Structural Dynamics with Applications in Earthquake and Wind Engineering

Konstantin Meskouris · Christoph Butenweg
Klaus-G. Hinzen · Rüdiger Höffer

Structural Dynamics with Applications in Earthquake and Wind Engineering

Second Edition

With contributions from Ana Cvetkovic, Linda Giresini, Britta Holtschoppen, Francesca Lupi, Hans-Jürgen Niemann

Springer

Konstantin Meskouris
Lehrstuhl für Baustatik und Baudynamik
RWTH Aachen University
Aachen, Germany

Christoph Butenweg
FH Aachen—University
of Applied Sciences
Aachen, Germany

Klaus-G. Hinzen
Erdbebenstation Bensberg, Institut
für Geologie und Mineralogie
Universität zu Köln
Bergisch Gladbach, Nordrhein-Westfalen
Germany

Rüdiger Höffer
Fakultät für Bau- und
Umweltingenieurwissenschaften
Ruhr-Universität Bochum
Bochum, Nordrhein-Westfalen
Germany

ISBN 978-3-662-57548-2 ISBN 978-3-662-57550-5 (eBook)
https://doi.org/10.1007/978-3-662-57550-5

Library of Congress Control Number: 2018949059

Originally published by Ernst & Sohn Verlag, Berlin, 2000

This Springer imprint is published by the registered company Springer-Verlag GmbH, DE part of Springer Nature.
The registered company address is: Heidelberger Platz 3, 14197 Berlin, Germany

Preface

The traditional design philosophy for buildings and structures, which takes almost exclusively only statics considerations into account, is being steadily supplemented by the need for carrying out additional verifications concerning their response and safety under dynamic loads. A reason for this may lie in the proliferation of modern bold architectural design forms favouring unorthodox and/or very slender structures, which are often susceptible to vibration under dynamic excitation. In addition, satisfying higher safety demands is increasingly required for buildings serving important societal needs (e.g. hospitals), or structures with high intrinsic risk potential (e.g. large industrial units). A prerequisite for carrying out complex dynamic analyses is a familiarity with the theoretical foundations and numerical methods of structural dynamics together with experience in the application of the latter and an insight into the nature of dynamic loads. The present book addresses both students and practising civil engineers offering an overview of the theoretical basics of structural dynamics complete with the relevant software for analysing the response of structures subject to earthquake and wind loads and illustrating its use by means of many examples worked out in detail, with input files for the programmes included. In the spirit of "learning by doing", it thus encourages readers to apply the tools described to their own problems, allowing them to become familiar with the broad field of structural dynamics in the process. Chapter 1 deals with the basic theory of structural dynamics followed by chapters on wind and earthquake loads. Chapters 4 and 5 deal with the behaviour of buildings and industrial units under seismic loading, respectively, while the final chapter is devoted to the application of wind engineering methods to slender tower-like structures. May this book contribute to a deeper understanding and familiarity of civil engineering students and practising engineers with the standard structural dynamics methods, enabling them to confidently carry out all necessary calculations for evaluating and verifying the safety of buildings and structures!

Aachen, Germany — Konstantin Meskouris
Aachen, Germany — Christoph Butenweg
Bergisch Gladbach, Germany — Klaus-G. Hinzen
Bochum, Germany — Rüdiger Höffer

Contents

Authors and Contributors

About the Authors

Prof. Dr.-Ing. Konstantin Meskouris was born in Athens, Greece. He studied Civil Engineering in Vienna and Munich, graduating in 1970 at the Technical University (TU) Munich, from which he also received his doctoral degree. From 1996 until his retirement in 2012, he served as Head of the Institute for Structural Statics and Dynamics in the Civil Engineering Department of RWTH Aachen University.

Prof. Dr.-Ing. Christoph Butenweg was born in Borken, Germany. He graduated in Civil Engineering at the Ruhr-Universität Bochum. From 1994 to 1999, he served as Research Assistant at Essen University and subsequently as Senior Engineer at the Institute of Structural Statics and Dynamics of RWTH Aachen University. Since 2006, he has been Manager of the SDA-engineering GmbH in Herzogenrath and since 2016 Full Professor for Technical Mechanics and Structural Engineering at FH Aachen – University of Applied Sciences.

Prof. Dr. Klaus-G. Hinzen was born in Hagen, Germany. He studied Geophysics at the Ruhr-Universität Bochum, graduating in 1979 and receiving his doctorate there in 1984. From 1984 to 1995, he served as Researcher at the Federal Institute for Geosciences and Natural Resources, Hanover. Since 1995, he has been Head of the Seismological Station Bensberg of Cologne University.

Prof. Dr.-Ing. Rüdiger Höffer was born in Attendorn, Germany. He graduated in Civil Engineering at Ruhr-Universität Bochum where he afterwards served as a research assistant. In 1996, he earned his doctoral degree in Bochum and subsequently spent two years as a postdocoral researcher abroad. He gained practical experience as a project engineer and managing director of consulting offices at Düsseldorf and Bochum, Germany. In 2003, he was appointed Full Professor for Wind Engineering at the Ruhr-Universität Bochum.

Contributors

Ana Cvetkovic M.Sc., born at Pozarevac, Serbia, earned her bachelor diploma in Structural Engineering at Belgrade University and graduated master's degree in Computational Engineering at the Ruhr-Universität Bochum in 2017. Since 2017, she works as Project Engineer in structural engineering.

Linda Giresini was born in Tempio, Italy. She earned her doctoral degree in Civil Engineering at the University of Pisa. She worked as Researcher at the Institute of Structural Statics and Dynamics, RWTH Aachen University and at the Institute for Sustainability and Innovation in Structural Engineering, University of Minho (Portugal) in 2013–2014. From 2014 to 2015, she was Research Assistant at the University of Sassari, Italy. Since 2016, she has been working as Assistant Professor of Structural Design at the University of Pisa.

Dr.-Ing. Britta Holtschoppen studied Civil Engineering at RWTH Aachen University and graduated in 2004. After a research stay at the University of Bristol, Great Britain, she returned to the Chair of Structural Statics and Dynamics of RWTH Aachen University and focused her research on the seismic design of industrial facilities with special emphasis on secondary structures, tank structures and probabilistic seismic performance assessment. Since 2016, she has worked at SDA-engineering GmbH, Herzogenrath.

Dr.-Ing. Francesca Lupi was born in Prato, Italy. She graduated master's degree in Civil Engineering at the Universita Degli Studi di Firenze, Italy, and earned her doctoral degree in 2013 in a joint programme of her home university and the Technische Universität Braunschweig. In 2015, she received a research scholarship for postdocs from the Alexander-von-Humboldt Foundation, Germany. Since 2018, she is employed as Senior Research Assistant at the Ruhr-Universität Bochum.

Prof. Dr.-Ing. habil. Hans-Jürgen Niemann was born 1935 in Lüneburg, Germany. He studied Civil Engineering at the Universität Hannover and graduated as Diplomingenieur. He worked as Research Assistant at the Universität Hannover and the then newly established Ruhr-Universität Bochum, graduated as Doctor of Civil Engineering and obtained the habilitation in Structural Engineering from the Ruhr-Universität Bochum. He was appointed Full Professor of Wind Engineering and Fluid Mechanics and served at the Department of Civil Engineering at Ruhr-Universität Bochum until 2001. In the same year, he founded the Engineering Company Niemann und Partners, Bochum.

Chapter 1
Basic Theory and Numerical Tools

Konstantin Meskouris

Abstract This chapter offers an overview of the theoretical foundations and the standard numerical methods for solving structural dynamics problems, with emphasis placed firmly on the latter. Starting with the analysis of single degree of freedom (SDOF) systems both in the time and in the frequency domain, it includes sections on the computation of elastic and inelastic response spectra, filtering in the frequency domain, the analysis of nonlinear SDOF systems and the generation of spectrum compatible ground motion time histories. Discrete multi-degree of freedom (MDOF) systems, condensation techniques and damping models are considered next. Both modal analysis ("response modal analysis") and direct integration methods are employed, focussing especially on the behaviour of MDOF systems subject to seismic excitations described by response spectra or sets of specific ground motion time histories. Detailed descriptions of the software used for solving the numerous examples presented complete with full input-output parameter lists conclude the chapter.

Keywords SDOF system · Seismic excitation · Response spectrum
Spectrum compatible accelerogram · Damping · MDOF system · Modal analysis
Direct integration

In this section, the most important basics of structural dynamics are introduced, which are needed in the further chapters of this book. The explanation of the theoretical derivation is kept to a minimum, while the emphasis is set on practical applications. For most algorithms, easy to use computing programs are provided, which application is illustrated by several examples.

Electronic supplementary material The online version of this chapter (https://doi.org/10.1007/978-3-662-57550-5_1) contains supplementary material, which is available to authorized users.

K. Meskouris et al., *Structural Dynamics with Applications in Earthquake and Wind Engineering*, https://doi.org/10.1007/978-3-662-57550-5_1

1.1 Single-Degree-of-Freedom Systems

Single-degree-of-freedom (SDOF) systems are the simplest oscillators. In spite of their simplicity, they are being successfully used as numerical models in many real-life cases. They are discussed in some detail in this chapter because of their wide application range and also because due to their "straightforwardness" they are eminently suitable for introducing basic structural dynamics methods and concepts.

1.1.1 Linear SDOF Systems in the Time Domain

Figure 1.1 depicts the standard case of a viscously damped SDOF system subject to a time-varying external load F(t). From the free-body diagram we obtain by D'ALEMBERT'S principle

$$\underline{F}_I + \underline{F}_D + \underline{F}_R = \underline{F}(t) \tag{1.1.1}$$

with $\underline{F}_I$, $\underline{F}_D$ and $\underline{F}_R$ as inertia, damping and restoring force, respectively. Setting the inertia force equal to mass times acceleration, the damping force equal to a coefficient c times velocity (linear viscous damping model) and the restoring force equal to displacement times the spring stiffness k yields the following 2nd order inhomogeneous linear ordinary differential equation (ODE) with constant coefficients:

$$m \cdot \ddot{\underline{u}} + c \cdot \dot{\underline{u}} + k \cdot \underline{u} = \underline{F}(t) \tag{1.1.2}$$

In order to avoid numerical errors in practice, it is advisable to use a consistent system of units, in which mass/force conversions are taken care of automatically. This is e.g. the case when masses are expressed in tons (1 ton = 1000 kg), forces in kN, lengths in m and time in s. Accordingly, c in (1.1.2) might be given in units of kN s/m, k in kN/m, F in kN and u in m.

The general solution of (1.1.2) is equal to the sum of the solution of the homogeneous equation ($\underline{F}(t) = 0$) and a "particular integral". The homogeneous ODE

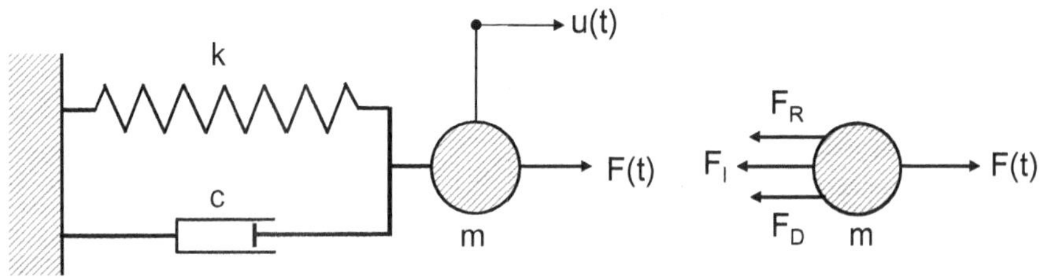

Fig. 1.1 SDOF system with free-body diagram

$$\underline{\ddot{u}} + \frac{c}{m} \cdot \underline{\dot{u}} + \frac{k}{m} \cdot \underline{u} = 0 \tag{1.1.3}$$

is satisfied by the function

$$u = e^{\lambda t};\ \dot{u} = \lambda e^{\lambda t};\ \ddot{u} = \lambda^2 e^{\lambda t} \tag{1.1.4}$$

leading to the characteristic equation

$$\lambda^2 + \frac{c}{m}\lambda + \omega_1^2 = 0;\quad \omega_1^2 = \frac{k}{m} \tag{1.1.5}$$

with ω_1 as circular natural frequency of the system. Its solutions are given by

$$\lambda_{1,2} = -\frac{c}{2m} \pm \sqrt{\left(\frac{c}{2m}\right)^2 - \omega_1^2} \tag{1.1.6}$$

The behaviour of the solution of the ODE depends on whether the radicand in Eq. (1.1.6) is less than, equal to or greater than zero, corresponding to the underdamped, critically damped and overdamped case, respectively. In the latter case, λ_1 and λ_2 are real and no vibration occurs. The critical damping is defined as the value of c for which the radicand is equal to zero:

$$\frac{c}{2m} = \omega_1 \rightarrow c_{krit} = 2m\,\omega_1 = 2\sqrt{km} \tag{1.1.7}$$

The dimensionless ratio D or ξ of the actual damping coefficient, c, to the critical damping, c_{krit}, called "damping ratio", is regularly used for quantifying damping. The following expressions hold:

$$D = \xi = \frac{c}{c_{krit}} = \frac{c}{2m\omega_1};\ \frac{c}{m} = 2\xi\omega_1 \tag{1.1.8}$$

Table 1.1 summarizes some typical values for the damping ratio for low-amplitude building vibrations.

Introducing the damping ratio ξ and the natural circular frequency ω_1, the differential equation (1.1.3) can also be written as:

$$\ddot{u} + 2\xi\omega_1\dot{u} + \omega_1^2 u = 0 \tag{1.1.9}$$

Table 1.1 Damping ratios for different structural types

Type of structure	Damping ratio D or ξ (%)
Steel structure, welded	0.2–0.3
Steel structure, bolted	0.5–0.6
Reinforced concrete	1.0–1.5
Masonry	1.5–2

Its solution is

$$u(t) = e^{-\xi\omega_1 t}\left(C_1 \cos\sqrt{1-\xi^2}\omega_1 t + C_2 \sin\sqrt{1-\xi^2}\omega_1 t\right) \tag{1.1.10}$$

where C_1, C_2 are integration constants. For general initial conditions $u(0) = u_0, \dot{u}(0) = \dot{u}_0$ this leads to

$$u(t) = e^{-\xi\omega_1 t}(u_0 \cos\sqrt{1-\xi^2}\omega_1 t + \frac{(\dot{u}_0 + \xi\omega_1 u_0)}{\omega_1\sqrt{1-\xi^2}} \sin\sqrt{1-\xi^2}\omega_1 t) \tag{1.1.11}$$

This expression can be further simplified by introducing the damped natural circular frequency ω_D (corresponding damped natural period T_D):

$$\omega_D = \omega_1\sqrt{1-\xi^2}; \quad T_D = 2\pi/\omega_D \tag{1.1.12}$$

In the case of forced vibrations, Eq. (1.1.9) reads

$$\ddot{u} + 2\xi\omega_1\dot{u} + \omega_1^2 u = \frac{F(t)}{m} = f(t) \tag{1.1.13}$$

The general solution of (1.1.13) is given by the sum of the homogeneous solution Eq. (1.1.11) and the particular integral (DUHAMEL integral)

$$u_p(t) = \frac{1}{\omega_D}\int_0^t f(\tau)e^{-\xi\omega_1(t-\tau)} \sin\omega_D(t-\tau)d\tau \tag{1.1.14}$$

The DUHAMEL integral can be evaluated numerically for arbitrary forcing functions F(t). Alternatively, Eq. (1.1.13) can be solved by various Direct Integration algorithms as explained later.

Another widely used damping parameter, in addition to the critical damping ratio D or ξ according to Eq. (1.1.8), is the so-called logarithmic decrement Λ. It is defined as the natural logarithm of the ratio of the amplitudes of two successive positive (or negative) peaks:

$$\begin{aligned}\Lambda &= \ln\frac{u_i}{u_{i+1}} = \ln\frac{e^{-\xi\omega_1 t_i}\cos\omega_D t_i}{e^{-\xi\omega_1 t_{i+1}}\cos\omega_D t_{i+1}} = \ln e^{-\xi\omega_1(t_i - t_{i+1})} = \xi\omega_1(t_{i+1} - t_i) \\ &= \xi\omega_1\frac{2\pi}{\omega_D} = \xi\,\omega_1 T_D = \xi\,\omega_1\frac{2\pi}{\omega_1\sqrt{1-\xi^2}} = \xi\frac{2\pi}{\sqrt{1-\xi^2}}\end{aligned} \tag{1.1.15}$$

For the lightly damped systems normally encountered in structural dynamics, it is sufficiently accurate to write

$$\xi = D \approx \frac{\Lambda}{2\pi} \tag{1.1.16}$$

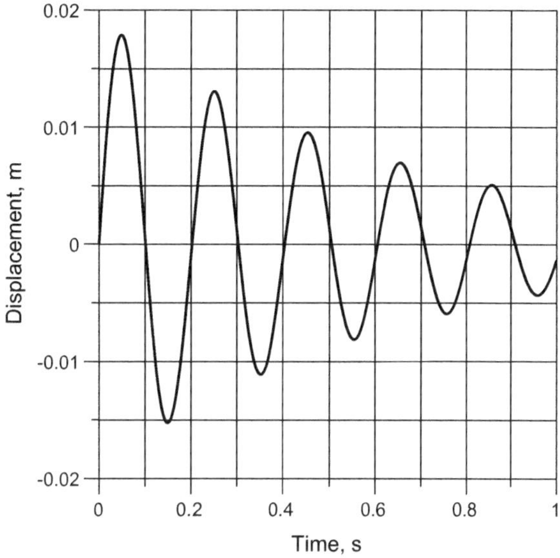

Fig. 1.2 Free vibration with viscous damping

The logarithmic decrement can be experimentally determined from time-history measurements of free vibrations. Normally, two peaks u_1 and u_{n+1} occurring at times t_1 and t_{n+1} and spanning n vibration cycles are considered, in which case we obtain

$$\Lambda = \frac{1}{n} \ln \frac{u_1}{u_{n+1}} \qquad (1.1.17)$$

As an example, Fig. 1.2 shows the (calculated) displacement time history for a SDOF system with a natural period of $T_1 = 0.20$ s, an initial velocity at $t = 0$ of 0.6 m/s and a damping value of $D = 5\%$. The positive peak amplitudes of the first four cycles are given as 0.01785, 0.01304, 0.009518, 0.006949 and 0.005071 m, leading to a logarithmic decrement of

$$\Lambda = \ln \frac{0.01785}{0.01304} \approx \frac{1}{4} \ln \frac{0.01785}{0.005071} = 0.315; \quad D \approx \frac{\Lambda}{2\pi} = 0.05 \qquad (1.1.18)$$

As mentioned above, the differential equation of motion for the linear SDOF oscillator given by Eq. (1.1.2) with the general initial conditions $u(0) = u_0$, $\dot{u}(0) = \dot{u}_0$ can also be solved by Direct Integration in the time domain. There exist many suitable algorithms for solving this classical initial-value problem.

Two issues are of central importance for the choice of an integration scheme, namely its stability and its accuracy. An unconditionally stable algorithm is present if the solution u(t) remains finite for arbitrary initial conditions and arbitrarily large $\Delta t/T$ ratios, Δt being the time step employed in the integration and T the natural period of the SDOF system, $T = 2\pi\sqrt{m/k}$. A conditionally stable algorithm (which is generally more accurate than an unconditionally stable one) implies that the solution remains finite only if the ratio $\Delta t/T$ does not exceed a certain value. For SDOF systems with known T it is easy to choose a suitable integration time step Δt; however, unconditionally stable algorithms are generally preferable, especially if nonlinearities are to be considered.

The accuracy of a Direct Integration algorithm depends on the loading function f(t), the system's properties and especially on the ratio of the time step Δt to the period T. The deviation of the computed solution from the true one makes itself felt as an elongation of the period and a decay of the amplitude of the former, corresponding to a fictitious additional damping.

Integration algorithms may also be divided into single-step and multi-step methods, which can also be implicit or explicit. Single-step methods, which are quite popular in structural dynamics, furnish the values of u, $\dot{u}$ and $\ddot{u}$ at time $t + \Delta t$ as functions of the same variables at time t alone, while multi-step methods require additional values at times $t - \Delta t$, $t - 2\Delta t$ etc. Multi-step methods therefore involve additional initial computations (e.g. by a single-step algorithm), while single-step methods are "self-starting". Explicit algorithms furnish the solution at time $t + \Delta t$ directly, while in implicit methods the unknowns appear on both sides of algebraic equations and must be determined by solving the corresponding equation system (or just one equation for a SDOF system). This shortcoming of implicit algorithms is offset by their better stability properties.

The well-known NEWMARK β-γ-algorithm, to be used here, is an implicit, single-step scheme with two parameters β and γ which determine its stability and accuracy properties. Considering the time points t_1 and t_2, with $t_2 = t_1 + \Delta t$, the dynamic equilibrium of the SDOF system at time t_2 is given by

$$m\ddot{u}_2 + c\dot{u}_2 + ku_2 = F(t_2) = F_2 \tag{1.1.19}$$

Introducing increments of the displacement, velocity, acceleration and external force according to $\Delta u = u_2 - u_1$, $\Delta\dot{u} = \dot{u}_2 - \dot{u}_1$ etc. leads to the incremental version of Eq. (1.1.2)

$$m\,\Delta\ddot{u} + c\,\Delta\dot{u} + k\,\Delta u = \Delta F \tag{1.1.20}$$

The increments $\Delta\ddot{u}$, $\Delta\dot{u}$ can be given as functions of the displacement increment Δu and the known values of velocity and acceleration at time t_1:

$$\begin{aligned}\Delta\dot{u} &= \frac{\gamma}{\beta\Delta t}\Delta u - \frac{\gamma}{\beta}\dot{u}_1 - \Delta t\left(\frac{\gamma}{2\beta} - 1\right)\ddot{u}_1\\ \Delta\ddot{u} &= \frac{1}{\beta(\Delta t)^2}\Delta u - \frac{1}{\beta\Delta t}\dot{u}_1 - \frac{1}{2\beta}\ddot{u}_1\end{aligned} \tag{1.1.21}$$

Values of $\beta = 1/4$ and $\gamma = 1/2$ correspond to an unconditionally stable scheme which assumes a constant acceleration $\ddot{u}$ between t_1 and t_2. For $\beta = 1/6$ and $\gamma = 1/2$ the integrator is only conditionally stable and the acceleration varies linearly between t_1 and t_2. The displacement increment Δu is given by

$$\Delta u = \frac{f^*}{k^*} \tag{1.1.22}$$

with

$$k^* = m\frac{1}{\beta \Delta t^2} + c\frac{\gamma}{\beta \Delta t} + k \tag{1.1.23}$$

and

$$f^* = \Delta F + m(\frac{\dot{u}_1}{\beta \Delta t} + \frac{\ddot{u}_1}{2\beta}) + c\left(\frac{\gamma \dot{u}_1}{\beta} + \ddot{u}_1 \Delta t(\frac{\gamma}{2\beta} - 1)\right) \tag{1.1.24}$$

The NEWMARK algorithm is used in the programs SDOF1 and SDOF2, details on which can be found in Appendix. The SDOF1 program deals with the case when F(t) is an arbitrary piecewise linear function, while SDOF2 considers steady-state excitations of the type $F(t) = A \cdot \sin(\Omega_1 t) + B \cos(\Omega_2 t)$. For SDOF1 it is usually necessary to first use the program LININT, also described in Appendix, for determining additional values of the forcing function F(t) for the chosen time step Δt by linear interpolation.

Example 1.1
The task is to determine the maximum displacement u of the girder and also the maximum bending moment at the base of the central column for the frame shown in Fig. 1.3. Further data: Mass m = 12 t (assumed to be concentrated in the girder), damping value D = 1%, bending stiffness values $EI_G = 1.25 \times 10^5$ kNm², $EI_1 = 0.75 \times 10^5$ kNm², all beams and columns are considered to be inextensional ($EA \to \infty$).

The single-story two-bay frame can be modelled using a SDOF idealization as depicted in Fig. 1.1, with mass m = 12 t and spring stiffness k in kN/m. The latter can be determined from statics as the reciprocal of the horizontal girder displacement u due to a unit force F = 1.0 kN (program FRAME) or by carrying out a static condensation for the horizontal girder displacement as a master DOF (program CONDEN, see Sect. 1.2.1). Here, the first approach will be used based on the discretization

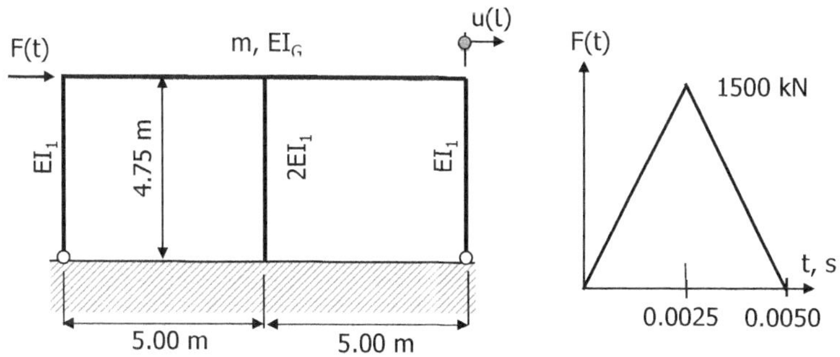

Fig. 1.3 Plane frame with triangular loading function F(t)

Discretized system	Input file EFRAM.txt

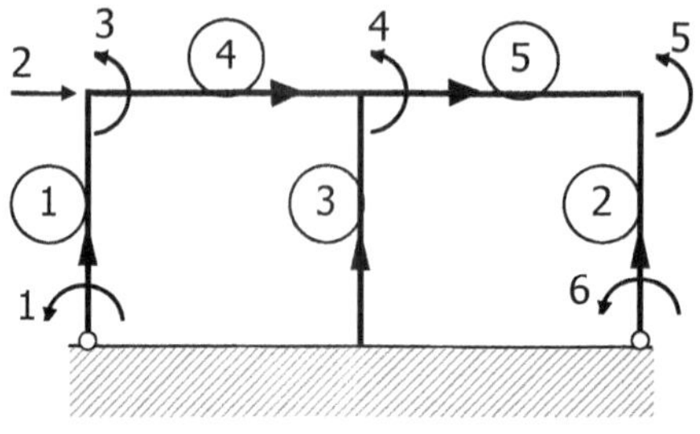


```
75000.,  4.75, 0., 90.
75000.,  4.75, 0., 90.
150000., 4.75, 0., 90.
125000., 5.0, 0., 0.
125000., 5.0, 0., 0.
0,0,1, 2,0,3,
0,0,6, 2,0,5,
0,0,0, 2,0,4,
2,0,3, 2,0,4,
2,0,4, 2,0,5,
0.,1.0, 0., 0., 0., 0.
```

Fig. 1.4 Discretization and input file for a unit load at DOF no. 2

of Fig. 1.4 (see input description for the program FRAME in Appendix), using the 6-DOF beam element shown in Fig. 1.46. For more details on the use of the program FRAME and the matrix deformation method of analysis employed here see Example 1.5.

The program FRAME yields a horizontal displacement of the girder due to the unit load (1 kN) equal to 6.292×10^{-5} m. The reciprocal is the spring stiffness, which in this case equals k = 15,893 kN/m. With k and m known, the undamped circular natural frequency of the frame is

$$\omega_1 = \sqrt{\frac{k}{m}} = \sqrt{\frac{15{,}893}{12}} = 36.39\,\frac{\text{rad}}{\text{s}} \tag{1.1.25}$$

the corresponding natural period being equal to

$$T_1 = \frac{2\pi}{\omega_1} = 0.173\,\text{s} \tag{1.1.26}$$

In view of the characteristics of the load function, a time step Δt of 0.5×10^{-3} s is chosen and the program LININT used to create the file RHS.txt as input file for the program SDOF1. For 600 time points (from zero to 0.3 s) Fig. 1.5 shows the computed time history for the horizontal displacement of the girder (solid line). The maximum displacement is reached at time t = 0.046 s and is equal to 0.008445 m or 8.4 mm. In view of the short duration of the excitation, a simple check on the validity of this result can be carried out by using the principle of impulse and momentum, which states that the final momentum of a mass m may be obtained by adding its initial momentum (which, in this case, is zero) to the time integral of the force during the interval considered. This allows the velocity at time t = 0.005 s to be determined as follows:

$$m \cdot \dot{u}_1 + \int_0^{0.005} F(t)dt = m \cdot \dot{u}_2$$

$$\rightarrow \dot{u}(0.005) = \frac{\frac{1}{2}1500 \cdot 0.005\,\text{kN s}}{12\,\text{t}} = 0.3125\frac{\text{m}}{\text{s}} \tag{1.1.27}$$

Equation (1.1.11) yields the following expression for the displacement u(t):

$$u(t) = e^{-0.01 \cdot 36.39 \cdot t}\frac{0.3125}{36.39\sqrt{1-0.01^2}}\sin\sqrt{1-0.01^2} \cdot 36.39 \cdot t \tag{1.1.28}$$

This function, depicted as a dashed line in Fig. 1.5, is in almost perfect agreement with the solution obtained using Direct Integration (solid line). From the maximum displacement $u_{max} = 0.00845$ m at $t = 0.046$ s the maximum restoring force is determined as $F_{R,max} = u_{max} \cdot k = 0.00845 \cdot 15{,}893 = 134$ kN. Using the program FRAME with 134 kN as the load corresponding to DOF no. 2 yields a bending moment at the base of the central column equal to 276 kNm. The bending moment diagram for the entire structure is shown in Fig. 1.6.

For a quick assessment of the maximum response of linear undamped SDOF systems subject to impulsive loading, shock or response spectra are quite useful. They present dynamic magnification factors, defined as ratios of maximum dynamic displacements $u_{dyn,max}$ to their static counterparts u_{stat} as functions of the impulse length ratio t_1/T, that is the duration t_1 of the impulse divided by the natural period of the SDOF system. Figure 1.7 shows shock spectra for three impulsive loading shapes, namely rectangular, trapezoidal and triangular.

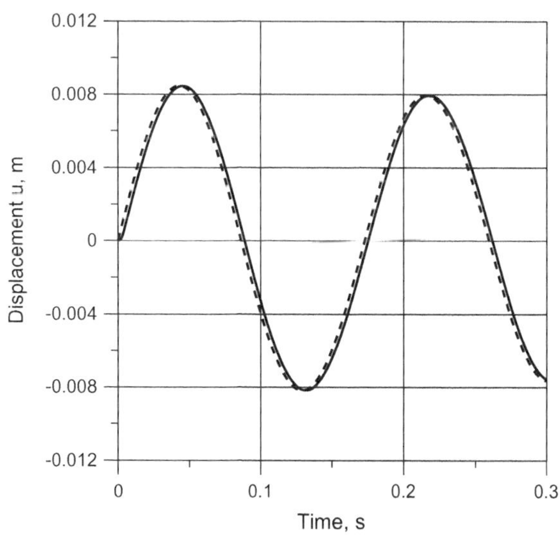

Fig. 1.5 Time histories for the girder horizontal displacement

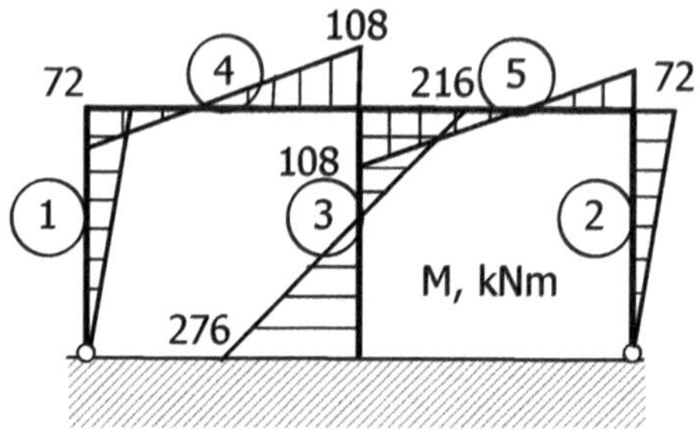

Fig. 1.6 Bending moment diagram at t = 0.00455 s

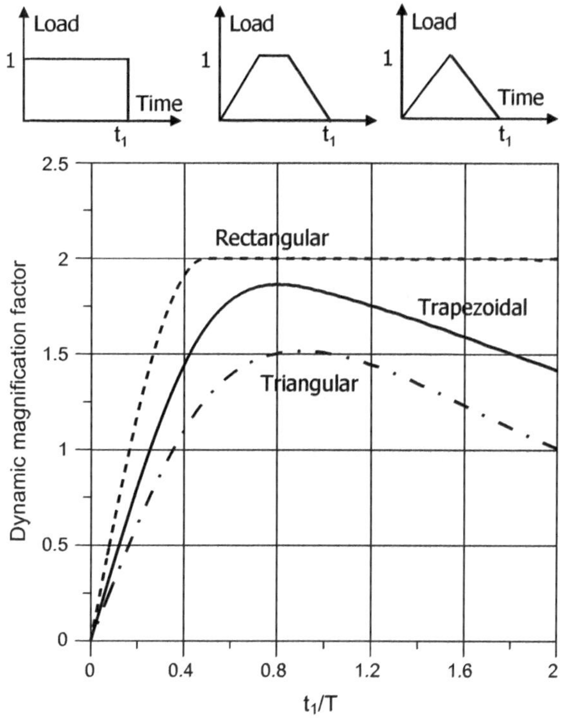

Fig. 1.7 Shock spectra for different impulsive loading shapes

A cursory look at Fig. 1.7 would seem to suggest that dynamic magnification factors do not exceed 2.0, which, however, is not the case. As an example, Figs. 1.8 and 1.9 show some more shock spectra for piecewise linear and sinusoidal impulse shapes. In Fig. 1.8 the solid line corresponds to the positive/negative impulse shown to the left and the dashed line to the positive/positive one shown to the right, while in Fig. 1.9 the solid line corresponds to the single half-sine and the dashed line to the double half-sine impulse.

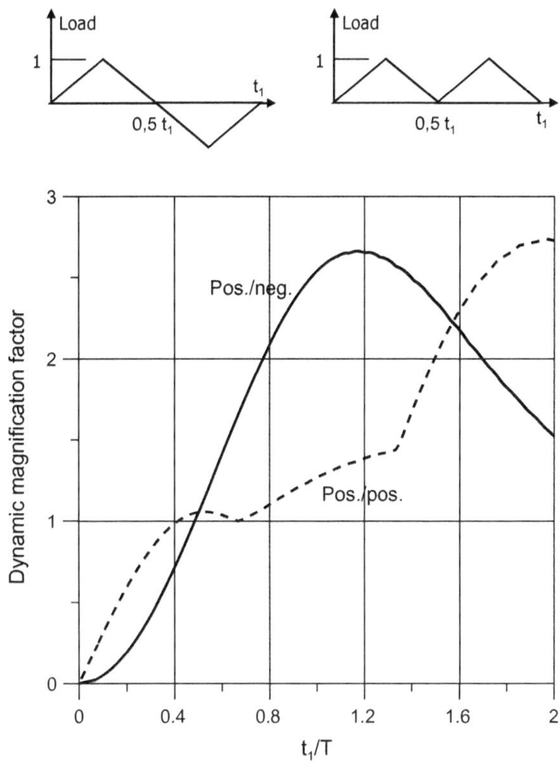

Fig. 1.8 Additional shock spectra for further polygonal impulses

A special case with significant practical importance is the linear SDOF system subject to stationary harmonic excitation as shown schematically in Fig. 1.10.

Its equation of motion is given by

$$\ddot{u} + 2\xi\omega_1\dot{u} + \omega_1^2 u = \frac{F_o}{m}\sin\Omega t \tag{1.1.29}$$

It has the general solution

$$\begin{aligned} u(t) &= \exp(-\xi\omega_1 t)(A\ \sin\omega_D t + B\ \sin\omega_D t) \\ &+ \frac{F_o}{k}\frac{1}{(1-\beta^2)^2 + (2\xi\beta)^2}\left[(1-\beta^2)\sin\Omega t - 2\xi\beta\cos\Omega t\right] \end{aligned} \tag{1.1.30}$$

where β is the ratio of the excitation frequency to the natural frequency of the system, that is

$$\beta = \frac{\Omega}{\omega_1} \tag{1.1.31}$$

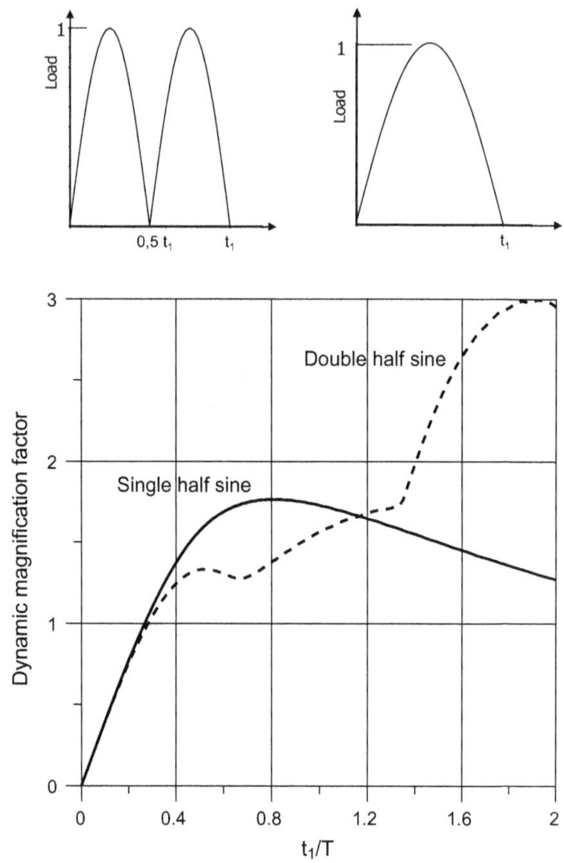

Fig. 1.9 Shock spectra for sinusoidal impulses

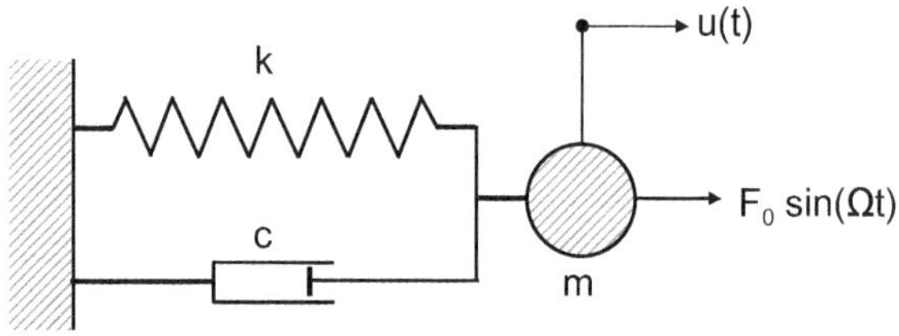

Fig. 1.10 SDOF system under harmonic excitation

The first part of the expression in Eq. (1.1.30) is the solution of the homogeneous differential equation; its constants A and B must be determined from the initial conditions. The second part is the particular integral which depends on the loading; this is the most important part of the system response, since the first part is eventually damped out, as is evident from the factor $e^{-\xi\omega_1 t}$. The second part of the solution can be written down in the form

$$u(t) = u_R \sin(\Omega t - \varphi) \tag{1.1.32}$$

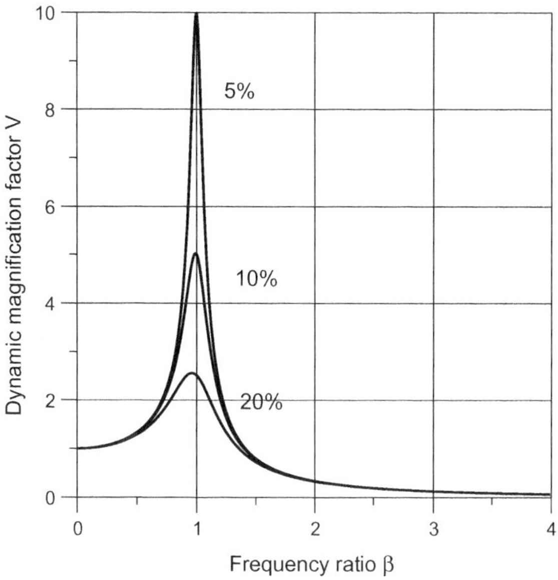

Fig. 1.11 Magnification factor V for different damping ratios D

with

$$u_R = \frac{F_o}{k}[(1-\beta^2)^2 + (2\xi\beta)^2]^{-0.5} \tag{1.1.33}$$

and

$$\varphi = \arctan\frac{2\xi\beta}{1-\beta^2} \tag{1.1.34}$$

From Eq. (1.1.33) the dynamic magnification factor V defined as

$$V = [(1-\beta^2)^2 + (2\xi\beta)^2]^{-0.5} \tag{1.1.35}$$

can be extracted. It is seen to be equal to the ratio of the dynamic to the static response of the harmonically excited SDOF system, where the maximum dynamic response is given by

$$\max u_p(t) = u_R = \frac{F_0}{k}V \tag{1.1.36}$$

Figure 1.11 shows V as a function of the frequency ratio for three damping ratios, namely D = 5, 10 and 20%; clearly, for undamped systems ($\xi = 0$), V tends to infinity.

Peak values of the magnification factor occur at the frequency ratio

$$\beta = \sqrt{1-2\xi^2} \tag{1.1.37}$$

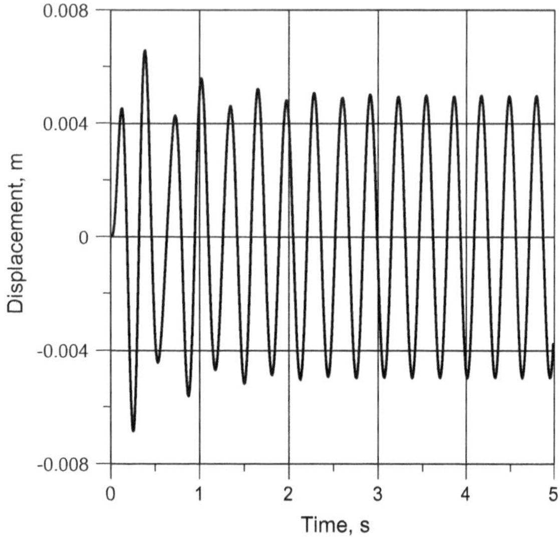

Fig. 1.12 Example 1.2, time history of displacement

They amount to

$$\max\ V = \frac{1}{2}\frac{1}{\xi\sqrt{1-\xi^2}} \tag{1.1.38}$$

For a damping ratio of 1% this gives a peak value of 50, for 5% of about 10 and even for a highly damped system with D = 20% we obtain V = 2.55.

Example 1.2
Consider the system of Fig. 1.10 with the following data: Spring stiffness k = 9000 kN/m, m = 10 t, F_0 = 25 kN, Ω = 20 rad/s and D = 5%. Determine the displacement and velocity time histories u(t) and $\dot{u}(t)$ as well as the maxima of the restoring force and the damping force.

The circular natural frequency ω_1 of the SDOF system is equal to $\sqrt{\frac{k}{m}} = \sqrt{\frac{9000}{10}} = 30\ \frac{rad}{s}$, corresponding to T_1 = 0.21 s. The program SDOF2 produces the results shown for the time histories of the displacement (Fig. 1.12) and the velocity (Fig. 1.13) using a time step of 0.005 s. The maximum restoring force F_R occurs at time 0.25 s, corresponding to a maximum displacement of −0.00688 m, and is equal to $k \cdot u_{max}$ = −61.9 kN, while the maximum velocity of 0.161 m/s, occurring at t = 0.32 s, produces a damping force equal to

$$2 \cdot \xi \cdot \omega_1 \cdot m \cdot \dot{u} = 2 \cdot 0,05 \cdot 30 \cdot 10 \cdot 0.161 = 4.83\,\text{kN} \tag{1.1.39}$$

Both these values were reached during the initial vibration stage, before the viscous damping mechanism eliminated the contribution of the "homogeneous" part

Fig. 1.13 Example 1.2, time history of velocity

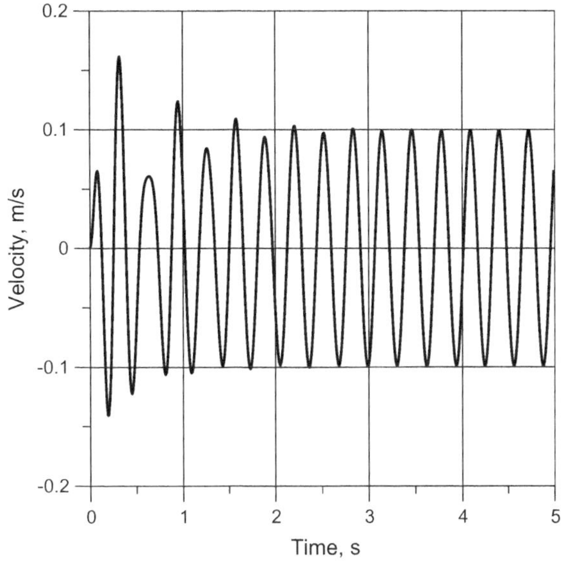

of the solution. In the subsequent steady-state harmonic vibration stage, with a frequency ratio of

$$\beta = \frac{\Omega}{\omega_1} = \frac{20}{30} = 0.667 \tag{1.1.40}$$

the maximum displacement amounts to

$$\max u_p(t) = \frac{F_0}{k} V = \frac{25}{9000} V(0.667) = \frac{25}{9000} 1.79 = 0.005\,\mathrm{m} \tag{1.1.41}$$

Since here u(t) is a sine wave with circular frequency Ω, the maximum velocity is readily determined as

$$\max \dot{u} = \Omega \cdot u_{max} = 20 \cdot 0.005 = 0.1\,\mathrm{m/s} \tag{1.1.42}$$

1.1.2 Linear SDOF Systems in the Frequency Domain

It can be shown that any real periodic function of time with period T

$$f(t + T) = f(t) \tag{1.1.43}$$

can be expressed in the form

$$f(t) = a_0 + \sum_{k=1}^{\infty} a_k \cos \omega_k t + \sum_{k=1}^{\infty} b_k \sin \omega_k t \tag{1.1.44}$$

with the coefficients

$$a_0 = \frac{\omega}{2\pi} \int_0^{\frac{2\pi}{\omega}} f(t)dt = \frac{1}{T} \int_0^{T} f(t)\, dt \tag{1.1.45}$$

and

$$a_k = \frac{2}{T} \int_{-T/2}^{T/2} f(t) \cos \omega_k t\, dt \tag{1.1.46}$$

$$b_k = \frac{2}{T} \int_{-T/2}^{T/2} f(t) \sin \omega_k t\, dt \tag{1.1.47}$$

Here, $\omega_k = k\frac{2\pi}{T}$, $k = 1, 2, \ldots\infty$. The coefficients a_k and b_k can be regarded as the real and imaginary part, respectively, of the harmonic component associated with the circular frequency ω_k. They can be displayed along a frequency axis at discrete points ω_k with an increment in rad/s equal to

$$\Delta\omega = \frac{2\pi}{T}; \quad \frac{2}{T} = \frac{\Delta\omega}{\pi} \tag{1.1.48}$$

Figure 1.14 shows such "comb spectra" consisting of discrete values of the coefficients a_k and b_k every $(2\pi/T)$ rad/s.

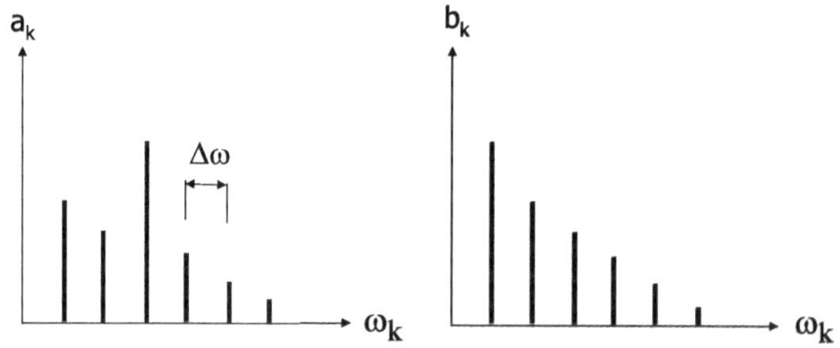

Fig. 1.14 "Comb spectra" for periodic functions

For $a_0 = 0$ we obtain

$$f(t) = \sum_{k=1}^{\infty} \left(\frac{2}{T} \int_{-T/2}^{T/2} f(t) \cdot \cos \omega_k t dt \right) \cos \omega_k t + \sum_{k=1}^{\infty} \left(\frac{2}{T} \int_{-T/2}^{T/2} f(t) \cdot \sin \omega_k t dt \right) \sin \omega_k t \tag{1.1.49}$$

or, using Eq. (1.1.44)

$$f(t) = \sum_{k=1}^{\infty} \left(\frac{\Delta\omega}{\pi} \int_{-T/2}^{T/2} f(t) \cdot \cos \omega_k t dt \right) \cos \omega_k t + \sum_{k=1}^{\infty} \left(\frac{\Delta\omega}{\pi} \int_{-T/2}^{T/2} f(t) \cdot \sin \omega_k t dt \right) \sin \omega_k t \tag{1.1.50}$$

For an aperiodic function we can assume $T \to \infty$, $\Delta\omega \to d\omega$ and

$$f(t) = \int_{\omega=0}^{\infty} \frac{1}{\pi} \left(\int_{-\infty}^{+\infty} f(t) \cdot \cos \omega t \, dt \right) \cdot \cos \omega t \, d\omega + \int_{\omega=0}^{\infty} \frac{1}{\pi} \left(\int_{-\infty}^{+\infty} f(t) \cdot \sin \omega t \, dt \right) \cdot \sin \omega t \, d\omega \tag{1.1.51}$$

Introducing

$$A(\omega) = \frac{1}{2\pi} \int_{-\infty}^{\infty} f(t) \cos \omega t \, dt, \quad B(\omega) = \frac{1}{2\pi} \int_{-\infty}^{\infty} f(t) \sin \omega t \, dt \tag{1.1.52}$$

leads to

$$f(t) = 2 \int_{\omega=0}^{\infty} A(\omega) \cdot \cos \omega t \, d\omega + 2 \int_{\omega=0}^{\infty} B(\omega) \cdot \sin \omega t \, d\omega \tag{1.1.53}$$

and, considering that

$$\begin{aligned} A(\omega) \cdot \cos \omega t &= A(-\omega) \cdot \cos(-\omega t) \\ B(\omega) \cdot \sin \omega t &= B(-\omega) \cdot \sin(-\omega t) \end{aligned} \tag{1.1.54}$$

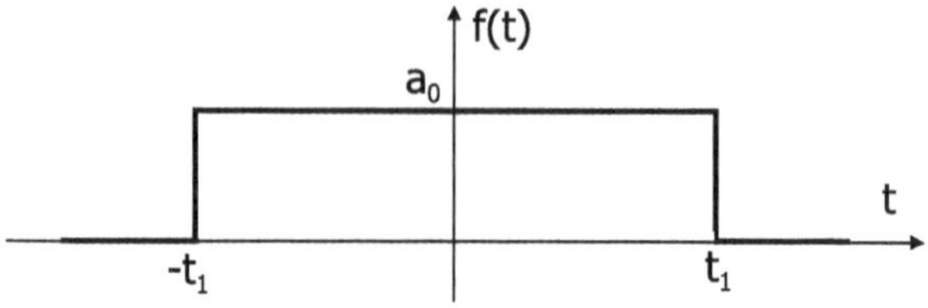

Fig. 1.15 Time domain function

to

$$f(t) = \int_{-\infty}^{\infty} A(\omega) \cdot \cos \omega t \, d\omega + \int_{-\infty}^{\infty} B(\omega) \cdot \sin \omega t \, d\omega \tag{1.1.55}$$

With complex coefficients

$$F(\omega) = A(\omega) - i \cdot B(\omega) \tag{1.1.56}$$

we finally obtain

$$F(\omega) = \frac{1}{2\pi}\left(\int_{-\infty}^{\infty} f(t) \cos \omega t \, dt\right) - i\frac{1}{2\pi}\left(\int_{-\infty}^{\infty} f(t) \sin \omega t \, dt\right)$$

$$F(\omega) = \frac{1}{2\pi}\left(\int_{-\infty}^{\infty} f(t)[\cos \omega t - i \, \sin \omega t]dt\right) \tag{1.1.57}$$

This is the formal definition of the FOURIER transform $F(\omega)$ of the time domain function $f(t)$:

$$F(\omega) = \frac{1}{2\pi}\int_{-\infty}^{\infty} f(t)e^{-i\omega t}dt \tag{1.1.58}$$

The inverse transform is given by

$$f(t) = \int_{-\infty}^{\infty} F(\omega)e^{i\omega t}d\omega \tag{1.1.59}$$

$F(\omega)$ and $f(t)$ form a "FOURIER transform pair". As a simple practical example for transforming a time domain function into the frequency domain, consider the "boxcar" function shown in Fig. 1.15.

The function is even, $f(t) = f(-t)$, so that in Eq. (1.1.52) $B(\omega) = 0$. We obtain

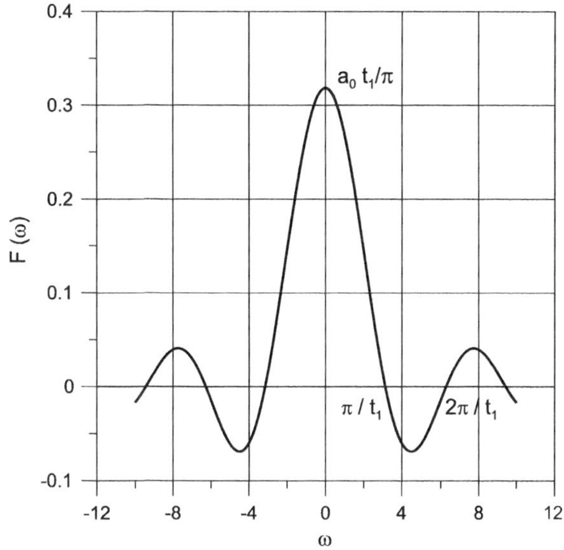

Fig. 1.16 Frequency domain counterpart of the time domain boxcar function for $a_0 = 1$, $t_1 = 1$ s

$$F(\omega) = A(\omega) = \frac{1}{2\pi}\int_{-t_1}^{t_1} a_0 \cos\omega t\,dt = \frac{a_0}{2\pi}\cdot\left[\frac{\sin\omega t}{\omega}\right]_{-t_1}^{t_1} = \frac{a_0\ t_1}{\pi}\ \frac{\sin\omega t_1}{\omega t_1} \qquad (1.1.60)$$

and considering that

$$\lim_{\omega\to 0}\ \frac{\sin\omega\, t_1}{\omega\, t_1} = 1 \qquad (1.1.61)$$

also

$$F(0) = a_0 t_1/\pi \qquad (1.1.62)$$

The resulting curve is plotted in Fig. 1.16, with zero crossings at $\pm k\cdot\pi/t_1,\ k = 1, 2, \ldots$. Obviously broad-band time domain functions (large t_1) correspond to narrow-band frequency domain functions and vice versa. To further illustrate this fact, consider that for a static load the frequency domain representation consists of a single spike at $\omega = 0$, while "white noise" in the frequency domain is represented by a DIRAC δ—function in the time domain.

A closed-form evaluation of the transform integrals in (1.1.58) and (1.1.59) is advisable only in special cases. Usually, the function f(t) is given in the form of a discrete time series, which must be handled numerically by a so-called "Discrete FOURIER Transform" (DFT) algorithm. This method will be applied to the time series f_r, r = 0, 1, 2, …N − 1, consisting of N points with a constant time step (sampling interval) Δt as shown in Fig. 1.17. It is assumed that the series is periodic, that is, $f_N = f_0$, $f_{N+1} = f_1$ etc. This is not as serious a restriction as it might seem at first glance, because any general non-periodic function can be practically considered as a periodic function with a sufficiently large period $T = N \cdot \Delta t$.

The DFT expression for the forward transform is given by

$$F_k = \frac{1}{N\,\Delta t}\sum_{r=0}^{N-1} f_r e^{-i\omega_k r \Delta t}\Delta t = \frac{1}{N}\sum_{r=0}^{N-1} f_r\, e^{-i\frac{2\pi kr}{N}} \tag{1.1.63}$$

or

$$F_k = \frac{1}{N}\sum_{r=0}^{N-1} f_r\left(\cos\frac{2\pi kr}{N} - i\,\sin\frac{2\pi kr}{N}\right) \tag{1.1.64}$$

This equation yields N complex coefficients F_k, $k = 0, 1, 2, \ldots N-1$ which correspond to the circular frequency values $\omega_k = (2\pi/T)k$. As shown in Fig. 1.17, $T = N \cdot \Delta t$ is the fictitious period of the time series, since the algorithm assumes that it is periodic. As mentioned, however, this assumption does not overly restrict its general applicability, because "spill-over" phenomena can be avoided by assuming a sufficiently long period T and padding the actual function with zeroes. Of the N complex coefficients F_k only those for $k = 0, 1, 2, \ldots N/2$, describing the frequency content of the time series for circular frequency values up to the so-called NYQUIST circular frequency are important:

$$\omega_{NYQ} = \frac{\pi}{\Delta t}, \quad f_{NYQ} = \frac{1}{2 \cdot \Delta t} \tag{1.1.65}$$

For k greater than N/2 the coefficients F_k are repeated, being, in fact, mirror images of the first N/2 coefficients. The real parts of complex coefficients at the same distance from the NYQUIST frequency (also designated as "folding frequency" for obvious reasons) are identical, while the imaginary parts have the same amplitudes but opposite signs. Clearly, the NYQUIST frequency is the highest frequency for which information can be gained from the harmonic analysis of the time series ("signal") in question, with Δt as sampling interval.

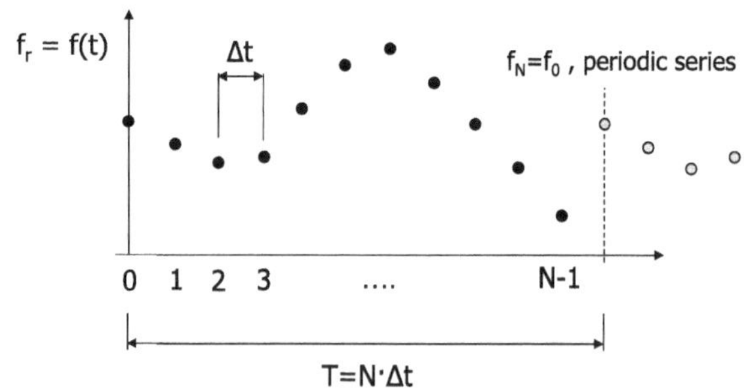

Fig. 1.17 Time series representing a periodic function

If the time series contains harmonic components with frequencies higher than the NYQUIST frequency, so-called "aliasing" occurs during harmonic analysis, leading to more or less erroneous results for the forward transform. If the presence of unwanted high-frequency components is detected by observing that the corresponding coefficients F_k near the NYQUIST frequency are not negligibly small, it might be advisable to eliminate these high-frequency components from the original signal (e.g. by low-pass filtering) before embarking on harmonic analysis. This is particularly important because the results for lower frequencies can be seriously distorted through aliasing.

The inverse transform to Eq. (1.1.64) is given by

$$f_r = \sum_{k=0}^{N-1} F_k \, e^{i \frac{2\pi kr}{N}}, \quad r = 0, 1, 2, \ldots N - 1 \tag{1.1.66}$$

Equations (1.1.64) and (1.1.66) are very easy to program. However, they are much too time-consuming for practical purposes if the number of data points, N, equals several tens or hundreds of thousands. So-called Fast-FOURIER transform (FFT) algorithms, of which the classical COOLEY-TUKEY method is an example, are much more efficient. For the COOLEY-TUKEY method the number of data points N must be equal to 2^n with n integer; this is achieved in practice by padding the function in question with zeros.

Two FFT-based programs, FFT1 and FFT2, are available for carrying out forward (FFT1) and inverse (FFT2) transforms (descriptions see Appendix). While FFT1 is used for transformations from the time domain to the frequency domain, FFT2 allows inverse transformations from the frequency domain to the time domain.

As an example, consider the time series consisting of $N = 8$ points representing the sine wave $f(t) = \sin \omega_1 t$ depicted in Fig. 1.18.

For a time step of 1 s the increment in the frequency domain amounts to

$$\Delta\omega = \frac{2\pi}{T} = \frac{2\pi}{8 \cdot 1.0} = 0.785 \, \frac{\text{rad}}{\text{s}} \tag{1.1.67}$$

and the NYQUIST circular frequency is equal to $4 \cdot 0.785 = 3.14$ rad/s.

Figure 1.19 shows the results produced by the program FFT1 for the real and imaginary parts of the eight FOURIER coefficients. The former are all zero, while for the latter only two coefficients (corresponding to ω_1 and to $7\omega_1$) are different from zero, their values being -0.5 and 0.5, respectively. The point symmetry about the NYQUIST frequency is obvious.

The differential equation of motion for a SDOF system, namely

$$\ddot{u} + 2\,\xi\,\omega_1\,\dot{u} + \omega_1^2\,u = \frac{F}{m} = f(t) \tag{1.1.13}$$

can be transformed into an algebraic equation by utilizing integral transform methods such as the FOURIER or the LAPLACE transform. The solution $U(\omega)$ of the algebraic

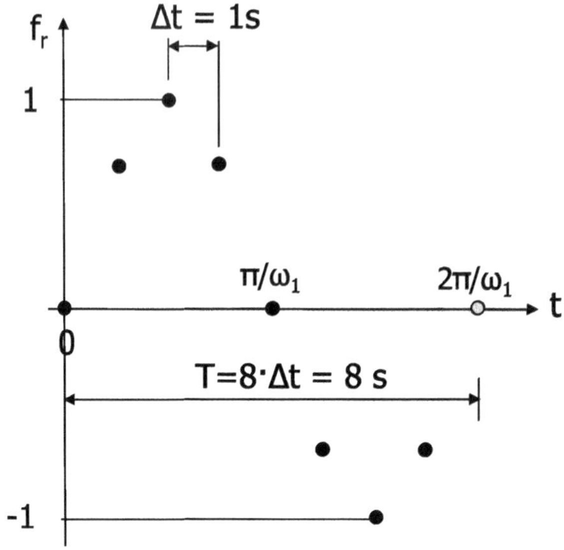

Fig. 1.18 Eight-point time series representing a sine wave

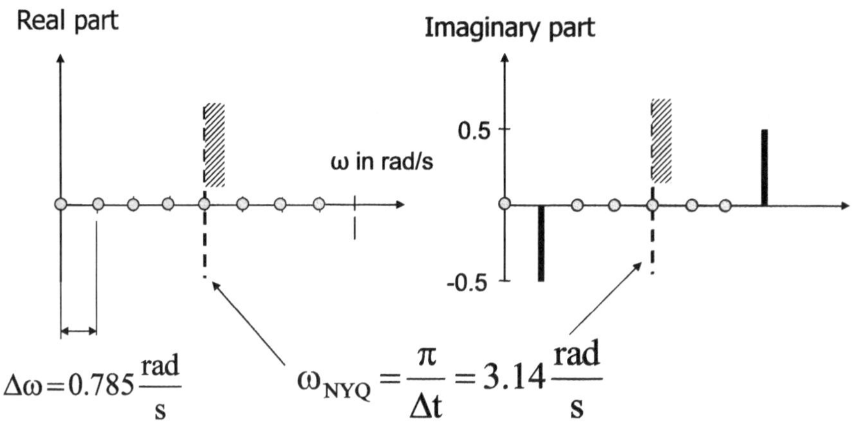

Fig. 1.19 FOURIER coefficients representing a sine wave

equation in the frequency domain can then be transformed back to u(t) in the time domain using the inverse transform algorithm.

The DUHAMEL integral in the time domain as given by Eq. (1.1.14) was

$$u_p(t) = \frac{1}{\omega_D} \int_0^t f(\tau) e^{-\xi\omega_1(t-\tau)} \sin \omega_D(t-\tau) d\tau \tag{1.1.14}$$

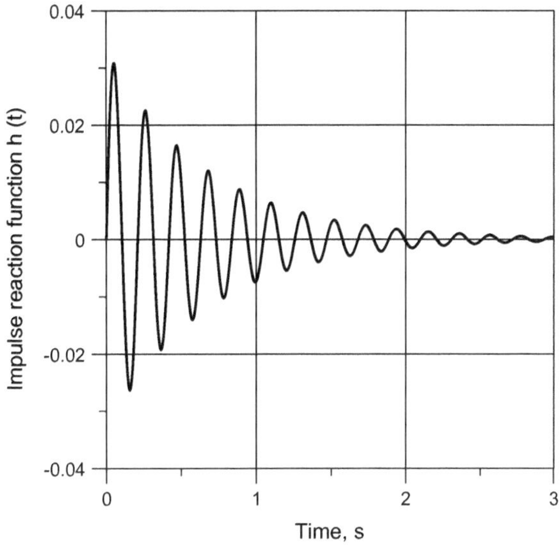

Fig. 1.20 Impulse response function

which can formally be written down as

$$u_p(t) = f(t) * h(t) = \int_{-\infty}^{\infty} f(\tau)h(t-\tau)d\tau \tag{1.1.68}$$

with the star denoting a "folding operator" and

$$h(t) = \frac{e^{-\xi\omega_1 t}}{\omega_D} \sin \omega_D t \tag{1.1.69}$$

h(t) is referred to as "impulse response function". It furnishes the response of a SDOF system to a unit impulse excitation at time t = 0, with the initial conditions

$$u_0 = 0, \quad \dot{u}_0 = 1 \tag{1.1.70}$$

As an example, Fig. 1.20 shows h(t) for $\omega_1 = 30$ rad/s and $\xi = 5\%$.

It can be shown that the equivalent of the "folding operation" in the time domain

$$u(t) = f(t) * h(t) \tag{1.1.71}$$

is a (complex) multiplication of the corresponding FOURIER transforms in the frequency domain:

$$U(\omega) = F(\omega) \cdot H(\omega) \tag{1.1.72}$$

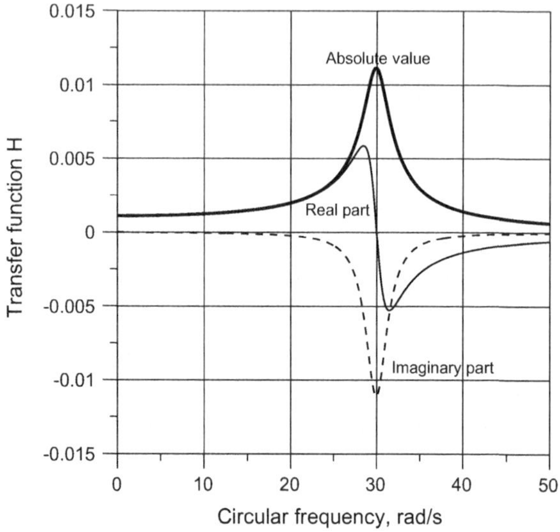

Fig. 1.21 Transfer function of a SDOF system (absolute value)

$H(\omega)$ as FOURIER transform of the impulse reaction function h(t) is denoted as the transfer function of the system. It can be visualised as the output/input ratio for the relevant response quantity for a stationary and harmonic input.

For the SDOF system of Eq. (1.1.13) with the unit harmonic excitation (input)

$$f(t) = 1 \cdot e^{i\omega t} \tag{1.1.73}$$

the response (output) is by definition:

$$u(t) = H(\omega)e^{i\omega t} \tag{1.1.74}$$

which implies

$$\begin{aligned} \dot{u}(t) &= i\omega\, H(\omega)e^{i\omega t} \\ \ddot{u}(t) &= -\omega^2 H(\omega)e^{i\omega t} \end{aligned} \tag{1.1.75}$$

The complex transfer function $H(\omega)$ is here given by

$$H(\omega) = \frac{1}{\omega_1^2 - \omega^2 + i\,2\,\xi\omega_1\,\omega} \tag{1.1.76}$$

As an example, Fig. 1.21 shows the complex transfer function $H(\omega)$ for $\omega_1 =$ 30 rad/s and $\xi = 5\%$.

Generally speaking, the transformation of an "input signal" f(t) into an "output signal" u(t) can be considered as a filtering process, which modifies the frequency

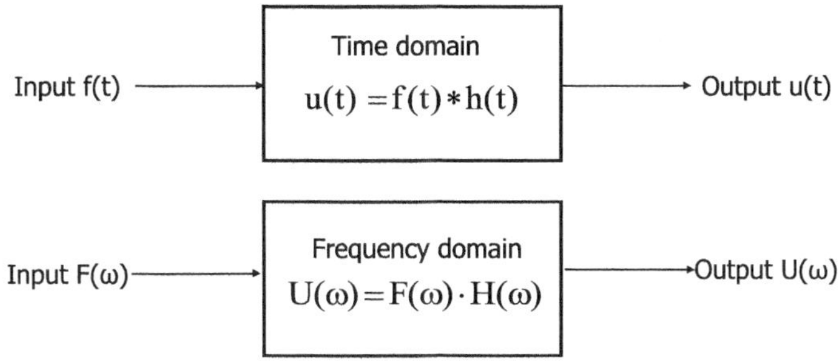

Fig. 1.22 Schematic input/output relationships in the time and frequency domains

content of the original signal according to the impulse reaction function (time domain) or the transfer function (frequency domain) characterising the system (Fig. 1.22).

A well-known second order low-pass filter with various applications in earthquake engineering is the KANAI-TAJIMI filter, the complex transfer function of which is given by

$$H(\omega) = \frac{1 + \frac{\omega^2}{\omega_0^2}(4\xi_0^2 - 1) - i2\xi_0\frac{\omega^3}{\omega_0^3}}{\left[1 - \left(\frac{\omega^2}{\omega_0^2}\right)^2\right]^2 + 4\xi_0^2\left(\frac{\omega^2}{\omega_0^2}\right)^2} \tag{1.1.77}$$

It is characterized by the parameter ω_0 (which can be regarded as the fundamental natural circular frequency of the ground) and ξ_0 (which describes its damping properties). Equation (1.1.78) presents the transfer function of a simple first order high-pass filter, characterized by its corner frequency ω_H, which, in conjunction with the low-pass filter of Eq. (1.1.79), characterized by its corner frequency ω_T, can be used as a simple band-pass filter.

$$H(\omega) = \frac{\omega^2 + i\,\omega\,\omega_H}{\omega_H^2 + \omega^2} \tag{1.1.78}$$

$$H(\omega) = \frac{\omega_T^2 - i\omega\omega_T}{\omega_T^2 + \omega^2} \tag{1.1.79}$$

To illustrate, Fig. 1.23 shows a time function f(t) consisting of three sine waves as given by Eq. (1.1.80)

$$f(t) = 1 \cdot \sin\frac{2\pi t}{1.024} + 0.5 \cdot \sin\frac{2\pi t}{0.1024} + 0.25 \cdot \sin\frac{2\pi t}{0.01024} \tag{1.1.80}$$

Using the program FILTER (see Appendix) and applying a KANAI-TAJIMI filter with $\omega_0 = 10$ rad/s and $\xi_0 = 30\%$ results in the signal shown in Fig. 1.24.

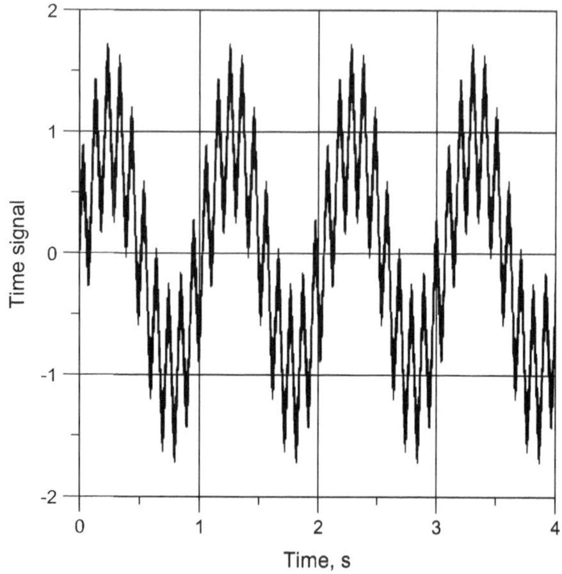

Fig. 1.23 Time signal for filtering in the frequency domain

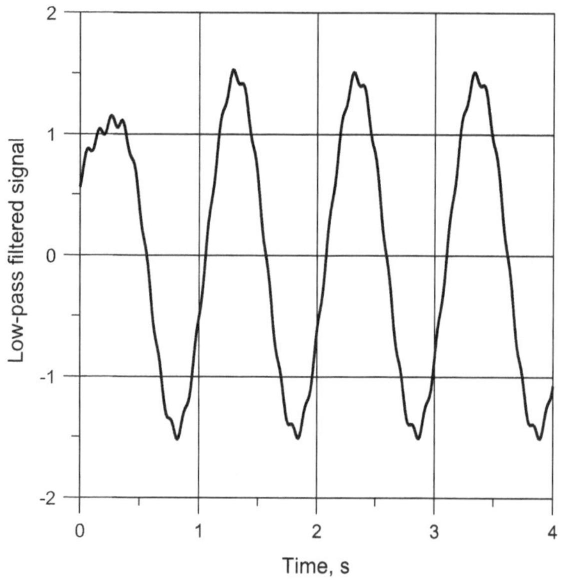

Fig. 1.24 The KANAI-TAJIMI filtered time signal Eq. (1.1.80)

1.1.3 Nonlinear SDOF Systems in the Time Domain

There are many types of nonlinear vibration phenomena associated with SDOF systems, e.g. if nonlinear damping mechanisms such as friction damping must be taken into account. Here, we will only consider the case of the restoring force F_R being a nonlinear function of the displacement u, as mirrored in the equation

$$m\ddot{u} + c\dot{u} + F_R(u) = F(t) \tag{1.1.81}$$

Only Direct Integration methods in the time domain are generally applicable for solving such equations. Standard frequency domain methods as well as the DUHAMEL integral approach cannot be employed because the superposition principle is not valid for nonlinear system behaviour and the methods mentioned depend upon the superposition of harmonic contributions in the frequency domain (the former) or impulsive contributions in the time domain (the latter). On the other hand, in employing time domain Direct Integration methods, it is a simple task to update the system configuration from time step to time step and to compute the currently valid restoring force corresponding to the actual displacement.

While there exist many nonlinear force-deformation laws of varying degrees of complexity, for standard applications simple multi-linear (e.g. bilinear) laws are usually sufficiently accurate for modelling the nonlinear restoring force, e.g. such as the elastic-perfectly plastic model shown in Fig. 1.25.

Other simple models are the bilinear model of Fig. 1.26a and the origin-oriented (UMEMURA) model shown in Fig. 1.26b.

The procedure for determining the relevant value of the restoring force for the elastic-perfectly plastic model for given displacement and velocity values of the SDOF system can be summarized as follows (see Fig. 1.25):

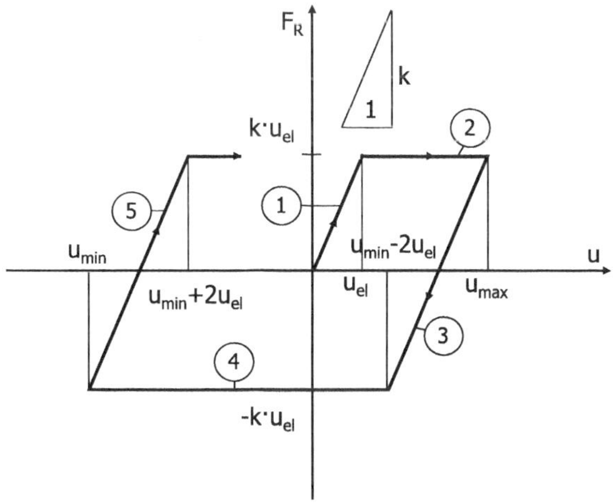

Fig. 1.25 Elastic-perfectly plastic model

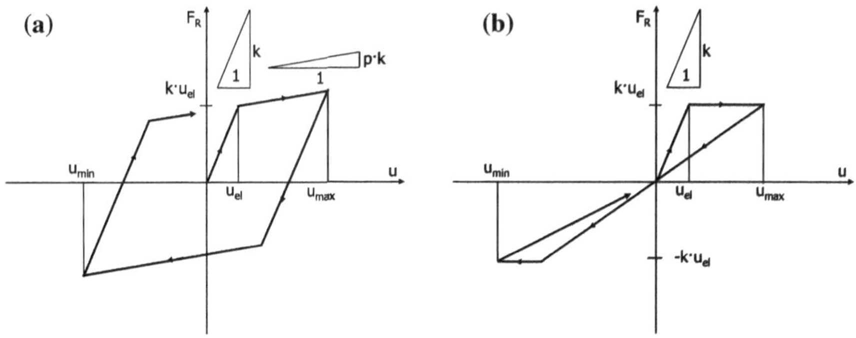

Fig. 1.26 **a** Bilinear model, **b** UMEMURA model

Region 1, elastic: $-u_{el} \leq u \leq u_{el}$, Restoring force $F_R = k\, u$

Region 2, plastic: $u > u_{el}$, $\dot{u} > 0$ (Increasing displacement), $F_R = k\, u_{el}$

As soon as the displacement stops increasing, $\dot{u} < 0$, Region 3 is entered and u_{max} is defined.

Region 3, elastic: $u_{max} - 2u_{el} \leq u \leq u_{max}$, $F_R = k \cdot (u_{el} - u_{max} + u)$

Region 4, perfectly plastic: $u < u_{max} - 2\,u_{el}$, $\dot{u} < 0$ and $F_R = -k \cdot u_{el}$

As soon as $\dot{u} > 0$, Region 5 is entered and the value of u_{min} is defined.

Region 5, elastic: $u_{min} \leq u \leq u_{min} + 2\,u_{el}$ and $F_R = k \cdot (u - u_{el} - u_{min})$

The computational procedure may be described as follows: Assume that a change in the restoring force has occurred during the time step between t_1 and t_2, e.g. because u has exceeded u_{el}. The equilibrium equation involving inertia, damping and restoring forces as well as the external loading at time t_2 is given by

$$F_I + F_D + F_R = F(t_2) \tag{1.1.82}$$

This equation is no longer satisfied, because the actual restoring force $F_R(t_2)$ is now different from the value it would have had if no system change had taken place. This leads to an unbalanced force R:

$$R = F(t_2) - [m\,\ddot{u}(t_2) + c\,\dot{u}(t_2) + F_R(t_2)] \tag{1.1.83}$$

There exist various alternatives for taking the unbalanced force R into account:

1. R is simply ignored and the computed displacement $u(t_2)$ is not modified; only the stiffness of the system is updated before continuing the computation for the next time step.
2. The computed solution $u(t_2)$ is not modified; however, R is considered in the external loading of the next time step.
3. The solution $u(t_2)$ is corrected iteratively, until the absolute value of R does not exceed a certain tolerance limit.

In the first two cases there is always the possibility that the computed solution and the correct solution will diverge by a certain degree. A simple method for checking the accuracy of the computed solution is to repeat the calculation with a smaller time step. The third alternative is generally the best, especially in the case of multilinear hysteretic laws. In these cases, the iteration process normally ends after a single cycle, since there is no further change in stiffness. However, if the hysteretic law contains curved branches, more than one iteration cycle is usually necessary until the absolute value of the unbalanced force R drops below the required threshold.

The incremental form of Eq. (1.1.81) is

$$m\,\Delta\ddot{u} + c\Delta\dot{u} + k\Delta u = \Delta F \tag{1.1.84}$$

This equation can be solved by the implicit NEWMARK integration scheme (with constant or linear acceleration during a time step). The displacement increment is computed as in the linear case, Eq. (1.1.22). At time $t = t_2$ we obtain

$$\begin{aligned} \ddot{u}_2 &= \ddot{u}_1 + c_1\,\Delta u - c_2\,\dot{u}_1 - c_3\ddot{u}_1 \\ \dot{u}_2 &= \dot{u}_1 + c_4\,\Delta u - c_5\,\dot{u}_1 - c_6\ddot{u}_1 \\ u_2 &= u_1 + \Delta u \end{aligned} \tag{1.1.85}$$

with

$$\begin{aligned} c_1 &= \frac{1}{\beta(\Delta t)^2}; \quad c_2 = \frac{1}{\beta\,\Delta t}; \quad c_3 = \frac{1}{2\beta}; \quad c_4 = \frac{\gamma}{\beta\,\Delta t}; \\ c_5 &= \frac{\gamma}{\beta}; \quad c_6 = \Delta t\left(\frac{c_5}{2} - 1\right) \end{aligned} \tag{1.1.86}$$

Next, the current restoring force $F_R(u)$ at time t_2 is determined from the assumed hysteretic law and the unbalanced force R evaluated. If it exceeds a given threshold, an iteration cycle is entered, else the computation proceeds as in the linear case. In the former case, a new generalised stiffness k^*_{new} is computed, based on the current system stiffness $k_{updated}$ at time t_2 and a correction δu of the displacement increment determined:

$$k^*_{new} = k_{updated} + c_1 m + c_4 c \tag{1.1.87}$$

$$\delta u = \frac{R}{k^*_{new}} \tag{1.1.88}$$

The updated values for $t = t_2$ are given by:

$$\begin{aligned} \ddot{u}_2 &= \ddot{u}_{2,old} + c_1 \cdot \delta u \\ \dot{u}_2 &= \dot{u}_{2,old} + c_4 \cdot \delta u \\ u_2 &= u_{2,old} + \delta u \end{aligned} \tag{1.1.89}$$

With these updated values a new unbalanced force R is computed. If the convergence criterion is met, the computation proceeds with the next time step, else a new iteration cycle is necessary.

The computer program SDOFNL (see Appendix) evaluates time responses of SDOF systems with nonlinear spring laws of elastoplastic, bilinear or UMEMURA type. The input file RHS.txt contains the ordinates of the loading function at a constant time interval, as in the program SDOF1. All further data are input interactively following corresponding prompts. Results are output in the file THNLM.txt and consist of time histories of the displacement, velocity and acceleration values as well as the restoring force.

Example 1.3

The plane frame of Fig. 1.27 is subject to the loading shown in Fig. 1.29. Further data: Mass m = 5 t, $EI_R = 0.70 \times 10^5$ kNm^2, $EI_1 = 0.25 \times 10^5$ kNm^2, $EI_2 = 0.50 \times 10^5$ kNm^2, damping D = 1%. Determine the time histories for u, $\dot{u}$ and $\ddot{u}$ as well as for the restoring force assuming (a) linear elastic behaviour throughout and (b) elastic-perfectly plastic behaviour with $u_{el} = 2$ cm.

Figure 1.28 shows the discretization of the frame and the input file for determining the horizontal displacement of the girder due to a unit load.

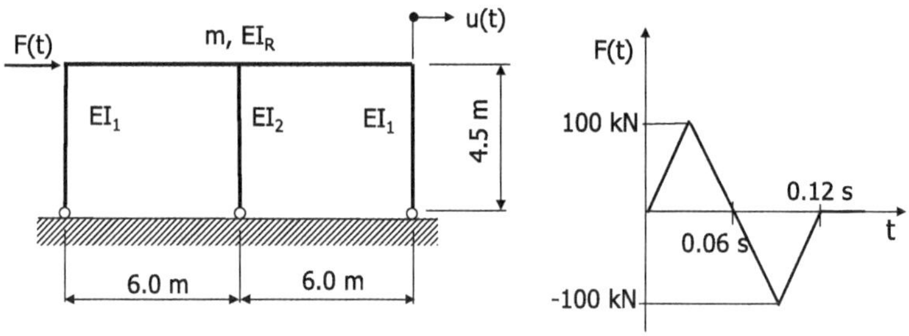

Fig. 1.27 Plane frame, Example 1.3

Discretized system	**Input file EFRAM.txt**
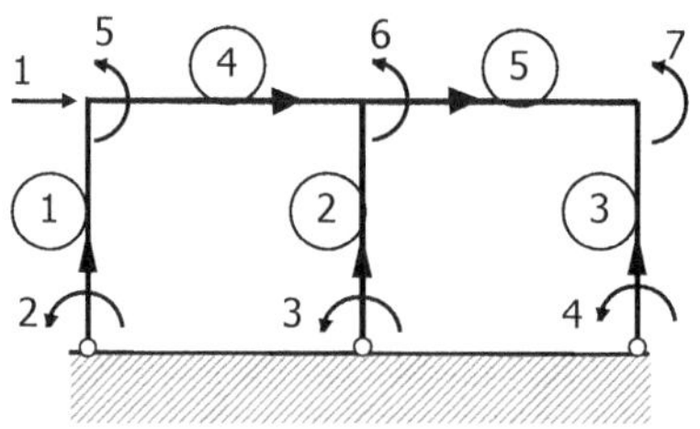	25000., 4.5, 0., 90. 50000., 4.5, 0., 90. 25000., 4.5, 0., 90. 70000., 6.0. 0., 0. 70000., 6.0. 0., 0. 0,0,2, 1,0,5 0,0,3, 1,0,6, 0,0,4, 1,0,7 1,0,5, 1,0,6, 1,0,6, 1,0,7, 1., 0., 0., 0., 0., 0., 0.,

Fig. 1.28 Discretization and input file for a unit load at DOF no. 1

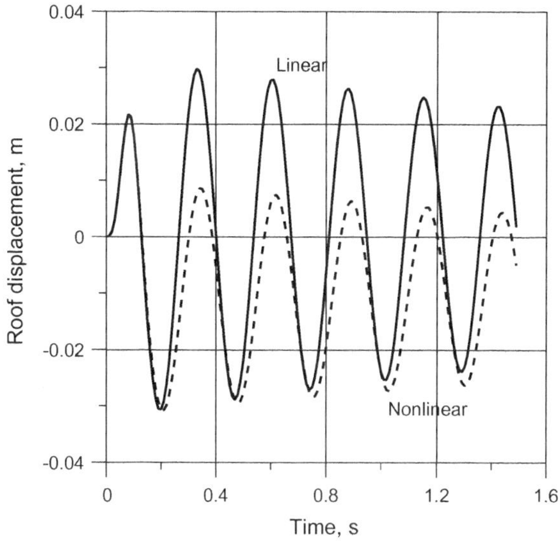

Fig. 1.29 Displacement time histories, Example 1.3

The resulting displacement at DOF no. 1 due to the unit load is equal to 0.00376 m, so that the spring stiffness amounts to k = 1/0.000376 = 2659 kN/m. This leads to a natural circular frequency of 23.06 rad/s or a natural period of 0.272 s. Discretizing the loading function by using 150 steps of 0.01 s each (program LININT) and applying the program SDOF1 for linear elastic behaviour yields a maximum absolute elastic roof displacement of 3.066 cm, corresponding to a restoring force of 81.5 kN. For the investigation of the nonlinear behaviour, the program SDOFNL is used and the elastic-perfectly plastic model option with u_{el} = 2 cm employed. Results yield a maximum absolute elastoplastic displacement of 3.08 cm (that is, practically equal to the elastic value) and a maximum absolute restoring force of 53.2 kN. Figures 1.29, 1.30, 1.31 and 1.32 compare time histories for displacements, velocities, accelerations and restoring forces for the linear and nonlinear models, with solid curves for the former case and dashed curves for the latter. Additionally, Fig. 1.33 shows the hysteretic loop (restoring force as a function of displacement) for the nonlinear case.

1.1.4 Applications of the Theory of SDOF Systems: Response Spectra

Figure 1.34 shows a SDOF system (point mass m on a massless cantilever beam with bending stiffness EI and length h) subject to a base excitation (ground motion) $\ddot{u}_g$.

The equation of motion is given by

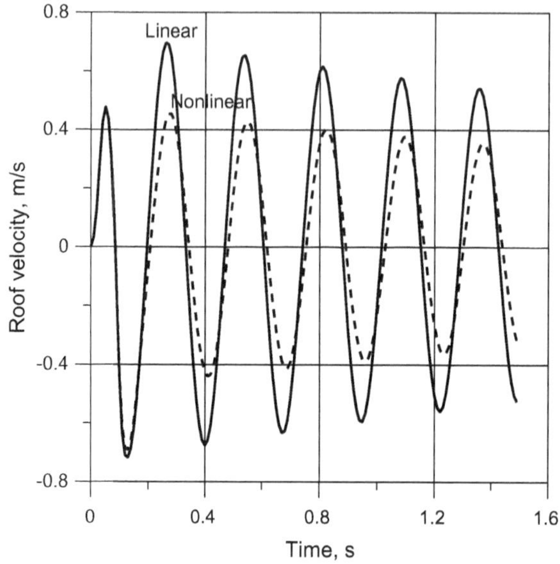

Fig. 1.30 Velocity time histories, Example 1.3

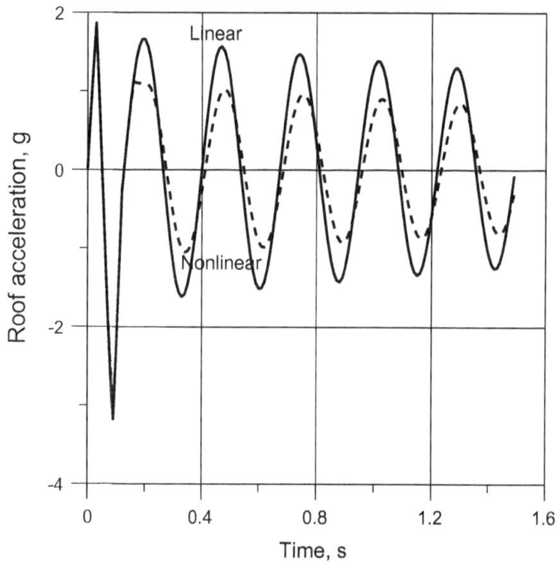

Fig. 1.31 Acceleration time histories, Example 1.3

$$m \cdot \left(\ddot{u} + \ddot{u}_g\right) + k \cdot u = 0 \qquad (1.1.90)$$

where u is the displacement relative to the base and k is the spring stiffness associated with the cantilever beam, that is $k = 3EI/h^3$. Taking additionally viscous damping into account, we obtain the expression

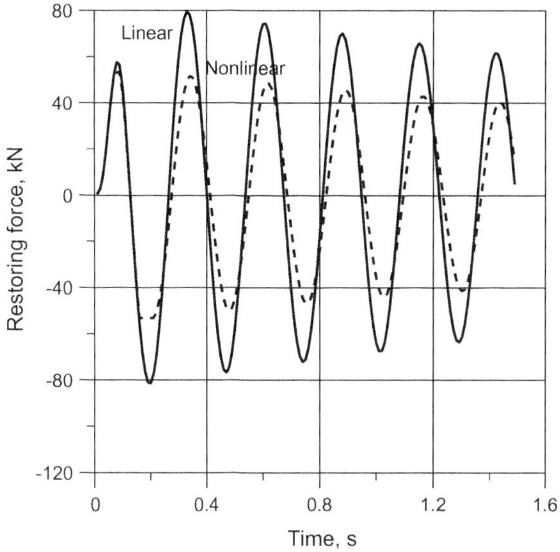

Fig. 1.32 Restoring force time histories, Example 1.3

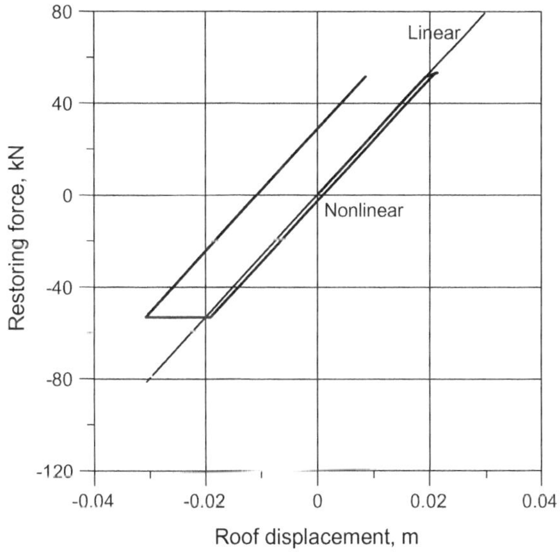

Fig. 1.33 Restoring force versus displacement, Example 1.3

$$m \cdot \ddot{u} + c \cdot \dot{u} + k \cdot u = (-)m \cdot \ddot{u}_g \tag{1.1.91}$$

The negative sign at the right-hand side is usually suppressed, since we are only interested in the absolute value of the maximum relative displacement max $|u| = S_d$ (spectral displacement). Dividing Eq. (1.1.91) by the mass m and introducing

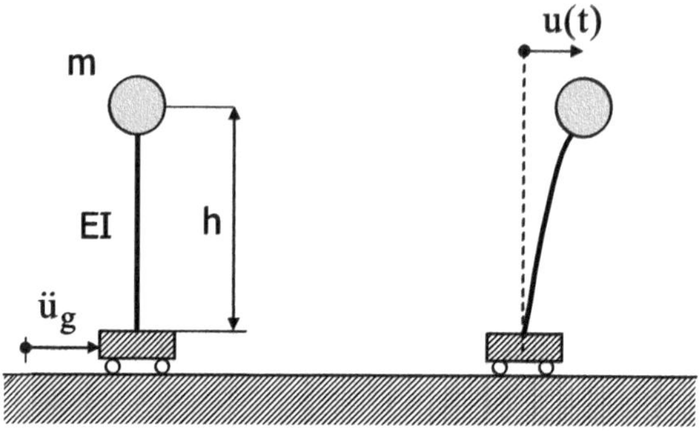

Fig. 1.34 SDOF system with base excitation

the natural circular frequency ω_1 of the SDOF system as well as its damping as percentage of critical damping (D or ξ), yields the ODE

$$\ddot{u} + 2 \cdot D \cdot \omega_1 \dot{u} + \omega_1^2 \cdot u = \ddot{u}_g(t) \tag{1.1.92}$$

The computed max $|u| = S_d$ values as function of the damping ratio D and the natural frequency ω_1 are the ordinates of the so-called displacement spectrum. Pseudo relative velocity (PSV) ordinates S_v and pseudo absolute acceleration (PAA) ordinates S_a are additionally defined as functions of S_d according to

$$S_d = (1/\omega_1) \cdot S_v = \left(1/\omega_1^2\right) \cdot S_a \tag{1.1.93}$$

The prefix "pseudo" for S_v and S_a reflects the fact that they do not necessarily represent maximum values for the velocity and acceleration of the SDOF system but are computed assuming a purely harmonic vibration.

For a given base excitation, a computer program such as SDOF1 can easily determine the maximum relative displacement S_d for a fixed D and a series of ω_1 values, and also furnish the (relative) displacement, (pseudo relative) velocity and (pseudo absolute) acceleration spectrum. It is meaningful to plot the results in a tripartite logarithmic diagram as shown schematically in Fig. 1.35, with the pseudo relative velocity spectral ordinates S_v plotted over the natural periods. From such tripartite logarithmic diagrams one can directly obtain the S_a and S_d ordinates along orthogonal axes rotated by 45° with respect to the (T, S_v) co-ordinate system, as shown in e.g. in Fig. 1.35 or Fig. 1.38. For very small periods (rigid systems), the spectral acceleration is constant and equal to the peak ground acceleration (PGA), for very large periods (very flexible systems) the diagram shows a constant spectral displacement equal to the maximum ground displacement.

For computing response spectra of given ground motion records the program SPECTRUM (see Appendix) can be used. It furnishes S_d, S_V, S_a and $S_{a,abs}$ (absolute acceleration relative to a fixed frame of reference) values as functions of natural periods T and a constant damping ratio, usually D = 5%. Additionally, it evaluates the so-called HOUSNER intensity SI, defined as

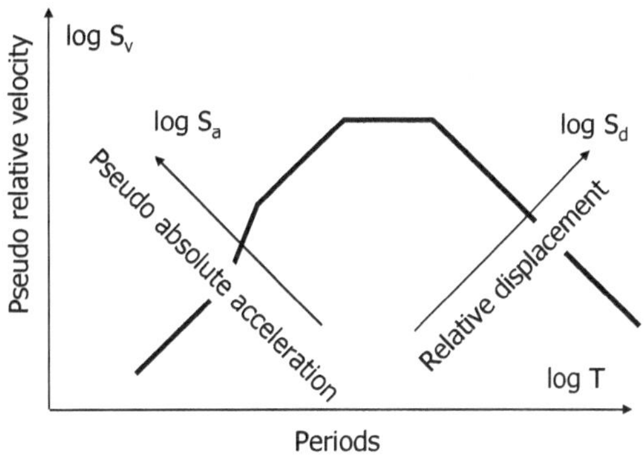

Fig. 1.35 Logarithmic plot of a response spectrum (schematically)

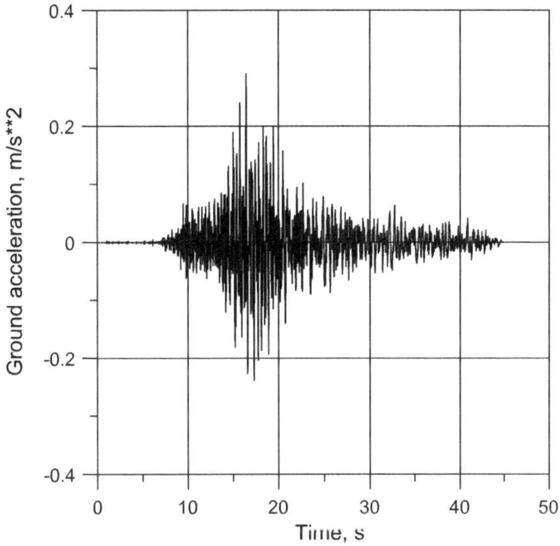

Fig. 1.36 Acceleration record of the Roermond 1992 earthquake

$$SI(D) = \int_{T=0.1}^{2.5} S_v(T, D)dT \tag{1.1.94}$$

As an example, consider the ground motion record shown in Fig. 1.36. It is the NS component of the 1992 Roermond earthquake measured in Bergheim, with a spectral intensity of 3.28 cm.

Figure 1.37 depicts its pseudo absolute acceleration spectrum for D = 5%, while Fig. 1.38 shows the logarithmic plot of the (S_V, T) relationship for the same motion.

In addition to the elastic response spectra, it is also possible to compute inelastic response spectra based on the elastic-perfectly plastic law of Fig. 1.25 and assuming a certain target ductility μ, defined by:

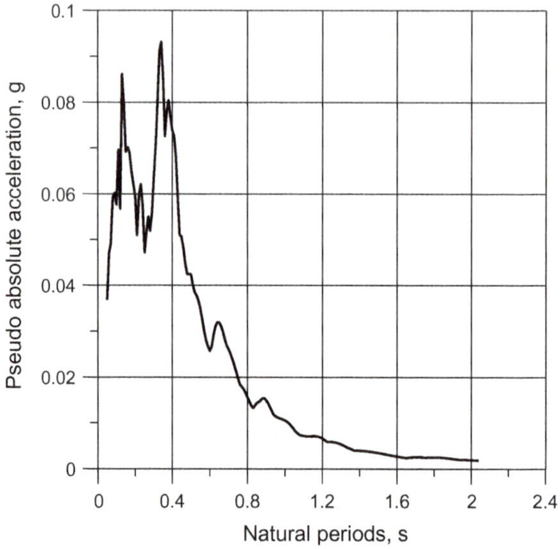

Fig. 1.37 Pseudo absolute acceleration spectrum, Roermond record

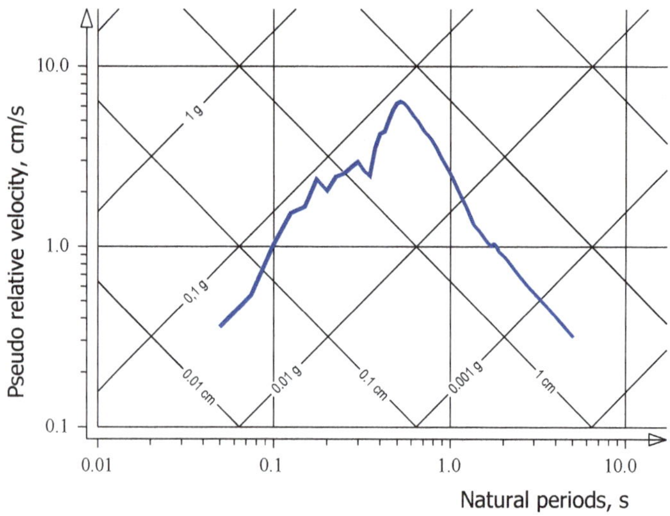

Fig. 1.38 Logarithmic tripartite spectrum (D = 5%), Roermond record

$$\mu = \frac{u_{max}}{u_{el}} \tag{1.1.95}$$

This ductility value is the additional parameter needed to characterize the inelastic response spectrum. Its ordinates refer to the maximum elastic displacement u_{el}, with $S_d = u_{el}$, $S_v = u_{el} \cdot \omega$ and $S_a = u_{el} \cdot \omega^2$. Inelastic response spectra of given accelerograms for arbitrary target ductilities μ can be computed by means of the program

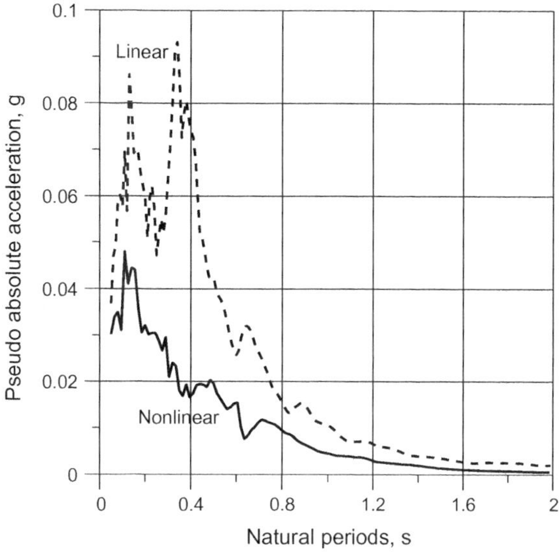

Fig. 1.39 Inelastic ($\mu = 2.5$) and elastic spectrum, Roermond record

NLSPEC (see Appendix). Its output file contains the natural periods (in s), the spectral ordinates $S_d = u_{el}$ for the displacement (in cm), S_v for the pseudo relative velocity (in cm/s) and S_a for the pseudo absolute acceleration (in g) in the first four columns, and, in the fifth column, the maximum ductility μ that was actually reached, which does not always coincide exactly with the target ductility that has been specified. The program's CPU time requirements are quite noticeable, especially for long records, due to the iterative process used for the determination of the correct spectral ordinates. It is also possible that no satisfactory solution can be found for a combination of target ductility, natural period and acceleration record, since not every ductility value is physically attainable for a certain system and excitation record. In this case, after having reached the internally determined maximum iteration cycle number, the program simply uses the best approximation found so far and proceeds to the next period value.

To illustrate, Fig. 1.39 depicts the inelastic acceleration response spectrum for the Roermond record of Fig. 1.36 for a target ductility $\mu = 2.5$ (solid line). The corresponding elastic response spectrum of Fig. 1.37 is included as a dashed line; it is obvious that by allowing a certain degree of inelastic action, it is possible to drastically reduce structural seismic demands.

Analyzing the nonlinear behaviour of seismically excited structures is usually carried out by means of time-domain direct integration methods, which furnish the system's response to explicit ground motion records. It is not always possible to obtain suitable accelerograms measured near the site in question to be used directly in such an investigation, in which case numerically generated records must be computed, taking into account all available site information (e.g. soil type and distance from the nearest seismic fault). The simplest approach lies in utilising an elastic response

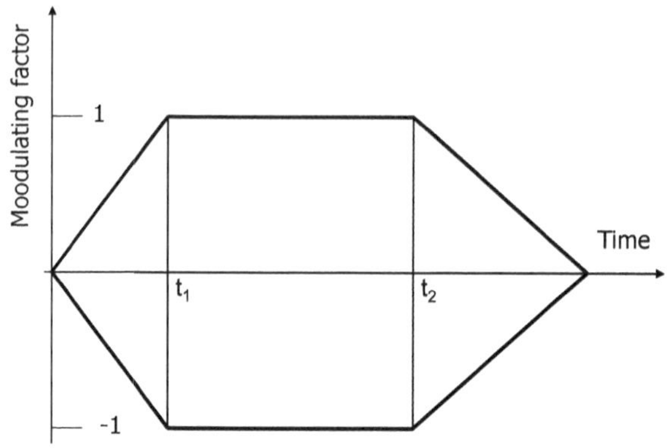

Fig. 1.40 Trapezoidal intensity function for modulating synthetic accelerograms

spectrum for the site and generating an acceleration record the response spectrum of which closely matches the original "target spectrum" in the period range of interest. A suitable value for the duration of the record must be chosen additionally, since this information is not directly contained in the target spectrum.

The program SYNTH (see Appendix) produces such synthetic records as sums of sine waves with random phase angles (uniformly distributed between zero and 2π) and iteratively adjusts it so that it more or less fits the target spectrum; the deterministic trapezoidal intensity function of Fig. 1.40 is used for modulating the stationary signal resulting from superposing the single harmonic components.

In SYNTH, the target spectrum is described by NK pairs of (T, Sv) values in s and cm/s, and it is assumed to be piecewise linear in a tripartite logarithmic plot. The period range of interest, where the target spectrum is to be approximated by the spectrum of the generated record, is defined by entering the two periods TANF and TEND (TANF < TEND) bracketing this range. The interval (TEND – TANF) should lie within the spectral polygon defined by the NK (T, Sv) values; standard values are TANF = 0.1 s and TEND = 2.5 s. Alternative ways of creating artificial earthquake records are:

1. Modulating white noise records with variable filters (e.g. KANAI-TAJIMI).
2. Computing bedrock response spectra and transforming corresponding records for vertically propagating shear waves all the way to the surface using one-dimensional models (e.g. SHAKE 91 program).
3. Using advanced geophysical simulation models which can also consider 2D or 3D effects.

Example 1.4

Determine a spectrum-compatible accelerogram for the elastic response spectrum of Eurocode 8 defined by the expressions in Eq. (1.1.96) as the target spectrum. In Eq. (1.1.96), $S_e(T)$ is the spectral acceleration, a_g the design ground acceleration on type A ground (rock), T_B, T_C and T_D the corner periods (here assumed as 0.05, 0.20 and 2.0 s, respectively), S is the soil factor and η the damping correction factor

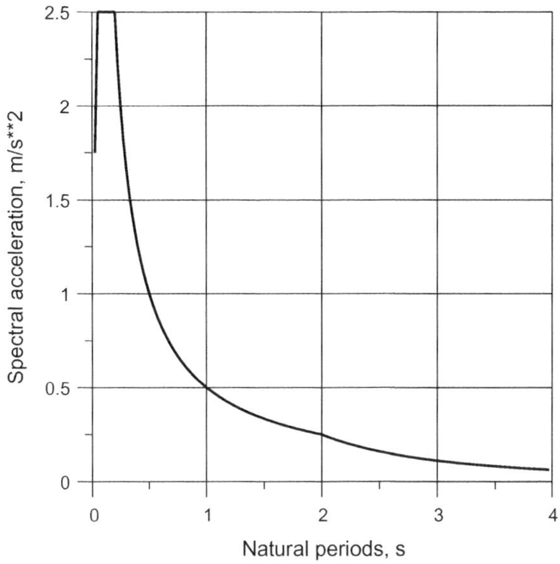

Fig. 1.41 Elastic EC 8 spectrum for ground type A, linear (S_a-T)-plot

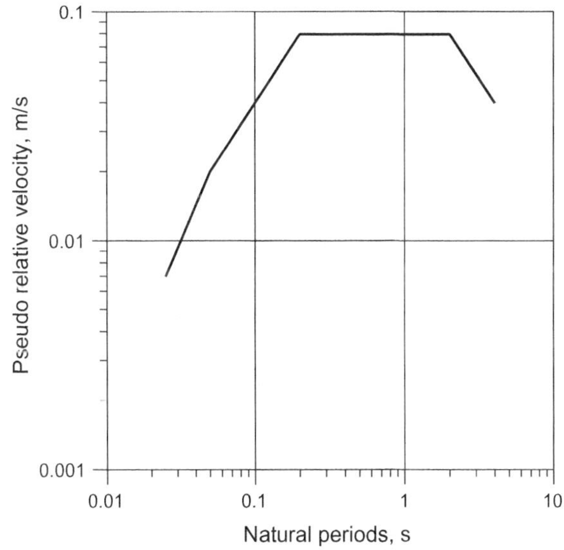

Fig. 1.42 Elastic EC 8 spectrum for ground type A, logarithmic (S_V-T)-plot

(assumed equal to 1 for 5% damping, as is here the case). This code spectrum, normalized to $a_g \cdot S = 1.0$ m/s^2, is shown in Figs. 1.41 and 1.42.

$$
\begin{aligned}
&0 \le T \le T_B: \quad S_e(T) = a_g \cdot S \cdot \left[1 + \frac{T}{T_B}(\eta \cdot 2.5 - 1)\right] \\
&T_B \le T \le T_C: \quad S_e(T) = a_g \cdot S \cdot \eta \cdot 2.5 \\
&T_C \le T \le T_D: \quad S_e(T) = a_g \cdot S \cdot \eta \cdot 2.5 \cdot (T_C/T) \\
&T_D \le T \le 4s: \quad S_e(T) = a_g \cdot S \cdot \eta \cdot 2.5 \cdot (T_C \cdot T_D/T^2)
\end{aligned} \tag{1.1.96}
$$

A nominal duration of 15 s has been chosen for the synthetic accelerogram to be computed, with the trapezoidal intensity function of Fig. 1.40 and $t_1 = 2$ s, $t_2 = 11$ s. The target spectrum, described by straight lines in the logarithmic (S_v-T) diagram of Fig. 1.42 is described by NK = 5 points as follows:

Natural period (s)	Pseudo-relative velocity (cm/s)
0.025	0.6963
0.050	1.98944
0.20	7.95775
2.0	7.95775
4.0	3.97887

The program SYNTH delivers the record shown in Fig. 1.43; its response spectrum is depicted in Fig. 1.44 together with the target spectrum (dashed line). The prescribed natural period range to be approximated by the synthetic record was between 0.1 and 2.5 s.

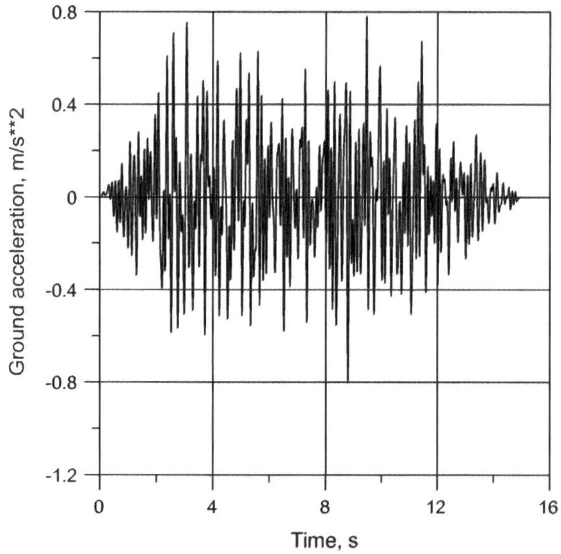

Fig. 1.43 Artificial spectrum compatible accelerogram

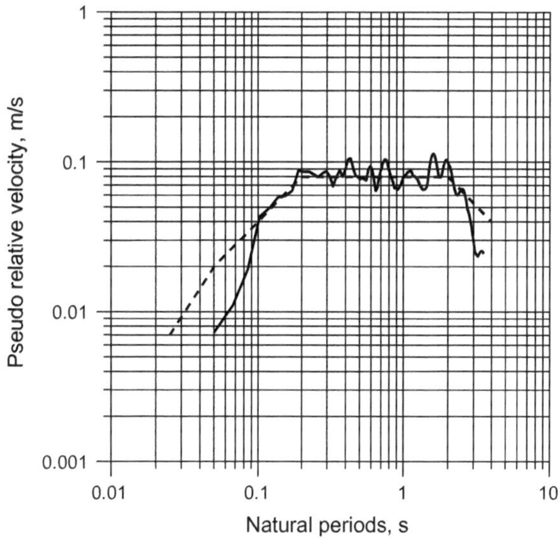

Fig. 1.44 Response spectrum of the accelerogram Fig. 1.44 and corresponding target spectrum (dashed line)

1.2 Discrete Multi-degree of Freedom Systems

Multi-degree-of freedom (MDOF) systems typically consist of discrete masses, springs and dashpots and the resulting differential equation system has the form

$$\underline{F}_I + \underline{F}_D + \underline{F}_R = \underline{F} \tag{1.2.1}$$

In this equation $\underline{F}$ is the loading vector, $\underline{F}_I$ the vector of inertial forces, $\underline{F}_D$ the vector of damping forces and $\underline{F}_R$ the vector of restoring forces. Equation (1.2.1) is analogous to the equilibrium condition of the SDOF system with vectors replacing scalar quantities. The actual differential equation system expressing the equilibrium between inertia, damping and restoring forces and the external loading is given by

$$\underline{M}\,\ddot{\underline{V}} + \underline{C}\,\dot{\underline{V}} + \underline{K}\,\underline{V} = \underline{F} \tag{1.2.2}$$

The (n, 1) column vector $\underline{V}$ contains the displacements in the n degrees of freedom of the MDOF system while $\dot{\underline{V}}$, $\ddot{\underline{V}}$ denote the corresponding velocities and accelerations. $\underline{K}$ is the stiffness matrix of the system, $\underline{M}$ its mass matrix and $\underline{C}$ the damping matrix, all matrices being square (n, n) matrices. For systems consisting of discrete springs, masses and (viscous) dashpots the derivation of these matrices is, at least in principle, a straightforward task. However, it is not immediately obvious how to meaningfully model real-life structures which possess distributed mass, stiffness and damping properties and an infinite number of degrees of freedom. Generally, these properties can be expressed by a finite number of suitably chosen generalised coordinates (e.g. by discretizing the continuum using finite element formulations) or on a

Fig. 1.45 Plane frame

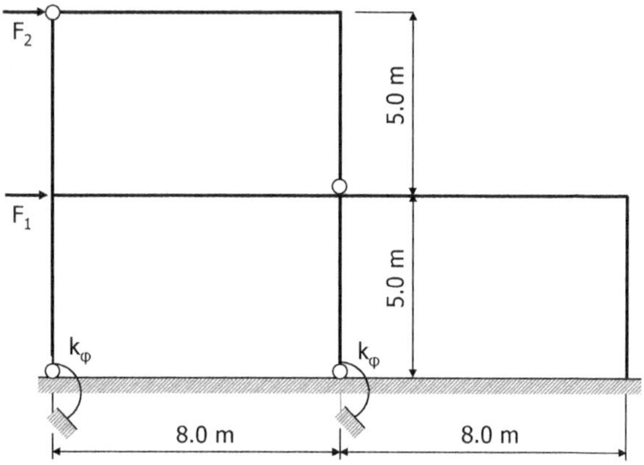

semi-empirical basis by introducing discrete concentrated masses, stiffness matrices derived from analytical or experimental results and meaningful damping models. In the interest of reducing the work load associated with dynamic analyses, is important to minimise the number n of the system's degrees of freedom. This can be done with the aid of condensation and substructuring techniques as described later on. In the following, only MDOF systems with discrete point masses ("lumped-mass" idealisations) are considered because of their versatility and widespread practical applicability. For these models the mass matrix is a diagonal matrix and if the additional assumption of "modal damping" is made, the solution of Eq. (1.1.2) becomes quite straightforward.

Before dealing with condensation techniques, a short recap of the matrix structural analysis procedure by the displacement method will be given in the following example.

Example 1.5

Consider the plane frame under static loading ($F_1 = 20$ kN, $F_2 = 10$ kN) shown in Fig. 1.45, for which the resulting deformations and internal forces (bending moments, shear and axial forces) are to be determined. The columns have a bending stiffness $EI = 3.24 \times 10^5$ kNm^2 and an axial stiffness $EA = 1.08 \times 10^7$ kN; the corresponding values for the ground floor girders are $EI = 1.09 \times 10^6$ kNm^2 and $EA = 1.62 \times 10^7$ kN and for the roof girder $EI = 5.15 \times 10^5$ kNm^2 and $EA = 1.26 \times 10^7$ kN. Finally, the rotational spring constant k_φ is equal to 5.0×10^6 kNm/rad.

The determination of the internal forces and displacements is carried out by first solving the algebraic equation system Eq. (1.2.3) for the system deformations $\underline{V}$:

$$\underline{K}\,\underline{V} = \underline{F} \tag{1.2.3}$$

With $\underline{V}$ known, the end section deformations of each beam element corresponding to the DOF shown in Fig. 1.46 are readily available (vector $\underline{v}$) and the internal forces (vector $\underline{s}$) can be computed through

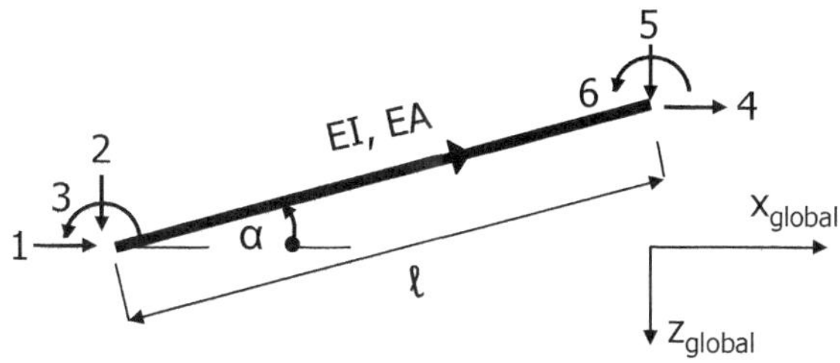

Fig. 1.46 6 DOF plane beam element with global coordinate system

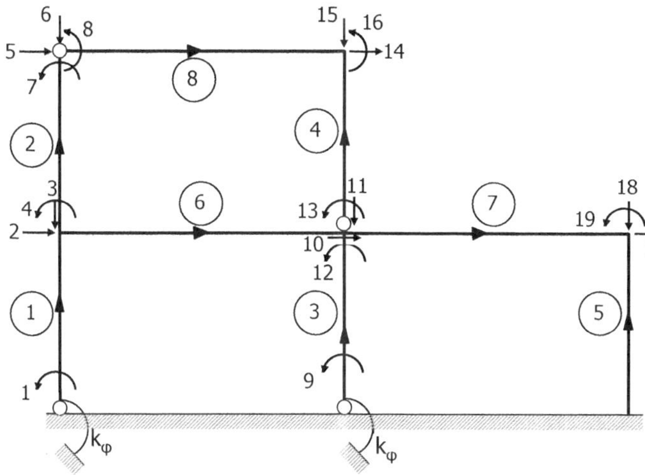

Fig. 1.47 Discretization of the plane frame of Fig. 1.1

$$\underline{k}\,\underline{v} = \underline{s} \tag{1.2.4}$$

where k is the (6, 6) stiffness matrix of the beam element in question.

A possible discretization of the plane frame Fig. 1.45 using the standard beam element with the six DOF of Fig. 1.46 is shown in Fig. 1.47. The model consists of 8 beam elements with 19 system DOF in total.

In the input file EFRAM.txt of the program FRAME, the first 8 rows contain the (bending and axial) stiffness values, the lengths and the angles α for the 8 beam elements, followed by the corresponding incidence vectors and the system loading vector (see description in Appendix). The (2, 2) spring support matrices corresponding to the foundation springs are entered in the file FEDMAT.txt and their incidence vectors in INZFED.txt. Here, FEDMAT.txt consists of the following two rows

5e6, −5e6, −5e6, 5e6
5e6, −5e6, −5e6, 5e6

and INZFED.txt of the rows

0, 1
0, 9

Results are output in the file AFRAM.txt and (partly) reproduced below. Given are the internal forces and displacements in the global (x, z) coordinate system according to sign convention II, meaning that positive values at both end sections are defined in the direction of the positive global (x, z) axes, rotations and bending moments acting counter-clockwise when positive. For each beam element, the first row contains the deformation values corresponding to the DOF 1–6 of Fig. 1.46, the second row the associated internal forces (x- and z-force components followed by bending moments).

Beam no. 1

0.0000E + 00 0.0000E + 00 −0.4555E−05 0.4373E−03 −0.4864E−05 −0.7755E−04
−0.7218E + 01 0.1051E + 02 0.2277E + 02 0.7218E + 01 −0.1051E + 02 0.1331E + 02

Beam no. 2

0.4373E−03 −0.4864E−05 −0.7755E−04 0.1554E−02 −0.6117E−05 −0.2963E−03
−0.5671E + 01 0.2705E + 01 0.2836E + 02 0.5671E + 01 −0.2705E + 01 0.1817E−13

Beam no. 3

0.0000E + 00 0.0000E + 00 −0.6313E−05 0.4282E−03 0.2900E−05 −0.7358E−06
−0.1277E + 02 −0.6263E + 01 0.3157E + 02 0.1277E + 02 0.6263E + 01 0.3229E + 02

Beam no. 4

0.4282E−03 0.2900E−05 −0.2804E−03 0.1552E−02 0.4152E−05 −0.1134E−03
−0.4329E + 01 −0.2705E + 01 0.1943E−13 0.4329E + 01 0.2705E + 01 0.2164E + 02

Beam no. 5

0.0000E + 00 0.0000E + 00 0.0000E + 00 0.4233E−03 0.1965E−05 −0.4056E−04
−0.1001E + 02 −0.4244E + 01 0.2766E + 02 0.1001E + 02 0.4244E + 01 0.2240E + 02

Beam no. 6

0.4373E−03 −0.4864E−05 −0.7755E−04 0.4282E−03 0.2900E−05 −0.7358E−06
0.1845E + 02 0.7801E + 01 −0.4167E + 02 −0.1845E + 02 −0.7801E + 01 −0.2074E + 02

Beam no. 7

0.4282E−03 0.2900E−05 −0.7358E−06 0.4233E−03 0.1965E−05 −0.4056E−04
0.1001E + 02 0.4244E + 01 −0.1155E + 02 −0.1001E + 02 −0.4244E + 01 −0.2240E + 02

Beam no. 8

0.1554E−02 −0.6117E−05 0.5475E−04 0.1552E−02 0.4152E−05 −0.1134E−03
0.4329E + 01 0.2705E + 01 0.1144E−14 −0.4329E + 01 −0.2705E + 01 −0.2164E + 02

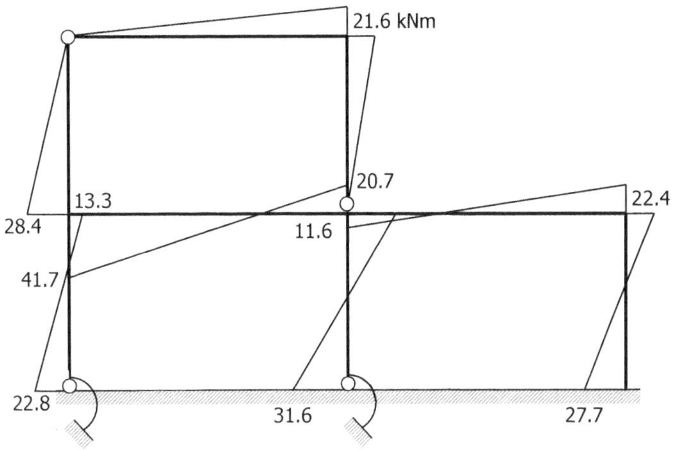

Fig. 1.48 Bending moment diagram in kNm

The resulting bending moment diagram is shown in Fig. 1.48.

1.2.1 Condensation Techniques

The dimension n of the matrices in Eq. (1.2.2), that is the number of components of the vectors $\underline{V}$, $\dot{\underline{V}}$ and $\ddot{\underline{V}}$, should be as small as possible in order to minimise the solution effort. One way to achieve this is to consider only the structural degrees of freedom associated with large, or at any rate substantial inertia forces, which are denoted as "master" degrees of freedom. This does not mean that the displacements, velocities and accelerations in the remaining "slave" degrees of freedom are neglected but simply that they are expressed as functions of the corresponding values of the master degrees of freedom. The concept of expressing a group of degrees of freedom as functions of other degrees of freedom ("condensation") is also quite useful when different parts of a system (substructures) are to be investigated separately before being integrated in the overall structure. In this case, the "coupling" degrees of freedom connecting the different substructures are considered as master degrees of freedom and the substructure itself can be regarded as a "macroelement" with "master" coupling DOF which express the contribution of the particular substructure to the whole assemblage. Formally, the first step lies in distinguishing between master and slave degrees of freedom, with the latter, $\underline{V}_{\varphi}$, to be expressed as functions of the former, $\underline{V}_u$. The corresponding matrix equation for the undamped system is:

$$\begin{pmatrix} \underline{M}_{uu} & \underline{M}_{u\varphi} \\ \underline{M}_{\varphi u} & \underline{M}_{\varphi\varphi} \end{pmatrix} \cdot \begin{pmatrix} \ddot{\underline{V}}_u \\ \ddot{\underline{V}}_{\varphi} \end{pmatrix} + \begin{pmatrix} \underline{K}_{uu} & \underline{K}_{u\varphi} \\ \underline{K}_{\varphi u} & \underline{K}_{\varphi\varphi} \end{pmatrix} \cdot \begin{pmatrix} \underline{V}_u \\ \underline{V}_{\varphi} \end{pmatrix} = \begin{pmatrix} \underline{F}_u \\ \underline{F}_{\varphi} \end{pmatrix} \tag{1.2.5}$$

The n master DOF $\underline{V}_u$ are those associated with significant inertia forces. The remaining DOF are given as

$$\begin{aligned}\underline{V}_\varphi &= \underline{a}\,\underline{V}_u,\\ \dot{\underline{V}}_\varphi &= \underline{a}\,\dot{\underline{V}}_u,\\ \ddot{\underline{V}}_\varphi &= \underline{a}\,\ddot{\underline{V}}_u\end{aligned} \tag{1.2.6}$$

leading to

$$\underline{V} = \begin{bmatrix}\underline{V}_u\\ \underline{V}_\varphi\end{bmatrix} = \begin{bmatrix}\underline{V}_u\\ \underline{a}\,\underline{V}_u\end{bmatrix} = \begin{bmatrix}\underline{I}\\ \underline{a}\end{bmatrix}\underline{V}_u = \underline{A}\,\underline{V}_u \tag{1.2.7}$$

with similar expressions for the velocity and acceleration vectors $\dot{\underline{V}}$, $\ddot{\underline{V}}$. The original system of differential equations (still without damping) is reduced to

$$\begin{aligned}&\underline{M}\,\underline{A}\,\ddot{\underline{V}}_u + \underline{K}\,\underline{A}\,\underline{V}_u = \underline{F}\\ &\underline{A}^T\underline{M}\,\underline{A}\,\ddot{\underline{V}}_u + \underline{A}^T\underline{K}\,\underline{A}\,\underline{V}_u = \underline{A}^T\underline{F}\end{aligned} \tag{1.2.8}$$

or

$$\tilde{\underline{M}}\,\ddot{\underline{V}}_u + \tilde{\underline{K}}\,\underline{V}_u = \tilde{\underline{F}} \tag{1.2.9}$$

with

$$\begin{aligned}\tilde{\underline{M}} &= \underline{A}^T\underline{M}\,\underline{A}\\ \tilde{\underline{K}} &= \underline{A}^T\underline{K}\,\underline{A}\\ \tilde{\underline{F}} &= \underline{A}^T\underline{F}\end{aligned} \tag{1.2.10}$$

The matrix $\tilde{\underline{K}}$ is the condensed or reduced stiffness matrix in the master degrees of freedom and $\tilde{\underline{F}}$ is the corresponding reduced load vector. Once the vectors $\underline{V}_u$, $\dot{\underline{V}}_u$, $\ddot{\underline{V}}_u$ have been determined, the unknown variables in the slave degrees of freedom can be readily computed through Eq. (1.2.6).

In the case of the standard "static condensation" procedure, static equations are used for eliminating the slave degrees of freedom, that is, expressing them as functions of a subset of master degrees of freedom. Writing down the second row of Eq. (1.2.5) in full and noting that $\underline{K}_{u\varphi}^T = \underline{K}_{\varphi u}$ leads to

$$\underline{F}_\varphi = \underline{M}_{\varphi u}\ddot{\underline{V}}_u + \underline{M}_{\varphi\varphi}\ddot{\underline{V}}_\varphi + \underline{K}_{u\varphi}^T\underline{V}_u + \underline{K}_{\varphi\varphi}\underline{V}_\varphi \tag{1.2.11}$$

Assuming that the inertia forces in the slave degrees of freedom are small, it is permissible to retain only the "static" part of this expression:

$$\begin{aligned} \underline{F}_{\varphi} &= \underline{K}_{u\varphi}^{T}\underline{V}_{u} + \underline{K}_{\varphi\varphi}\underline{V}_{\varphi} \\ \underline{V}_{\varphi} &= \underline{K}_{\varphi\varphi}^{-1}(\underline{F}_{\varphi} - \underline{K}_{u\varphi}^{T}\underline{V}_{u}) \end{aligned} \tag{1.2.12}$$

If no external forces act along slave degrees of freedom, so that all "loaded" DOF are master DOF, we obtain

$$\underline{V}_{\varphi} = -\underline{K}_{\varphi\varphi}^{-1}\underline{K}_{u\varphi}^{T}\underline{V}_{u} \tag{1.2.13}$$

leading to

$$\underline{V} = \begin{bmatrix} \underline{V}_{u} \\ \underline{V}_{\varphi} \end{bmatrix} = \begin{bmatrix} \underline{I} \\ -\underline{K}_{\varphi\varphi}^{-1}\underline{K}_{u\varphi}^{T} \end{bmatrix} \underline{V}_{u} = \underline{A}\,\underline{V}_{u} \tag{1.2.14}$$

The reduced (or condensed) mass matrix is given by

$$\begin{aligned} \underline{\tilde{M}} = \underline{M}_{uu} &- \underline{K}_{u\varphi}\underline{K}_{\varphi\varphi}^{-1}\underline{M}_{u\varphi}^{T} - \underline{M}_{u\varphi}\underline{K}_{\varphi\varphi}^{-1}\underline{K}_{u\varphi}^{T} \\ &+ \underline{K}_{u\varphi}\underline{K}_{\varphi\varphi}^{-1}\underline{M}_{\varphi\varphi}\underline{K}_{\varphi\varphi}^{-1}\underline{K}_{u\varphi}^{T} \end{aligned} \tag{1.2.15}$$

However, when dealing with lumped mass models, only the master DOF are usually endowed with masses, in which case $\underline{\tilde{M}} = \underline{M}_{uu}$. The reduced stiffness matrix is given by

$$\underline{\tilde{K}} = \underline{K}_{uu} - \underline{K}_{u\varphi}\underline{K}_{\varphi\varphi}^{-1}\underline{K}_{u\varphi}^{T} \tag{1.2.16}$$

and the load vector is simply

$$\underline{\tilde{F}} = \underline{F}_{u} \tag{1.2.17}$$

A straightforward method for obtaining the condensed stiffness matrix for $\underline{F}_{\varphi} = 0$ is the following: Unit loads corresponding to the n master degrees of freedom are applied singly and sequentially, and the resulting deformation patterns in the n master DOF stored as columns of the (n, n) flexibility matrix of the structure. The inverse of the flexibility matrix thus obtained is the condensed stiffness matrix $\underline{\tilde{K}}$.

The numerical determination of the condensed stiffness matrix and the corresponding transformation matrix A after Eq. (1.2.7) for arbitrary plane frames is carried out by the program CONDEN. Its main input file ECOND.txt is similar to EFRAM.txt, with the difference that instead of the loading vector in EFRAM.txt, in ECOND.txt it is the numbers of the master DOF that must be entered.

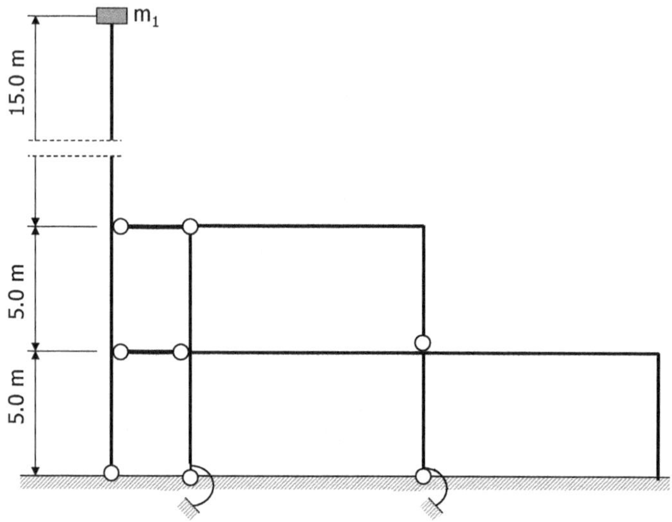

Fig. 1.49 Mast attached to the two-story building of Example 1.5

As already mentioned, once the deformations in the master DOF are known, the deformations in the remaining DOF can be computed using Eq. (1.2.7) and the internal forces $\underline{s}$ evaluated using Eq. 1.2.3. This can be done by using the program SECSYS (Appendix), which yields the complete deformations and internal forces in the "secondary structure" once the deformations in the DOF coupling it to the primary structure have been determined.

Example 1.6
The 25 m mast of Fig. 1.49 with a bending stiffness $EI = 1.9 \times 10^5$ kNm^2 and an axial stiffness $EA = 4.4 \times 10^6$ kN is attached to the two-story building from Example 1.5 by inextensional links as shown. Determine the static horizontal displacement of the mast top due to a unit horizontal load acting at m_1 and the corresponding bending moments at the base of the three columns of the building considering the latter as a "macroelement" (secondary structure) attached to the mast (primary structure).

Using the program CONDEN with the degrees of freedom 2 and 5 in Fig. 1.47 as master DOF yields the (2, 2) stiffness matrix ("coupling matrix")

$$\underline{K} = \begin{pmatrix} 0.9057E+05 & -0.1262E+05 \\ -0.1262E+05 & 0.9983E+04 \end{pmatrix} \tag{1.2.18}$$

which characterizes the springs coupling the DOF 2 and 5 of the primary structure of Fig. 1.50. Employing this spring matrix and based on the discretization of Fig. 1.50, the program FRAME yields a displacement at the top of the mast due to the horizontal load of 1 kN equal to 0.00955 m, that is 9.55 mm. The displacements in the coupling DOF no. 2 and 5 were computed as 0.000028465 and 0.00042722 m, respectively.

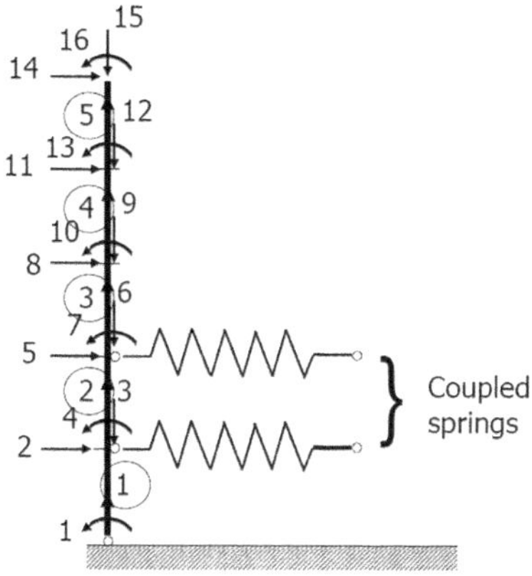

Fig. 1.50 Idealized primary structure

Entering these two values in the input file VU of the program SECSYS yields for the bending moments at the base of the columns (elements 1, 3 and 5 according to Fig. 1.47) values of −0.206 kNm, 2.447 kNm and 1.727 kNm, respectively. Of course, the same values can be arrived at by analyzing the complete system "in one go" using the program FRAME.

1.2.2 Lumped-Mass Models of MDOF Systems

Lumped-mass idealisations of multi-degree-of-freedom systems are quite popular because of their simplicity. Their central feature is that the mass matrix $\underline{M}$ is a diagonal matrix, with each coefficient on the main diagonal giving the mass associated with the corresponding active kinematic degree of freedom. These masses are usually determined "by hand" and the corresponding DOF are the master DOF associated with medium or large inertia forces. The first step in setting up the lumped-mass idealisation consists in identifying the n master degrees of freedom to be retained in the equation and eliminating the remaining DOF, e.g. by static condensation. For plane frame structures, the determination of the pertinent (n, n) condensed stiffness matrix can be carried out using the program CONDEN as demonstrated in Example 1.6 and in the following Example 1.7 for a typical moment-resisting frame.

Example 1.7
Determine the condensed stiffness matrix of the frame shown in Fig. 1.51 for the horizontal floor displacements as master DOF. All girders and columns may be considered as inextensional ($EA \rightarrow \infty$); the bending stiffness values are $EI_G = 2.50 \times 10^5$ kNm^2 for the girders and $EI_C = 2.0 \times 10^5$ kNm^2 for the columns, respectively. Masses are considered to be concentrated at the floor levels.

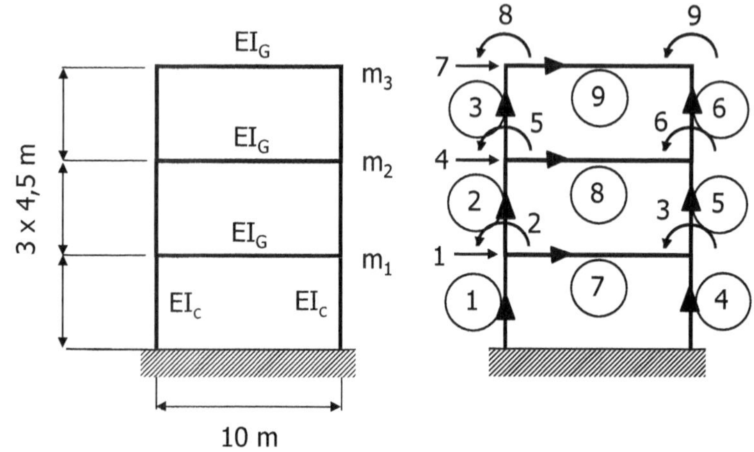

Fig. 1.51 Moment resisting frame and its discrete model

The discretization, based on the assumption that all elements are inextensional and thus dispensing with the corresponding DOF, is also shown in Fig. 1.51, the master DOF being 1, 4 and 7. EA values are not needed, all nine beam elements being inextensional, so they may be arbitrarily put equal to zero. CONDEN yields as a result the following (3, 3) condensed stiffness matrix (in the output file KMATR.txt):

$$\begin{array}{ccc} 0.9027251E+05 & -0.5123709E+05 & 0.1098853E+05 \\ -0.5123709E+05 & 0.6989332E+05 & -0.3229571E+05 \\ 0.1098853E+05 & -0.3229571E+05 & 0.2323923E+05 \end{array}$$

The (9, 3) matrix $\underline{A}$ from Eq. (1.2.7), which is needed for the subsequent evaluation of the deformations in the slave DOF, is output in the file AMAT.txt. The corresponding lumped-mass matrix in this case would be, taking into account the sequence of the master DOF used in CONDEN:

$$\underline{M} = \begin{pmatrix} m_1 & 0 & 0 \\ 0 & m_2 & 0 \\ 0 & 0 & m_3 \end{pmatrix} \tag{1.2.19}$$

1.2.3 Modal Analysis for Lumped-Mass Systems

Starting point is the ordinary differential equation system without damping

$$\underline{M}\,\ddot{\underline{V}} + \underline{K}\,\underline{V} = \underline{F} \tag{1.2.20}$$

While the mass matrix $\underline{M}$ is a diagonal matrix due to the lumped-mass assumption, this is certainly not the case for the (condensed) stiffness matrix $\underline{K}$. This leads to a "stiffness coupling" of the responses in the single degrees of freedom.

In order to arrive at independent (uncoupled) differential equations of motion, new generalized co-ordinates $\underline{\eta}$ may be introduced to replace the physical coordinates $\underline{V}$ (concept of Modal Analysis). These "generalized coordinates" may be visualized as amplitudes of mutually orthogonal system deflection shapes, the latter being usually chosen to be the "free vibration shapes" or eigenvectors of the undamped system as determined by solving the eigenvalue problem. The single steps are as follows:

We consider the system Eq. (1.2.2) with the initial conditions

$$\begin{aligned} \underline{V}(0) &= \underline{V}_0 \\ \dot{\underline{V}}(0) &= \dot{\underline{V}}_0 \end{aligned} \tag{1.2.21}$$

The modal coordinates are introduced through the linear transformations

$$\begin{aligned} \underline{V} &= \underline{\Phi}\underline{\eta} \\ \dot{\underline{V}} &= \underline{\Phi}\,\dot{\underline{\eta}} \\ \ddot{\underline{V}} &= \underline{\Phi}\,\ddot{\underline{\eta}} \end{aligned} \tag{1.2.22}$$

where the (n, r) matrix $\underline{\Phi}$ is termed the "modal matrix". Its coefficients are time-independent and its r columns, with r typically much smaller than the number n of rows of $\underline{V}$, are the eigenvectors of the system. Introducing Eq. (1.2.22) into Eq. (1.2.2) leads to

$$\underline{M}\,\underline{\Phi}\,\ddot{\underline{\eta}} + \underline{C}\,\underline{\Phi}\,\dot{\underline{\eta}} + \underline{K}\,\underline{\Phi}\,\underline{\eta} = \underline{F}(t) \tag{1.2.23}$$

and also

$$\underline{\Phi}^T\,\underline{M}\,\underline{\Phi}\ddot{\underline{\eta}} + \underline{\Phi}^T\,\underline{C}\,\underline{\Phi}\,\dot{\underline{\eta}} + \underline{\Phi}^T\,\underline{K}\,\underline{\Phi}\,\underline{\eta} = \underline{\Phi}^T\,\underline{F}(t) \tag{1.2.24}$$

In order to obtain uncoupled differential equations, we demand that the transformations in Eq. (1.2.24) should lead to diagonal matrices. For simplicity, we also demand that the transformation of the diagonal mass matrix by the modal matrix lead to a unit matrix, thus obtaining unit "modal masses" for the r "modal contributions":

$$\underline{\Phi}^T\underline{M}\,\underline{\Phi} = \underline{I} \tag{1.2.25}$$

$$\underline{\Phi}^T\underline{K}\,\underline{\Phi} = \underline{\omega}^2 = \mathrm{diag}\left[\omega_i^2\right] \tag{1.2.26}$$

The diagonalisation of the damping matrix $\underline{C}$ is deferred until later. Equations (1.2.25) and (1.2.26) can be combined by pre-multiplying the latter with the unit matrix which is subsequently replaced on the right-hand side by the left-hand side of Eq. (1.2.25):

$$\underline{\Phi}^{T}\underline{K}\,\underline{\Phi} = \underline{\Phi}^{T}\,\underline{M}\,\underline{\Phi}\,\underline{\omega}^{2} \tag{1.2.27}$$

This leads to the general eigenvalue problem

$$\underline{K}\,\underline{\Phi} = \underline{M}\,\underline{\Phi}\,\underline{\omega}^{2} \tag{1.2.28}$$

the solution of which (modal matrix $\underline{\Phi}$) can be used to diagonalise the stiffness matrix $\underline{K}$. In this equation,

- $\underline{\omega}^2$ is a diagonal matrix with the r eigenvalues ω_i^2 (squares of the natural circular frequencies) as coefficients, and
- $\underline{\Phi}$ is the (n, r) modal matrix, the columns of which are the r eigenvectors or mode shapes.

There exist many powerful solution algorithms for solving the eigenvalue problem; one of the simplest, well-suited for small problems, is the Jacobi algorithm (program JACOBI) employed here. Its input simply consists of the (n, n) stiffness matrix $\underline{K}$ (found in the file KMATR.txt, as created by CONDEN) and the n diagonal elements of the mass matrix $\underline{M}$ (file MDIAG.txt), while its output yields all n natural periods of the system with the corresponding mode shapes (eigenvectors), the latter normalized to unit mass as per Eq. (1.2.25).

The easiest way to include damping is to assume that the damping matrix $\underline{C}$ can also be brought to diagonal form by transforming it with the modal matrix $\underline{\Phi}$ according to

$$\underline{\tilde{C}} = \underline{\Phi}^{T}\,\underline{C}\,\underline{\Phi} = \mathrm{diag}\left[\tilde{c}_{ii}\right] \tag{1.2.29}$$

In analogy to the single-degree-of-freedom system, the diagonal term $\tilde{c}_{ii}$ is set equal to

$$\tilde{c}_{ii} = 2D_i\,\omega_i \tag{1.2.30}$$

D_i is the modal damping ratio and ω_i the circular natural frequency of the *i*th mode. This leads to the uncoupled system of differential equations

$$\underline{\ddot{\eta}} + \underline{\tilde{C}}\,\underline{\dot{\eta}} + \underline{\omega}^{2}\,\underline{\eta} = \underline{\Phi}^{T}\underline{F} \tag{1.2.31}$$

It consists of r 2nd order differential equations of the general form

$$\ddot{\eta}_i + 2D_i\,\omega_i\dot{\eta}_i + \omega_i^2\eta_i = \underline{\Phi}_i^T\underline{F}, \quad i = 1, 2, \ldots r \tag{1.2.32}$$

Each equation can be solved separately by means of the methods discussed earlier. However, we still need the correct initial conditions for the velocity and displacement values of the modal co-ordinates. To this effect, Eq. (1.2.25) is formally rewritten for a square modal matrix as

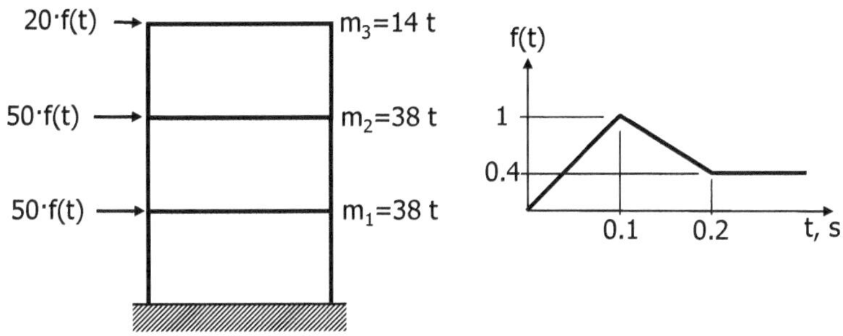

Fig. 1.52 Frame with loading

$$\underline{\Phi}^{-1} = \underline{\Phi}^{T}\underline{M} \tag{1.2.33}$$

which, together with the definitions of the modal co-ordinates of Eq. (1.2.22), leads to

$$\underline{\eta}(0) = \underline{\eta}_0 = \underline{\Phi}^{T}\underline{M}\,\underline{V}_0 \tag{1.2.34}$$

$$\dot{\underline{\eta}}(0) = \dot{\underline{\eta}}_0 = \underline{\Phi}^{T}\underline{M}\,\dot{\underline{V}}_0 \tag{1.2.35}$$

The special advantage of the modal analysis approach described here lies in the fact that quite accurate solutions can be arrived at using only a few modal contributions. The relative importance of the ith modal contribution can be judged by the magnitude of the "generalised load" $\underline{\Phi}_i^T\ \underline{F}$ and also by the closeness of the frequencies contained in the loading to the natural frequency ω_i of the ith mode. On the other hand, the solution of the eigenvalue problem may be quite time-consuming for large systems. Due to the underlying superposition principle, modal analysis is, furthermore, strictly only applicable to linear systems, although some degree of nonlinearity may be accommodated by modifying the mode shapes accordingly. Once the time histories $\eta_i(t),\ i = 1, 2, \ldots r$ of the modal co-ordinates have been determined (together with their time derivatives), the displacements, velocities and accelerations in the original degrees of freedom can be computed through Eq. (1.2.22), with the number r of modal contributions generally significantly smaller than the number n of the master degrees of freedom of the system, as mentioned.

Example 1.8

Determine the three natural periods and mode shapes of the moment-resisting frame shown in Figs. 1.51 and 1.52, also set up and solve the modal differential equation for the fundamental mode assuming a modal damping value of $D = 1\%$. With KMATR.txt as determined in Example 1.7, the program JACOBI produces the following results (file OUTJAC.txt) for the mode shapes normalized to unit modal masses and shown in Fig. 1.53 (mode shapes 1, 2 and 3 as solid, dashed and dash-dot lines, respectively):

No.	1	2	3
Periods (s)	0.5179	0.1588	0.0974
Mode shape ordinates	0.05262	−0.11758	0.09860
	0.12052	−0.03285	−0.10350
	0.15648	0.17599	0.12637

The circular natural frequencies and the corresponding mode shapes are also output to the files OMEG.txt and PHI.txt for use by other programs (e.g. MODAL, see later).

In determining the decoupled equations in the modal coordinates η_1 to η_3, the first step is to compute the modal participation factors. For the fundamental mode we obtain

$$\underline{\Phi}_1^T \underline{F} = \begin{pmatrix} 0.0526 & 0.1205 & 0.1565 \end{pmatrix} \begin{pmatrix} 50.0 \\ 50.0 \\ 20.0 \end{pmatrix} = 11.79 \tag{1.2.36}$$

and the corresponding uncoupled differential equation for the modal coordinate η_1 becomes:

$$\ddot{\eta}_1 + 2 \cdot 0.01 \cdot \frac{2\pi}{0.518} \dot{\eta}_1 + 12.13^2 \eta_1 = 11.79 \cdot f(t) \tag{1.2.37}$$

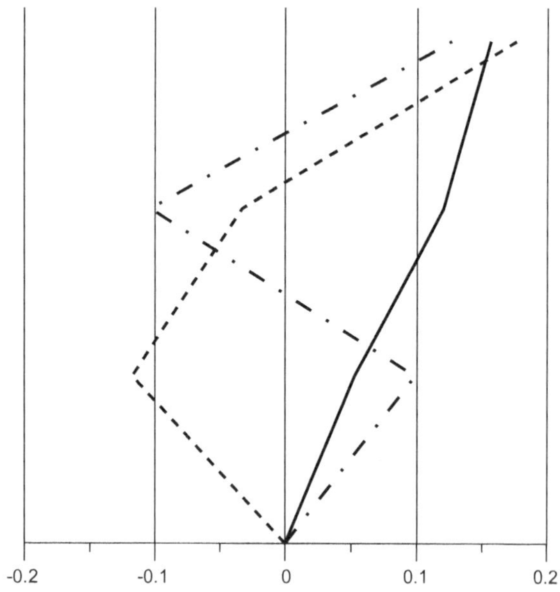

Fig. 1.53 Mode shapes of the three-story frame

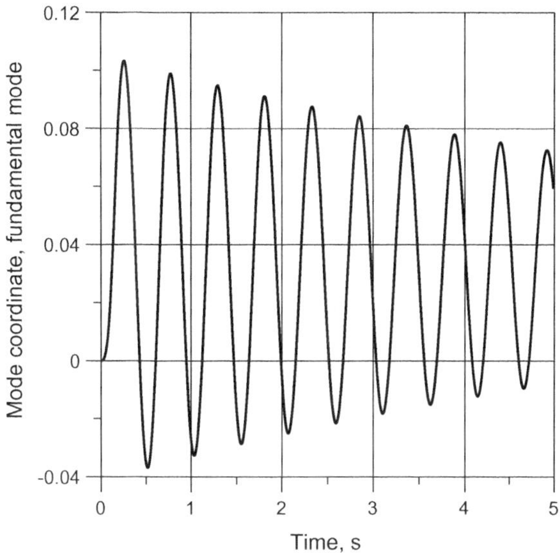

Fig. 1.54 Time history of the fundamental mode coordinate η_1

Its solution, computed by the program SDOF1, is shown in Fig. 1.54. The maximum attained is $\eta_1 = 0.1033$, corresponding to a maximum roof displacement of $0.1033 \cdot 0.1565 = 0.0162$ m or 16 mm.

For taking more modal contributions into account, the program MODAL can be used. It requires (in the input file LOADV.txt) the right-hand side $\underline{F}$ of Eq. (1.2.2) consisting of n load components at every time step (assuming a constant time increment). In the special case when all n components of $\underline{F}$ share the same piecewise linear time function, LOADV.txt can be set up using linear interpolation by the program INTERP. MODAL produces displacement time histories in the master DOF which can be used by another program, INTFOR, to determine maxima and time histories of internal forces and displacements for plane frame systems. The procedure is illustrated by Example 1.9.

Example 1.9

Determine time histories and maximum values for the horizontal roof displacement and the bending moment at the base of the ground floor columns of the three-story frame of Example 1.8 taking into account all three modal contributions.

In the program MODAL all three modal contributions are considered, with D = 1% for each. For computing maxima of the internal forces and the time history of the bending moment at the base of the ground floor columns, the program INTFOR is used (output files MAXMIN.txt and THHVM.txt). It furnishes a maximum bending moment of 218.3 kNm occurring at time t = 0.28 s. If only the contribution of the fundamental natural mode is considered, the corresponding value is 207.3 kNm, that is approx. 5% off. On the other hand, the maximum roof displacement taking into account all three modal contributions is computed as 16 mm, that is essentially the value determined in Example 1.8 by considering the fundamental mode contribution

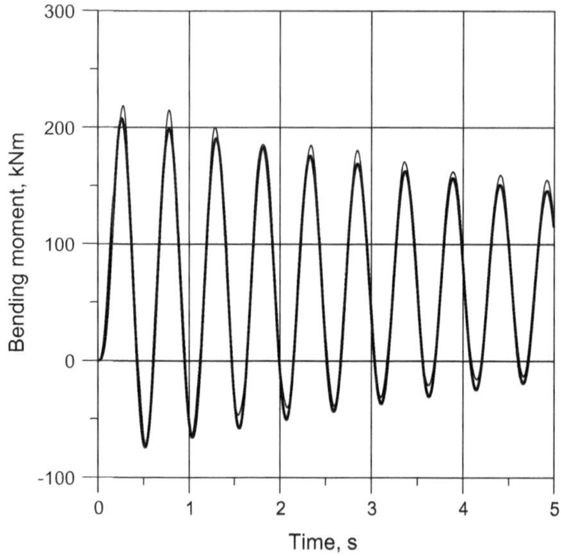

Fig. 1.55 Time histories of the bending moment at the base of the ground floor columns considering the fundamental mode alone (bold line) and all three modal contributions (normal line)

alone. This confirms the well-known fact that for internal forces such as bending moments, which depend on the second derivatives of displacements, the influence of higher modal contributions is more pronounced than for the displacements themselves. Figure 1.55 shows time histories for the bending moment at the base of the ground floor columns, considering all three modal contributions (normal line) and the fundamental mode alone (bold line).

As an independent check on the validity of the dynamic solution, we compute the response of the frame for static loads $F_1 = F_2 = 20$ kN and $F_3 = 8$ kN (that is, for $f(t) = 0.4$, see Fig. 1.52) by solving the algebraic equation system $\underline{F} = \underline{\underline{K}} \cdot \underline{V}$, with K as given in Example 1.7 (e.g. using program EQSOLV, see Appendix). We obtain for the static displacements of the three floors (from base to top) 1.83, 3.87 and 4.86 mm, respectively. The latter agrees well with the approximate value corresponding to an estimated $\eta_1 = 0.03$ (see Fig. 1.54) for large values of t, which yields $0.03 \cdot 0.1565 = 0.0047$ m or 4.7 mm.

1.2.4 The Linear Viscous Damping Model

The correct consideration of damping is of prime importance for the validity and accuracy of the results obtained in any structural dynamics problem. However, in view of the complexity of the different energy dissipation mechanisms contributing to damping, one is usually quite satisfied in being able to reproduce the correct order of magnitude of the overall energy dissipation, mostly by referring to past experience with similar structures. The mathematically simple linear viscous damping model is widely used for this purpose because of its ease of application, even if it cannot describe the damping phenomenon in all its complexity and heterogeneity.

Referring to Eq. (1.2.2), we are dealing with "proportional" damping if, for instance, the viscous damping matrix $\underline{C}$ is diagonalised by a transformation with the modal matrix $\underline{\Phi}$, the columns of which are the modal shapes of the undamped system:

$$\underline{\tilde{C}} = \underline{\Phi}^T \underline{C}\, \underline{\Phi} = \text{diag}\left[\tilde{c}_{ii}\right] = \text{diag}[2D_i\omega_i] \tag{1.2.38}$$

D_i is the damping ratio (in percent of critical damping) for the *i*th modal shape with the corresponding circular natural frequency ω_i. Because of the simplicity of this idealisation, other non-viscous damping mechanisms (such as material damping) are often approximated by an equivalent linear viscous damping model. If Eq. (1.2.33) holds, the system possesses "classical" mode shapes and its eigenvectors as columns of the modal matrix are real and mutually orthogonal. Physically, the existence of classical mode shapes implies that the damping mechanisms over the whole system are somehow similar. Strong discrepancies in the spatial distribution of the energy dissipation rate (e.g. if soil-structure interaction effects are considered) cause a distribution of damping forces that varies significantly from those of the restoring and inertia forces, thus leading to the general case of non-proportional viscous damping.

Usually, no explicit viscous damping matrix is known, and one can only assume modal damping ratios D_i ($i = 1, \ldots, r$) for the r modal contributions to be considered in the context of a modal analysis approach. For the modal analysis itself, it is sufficient to supply the damping ratio D_i alone and no explicit damping matrix $\underline{C}$ is needed. If, on the other hand, $\underline{C}$ itself is needed, e.g. for solving Eq. (1.2.2) by Direct Integration, two methods are recommended for setting up $\underline{C}$ based on one or more known modal damping ratios.

1.2.4.1 RAYLEIGH Damping

In this case $\underline{C}$ is computed as a linear combination of the system's stiffness and mass matrix according to

$$\underline{C} = \alpha\underline{M} + \beta\underline{K} \tag{1.2.39}$$

With

$$2D_i\omega_i = \underline{\Phi}_i^T \underline{C}\, \underline{\Phi}_i = \underline{\Phi}_i^T \left(\alpha\underline{M} + \beta\underline{K}\right)\underline{\Phi}_i = \alpha + \beta\,\omega_i^2 \tag{1.2.40}$$

we obtain

$$D_i = \frac{\alpha}{2\,\omega_i} + \beta\frac{\omega_i}{2} = \alpha\frac{T_i}{4\pi} + \beta\frac{\pi}{T_i} \tag{1.2.41}$$

The two coefficients α and β can be determined in such a manner that prescribed damping ratios D_1 and D_2 appear at two period values T_1 and T_2 (which are not necessarily natural periods of the structure). The coefficients can be determined from the expressions:

$$\alpha = 4\pi \frac{T_1 D_1 - T_2 D_2}{T_1^2 - T_2^2} \tag{1.2.42}$$

$$\beta = T_1 T_2 \frac{T_1 D_2 - T_2 D_1}{\pi \left(T_1^2 - T_2^2\right)} \tag{1.2.43}$$

For the special case of stiffness proportional damping, we have $\alpha = 0$ and

$$\beta = \frac{D_1 T_1}{\pi} \tag{1.2.44}$$

For mass proportional damping ($\beta = 0$), α is given by

$$\alpha = \frac{4\pi D_1}{T_1} \tag{1.2.45}$$

For these special cases (mass or stiffness proportional damping), only one damping ratio D_1 for the period T_1 can be prescribed. Generally, the stiffness proportional assumption is more realistic than the mass proportional one, because the latter yields smaller damping values for higher natural frequencies (lower periods), contrary to experience. To illustrate, Fig. 1.56 shows curves for the damping ratio D against periods for all three RAYLEIGH damping variants just mentioned, these being the stiffness proportional ($\beta\,\underline{K}$), the mass proportional ($\alpha\,\underline{M}$) and the general ($\alpha\,\underline{M} + \beta\,\underline{K}$) variant. Once α and/or β have been determined, the damping ratio D is fixed for all periods, as illustrated by Example 1.10.

Example 1.10

Determine the damping ratio D as a function of natural period for the plane frame of Example 1.8 for the following three cases:

- General RAYLEIGH damping with $D_1 = 1\%$ for $T_1 = 0.5179$ s and $D_2 = 2\%$ for $T_2 = 0.1588$ s
- Stiffness proportional RAYLEIGH damping with $D_1 = 1\%$ for $T_1 = 0.5179$ s
- Mass proportional RAYLEIGH damping with $D_1 = 1\%$ for $T_1 = 0.5179$ s

Using the expressions (1.2.42) to (1.2.45) we obtain the following coefficients for the three cases:

	$\alpha\,\underline{M} + \beta\,\underline{K}$	$\beta\,\underline{K}$	$\alpha\,\underline{M}$
α	0.1036	–	0.2426
β	0.000945	0.00165	–

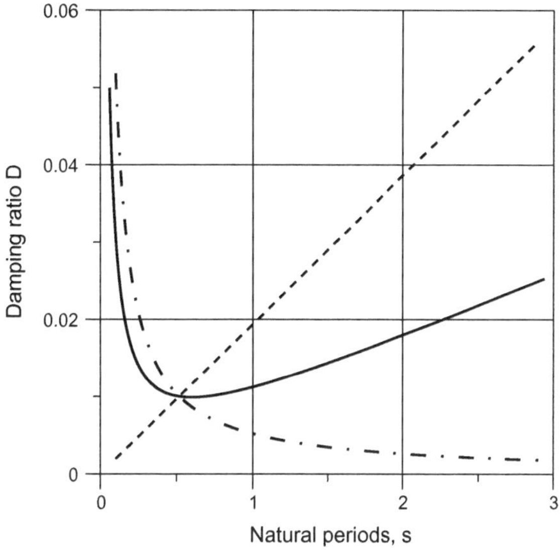

Fig. 1.56 Modal damping ratios for different variants of RAYLEIGH damping, Example 1.10

Figure 1.56 shows the corresponding functions D(T) for all three cases. Therein, the general $\alpha\, \underline{M} + \beta\, \underline{K}$ case is represented by the solid line, the mass proportional case by the dashed line and the stiffness proportional case by the dash-dot curve.

The damping matrix $\underline{C}$ can be computed by the program CRAY (Appendix). It furnishes $\underline{C}$ for mass proportional, stiffness proportional or general RAYLEIGH damping, with the stiffness matrix (file KMATR.txt) and the diagonal of the lumped-mass matrix (file MDIAG.txt) as input files. The necessary data for damping ratio(s) and corresponding period(s) are entered interactively, and the resulting damping matrix is output in the file CMATR.txt.

If more than two given modal damping ratios must be taken into consideration in computing the viscous damping matrix $\underline{C}$, the following well-known approach after CLOUGH and PENZIEN is recommended.

1.2.4.2 Complete Modal Damping

Equation (1.2.33) may be rewritten as

$$\underline{\Phi}^T \underline{C}\, \underline{\Phi} = \text{diag}[2 D_i\, \omega_i] \tag{1.2.46}$$

$$\underline{C}\, \underline{\Phi} = \left(\underline{\Phi}^T\right)^{-1} \text{diag}[2\, D_i\, \omega_i] \tag{1.2.47}$$

$$\underline{C} = \left(\underline{\Phi}^T\right)^{-1} \text{diag}[2\, D_i\, \omega_i] \underline{\Phi}^{-1} \tag{1.2.48}$$

and considering that

$$\left(\underline{\Phi}^{T}\right)^{-1} = \underline{M}\,\underline{\Phi} \tag{1.2.49}$$

we obtain

$$\underline{C} = \underline{M}\ \underline{\Phi}\ \text{diag}[2\ D_i\ \omega_i]\underline{\Phi}^{T}\ \underline{M} \tag{1.2.50}$$

Through this expression all assumed modal damping ratios D_i, $i = 1, \ldots, r$ may now be considered in computing the damping matrix. This is carried out by the program CMOD, which requires MDIAG.txt, PHI.txt and OMEG.txt as input files and also (in the file DAEM.txt) the r damping values D_i.

1.2.5 Direct Integration for Lumped-Mass Systems

Direct numerical integration methods require no solution of the eigenvalue problem and are also applicable to nonlinear systems. They generate numerically approximate solutions for a response process $\{\ \underline{V},\ \underline{\dot{V}},\ \underline{\ddot{V}}\ \}$ at discrete time points $t = 1 \cdot \Delta t, 2 \cdot \Delta t, \ldots, n \cdot \Delta t$. As a starting point, it is assumed that the response process is known at a defined time t (usually $t = 0$) and from these known initial values (and the known excitation process $\underline{P}$) the system response at time $t + \Delta t$ is calculated.

Using an unconditionally stable integration scheme is important because the time step Δt is normally larger than the highest natural period of the structure. In the case of "blow up", not only would the higher mode contributions to the solution be lost, but the low-frequency solution components would also be affected. Consequently, unconditionally stable implicit single-step integration algorithms such as the NEWMARK method are quite popular in practice. In the NEWMARK algorithm, the solution at time $t + \Delta t$ is given as follows:

$$\underline{\dot{V}}_{t+\Delta t} = \underline{\dot{V}}_t + \Delta t \cdot (1 - \gamma)\underline{\ddot{V}}_t + \Delta t \cdot \gamma \cdot \underline{\ddot{V}}_{t+\Delta t} \tag{1.2.51}$$

$$\underline{V}_{t+\Delta t} = \underline{V}_t + \underline{\dot{V}}_t \Delta t + (\Delta t)^2 \cdot \left(\frac{1}{2} - \beta\right) \cdot \underline{\ddot{V}}_t + (\Delta t)^2 \cdot \beta \cdot \underline{\ddot{V}}_{t+\Delta t} \tag{1.2.52}$$

with β and γ equal to 0.25 and 0.50 for the unconditionally stable "constant acceleration" scheme (and $\beta = 1/6$, $\gamma = 0.50$ for the only conditionally stable "linear acceleration" variant).

The computer program NEWMAR (Appendix) can be used for the task. It requires the square system stiffness and damping matrices $\underline{K}$ and $\underline{C}$ as input, to be read from the files KMATR.txt and CMATR.txt, while the diagonal of the mass matrix $\underline{M}$ is read from the file MDIAG.txt. The determination of CMATR can be carried out, for example, by using the programs CRAY (for RAYLEIGH damping) or CMOD (for complete modal damping), as described in the last section.

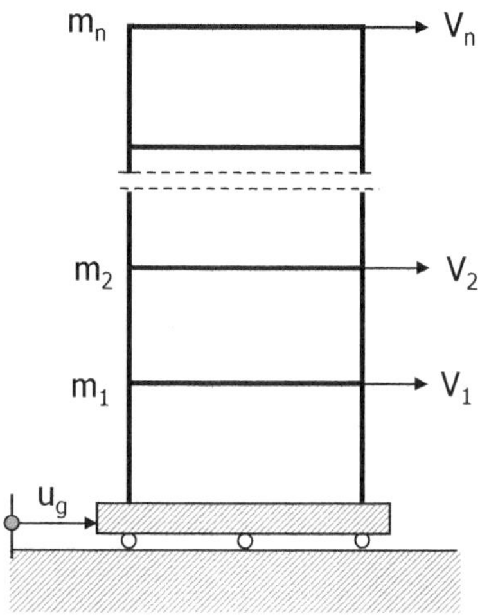

Fig. 1.57 n-story building subject to base excitation

1.2.6 Application to Base Excitation

Consider the N story building shown schematically in Fig. 1.57 subject to the support motion u_g.

The n horizontal story displacements relative to the support are the master degrees of freedom associated with the n story masses of the building; they are the elements of the vector $\underline{V}^T = (V_1, V_2, \ldots, V_n)$. The system of differential equations of motion of this cantilever system is usually expressed as

$$\begin{aligned} &\underline{M}\left(\ddot{\underline{V}} + \underline{r}\ddot{u}_g\right) + \underline{C}\,\dot{\underline{V}} + \underline{K}\,\underline{V} = 0 \\ &\underline{M}\,\ddot{\underline{V}} + \underline{C}\,\dot{\underline{V}} + \underline{K}\,\underline{V} = -\underline{M}\,\underline{r}\ddot{u}_g \end{aligned} \qquad (1.2.53)$$

Herein $\underline{K}$ is the condensed (n, n) stiffness matrix in the n master degrees of freedom, $\underline{M}$ the diagonal mass matrix and $\underline{C}$ the viscous damping matrix (which is needed in an explicit form only if the equation is to be solved by Direct Integration). The column vector $\underline{r}$ contains the displacements in the n master degrees of freedom due to a unit displacement of the support in the direction of the seismic action. In the standard case of the building shown in Fig. 1.57 subject to horizontal seismic excitation, it consists of n coefficients equal to 1; for the structure shown in Fig. 1.58 it would be equal to (1, 1, 0) for a horizontal seismic excitation and (0, 0, 1) for a vertical excitation.

Equation (1.2.53) can be solved in the time or in the frequency domain by standard methods. In earthquake engineering, the modal analysis response spectrum approach is by far the most common. In it, the modal decomposition of Eq. (1.2.53) with assumed proportional damping ($\underline{\Phi}^T \underline{C}\, \underline{\Phi} = \mathrm{diag}[2\,D_i\omega_i]$) yields n uncoupled equations in the modal coordinates η_i, $i = 1, 2, \ldots n$:

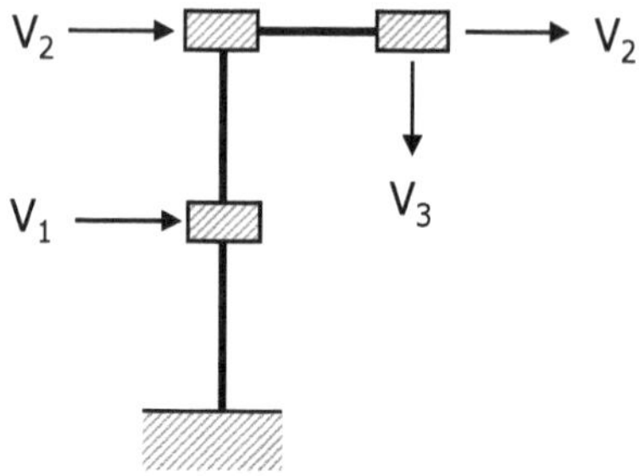

Fig. 1.58 Another system with three master DOF

$$\ddot{\eta}_i + 2D_i\ \omega_i\ \dot{\eta}_i + \omega_i^2\ \eta_i = \underline{\Phi}_i^T\ \underline{P} \tag{1.2.54}$$

Herein, $\underline{\Phi}_i$ is the *i*th eigenvector with the circular natural frequency ω_i in rad/s; the modal damping ratio D_i is usually set equal to 5%. For each modal contribution i, a participation factor β_i is introduced according to

$$\beta_i = (-)\frac{\underline{\Phi}_i^T\ \underline{M}\ \underline{r}}{\underline{\Phi}_i^T \underline{M}\ \underline{\Phi}_i} = \frac{L_i}{M_i} = \frac{L_i}{1} = \underline{\Phi}_i^T \underline{M}\ \underline{r} \tag{1.2.55}$$

The eigenvectors computed by the program JACOBI have already been normalized to a unit modal mass so that $M_i = \underline{\Phi}_i^T\ \underline{M}\ \underline{\Phi}_i$ is equal to 1. The negative sign in Eq. (1.2.55) is usually suppressed, and Eq. (1.2.54) becomes

$$\ddot{\eta}_i + 2D_i\omega_i\dot{\eta}_i + \omega_i^2\eta_i = \beta_i\ddot{u}_g(t) \tag{1.2.56}$$

The solution of this ordinary differential equation is given (DUHAMEL integral) by

$$\eta_i(t) = \beta_i \cdot \overline{S}_{d,i} \tag{1.2.57}$$

where

$$\overline{S}_{d,i}(t) = \frac{1}{\omega_{Di}} \int_0^t \ddot{u}_g\ e^{-D_i\omega_i(t-\tau)}\ \sin\omega_{Di}(t-\tau)d\tau \tag{1.2.58}$$

The absolute maximum value of $\overline{S}_{d,i}$ is set equal to the ordinate S_d of the relative displacement spectrum according to Eq. (1.1.93):

$$\max\left|\overline{S}_d(t)\right| = S_d = (1/\omega_1) \cdot S_v = \left(1/\omega_1^2\right) \cdot S_a \tag{1.2.59}$$

The absolute maximum value max η_i of the modal co-ordinate η_i is given in Eq. (1.2.60) as a function of the ordinate $S_{d,i}$ of the displacement spectrum, $S_{v,i}$ of the pseudo relative velocity spectrum or $S_{a,i}$ of the pseudo absolute acceleration spectrum:

$$\max \eta_i = \beta_i \cdot S_{d,i} = \beta_i \cdot \frac{S_{v,i}}{\omega_i} = \beta_i \cdot \frac{S_{a,i}}{\omega_i^2} \tag{1.2.60}$$

The spectral ordinates themselves are functions of the circular natural frequency ω_i and damping value D_i of the *i*th mode. The modal maximum displacement values are finally given by

$$\begin{aligned} \max \underline{V}_i &= \max \eta_i \cdot \underline{\Phi}_i = \beta_i \cdot S_{d,i} \cdot \underline{\Phi}_i = \beta_i \cdot \frac{S_{v,i}}{\omega_i} \cdot \underline{\Phi}_i \\ &= \beta_i \cdot \frac{S_{a,i}}{\omega_i^2} \cdot \underline{\Phi}_i \end{aligned} \tag{1.2.61}$$

The internal forces in the *i*th mode can be computed from the corresponding modal displacements by using standard methods of matrix structural analysis. Alternatively, we can compute modal equivalent static loads to be applied to the structure as external loads. The maximum elastic restoring forces of the *i*th mode are given by

$$\max\left(\underline{K} \cdot \underline{V}\right)_i = \underline{K}\, \beta_i\, S_{d,i}\, \underline{\Phi}_i \tag{1.2.62}$$

They are equal to the modal inertia forces, as can be seen from the following derivation:

$$\begin{aligned} \max\left(\underline{K} \cdot \underline{V}\right)_i &= \underline{K}\beta_i S_{d,i} \underline{\Phi}_i = \omega_i^2 \underline{M} \beta_i S_{d,i} \underline{\Phi}_i \\ &= \underline{M}\, \beta_i\, S_{d,i} \omega_i^2\, \underline{\Phi}_i = \underline{M}\, \beta_i\, S_{a,i}\, \underline{\Phi}_i = \max\left(\underline{M} \cdot \ddot{\underline{V}}_{abs}\right)_i \end{aligned} \tag{1.2.63}$$

This leads to Eq. (1.2.64) for the equivalent static load H_E acting in the *i*th mode in the direction of the degree of freedom k

$$H_{E,k,i} = \beta_i\, S_{a,i}\, m_k\, \Phi_{i,k} \tag{1.2.64}$$

Here, m_k is the mass associated with the degree of freedom k and $\Phi_{i,k}$ is the corresponding ordinate in the *i*th mode shape.

An importance question concerns the number of modal contributions that are needed in order to achieve a certain accuracy of the results. In this context, the "effective modal mass" $M_{i,eff}$ associated with the *i*th mode shape is introduced as follows:

$$M_{i,eff} = \beta_i^2\, M_i = \beta_i^2 (\underline{\Phi}_i^T\, \underline{M}\, \underline{\Phi}_i) \tag{1.2.65}$$

It is shown in the following that the sum of all n effective modal masses is equal to the effective total mass $M_{Tot,eff}$, which is also given by

$$M_{Tot,eff} = \underline{r}^T \underline{M}\, \underline{r} \tag{1.2.66}$$

To that effect, we introduce

$$\underline{r} = \underline{\Phi}\, \underline{\beta} \tag{1.2.67}$$

where $\underline{\Phi}$ is the modal matrix with the n mode shapes as columns and $\underline{\beta}$ a column vector containing the n participation factors. Multiplying both sides of Eq. (1.2.67) with $\underline{\Phi}_i^T\, \underline{M}$ leads to

$$\underline{\Phi}_i^T \underline{M}\, \underline{r} = \underline{\Phi}_i^T\, \underline{M}\, \underline{\Phi}\, \underline{\beta} \tag{1.2.68}$$

or

$$L_i = M_i\, \beta_i; \quad \beta_i = \frac{L_i}{M_i} \tag{1.2.69}$$

as already given by Eq. (1.2.55). Introducing Eq. (1.2.67) in the expression (1.2.66) for the effective total mass yields

$$\begin{aligned} M_{Tot,eff} &= \underline{\beta}^T\, \underline{\Phi}^T\, \underline{M}\, \underline{\Phi}\, \underline{\beta} = \underline{\beta}^T\, \underline{M_i \beta} \\ &= \sum_{i=1}^{n} \beta_i^2\, M_i = \sum_{i=1}^{n} M_{i,eff} \end{aligned} \tag{1.2.70}$$

A useful rule states that in order to achieve a satisfactory accuracy one should consider enough modal contributions so that the sum of their effective modal masses (this sum being equal to the sum of the squares of the participation factors in the case of unit modal masses, $M_i = 1$) is equal to at least 90% of the effective total mass. The latter is equal to the sum of the story masses for building models like the one shown in Fig. 1.57, where all components of the vector $\underline{r}$ are equal to 1. Careful attention is required if rotations are also present among the master degrees of freedom, with the associated mass moments of inertia as entries on the diagonal of the mass matrix: The corresponding inertia forces are activated only for rotational support excitations, in which case the vector $\underline{r}$ contains the displacements of the translational degrees of freedom and the rotations of the rotary degrees of freedom due to a unit support rotation.

The product of the effective modal mass and the spectral acceleration S_a equals the total seismic base shear force F_i in the *i*th mode:

$$F_i = M_{i,eff} \cdot S_{a,i} \tag{1.2.71}$$

Analogously, the "total base shear" can be expressed as the product of the effective total mass and the spectral acceleration for the first natural period of the structure. It is denoted by F_b in Eurocode 8 (EC 8) and it is given by

$$F_b = \frac{W}{g} \cdot S_a(T_1) \tag{1.2.72}$$

Here W is the total weight of the structure. This expression is exploited further in the context of the simplified response spectrum method.

The program MDA2DE (Appendix) computes modal displacements and equivalent modal static loads for plane frame structures, the natural periods and mode shapes of which have been computed by the program JACOBI; the spectral ordinates corresponding to the single natural periods must be input interactively. MDA2DE requires the input files MDIAG.txt, OMEG.txt, PHI.txt and RVEKT.txt; the latter contains the displacements in all (master) degrees of freedom due to a unit support displacement in the direction of the seismic excitation.

Example 1.11

Determine maximum roof displacements, base shears and equivalent static loads for all three modal contributions for the three-story frame of Fig. 1.51 subject to a base excitation described by the elastic spectrum depicted in Fig. 1.41. Its natural periods and eigenvectors have been determined in Example 1.8 and the corresponding spectral ordinates given as pseudo-relative velocity values S_v are as follows:

Mode no.	1	2	3
Natural period (s)	0.5179	0.1588	0.0974
S_V (m/s)	0.07958	0.06318	0.03875

The program MDA2DE yields the following results for modal participation factors, base shear values, modal displacements and equivalent static loads:

Mode no.	1	2	3
Participation factor	8.77	−3.25	1.58
Modal displacements in mm (base to roof level)	3.027	0.61	0.094
	6.934	0.17	−0.098
	9.003	−0.91	0.120
Base shears in kN	74.25	26.45	6.26
Equivalent static loads in kN (base to roof level)	16.928	36.338	14.826
	38.774	10.151	−15.563
	18.548	−20.037	7.001

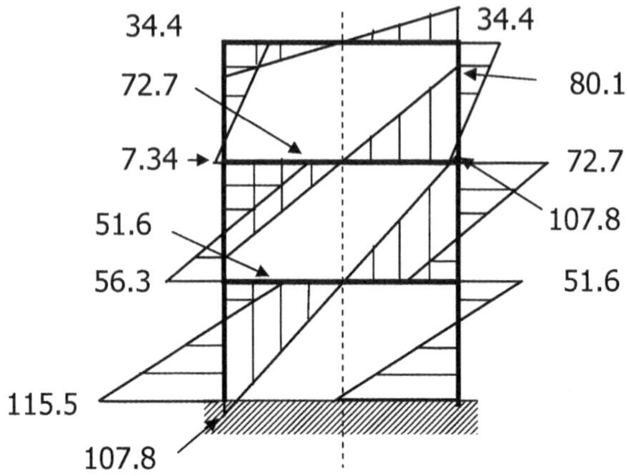

Fig. 1.59 Bending moments in the fundamental mode, in kNm

From these results, modal internal force distributions can be calculated e.g. by using the program FRAME as in Example 1.8. To illustrate, the bending moment diagram for the fundamental mode is shown in Fig. 1.59. The bending moments at the base of the ground floor columns in the 2nd and 3rd mode are, in comparison, 31.9 and 6.55 kNm, respectively.

For the combination of all modal responses (generally including translational and torsional modes when using 3D models), it is customary to employ the SRSS rule (square root of the sum of the squares), as long as the modes can be considered to be independent of each other. This can be assumed to be the case if the natural periods T_i and T_j of the modes i and j (with $T_j \leq T_i$) fulfil the condition (as is here the case)

$$T_j \leq 0.90 \cdot T_i \tag{1.2.73}$$

By the SRSS rule, the maximum roof displacement for this example is given as

$$\max w_{roof} = \sqrt{9.00^2 + (-0.91)^2 + 0.12^2} = 9.05 \ \text{mm} \tag{1.2.74}$$

The SRSS value for the bending moment at the base of the ground floor columns is equal to:

$$\max M = \sqrt{115.5^2 + 31.9^2 + 6.55^2} = 120.0 \ \text{kNm} \tag{1.2.75}$$

If the condition (1.2.73) is not satisfied, the combination of modal contributions must be carried out by more sophisticated procedures, such as the Complete Quadratic Combination (CQC). For a vector $\underline{S}^T$ consisting of p modal contributions, $\underline{S}^T = (E_1, E_2, \ldots, E_p)$, the CQC rule yields:

$$\underline{S}_E = \sqrt{\sum_{i=1}^{p}\sum_{j=1}^{p} E_i E_j \, \varepsilon_{ij} \, \alpha_{ij}} \tag{1.2.76}$$

with

$$\varepsilon_{ij} = \frac{8\sqrt{D_i D_j}(D_i + r\,D_j) r^{1.5}}{\left(1 - r^2\right)^2 + 4 D_i D_j\, r\left(1 + r^2\right) + 4\left(D_i^2 + D_j^2\right) r^2} \tag{1.2.77}$$

and

$$r = \frac{\omega_j}{\omega_i} \leq 1; \quad \alpha_{ij} = \frac{\beta_i \, \beta_j}{\left|\beta_i \, \beta_j\right|} = +1 \text{ or } -1 \tag{1.2.78}$$

with the modal participation factors β.

For the standard case of plane models subject only to a single ground acceleration component (horizontal acceleration), time histories of internal forces and displacements for given acceleration records can be computed using the programs MODBEN (on the basis of modal analysis) or NEWBEN (employing Direct Integration), both described in Appendix. Their use is illustrated by the following example.

Example 1.12

Determine time histories and maxima of the roof displacement and the bending moment at the base of the ground floor columns for the frame of Fig. 1.51 subject to the acceleration record determined in Example 1.4.

Using the program MODBEN with this record as input (in file ACC.txt) and 5% modal damping for all three modal contributions furnishes the time histories shown in Figs. 1.60 and 1.61, while maxima (as computed by the program INTFOR) are 0.9 cm for the roof displacement and 110 kNm for the bending moment at the ground floor column base. These results can be verified using the program NEWBEN after creating the damping matrix required by this program, e.g. by using the program CRAY for stiffness-proportional 5% damping at the fundamental period of the frame of 0.52 s.

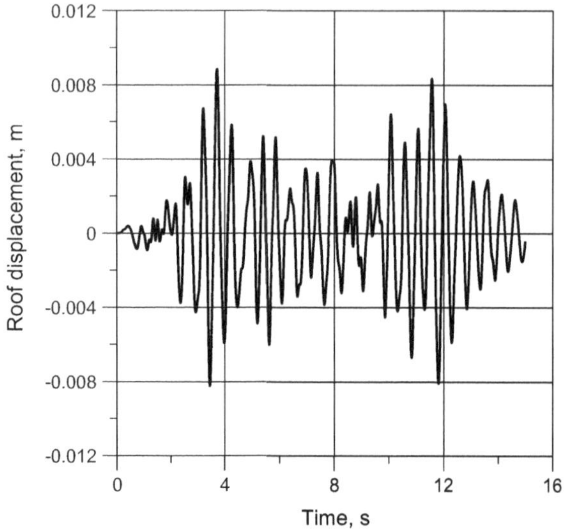

Fig. 1.60 Roof displacement time history

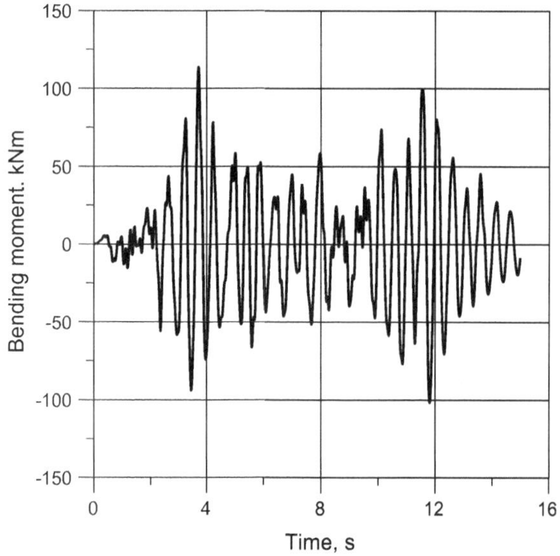

Fig. 1.61 Bending moment at the ground floor column base time history

Appendix: Descriptions of the Programs of Chapter 1, in Alphabetical Order

<table>
<tr><th colspan="2">BASCOR</th></tr>
<tr><td colspan="2">Linear base-line correction of an accelerogram and evaluation of the corresponding ground velocity and ground displacement time histories</td></tr>
<tr><th colspan="2">Interactive input</th></tr>
<tr><td>Number NANZ of points in the accelerogram?
Constant time step DT?
Factor FAKT to obtain acceleration ordinates in m/s**2 units?</td><td>The ordinates of the record are multiplied by FAKT (e.g. 9.81 if they are originally in g) in order to obtain m/s^2-units.</td></tr>
<tr><th colspan="2">Input file</th></tr>
<tr><td>ACC.txt</td><td>Contains the accelerogram in 2E16.7 format, with time points in the first column and ordinates in the second column (NANZ rows).</td></tr>
<tr><th colspan="2">Output files</th></tr>
<tr><td>ACCCOR.txt
VELCOR.txt
DISCOR.txt
VELUNC.txt
DISUNC.txt</td><td>Corrected acceleration time history, ordinates in m/s^2.
Corrected velocity time history, ordinates in m/s.
Corrected displacement time history, ordinates in m.
Uncorrected velocity time history.
Uncorrected displacement time history.

All output files consist of NANZ rows in 2E16.7 format, with time points in the first and ordinates in the second column.</td></tr>
</table>

CMOD	
Evaluation of the damping matrix $\underline{C}$ by the complete modal approach	
Interactive input	
Dimension NDU of the square matrices M, C, K? Number NMOD of modes to be considered?	
Input files	
MDIAG.txt	The diagonal of the mass matrix (NDU values in free format).
OMEG.txt	Natural circular frequencies ω_i of the system; this file is created by the JACOBI program
PHI.txt	Mode shapes $\underline{\Phi}_i$ of the system; this file is created by the JACOBI program
DAEM.txt	NMOD damping ratios for the NMOD modes, in free format.
Output file	
CMATR.txt	Resulting damping matrix (NDU*NDU values in free format).

<table>
<tr><td colspan="2">CONDEN</td></tr>
<tr><td colspan="2">Performs a static condensation for a plane frame</td></tr>
<tr><td colspan="2">Interactive input:</td></tr>
<tr><td>Number NDOF of the system degrees of freedom?
Number NDU of the master degrees of freedom?
Number NELEM of elements?
Number NFED of elastic spring matrices?

Dimensions of the square elastic spring matrices (NFED numbers)?</td><td>Each matrix couples a number of active kinematic degrees of freedom with one another and/or with the ground.</td></tr>
<tr><td colspan="2">Input files</td></tr>
<tr><td>ECOND.txt</td><td>The first NELEM rows contain EI, ℓ, EA and α for each beam element, in free format. EI is its constant bending stiffness (e.g. in kNm^2), ℓ is the element length (e.g. in m), EA is the axial stiffness (e.g. in kN) and α is the angle between the global x-axis and the element axis (in degrees, positive counter-clockwise). The next NELEM rows contain the incidence vectors for all elements in free format, indicating the 6 global degrees of freedom corresponding to the 6 local degrees of freedom (u_1, w_1, φ_1, u_2, w_2, φ_2) of each element. Next, the NDU numbers of the master kinematic degrees of freedom are entered, in free format.</td></tr>
<tr><td>INZFED.txt</td><td>NFED rows, one for each spring matrix, containing the numbers of the system degrees of freedom which are constrained by this elastic support (free format).</td></tr>
<tr><td>FEDMAT.txt</td><td>The coefficients of the NFED elastic support matrices, in free format.</td></tr>
<tr><td colspan="2">Output files</td></tr>
<tr><td>KMATR.txt</td><td>The condensed stiffness matrix (NDU * NDU coefficients, 6E16.7 format).</td></tr>
<tr><td>AMAT.txt</td><td>The matrix A (NDOF rows, NDU columns) serves for evaluating the displacements in all degrees of freedom ($\underline{V}$) from known displacements in the NDU master degrees of freedom $\underline{V}_u$ according to ($\underline{V} = \underline{A}\ \underline{V}_u$).</td></tr>
</table>

CRAY	
Evaluation of $\underline{C} = \alpha\underline{M} + \beta\underline{K}$, $\underline{C} = \alpha\underline{M}$ or $\underline{C} = \beta\underline{K}$	
Interactive input	
Dimension NDU of the square matrices M, C, K? Type of the desired damping matrix C: C = alfa * M + beta * K: IKN=1 C = alfa * M : IKN=2 C = beta * K : IKN=3 IKN=?	
Enter the first natural period T1 (T1>T2): Corresponding damping ratio? Enter the second natural period T2 (T2<T1): Corresponding damping ratio?	For the general case (IKN = 1)
Enter the natural period (in s): Corresponding damping ratio?	If IKN = 2 or 3
Input files	
MDIAG.txt	The diagonal of the mass matrix (NDU coefficients in free format).
KMATR.txt	Contains the condensed stiffness matrix of the system (NDU*NDU coefficients) created by CONDEN.
Output files	
CMATR.txt	Damping matrix, NDU*NDU coefficients in free format.
CTRLC.txt	Additional information (computed parameters α and β)

EQSOLV	
Solves the linear algebraic system $\underline{A}\ \underline{X} = \underline{B}$	
Interactive input	
Number of equations N?	
Input files	
COEFMAT.txt	Matrix $\underline{A}$, entered column wise in free format.
RHS.txt	Right-hand side (vector $\underline{B}$), in free format
Output files	
SOLUTION.txt	The solution vector $\underline{X}$.

FFT1	
Transformation of a time series from the time into the frequency domain	
Interactive input:	
Number NPKT of data points in the time series (max. 8192)? NANZ, desired number as a power of 2 with NANZ ≥ NPKT (max. 8192)? The power of 2 corresponding to NANZ (max. 13)? Time step DT?	The time series with NPKT values is padded by (NANZ-NPKT) zeros; the actual time series to be transformed thus consists of NANZ values at a constant time step DT.
Input file	
TIMSER.txt	NPKT rows containing the time t and the corresponding ordinate f(t) in a 2E16.7 format.
Output files	
OMCOF.txt OMSQU.txt	Contains the computed (NANZ/2) complex FOURIER coefficients, with the circular frequencies in the first, the real part of the coefficient in the second and its imaginary part in the third column (3E16.7 format). Contains the circular frequencies in the first column and the squares of the FOURIER coefficients in the second column (2E16.7 format).

<table>
<tr><th colspan="2">FFT2</th></tr>
<tr><td colspan="2">Transformation from the frequency domain into the time domain</td></tr>
<tr><td colspan="2">Interactive input:</td></tr>
<tr><td>Number NANZ of time series data points (must be a power of 2, max. 8192 values)?

Power of 2 corresponding to NANZ (max. 13)?

Circular frequency step
DOM = 2*pi/(NANZ*DT) ?</td><td>(NANZ/2) complex values are input from the file OMCOF.TXT. They are the FOURIER coefficients at a frequency step of
$\Delta\omega = \frac{2\pi}{NANZ \cdot DT}$</td></tr>
<tr><td colspan="2">Input file:</td></tr>
<tr><td>OMCOF.txt</td><td>Contains (NANZ/2) complex FOURIER coefficients, with the circular frequencies in the first, the real part of the coefficient in the second and its imaginary part in the third column (3E16.7 format).</td></tr>
<tr><td colspan="2">Output file:</td></tr>
<tr><td>RESULT.txt</td><td>Contains the computed time series (NANZ values), with the time points in the first and the ordinates in the second column (2E16.7 format).</td></tr>
</table>

FILTER	
Filtering of a time series in the frequency domain (high-, low- and band-pass filter)	
Interactive input:	
Number NANZ of the time series ordinates (max. 4096)? Next higher exponent of 2? (e.g. 12 for 4096 points) Constant time step? High-pass filtering desired? Y/N Circular corner frequency of the high-pass filter? Low-pass filtering desired? Y/N Circular frequency of the KANAI-TAJIMI filter? Damping ratio of the KANAI-TAJIMI filter? Band-pass filtering desired? Y/N Circular frequencies OMH and OMT of the band-pass filter?	The time series is padded with zeros in the program so that the number of values is equal to a power of 2. The filter parameters for the desired filters are entered.
Input file	
File name	
TIMSER.txt	The time series F(t) to be filtered, NANZ rows containing [t, F(t)] in a 2E16.7 format.
Output file	
FILT.txt	The computed filtered time series, with time points in the first and ordinates in the second column (2E16.7 format).

FRAME	
Evaluation of the internal forces and displacements in plane frames subject to static loading	
Interactive input:	
Total no. of DOF? Total no. of beams? No. of beams with trapezoidal load? Number NFED of spring matrices? Dimensions of spring matrices (NFED numbers)?	Each elastic support matrix couples a number of active kinematic degrees of freedom with one another and/or with the ground.

Input files:	
EFRAM.txt	The first NELEM rows contain EI, ℓ, EA and α for each beam element, in free format. EI is its constant bending stiffness (e.g. in kNm^2), ℓ is the element length (e.g. in m), EA is the axial stiffness (e.g. in kN) and α is the angle between the global x-axis and the element axis (in degrees, positive counter-clockwise). The next NELEM rows contain the incidence vectors for all elements in free format, indicating the 6 global degrees of freedom corresponding to the 6 local degrees of freedom (u_1, w_1, φ_1, u_2, w_2, φ_2) of each element. Next, the NDOF load components (nodal forces and nodal moments) corresponding to the active kinematic degrees of freedom are entered, in free format.
INZFED.txt	NFED rows, one for each elastic support matrix, containing the numbers of the system degrees of freedom constrained by this spring matrix (free format).
FEDMAT.txt	The coefficients of all NFED spring matrices, in free format.
TRAPNR.txt	Numbers of the beams with trapezoidal load (NTRAP numbers).
TRAP.txt	In NTRAP rows the ordinates of the trapezoidal load parallel and perpendicular to the beam axis at the beam end sections (4 values, in free format)
Output files:	
AFRAM.txt	Contains the computed displacements in all NDOF system degrees of freedom followed by the deformations and end forces/moments of each element in both the global and the local coordinate system.

HUSID		
Evaluation of the HUSID diagram, the ARIAS intensity and the strong motion duration of an accelerogram		
Interactive input:		
Number NANZ of points in the accelerogram? Constant time step DT? Factor FAKT to obtain acceleration ordinates in m/s**2 units?		The ordinates of the record in ACC.txt are multiplied by FAKT (e.g. 9.81 if they are originally in g) in order to obtain m/s^2 units.
Input file:		
ACC.txt	Contains the accelerogram in 2E16.7 format, with time points in the first and ordinates in the second column (NANZ rows).	
Output file:		
HUSDIAG.txt	NANZ rows containing the normalized HUSID diagram (with time points in the first and ordinates in the second column) in 2E16.7 format.	
ARIAS.txt	The peak ground acceleration (PGA) in m/s^2, the ARIAS intensity (INT * π /2g) in m/s and the effective strong motion duration in s (time duration between 5% and 95% of the HUSID diagram).	

INTERP	
Linear interpolation of a time function as in LININT, here carried out for several components of a load vector which have different amplitudes but the same time function	
Interactive in-/output:	
Number NPKT of points defining the function? Constant interpolation step? Number NANZ of ordinates to be computed? Number of rows of the load vector? Should the time be written in the first column of the output file (Y/N)?	
Input files:	
FKT.txt	NPKT rows containing the [t, f(t)] coordinates of the points defining the polygonal time function (free format).
AMPL.txt	The amplitudes of the NDU load vector components in free format.
Output file:	
LOADV.txt	If the time points are to be output, LOADV contains (NDU+1) columns in (E16.7, 30E12.4) format, with the time points in the first column and the components of the load vector in the next NDU columns. If no time points are to be output, LOADV consists of NDU columns and NANZ rows, giving the time histories of the NDU load vector components in free format.

<table>
<tr><th colspan="2">INTFOR</th></tr>
<tr><td colspan="2">Evaluation of internal forces and displacements in plane frames from known displacements in the master degrees of freedom</td></tr>
<tr><th colspan="2">Interactive input:</th></tr>
<tr><td colspan="2">Number NDU of the master degrees of freedom?
Number NELEM of elements?
Number NDPHI of the slave degrees of freedom?
Number NT of time steps?
Constant time step DT?
For which time step should the state variables be computed?
For which internal force component (H, V, M) should the extreme values be calculated?
Enter 1 for horizontal component
Enter 2 for vertical component
Enter 3 for bending moment

Should a state variable time history be output into the file THHVM ? Enter 1 for yes, 0 for no.
Corresponding element number?
End 1 or end 2 of the element?
Displacement (enter 1) or internal force (enter 2)?
Horizontal (1), vertical (2) or rotation (3)?
Horizontal (1), vertical (2) or moment (3)?</td></tr>
<tr><th colspan="2">Input files:</th></tr>
<tr><td>ECOND.txt</td><td>The first NELEM rows contain EI, ℓ, EA and α for each beam element, in free format. EI is its constant bending stiffness (e.g. in kNm^2), ℓ is the element length (e.g. in m), EA is the axial stiffness (e.g. in kN) and α is the angle between the global x-axis and the element axis (in degrees, positive counter-clockwise).
The next NELEM rows contain the incidence vectors for all elements in free format, indicating the 6 global degrees of freedom corresponding to the 6 local degrees of freedom (u_1, w_1, φ_1, u_2, w_2, φ_2) of each element. Next, the NDU numbers of the master kinematic degrees of freedom are entered, in free format.</td></tr>
<tr><td>THISDU.txt</td><td>Contains the displacements in the NDU master degrees of freedom in all NT time points in free format, without the time points.</td></tr>
<tr><td>AMAT.txt</td><td>The matrix A (NDOF rows, NDU columns) serves for evaluating the displacements in all degrees of freedom from known displacements in the NDU master degrees of freedom.</td></tr>
</table>

Output files:	
FORSTA.txt	The displacements and end forces/moments at a chosen time point, sorted by element number. All state variables refer to the global (x, z)- coordinate system and are arranged in the following sequence: Horizontal component, vertical component, rotation or bending moment at end 1 and end 2 of the beam element.
THHVM.txt	Time history of an internal force or displacement component, with the times in the first column and the ordinates in the second column (2E16.7 format).
MAXMIN.txt	This file contains for each element the evaluated maximum and minimum values of the internal force (H, V or M) at both element ends, with time of occurrence and the other internal force components acting simultaneously.
CTRLIF.txt	Echo print of input data

JACOBI	
Solution of the general linear eigenvalue problem	
Interactive input:	
Number of equations?	Number NDU of the master degrees of freedom, this being the order of the stiffness and mass matrices.
Input files:	
KMATR.txt	The condensed (NDU*NDU) stiffness matrix as computed by the program CONDEN.
MDIAG.txt	The diagonal of the mass matrix consisting of the NDU masses in the master degrees of freedom (free format).
Output files:	
File name	
OUTJAC.txt	All natural periods and eigenvectors, the latter normalized for unit modal masses.
OMEG.txt	Contains the natural circular frequencies of the system.
PHI.txt	Contains the modal shapes of the system.

LININT	
Linear interpolation of a polygonal function f(t)	
Interactive input	
Number NPKT of points defining the function? Constant interpolation time step? Number NANZ of ordinates to be computed?	
Input file:	
FKT.txt	NPKT rows containing the [t, f(t)] coordinates of the corner points defining the polygonal function.
Output file:	
RHS.txt	NANZ rows in 2E16.7 format, with times in the first and function ordinates in the second column.

<table>
<tr><th colspan="2">MDA2DE</th></tr>
<tr><td colspan="2">Response spectrum modal analysis of plane frame structures</td></tr>
<tr><td colspan="2">Interactive input:</td></tr>
<tr><td>Number NDU of the master degrees of freedom?
Number NMOD of modes to be considered?
Factor for the mass matrix?

Which type of spectral ordinate would you like to enter?
Displacement Sd (m): JKN = 1
Velocity Sv (m/s): JKN = 2
Acceleration Sa (g): JKN = 3
JKN = ?

Spectral displacement in m?
Spectral velocity in m/s?
Spectral acceleration in g?</td><td>The values of the discrete masses in MDIAG are multiplied by this factor.

These data must be entered for each of the NMOD modes:</td></tr>
<tr><td colspan="2">Input files:</td></tr>
<tr><td>MDIAG.txt</td><td>The diagonal of the mass matrix consisting of the NDU masses in the master degrees of freedom, in free format.</td></tr>
<tr><td>OMEG.txt</td><td>The circular natural frequencies of the system; this file is created by the program JACOBI</td></tr>
<tr><td>PHI.txt</td><td>The mode shapes of the system; this file is created by the program JACOBI.</td></tr>
<tr><td>RVEKT.txt</td><td>NDU values in free format representing the displacements in each master degree of freedom for a unit base displacement in the direction of the seismic excitation.</td></tr>
</table>

<table>
<tr><th colspan="2">Output file:</th></tr>
<tr><td>File name</td><td></td></tr>
<tr><td>EQUIV.txt</td><td>It contains the modal displacements and the equivalent static forces in the master degrees of freedom for NMOD modes. In addition, for each mode the modal excitation factor, the effective mass and the base shear force are output as well as the total seismically active mass and the mass actually taken into account by the modes considered.</td></tr>
</table>

MODAL	
Modal analysis of MDOF systems after solving the eigenvalue problem by direct integration of the uncoupled modes	
Interactive input:	
Number NDU of the master degrees of freedom? Number NDPHI of the slave degrees of freedom? Number NMOD of modes to be considered? Number NT of time steps (= ordinates of the load vector)? Constant time step DT? Damping of mode no. ... ?	
Input files:	
MDIAG.txt	The diagonal of the mass matrix consisting of the NDU masses in the master degrees of freedom, in free format.
OMEG.txt	The circular natural frequencies of the system; this file is created by the program JACOBI
PHI.txt	The mode shapes of the system; this file is created by the program JACOBI.
V0.txt	NDU values in free format indicating the displacements in the master degrees of freedom at time t=0.
VP0.txt	NDU values in free format indicating the velocities in the master degrees of freedom at time t=0.
LOADV.txt	Contains NT*NDU values corresponding to the NDU components of the load vector at all NT time points in free format. This file may be created by the INTERP program (without writing the time points in the output file).
AMAT.txt	The matrix A (NDOF * NDU) serves for evaluating the displacements in all degrees of freedom from known displacements in the NDU master degrees of freedom (created by the program CONDEN).

Output files:	
THISMOD.txt	Contains the time histories of displacements in the NDU master degrees of freedom (F7.4, 30E14.6 format), with the time points in the first column.
THISDU.txt	Contains the displacements in the NDU master degrees of freedom in all NT time points in free format, without the time points. This file serves as input for the program INTFOR.
THISDG.txt	Contains the displacements in all (NDU+NDPHI) degrees of freedom in free format, without the corresponding time points.
CTRLMD.txt	Echo print of input data, output of the computed maximum absolute displacement value.

<table>
<tr><th colspan="2">MODBEN</th></tr>
<tr><td colspan="2">Seismic analysis of plane frame structures by direct integration of the decoupled equations of motion</td></tr>
<tr><th colspan="2">Interactive in-/output:</th></tr>
<tr><td>Number NDU of the master degrees of freedom?
Number NDPHI of the slave degrees of freedom?
Number NMOD of modes to be considered?
Number NT of time steps?
Constant time step DT?
Factor for the accelerogram?

Enter damping ratio of mode no. ...</td><td>The resulting dimension of the acceleration multiplied by this factor should be (L/T^2), e.g. in m/s^2 units.</td></tr>
<tr><th colspan="2">Input files:</th></tr>
<tr><td>MDIAG.txt</td><td>The diagonal of the mass matrix consisting of the NDU masses in the master degrees of freedom, in free format.</td></tr>
<tr><td>OMEG.txt</td><td>The circular natural frequencies of the system; this file is created by the program JACOBI</td></tr>
<tr><td>PHI.txt</td><td>The mode shapes of the system; this file is created by the program JACOBI.</td></tr>
<tr><td>V0.txt</td><td>NDU values in free format indicating the displacements in the master degrees of freedom at time t=0.</td></tr>
<tr><td>VP0.txt</td><td>NDU values in free format indicating the velocities in the master degrees of freedom at time t=0.</td></tr>
<tr><td>ACC.txt</td><td>Contains the accelerogram in 2E16.7 format, with time points in the first and ordinates in the second column (NT rows).</td></tr>
<tr><td>AMAT.txt</td><td>The matrix A (NDOF * NDU values) is required for the evaluation of the displacements in all degrees of freedom from known displacements in the master degrees of freedom.</td></tr>
<tr><td>RVEKT.txt</td><td>NDU values in free format representing the displacements in each master degree of freedom for a unit base displacement in the direction of the seismic excitation.</td></tr>
</table>

Output files:	
THISMOD.txt	The file contains the displacement time histories in the NDU master degrees of freedom, with time points in the first column and displacements in the other NDU columns (F8.4, 30E16.7 format)
THISDG.txt	The displacements in all (NDU + NDPHI) degrees of freedom for all NT time points, in free format.
THISDU.txt	The displacements in the NDU master degrees of freedom for all NT time points, in free format.
CTRLMB.txt	Echo print of the input data. Additionally, the evaluated modal excitation factor for each mode and the (absolute) maximum displacement value with the corresponding time.

NEWBEN	
Seismic analysis of plane structures by direct integration of the coupled equations of motion	
Interactive in-/output:	
Number NDU of the master degrees of freedom? Number NT of time steps? Constant time step DT? Factor for the accelerogram? Should displacements, velocities or accelerations be output in THISNEW.txt? For displacements: IOUT = 1 For velocities: IOUT = 2 For accelerations: IOUT = 3 IOUT = ?	The resulting dimension of the acceleration multiplied by this factor should be (L/T^2), e.g. in m/s^2 units.

Input files:	
MDIAG.txt	The diagonal of the mass matrix consisting of the NDU masses in the master degrees of freedom, in free format.
KMATR.txt	The (NDU*NDU) stiffness matrix, e.g. as generated by the program CONDEN.
CMATR.txt	The (NDU*NDU) damping matrix, e.g. as generated by CRAY or CMOD
V0.txt	NDU values in free format indicating the displacements in the master degrees of freedom at time t=0.
VP0.txt	NDU values in free format indicating the velocities in the master degrees of freedom at time t=0.
ACC.txt	Contains the accelerogram in 2E16.7 format, with time points in the first and ordinates in the second column (NT rows).
RVEKT.txt	NDU values in free format representing the displacements in each master degree of freedom for a unit base displacement in the direction of the seismic excitation.
Output files:	
THISNEW.txt	Either the displacements, velocities or the accelerations of the NDU degrees of freedom at each time step with the time points in the first column (F8.4, 30E16.7 format).
THISDU.txt	The displacements in the NDU master degrees of freedom for all NT time points in free format.
CTRLNB.txt	Echo print of the input data. Additionally, the (absolute) maximum displacement, the corresponding degree of freedom number (FRH) and the corresponding time and time step number.

<table>
<tr><th colspan="2">NEWMAR</th></tr>
<tr><td colspan="2">Solution of the coupled differential equation system by implicit direct integration (NEWMARK's method)</td></tr>
<tr><th colspan="2">Interactive in-/output:</th></tr>
<tr><td>Number NDU of the master degrees of freedom ?
Number NT of time steps (= ordinates of the load vector)?
Constant time step DT?
Should displacements, velocities or accelerations be written in the output file THISNEW.txt?
Enter IOUT=1 for displacements
Enter IOUT=2 for velocities
Enter IOUT=3 for accelerations
IOUT = ?
The acceleration will be output in your units (e.g. in m/s**2), not in g!</td><td>This message will appear if IOUT=3.</td></tr>
</table>

Input files:	
File name	Notes
KMATR.txt	Contains the condensed stiffness matrix of the system (NDU*NDU values) as created by the program CONDEN.
MDIAG.txt	The diagonal of the mass matrix (NDU coefficients in free format).
CMATR.txt	(NDU*NDU) damping matrix, e.g. as created by CRAY or CMOD.
V0.txt	NDU values in free format indicating the displacements in the master degrees of freedom at time t=0.
VP0.txt	NDU values in free format indicating the velocities in the master degrees of freedom at time t=0.
LOADV.txt	Contains NT*NDU values corresponding to the NDU components of the load vector at all NT time points in free format. This file may be created by the INTERP program (without the time points in the output file).

Output files:	
THISNEW.txt	This file contains the time points in the first column, and (alternatively) the displacements, the velocities or the accelerations in the NDU master degrees of freedom in the following columns (F7.4, 30E14.6 format).
THISDU.txt	Contains the displacements in the NDU master degrees of freedom in all NT time steps in free format, without the time points. This file serves as input for the program INTFOR.
CTRLNM.txt	Echo print; additionally, the maximum absolute value of the displacement together with the time it occurred, the time step number and the corresponding degree of freedom.

<table>
<tr><th colspan="3">NLSPEC</th></tr>
<tr><td colspan="3">Evaluation of inelastic response spectra of an accelerogram for a given target maximum ductility μ</td></tr>
<tr><td colspan="3">Interactive input:</td></tr>
<tr><td colspan="2">Target ductility?
Input data for the nonlinear spring model:
For bilinear law JKN = 1
For elastic-perfectly plastic law JKN = 2
For UMEMURA law JKN = 3
JKN =?</td><td></td></tr>
<tr><td colspan="2">Strain hardening ratio p as a proportion pK of the initial stiffness K?</td><td>Only required for the bilinear law.</td></tr>
<tr><td colspan="2">Spectrum damping ratio D ?
Initial conditions: Displacement for t=0?
Velocity for t=0?
Number NANZ of points of the accelerogram? (<7000)
Constant time step DT?</td><td></td></tr>
<tr><td colspan="2">Factor FAKT to obtain acceleration ordinates in m/s**2 units?

Initial natural period?
Period increment?
Number of ordinates to be computed?</td><td>The ordinates of the record in ACC are multiplied by FAKT (e.g. 9.81 if they are originally in g) in order to obtain m/s²-units.</td></tr>
<tr><td colspan="3">Input file:</td></tr>
<tr><td>ACC.txt</td><td colspan="2">Contains the accelerogram in 2E16.7 format, with time points in the first and ordinates in the second column (NANZ rows).</td></tr>
<tr><td colspan="3">Output files:</td></tr>
<tr><td>File name</td><td colspan="2"></td></tr>
<tr><td>NLSPK.txt</td><td colspan="2">Five columns in 5E16.7 format containing the periods (s), the relative displacement spectrum (cm), the relative pseudo-velocity spectrum (cm/s), the absolute pseudo-acceleration spectrum (g) and the ductility actually present, which may deviate slightly from the target ductility.</td></tr>
<tr><td>CTRLNS.txt</td><td colspan="2">Echo print, in addition periods and the maximum ductility actually obtained.</td></tr>
</table>

SDOF1	
Direct integration for a SDOF system by NEWMARK's method	
Interactive input	
Circular natural frequency? Damping ratio? Mass?	For computing inertia, damping and restoring forces
Displacement for t=0? Velocity for t=0? Number NANZ of time steps?	For the loading function given in RHS.txt
Constant time step? Acceleration of gravity g in the units employed? (e.g. 9.81)	For outputting accelerations in g-units
Load factor?	If the ordinates of the load function have not already been divided by the mass of the system, (1/m) should be entered as load factor.
Input file:	
RHS.txt	NANZ rows with [t, f(t)] of the load function (right-hand side of the differential equation) in 2E16.7 format. The time step DT is constant. If the ordinates of the load function have not already been divided by the mass m of the SDOF system, (1/m) should be entered as load factor.
Output file:	
TIMHIS.txt	Time points and calculated values for displacement, velocity, acceleration (the latter in units of g) and restoring force (in kN) in 5E16.7 format.
CTRLS1.txt	Maxima of the absolute values of displacements, velocities, accelerations with the corresponding restoring, damping and inertia forces and times.

SDOF2	
Direct integration for a SDOF system by NEWMARK's method for the harmonic excitation (right hand side) A sin ω_1t + B cos ω_2t	
Interactive input	
Circular natural frequency?	
Damping ratio?	
Mass?	For computing inertia, damping and restoring forces
Displacement for t=0?	
Velocity for t=0?	
A, OM1, B, OM2	For describing the loading function
Number of time steps?	
Constant time step?	
Acceleration of gravity g in the units employed? (e.g. 9.81)	For outputting accelerations in g-units
Load factor?	If the ordinates of the load function have not already been divided by the mass of the system, (1/m) should be entered as load factor.
Output file:	
TIMHIS.txt	Time points and calculated values for displacement, velocity, acceleration (the latter in units of g) and restoring force in kN in a 5E16.7 format.
CTRLS2.txt	Maxima of the absolute values of displacements, velocities, accelerations with the corresponding restoring, damping and inertia forces and times.

<table>
<tr><th colspan="2">SDOFNL</th></tr>
<tr><td colspan="2">Time history analysis for a nonlinear SDOF system; elastoplastic, bilinear or UMEMURA-type constitutive laws possible</td></tr>
<tr><th colspan="2">Interactive input:</th></tr>
<tr><td>Data of the nonlinear SDOF system:
Bilinear constitutive law? Y/N
Elastoplastic constitutive law? Y/N
UMEMURA constitutive law? Y/N</td><td></td></tr>
<tr><td>Strain hardening ratio p as proportion pK of the initial stiffness?</td><td>Only required for the bilinear constitutive law.</td></tr>
<tr><td>Circular natural frequency?
Damping ratio?
Mass of the SDOF system?
Initial conditions:
Displacement for t=0?
Velocity for t=0?</td><td></td></tr>
<tr><td>Max. elastic displacement?</td><td>Limiting value for the validity of HOOKE's law.</td></tr>
<tr><td>Load function data:
Number NANZ of time steps?
Constant time step?</td><td></td></tr>
<tr><td>Acceleration of gravity g in the units employed? (e.g. 9.81)</td><td>Required for outputting accelerations in units of g.</td></tr>
<tr><td>Load factor?</td><td>If the ordinates of the load function have not already been divided by the mass m of the SDOF system, (1/m) should be entered as load factor.</td></tr>
<tr><th colspan="2">Input files:</th></tr>
<tr><td>RHS.txt</td><td>NANZ rows with [t, f(t)] of the load function (right-hand side of the differential equation) in 2E16.7 format. The time step DT is constant.</td></tr>
<tr><th colspan="2">Output files:</th></tr>
<tr><td>TIMHISNL.txt</td><td>The output file THNLM contains, in five columns (5E14.7 format), the times, the calculated values for displacement, velocity, and acceleration (the latter in units of g) and the restoring force $F_R(t)$.</td></tr>
<tr><td>CTRLNL.txt</td><td>Maxima for the displacement, velocity, acceleration (in g) and restoring force (in kN) with the corresponding times.</td></tr>
</table>

SECSYS	
Computation of internal forces and deformations in a substructure for given deformations in DOF coupling it to the primary structure	
Interactive input:	
Number of master DOF (coupling DOF)? Number of beam elements in the secondary structure? Number of slave DOF in the secondary structure?	
Input files	
ECOND.txt	Describes the primary structure; input file for the program CONDEN. The first NELEM rows contain EI, ℓ, EA and α for each beam element, in free format. EI is its constant bending stiffness (e.g. in kNm^2), ℓ is the element length (e.g. in m), EA is the axial stiffness (e.g. in kN) and α is the angle between the global x-axis and the element axis (in degrees, positive counter-clockwise). The next NELEM rows contain the incidence vectors for all elements in free format, indicating the 6 global degrees of freedom corresponding to the 6 local degrees of freedom (u_1, w_1, φ_1, u_2, w_2, φ_2) of each element. Next, the NDU numbers of the master kinematic degrees of freedom are entered, in free format.
AMAT.txt	The matrix A (NDOF rows, NDU columns) serves for evaluating the displacements in all degrees of freedom from known displacements in the NDU master degrees of freedom. Computed by the program CONDEN.
VU.txt	Deformations in the coupling DOF, in free format.
Output files	
RESULT.txt	The computed deformations and internal forces for all beam elements of the substructure.

SPECTRUM
Evaluation of the response spectrum of a given accelerogram
Interactive input:
Damping ratio D for the spectrum to be evaluated? Number of points in the accelerogram? (<7000) Constant time step DT? Factor FAKT to obtain acceleration ordinates in m/s**2 units? Initial natural period? Period increment? Number of spectral ordinates to be computed?

Input file:	
ACC.txt	Contains the record in 2E16.7 format, with time points in the first and ordinates in the second column (NANZ rows).
Output files:	
SPECTRUM.txt	Five columns in 6E16.7 format containing, respectively, the periods (s), the relative displacement spectrum (cm), the relative pseudo-velocity spectrum (cm/s), the absolute pseudo-acceleration spectrum (g) and the acceleration spectrum (g).
CTRLSP.txt	Echo print; additionally, the computed spectral intensity according to HOUSNER (in cm).

SYNTH	
Generation of spectrum compatible accelerograms	
Input file:	
ESYN.txt	This file contains the following data in free format: An arbitrary integer number IY (IY < 1024), Number NK of (T, S_v) – pairs to be read, defining the desired target spectrum, Number N of accelerogram ordinates to be generated, with the constant time step fixed at 0.01 s, Number of the time step at the end of the increasing part of the trapezoidal envelope function, Number of the time step at the beginning of the decreasing part of the trapezoidal envelope function, Number of iteration cycles, usually 10 to 20, Periods TANF and TEND, between which the target spectrum is to be matched, Damping ratio of the spectrum, The NK pairs (T, S_v) defining the target spectrum (T in s and S_v in cm/s); one pair per row.
Output files:	
File name	
CTRLSY.txt	Echo print of the input data.
ASYN.txt	The generated accelerogram in 2E16.7 format, with time points in the first and accelerations (in m/s^2) in the second column (N rows).

Chapter 2
Seismic Loading

Klaus-G. Hinzen

Abstract Since man has been erecting structures, earthquake ground motions have posed a threat to their stability. Only with the advent of and continual improvement to digital measurement, processing, and modeling techniques, have engineering seismologists been able to quantify the spreading of seismic waves in the Earth to facilitate site specific ground motions estimates for potentially damaging (i.e., 'future') earthquakes. These earthquake scenarios constitute the boundary conditions for earthquake-resistant design of buildings. The ground motions during an earthquake at a specific site are determined by the characteristics of the earthquake source, the travel path, and the local site conditions. Each of these links along the path of seismic waves influences the amplitudes, frequency content, and duration of the vibrations which ultimately influence the dynamic load of buildings. The chapter gives a brief introduction to the earthquake phenomenon and the spreading of seismic waves and the main characteristics and parameters of strong ground motions are explained. Further discussed are the effects of finite seismic sources and site effects due to the local geology on ground motions and spectra.

Keywords Seismic waves · Strong ground motions · Earthquake source
Site effects

The ground motions during an earthquake at a specific site are determined by the characteristics of the earthquake source, the travel path, and the local site conditions. Each of these links along the path of seismic waves influences the amplitudes, frequency content, and duration of the vibrations which ultimately influence the dynamic load of buildings. Section 2.1 gives a brief introduction to the earthquake phenomenon and the spreading of seismic waves. In Sect. 2.2 the main characteristics and parameters of strong ground motions are explained. The effect of finite seismic sources on ground motions and spectra are described in Sect. 2.3, followed by a closer look at the site effects due to local geology and building ground conditions. Sect. 2.5 provides some examples of designed or simulated ground motions. A highly quantitative presentation of elasticity theory, wave equations, and seismic source theory can be found in seismological textbooks (e.g. Aki and Richards 2002; Lay and Wallace 1995; Shearer 1999).

K. Meskouris et al., *Structural Dynamics with Applications in Earthquake and Wind Engineering*, https://doi.org/10.1007/978-3-662-57550-5_2

2.1 The Earthquake Phenomenon

Ever since mankind has erected buildings, natural forces, including ground shaking during earthquakes, affected these. With the development of instruments capable of recording the ground motion during an earthquake as function of time at the end of the 19th century, modern seismology progressed rapidly. The Great 1906 San Francisco Earthquake was a trigger for systematic studies of earthquake engineering problems (Fig. 2.1).

The Lawson Report (Lawson 1908) summarized the observations of earthquake phenomena and their analysis including source effects, wave spreading effects, site effects, and the structural behaviour of buildings. Reid (1910) also based his elastic rebound theory on the data recorded from this earthquake. This theory was the first comprehensive explanation of the connection between tectonic processes (plate movements) and the generation of seismic waves from a rupturing fault. The basis of Reid's theory in which the seismic source is modelled as a shear dislocation on the fault plane is still in use. Figure 2.2 shows an earthquake cycle in the elastic rebound theory. The cycle starts after an earthquake with the situation on the left panel. Over decades plates move in the directions indicated by the filled arrows while the rock in the vicinity of the fault is deformed and accumulating strain energy. When the stress on the fault exceeds the frictional forces, a rupture develops and opposite sites of the fault are accelerated to the new position (right panel) within seconds during an earthquake.

The vast majority of earthquakes occur at the major plate boundaries (interplate earthquakes). Even in a (geologically speaking) very short time window of 30 years,

Fig. 2.1 Structural damage of the San Francisco City Hall, which was opened shortly before the 18 April 1906 Earthquake (Photo USGS)

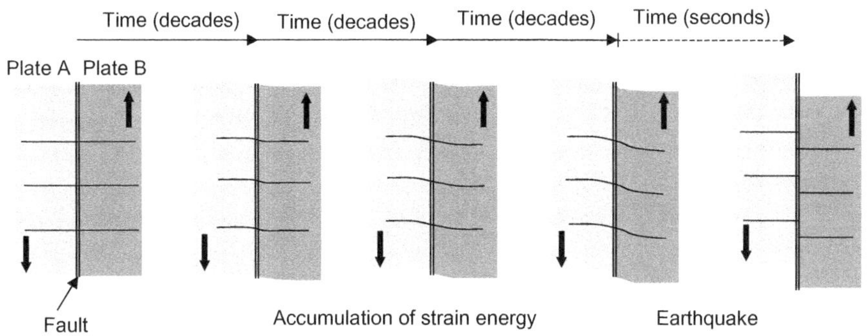

Fig. 2.2 Schematic drawing of an earthquake cycle in elastic rebound theory. Shown is a map-view of tectonic plates A and B, separated by a fault line

the epicentres of the larger earthquakes with magnitudes above 5 clearly trace the plate boundaries (Fig. 2.3). The general concept of plate tectonics was developed in the 1960s after advances in seismological instrumentation allowed precise earthquake locations and advanced methods of measuring the topography, and magnetic properties of the ocean basins became available. Driven by the internal heat of the Earth, plates move with respect to each other with velocities typically in the range of 3–5 cm/year and reaching up to 20 cm/year. However, in some areas low to moderate earthquake activity also exists inside the plate (intraplate seismicity), i.e., eastern North America, West India, Central Europe.

A rough classification of tectonic environments is: (1) shallow-crustal earthquakes in active tectonic regions, (2) shallow-crustal earthquakes in stable tectonic regions, (3) intermediate depth earthquakes in subducting plates, and (4) earthquakes along the interface of two subduction plates (Campbell 2003).

2.1.1 Earthquake Source Model

Though the earthquake source process can be very complex and is still a matter of research, the basic geometry can be summarized as shown in Fig. 2.4. The size of the earthquake is mainly determined by the source area, A the section of the fault, which is activated during the event. The rupture, which is assumed two dimensional in the model, nucleates at the hypocenter and spreads over the fault surface. Spreading velocity and amount of dislocation can vary over the source area. The second parameter determining the size of the earthquake dislocation averaged over the source area, $\overline{D}$. The orientation of the source can be described by three angles of orientation, strike, ϕ_s, and dip, δ, of the fault plane and rake, λ, giving the direction of the slip vector, $\underline{u}$, of the hanging wall (Fig. 2.5) with respect to the foot wall and measured in the fault plane. Usually the definitions by Aki and Richards (1980) are applied where ϕ_s is measured clockwise from north ($0 \leq \phi_s < 2\pi$) with the dip direction down

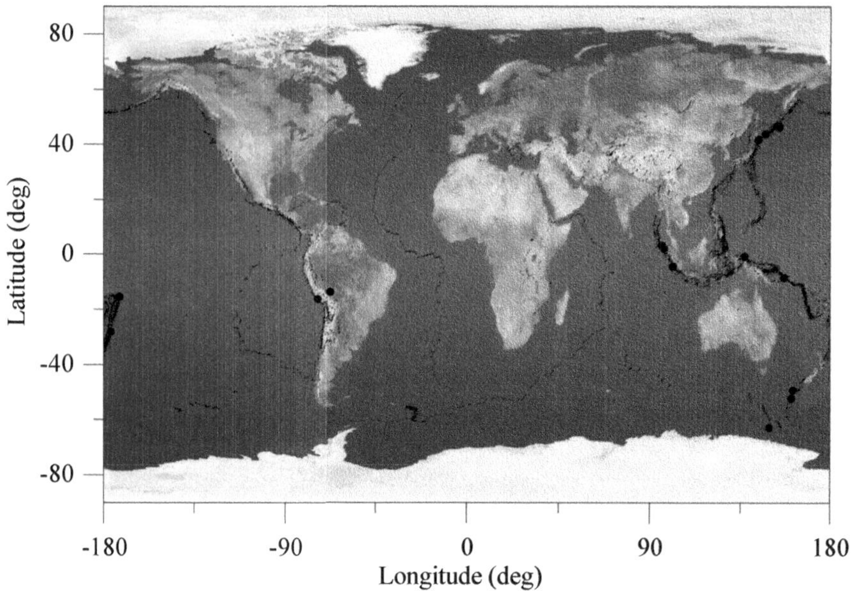

Fig. 2.3 Location of epicentres of 47,578 earthquakes with magnitudes larger than 5 that occurred between 1980 and 2009 (earthquake catalogue, USGS, plot https://gnugeneration.epfl.ch/)

to the right of the strike direction ($0 \leq \delta < \pi/2$). The zero direction for the rake is parallel to the strike direction and λ is measured counter clockwise. The seismic shear dislocation source is categorized using the slip-direction (Fig. 2.4).

The seismic wavefield radiated from such a shear dislocation can be modelled by a double couple source (Lay and Wallace 1995) incorporating the scalar seismic moment M_0 given by the product of the shear modulus of the material in the source region, μ, which is one of Lamé's constants of the medium, the source area, i.e. the fault rupture surface, and the displacement averaged over the source area (Aki 1966).

$$M_0 = \mu A\overline{D}. \tag{2.1.1}$$

For crustal rocks, usually relevant in engineering seismological problems, μ is often assumed to be on the order of 32 GPa. The size of the seismic moment can be determined from earthquake records as well as from the measurement of activated fault sections in the field. The physical unit is Nm, in seismological literature the non-SI unit of dyn cm (1 Nm = 10^7 dyn cm) is often used for traditional reasons.

A new unit of seismic moment, the Aki, was recommended by the International Association of Seismology and Physics of the Earth's Interior (IASPEI) in 2007. The recommendation was made by resolution, which states that 1 Aki (Ak), defined as 10^{18} Nm, be recognized as a standard unit of earthquake size. The unit is named after Professor Keiiti Aki, who was a pioneer in defining seismic moment and describing practical ways to measure it.

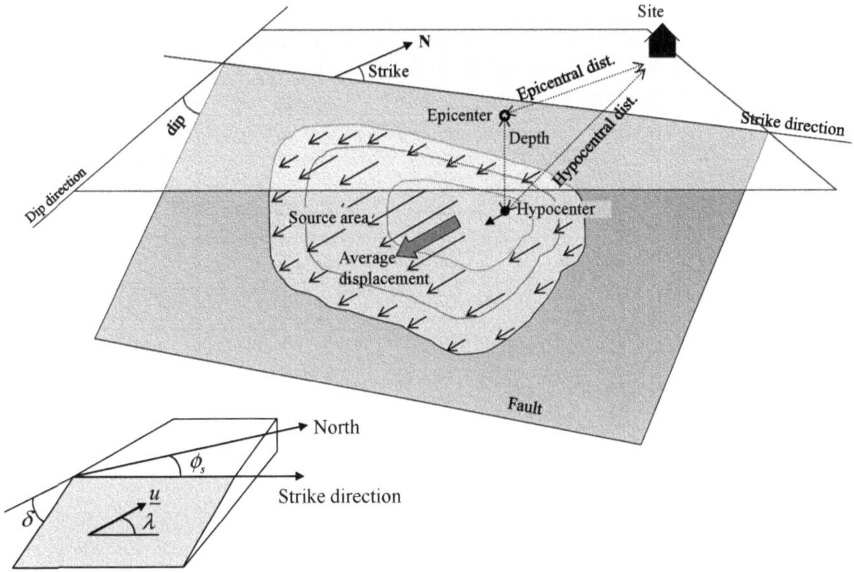

Fig. 2.4 Main elements of an earthquake source, the inset in the lower left shows the angles of orientation, strike, ϕ_s, dip, δ, and rake λ

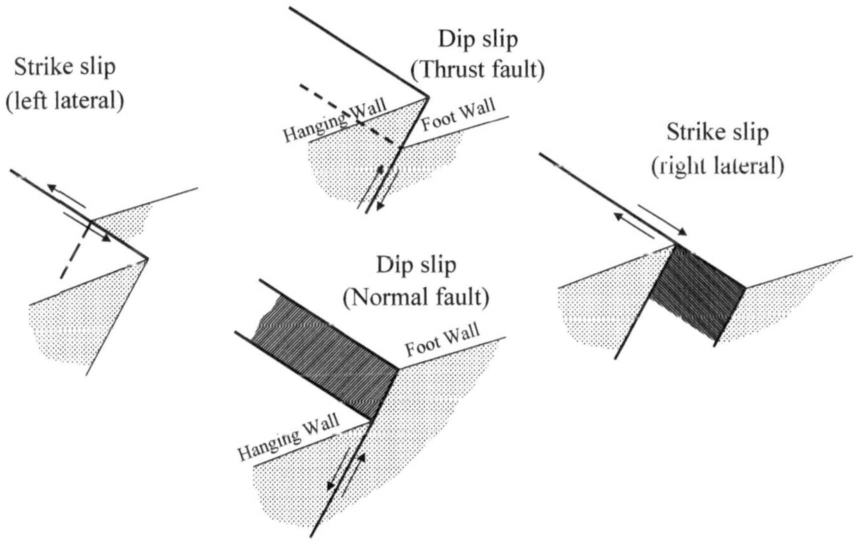

Fig. 2.5 Basic types of faulting mechanisms of a shear dislocation source

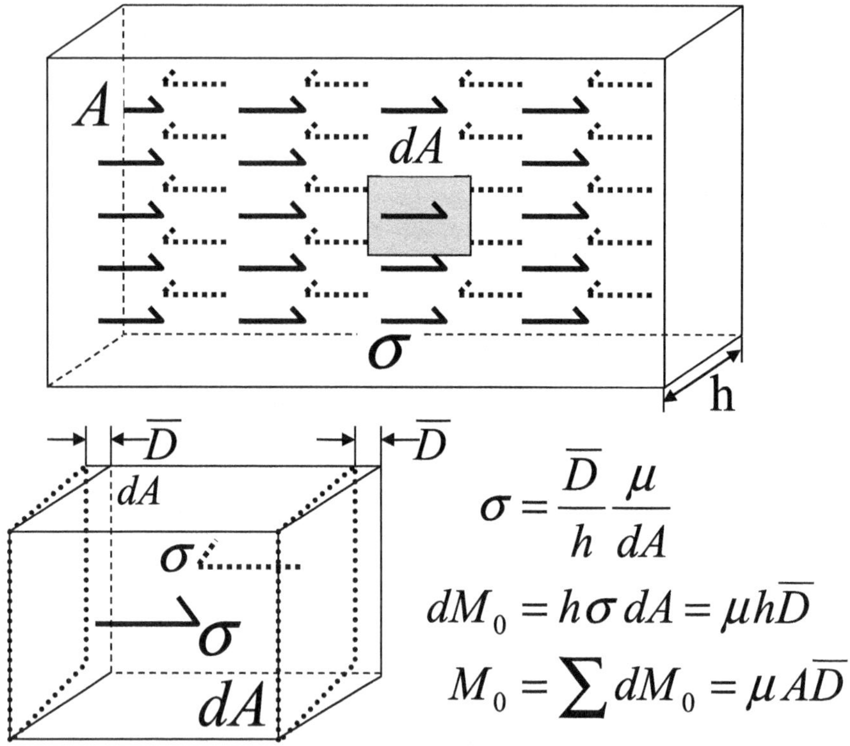

Fig. 2.6 Mechanical model for the explanation of the seismic moment (after Erdik and Durukal 2003)

Engineers often argue that M_0 is not really a moment; however, a mechanically oriented model for the explanation of the seismic moment is given by Erdik and Durukal (2003) in Fig. 2.6. Here, the relative slip on both sides of the dislocation source is caused by the acting stress, σ which exceeds the friction stress.

The stresses involved in the earthquake process are hard to determine. However, the stress drop, $\Delta\sigma$, which is defined as the difference between the initial static stress before an earthquake, σ_i and the final stress after the earthquake, σ_0, critically determines the nature of the ground motion.

$$\Delta\sigma = \sigma_i - \sigma_0. \tag{2.1.2}$$

Only under idealized condition $\sigma_0 = 0$; in actual earthquakes σ_0 will be finite, due to the friction on the fault surface. The average stress σ_m is defined as:

$$\sigma_m = (\sigma_i - \sigma_0)/2. \tag{2.1.3}$$

and the total seismic energy, E, released during the earthquake is (Erdik and Durukal 2003):

$$E = \sigma_m \overline{D} A = \frac{\sigma_m}{\mu} M_0. \tag{2.1.4}$$

Typical observed values of static stress drop are between 1 and 10 MPa (10–100 bar). Kanamori and Anderson (1975) used 3 MPa of static stress drop for intraplate earthquakes and 10 MPa for interplate earthquakes.

For a uniform shear rupture Aki (1972) gave the stress drop as:

$$\Delta\sigma = C\mu\left(\overline{D}/L_c\right). \tag{2.1.5}$$

where C is a constant value in the range of unity which depends on the source geometry and L_c is the so-called characteristic dimension (length) of the source. With another constant, C_1, and (2.1.1) the seismic moment can be related to stress drop and source dimension:

$$M_0 = C_1\,\Delta\sigma\,L_c. \tag{2.1.6}$$

These relations form the basis for a variety of earthquake source models with different geometries. For a circular fault of radius r, the relation between source dimension, stress drop, and average dislocation is (Keilis and Borok 1957; Kanamori and Anderson 1975):

$$\Delta\sigma = \frac{7\pi}{16}\mu\frac{\overline{D}}{r} = \frac{7}{16}\frac{M_0}{r^3} \tag{2.1.7}$$

The seismic moment is directly related to the physical source dimensions (2.1.1) and thus does not saturate for large earthquakes; an advantage in earthquake size determination compared to the traditional magnitude scales based on band limited parts of the seismic source spectrum. M_0 will never completely replace magnitudes and Kanamori (1977) defined the moment magnitude scale based on the logarithm of M_0:

$$M_W = \frac{2}{3}\left(\log M_0 - 9.05\right). \tag{2.1.8}$$

where M_0 is measured in Nm (when M_0 is given in dyn cm, the scaling factor is 16.05). A constant static stress drop of 3 MPa is assumed in this definition.

Example 2.1

Consider an earthquake of moment magnitude $M_W = 7.0$. Following the definition of (2.1.8), $\log M_0$ comes to 19.6 and the seismic moment is $M_0 = 4.0 \cdot 10^{19}$ Nm (or 40 Aki). Assuming the typical stress drop of an intraplate earthquake of 3 MPa, the source radius of a circular fault using (2.1.7) is 18 km.

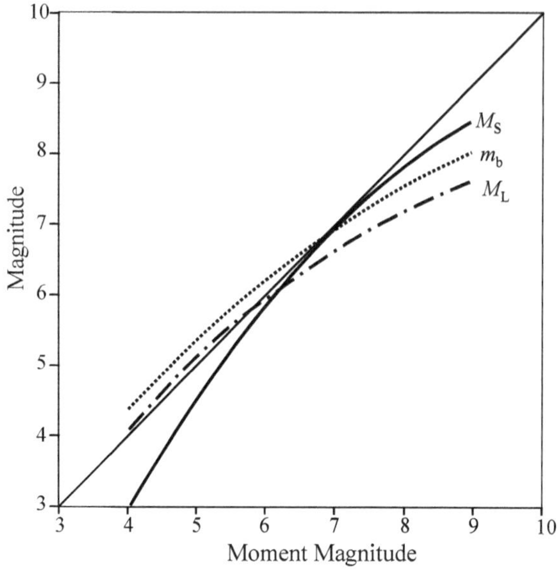

Fig. 2.7 Relation between moment magnitude (M_W), surface wave magnitude (M_S), body wave magnitude (m_b) and local magnitude (M_L) (after Boore and Joyner 1994)

The concept of the logarithmic measure of earthquake size was originally adopted from astronomy (measuring the brightness of stars) by Wadati and later defined by Richter (1935, 1958). Originally his local magnitude M_L was only defined for California earthquakes, due to the imbedded regional, empirical amplitude attenuation function. For earthquake observations at distances larger than 1000 km, the body wave magnitude m_b and the surface wave magnitude M_S were introduced. While m_b is based on the amplitudes of body waves (P-wave) at periods around 1 s, M_S is based on surface waves at periods of 20 s. In the middle range of magnitude values, these numbers on the different scales are close; however, at large magnitudes (above 6.5–7) saturation affects the classical magnitude measures (Fig. 2.7).

2.1.2 *Seismic Waves*

While the rupture processes and deformations in the vicinity of the fault are non-linear processes, the seismic source generates elastic waves, which can travel through the whole body of the Earth. These body waves are classified following the sense of motion of a particle with respect to the wave spreading direction (Fig. 2.8). In the early days of seismology the longitudinal waves were named primary waves (P waves), as they are the fastest travelling seismic waves. The transversal or shear waves (S waves) correspondingly were called secondary waves. For S waves the ground motion in the plane perpendicular to the spreading direction can be separated

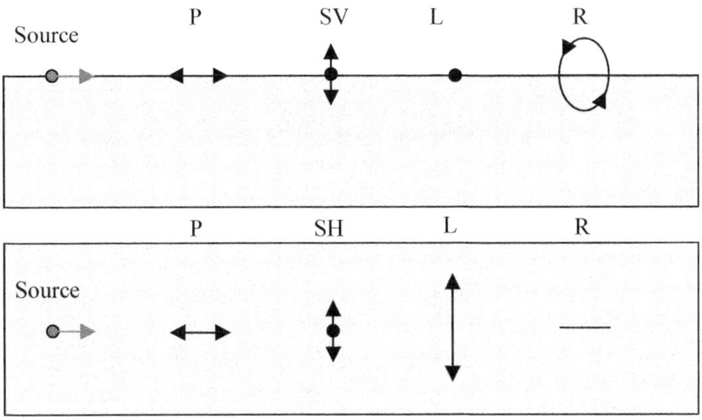

Fig. 2.8 Particle motion during the passage of seismic waves. Spreading direction in both panels is from left to right. Top and bottom panels show the motion in a vertical and horizontal plane, respectively. SV and SH are the vertically and horizontally polarized parts of S waves (after Bath 1973)

into the vertical (SV) horizontal (SH) component. The spreading velocity of P and S waves, v_P and v_S, respectively is:

$$v_P = \sqrt{(\lambda + 2\mu)/\rho}, \quad v_S = \sqrt{\mu/\rho}. \tag{2.1.8}$$

where λ is one of Lamé's constants of elasticity and ρ is the density. In case of a perfect elastic medium with a Poisson ratio of $\nu = 0.25$ the ratio between v_P and v_S is:

$$v_P/v_S = \sqrt{3} \approx 1.73. \tag{2.1.9}$$

Typical P-wave velocities in the Earth's crust are in the range of 5–7 km/s (18,000–25,000 km/h). In sedimentary basins and sediments on valley floors velocities v_P can be lower than 1 km/s.

While both body waves, P and S, exist in a whole space, the free surface of the Earth is a precondition for the formation of surface waves. Again, two types exist. Love waves (L) are formed by the interference of horizontally polarized S waves. Besides the free surface, at least one layer with a lower velocity than the underlying half space must be present. The ground motion is similar to that of SH waves (Fig. 2.7). Rayleigh waves are a consequence of interference of P and SV waves. The ground motion at the Earth's surface is elliptical with the larger half axis in the vertical direction and a retrograde sense of motion (Fig. 2.7). Surface wave amplitudes are in general larger than those of body waves; however, they exponentially decay with depth. The velocities of Love and Rayleigh waves are smaller than v_S. In Earth's crust the seismic velocities mostly increase with depth

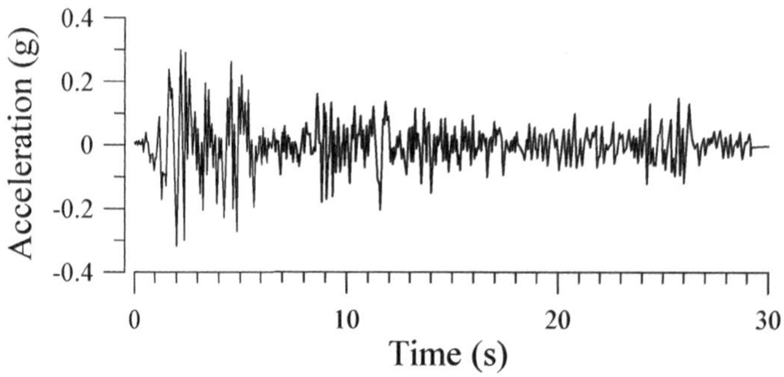

Fig. 2.9 Strong-motion record of the 1940 El Centro (Imperial Valley) earthquake. The seismogram shows the North-South component of motion. The instrument was attached to the El Centro Terminal Substation Building's concrete floor, and not in a free-field location

and this layering causes the dispersion of surface waves. In contrast to body waves, the spreading velocity becomes dependent of the wave's frequency.

2.2 Strong Ground Motion Characteristics

While in the early years of modern seismology the effects of earthquake ground shaking were studied thoroughly and the complete ground motion was recorded at large distances from the source, little was known about the ground motion characteristics close to the rupturing fault section until the construction of the first strong-motion instruments. Davis (1913) described the construction of the first strong-motion instrument and the first strong-motion record that was widely used in the aftermath was recorded during the 1940 El Centro (Imperial Valley) Earthquake (Fig. 2.9).

As more and more strong-motion instruments were installed worldwide, in recent time the amount of available records has dramatically increased. Several open sources for easy access to strong motion records exist (i.e., http://peer.berkeley.edu/smcat/; http://www.isesd.cv.ic.ac.uk/ESD/frameset.htm; http://www.cosmos-eq.org/). In most cases strong-motion recording is done with accelerometers. Figure 2.10 shows a typical installation of an accelerometric strong motion sensor in the free field. The sensor is tightly bolted to the foundation because at ground accelerations above some 0.2 g rattling of the sensor might otherwise occur.

Modern equipment with a high recording dynamic range allows for integration and double integration to deduce ground velocity and displacement, respectively. Figure 2.11 shows the three components of a strong ground motion recorded during the $M_W = 7.2$, 1999 Düzse Earthquake in Turkey at a distance of 39 km. Shown are the originally recorded acceleration, which reaches in maximum almost 80% of g, and the displacement determined by double integration after band-pass filtering of the accelerogram between 0.1 and 25 Hz.

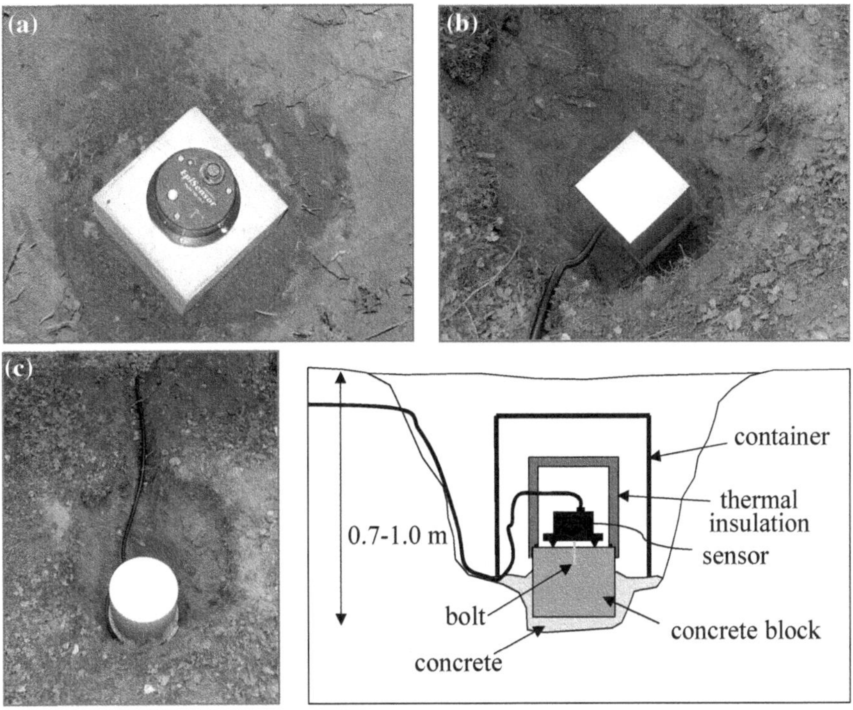

Fig. 2.10 Installation of a strong-motion sensor in free-filed conditions. **a** The 3-component sensor is bolted to a concrete foundation located in a 1 m deep ditch. **b** and **c** Sensor is covered by thermal and mechanical protection. The bottom right shows a schematic cut of the setup (after Hinzen and Fleischer 2007)

Such data resources are the basis for empirical attenuation relations, better known as ground motion models, which have become a standard in seismic hazard analysis to link earthquake size (magnitude), and distance between the point of interest to the seismic source to predict ground motion parameters. The source-to-site distance can be defined in different ways. For larger distances, the earthquake might by treated as a point source, however at closer distances, which are more important in engineering seismology, the source extension has to be taken into account. The point source distances are the epicentral distance, r_{epi}, and the hypocentral distance, r_{hyp}, as shown in Fig. 2.3 where the latter can be expressed with the depth of the focus, h as:

$$r_{hyp} = \sqrt{r_{epi}^2 + h^2}. \quad (2.2.1)$$

In cases where r_{epi} is not much larger than the source dimension three finite-source distances are of common use (Fig. 2.4). The closest horizontal distance of the site to the vertical projection of the fault to the Earth surface was introduced by Joyner and Boore (1981) and is referred to as Joyner–Boore distance, r_{jb}. In case of a dipping

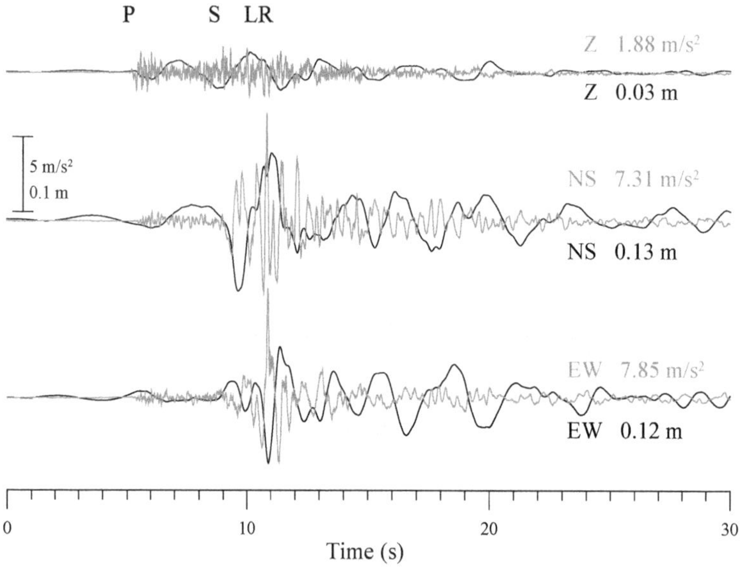

Fig. 2.11 Strong-motion record of the 1999, $M_W = 7.2$ Düzce earthquake, Turkey. Recording distance was 39 km from the fault. The gray lines show the acceleration and the black lines the displacement in the vertical (Z) and horizontal (EW, NS) components. The numbers at the end of the traces give the maximum ground acceleration and displacement, respectively. P, S and LR mark the arrivals of the P- and S-bodywaves and the Love and Raleigh surface waves (Acceleration data retrieved from http://www.isesd.cv.ic.ac.uk/)

fault, the projection to the surface encloses an area for which $r_{jb} = 0$. The closest distance between site and the rupture plane, introduced by Schnabel and Bolton Seed (1973) as rupture distance, r_{rup}, and the closest distance to the seismogenic part of the rupture plane, r_{seis} (Campbell 1987) are more difficult to determine, because the depth of the top of the seismogenic zone has to be known. In case of a vertical fault and a rupture reaching the surface $r_{jb} = r_{rup}$, however, in general $r_{seis} > r_{jb}$.

2.2.1 Prediction of Ground Motion Parameters

The simplest ground motion models only account for the distance measure and the earthquake size as parameters. More complex models include faulting mechanism (Fig. 2.4), local site conditions, stress drop (2.1.2), source directivity and radiation pattern, hanging- and footwall effects, and the tectonic environment. The relations are derived by regression calculations (Draper and Smith 1981) either from the strong motion databases, theoretical data, or a combination of both. In order to include the mentioned site and source effects, it is important that these parameters are included in the databases, so specific selection of datasets can be made. Also care has to be taken of the recording conditions because many instruments are installed in or at larger buildings and the records might partially include the building response, whereas the prediction models in general are aiming to give the free field ground motion parameters.

In its common logarithmic form the ground motion can be expressed by (Kramer 1996; Campbell 2003):

$$\underbrace{\ln Y}_{1} = \underbrace{c_1 + c_2 M + c_3 M^{c_4}}_{2} + c_5 \ln\left[R + c_6 \exp(c_7 M)\right] + \underbrace{c_8 R}_{5}$$
$$+ \underbrace{c_9 F}_{6} + \underbrace{c_{10} S}_{7} + \varepsilon \tag{2.2.2}$$

where Y is the strong-motion parameter of interest (i.e., peak ground acceleration or spectral acceleration response), M is the magnitude, F and S describe the faulting mechanism and site condition, respectively, ε is a random error term with zero mean and R is the distance term. The numbered brackets in (2.2.2) relate the terms to the individual effects (Kramer 1996):

① The peak values of strong-motion parameters are often lognormally distributed. Therefore the regression calculations are commonly made for log Y rather than Y itself.
② Magnitudes are typically determined from the logarithm of some peak ground motion parameter. Therefore, $\ln Y$ should be proportional to M. $Y \propto c_2 M$ is compatible with the original definition of magnitude by Richter (1935).
③ Amplitudes of body waves P and S decrease due to geometrical spreading typically with $1/R$, those of surface waves with $1/\sqrt{R}$.
④ As the source area increases with magnitude, the distances from which waves arrive at a site and cause strong ground motions can be larger than R (Fig. 2.12). This term allows for an effective distance that increases with magnitude.
⑤ Inelastic attenuation converts part of the seismic wave energy to heat and causes an exponential decay of the amplitude with R.
⑥ With the increase in strong-motion observations influence between source characteristics such as type of faulting (Fig. 2.4) and rupture process could be included in the regression analysis of some models.
⑦ At least a differentiation between hard rock and sediment sites is commonly regarded in the models.

More details are given in the *Earthquake Engineering Handbook* by Campbell (2003) who also summarizes the content of a large variety of models that cover different tectonic environments (i.e., shallow active and stable crust, extensional crust, subduction environment) and geographical regions (i.e., western and eastern North America, Europe, Japan and worldwide).

Example 2.2
Figure 2.13 compares the predicted variation of the maximum horizontal ground acceleration, *PHA*, with distance from the source for two models. The model by

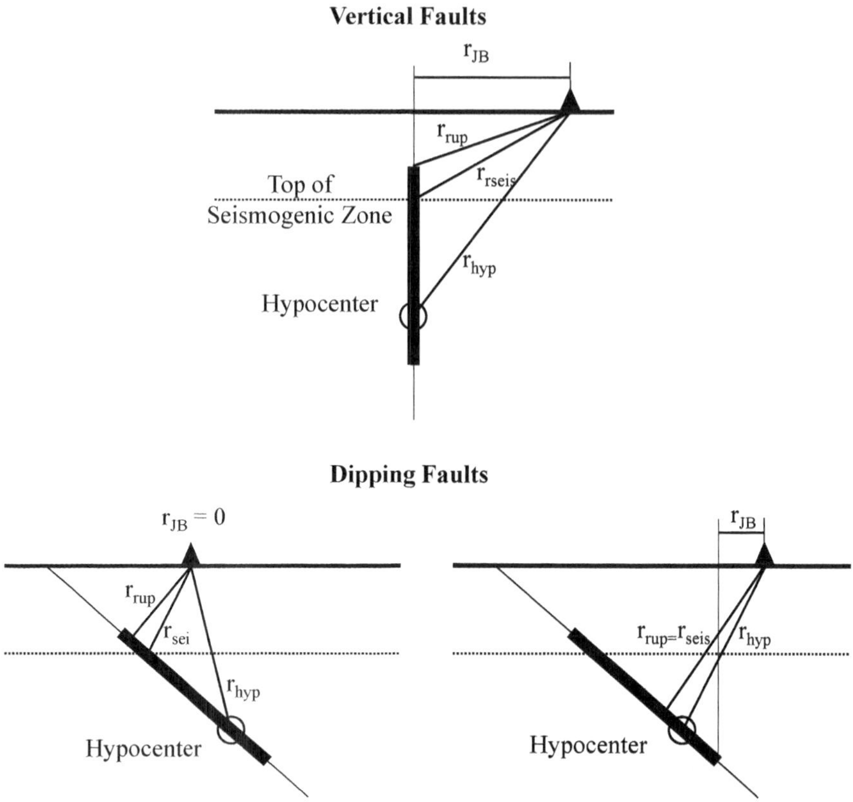

Fig. 2.12 Distance measures commonly used in engineering seismology (after Abrahamson and Shedlock 1997)

Campbell (1981) is based on a worldwide database and uses a distance range of maximum 50 km. The distance measure is the closet distance to the fault rupture and magnitudes below 6 are local magnitude and surface wave magnitude. This early model predicts the mean *PHA* and only depends on the terms ② and ③ in Eq. 2.2.2. The second model by Boore et al. (1993) is more complex. Here the regression is based only on earthquakes from western North America of magnitudes 5.0–7.7 and for distances up to 100 km from the surface projection of the fault. The curves shown in Fig. 2.13 are for a site with shear wave velocities of 360–750 m/s [site class B, term ⑦ in (2.2.2)] and the larger *PGA* of the two horizontal components.

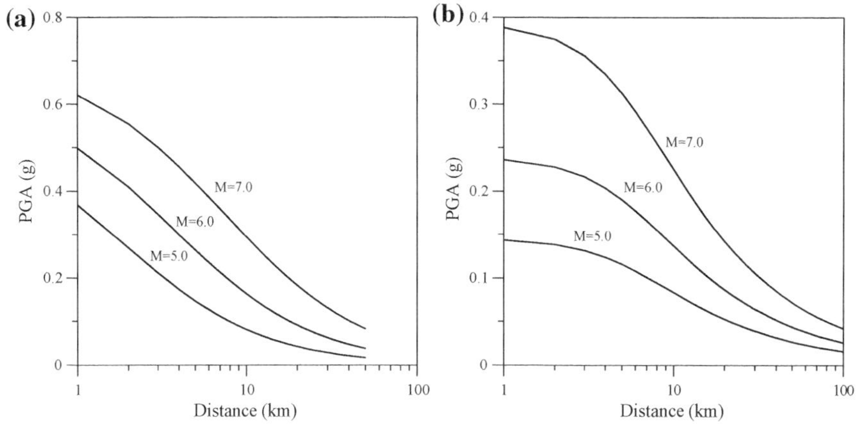

Fig. 2.13 Variation of peak ground acceleration, *PGA*, with distance for magnitudes 5.0, 6.0, and 7.0 predicted by the ground motion model of **a** Campbell (1981), and **b** Boore et al. (1993)

2.2.2 Site Specific Ground Motion Parameters

A broad spectrum of ground motion parameters has been defined in the field of engineering seismology including time- and frequency domain values.

Peak amplitudes of ground acceleration (PGA) are the most commonly used parameters. As the horizontal ground movements are usually more important than the vertical component in civil engineering, the maximum horizontal acceleration is addressed as PHA and the vertical component as PVA. The corresponding abbreviations for the ground velocity and displacement are PGV and PGD, respectively.

These peak values have in common that they are the maximum of a single half oscillation of the entire seismogram and the actual value a product of chance. This is often observed in records from closely spaced stations during the same earthquake. Even for conditions that appear identical, peak amplitudes can differ significantly. PGA may correlate with the damage caused by an earthquake, however, repeated moderate ground motion amplitudes can have a higher damage potential than a single large peak in the accelerogram.

Newmark and Hall (1982) introduced the concept of an effective acceleration that is supposed to be the acceleration, which is most closely related to the structural response and to the damage potential of an earthquake. This effective acceleration is smaller than the PGA measured in free-field outside of structures.

The total intensity I_T is defined as the area underneath the squared acceleration over the time interval T_d of the strong ground motion:

$$I_T = \int_0^{T_d} [a(t)]^2 dt. \tag{2.2.3}$$

A modification of (2.2.3) is the *rms* acceleration:

$$a_{rms} = \sqrt{\frac{1}{T_d}\int_0^{T_d} [a(t)]^2 dt}. \tag{2.2.4}$$

The square of a_{rms} is known as the *mean squared acceleration.* The influence of a large acceleration which lasts only for a short moment is small; most influential is the length of T_d, which can be defined in different ways and thus will produce different numerical values of a_{rms}.

If ground velocity $v(t)$ or displacement $d(t)$ is used in (2.2.4) instead of acceleration the corresponding *rms velocity* and *rms displacement* results:

$$v_{rms} = \sqrt{\frac{1}{T_d}\int_0^{T_d} [v(t)]^2 dt}, \quad d_{rms} = \sqrt{\frac{1}{T_d}\int_0^{T_d} [d(t)]^2 dt}. \tag{2.2.5}$$

A similar approach, however without the necessity to determine T_d was made by Arias (1970). Theoretically the integration for the Arias intensity, I_A, runs until infinity, but in practice a reasonable length of the time window has to be chosen:

$$I_A = \frac{\pi}{2g}\int_0^{\infty} [a(t)]^2 dt. \tag{2.2.6}$$

I_A has the unit of a velocity and has among other applications been utilized to quantify slope failure hazard (Wilson 1993).

The *characteristic intensity* is defined as:

$$I_c = (a_{rms})^{\frac{2}{3}} \sqrt{T_d}. \tag{2.2.6}$$

and the *specific energy density* is:

$$SED = \int_0^{T_d} [v(t)]^2 dt. \tag{2.2.7}$$

and the *cumulative absolute velocity* is defined as:

$$CAV = \int_0^{T_d} |a(t)|^2 dt. \tag{2.2.8}$$

Several modifications of the (2.2.3)–(2.2.8) have been described. Benjamin and Associates (1988) for example introduced the *effective design acceleration* (*EDA*) which is the PGA of a seismogram after it has been limited to frequencies below 9 Hz, avoiding that singular high frequency peak determines the whole design.

The *sustained maximum acceleration* (*SMA*) and *velocity* (*SMV*) was introduced by Nuttli (1979). These parameters give the sustained maximum acceleration/velocity during three cycles, and are defined as the third highest absolute value of acceleration in the time history.

Besides the amplitudes of strong ground motions, their duration determines the damage potential. Sites on soft sediments show a clear tendency for longer durations than rock sites at the same distance. While the first onset of P wave ground motion is easy to identify, the cessation of ground motions can be defined in different ways.

Bolt (1969) defined the duration of the strong ground motion as the time interval between the first and last exceedance of an acceleration threshold, which he choose at 0.05 g. This time was later called *bracketed duration*, D_b. Alternatively Trifunac and Brady (1975) suggested to use the time interval between the arrival of 5 and 95% of the total seismic energy at the site.

The *significant duration*, D_s is bound to the Arias Intensity, I_A, often also the time between 5 and 95% of I_A is used.

The *uniform duration*, D_u, is the sum of time intervals during which the acceleration exceeds a certain level. If the same threshold is chosen, $D_u < D_b$. More details about strong motion duration are given by Bommer and Martinez-Pereira (1999).

The *effective duration*, D_e is based on the D_s, however, both the start and end of the strong shaking phase are identified by absolute criteria.

Alternative to the time domain parameters, similar values can be defined in frequency domain. Parseval's theorem links the *total intensity*, I_T to the corresponding parameter in frequency domain:

$$I_T = \frac{1}{\pi} \int_0^{\omega_N} c_n^2 \, d\omega. \tag{2.2.9}$$

where ω_N is the Nyquist frequency and c_n are the Fourier coefficients of the frequency transformed seismogram.

The *predominant period*, T_p, is the period of the oscillations of the signal with the largest spectral amplitudes and can serve as a simple frequency characteristic of the ground motion. It is usually determined from smoothed amplitude spectra so that the values depend on the type of smoothing applied. The dominant period has a disadvantage similar to that of *PGA* as it does not say anything about the shape of the spectrum. Rather, different ground motions can result in the same dominant period.

Rathje et al. (1998) introduced the *mean period*, T_m, to characterize the frequency content on the ground motion:

$$T_m = \frac{\sum c_i^2/f_i}{\sum c_i^2}. \tag{2.2.10}$$

c_i are the Fourier amplitudes, and f_i represent the discrete Fourier transform frequencies between 0.25 and 20 Hz.

The form of the spectrum can be parameterized by its bandwidth, i.e., width of the frequency band in which the amplitudes are larger than half the maximum spectral amplitude.

The *acceleration spectrum intensity*, *ASI* (von Thun et al. 1988), is determined by integration of the response spectrum, S_a, with a critical damping, ξ of 5% within a fixed interval of periods:

$$ASI = \int_{0.1}^{0.5} S_a(\xi = 0.5, T)\, dT. \tag{2.2.11}$$

For the velocity the correspondent parameter is the *velocity spectrum intensity*:

$$VSI = \int_{0.1}^{2.5} S_v(\xi = 0.5, T)\, dT. \tag{2.2.12}$$

where S_v is the velocity response spectrum.

A simple parameter is the ratio of maximum velocity and maximum acceleration, $v_{\max}/a_{\max}$. These values are usually observed at parts of the signal with different frequencies. In case of a simple harmonic ground motion this would be $v_{\max}/a_{\max} = T/2\pi$. For the more irregular oscillations during an earthquake the ratio $v_{\max}/a_{\max}$ indirectly informs about the significant period of the signal.

Example 2.3

Figure 2.13 shows several parameters derived from a strong-motion seismogram recorded during the M 7.6 Chi-Chi earthquake which struck Taiwan September 21st in 1999. A baseline correction and a bandpass filter were applied to the raw acceleration data before single and double integration to calculate ground velocity, $v(t)$ and displacement $d(t)$:

$$v(t) = \int_{0}^{\infty} a(t)\, dt, \quad d(t) = \int_{0}^{\infty} v(t)\, dt. \tag{2.2.13}$$

While the calculation of displacement from the acceleration record is mathematically straight forward, in practice this is not a simple task. On one hand, acceleration records for technical reasons are often affected by an offset of the trace. In its simplest form this is a constant deviation of the zero line. Integration of an offset trace results in a ramp function which is steeper the larger the offset is. In the simplest form of baseline correction a polynomial function of grade 0 to 3 is fitted to the data and subsequently subtracted. In case of the example in Fig. 2.13 a linear baseline correction and a bandpass filter between 0.1 and 25 Hz was applied. Problematic is the proper correction of the offset in case of a static displacement at the recording site, which may happen close to the source. In case of a high pass filtering of the accelerogram, a possible final displacement will be removed from the data. The uncorrected and the filtered acceleration seismograms in Fig. 2.14 show only small differences. However, in the velocity and displacement seismograms, large differences between the filtered and unfiltered data are obvious. In addition to the ground motion data, the time dependent values of the Aria intensity so called Husid plot (Husid 1969) are shown. Here the *significant duration*, D_s between 5 and 95% of total A_I is indicated. The bottom trace shows the energy flux, *EF* which is the build up of the *SED* (2.2.7).

The acceleration response spectrum of the example record from Fig. 2.13 is shown in Fig. 2.14 for four levels of damping of the structure (Fig. 2.15).

The parameters introduced in Sect. 2.2.2 are given for the example record shown in Fig. 2.14 in Table 2.1. The list of parameters is by far not complete and the abundance of different definitions in some ways reflects the complexity of seismic ground motion. It also has to be kept in mind, that all these parameters are values derived from single traces, but seismic ground motion is a time varying vector with three components of translational and possibly rotational movement. The latter have long been ignored in engineering seismology; however, the advancing field of rotational seismology (i.e., special issue Vol. 99 of the Bulletin of the Seismological Society of America) shows that these components of ground motion might be significant for engineering problems as well.

2.3 Source Effects

Numerous international and national seismological networks and agencies perform routine localization of earthquakes and publish the results in catalogs, i.e. International Seismological Centre (ISC): http://www.isc.ac.uk/; U.S. Geological survey: http://earthquake.usgs.gov/earthquakes/; GeoForschungszentrum Potsdam (GFZ): http://geofon.gfz-potsdam.de/db/eqinfo.php. In these catalogs the location is given by the coordinates of the hypocenter (Fig. 2.3). In the location procedures the source is commonly treated as a point source. In the following we use the point source model to describe the general pattern of seismic wave radiation.

2.3.1 *Point Source Approximation and Equivalent Forces*

The point source model can be used to study the radiation pattern of seismic waves from the source. The displacement field of a shear dislocation source is hard to describe mathematically. Instead, the displacement field of force systems can be quantified which create the same displacements as a shear dislocation. The concept of these equivalent body forces is shown in Fig. 2.16.

In the actual earthquake source, complex fracturing and frictional processes arise on an extended fracture plane. These processes produce a dislocation history in space

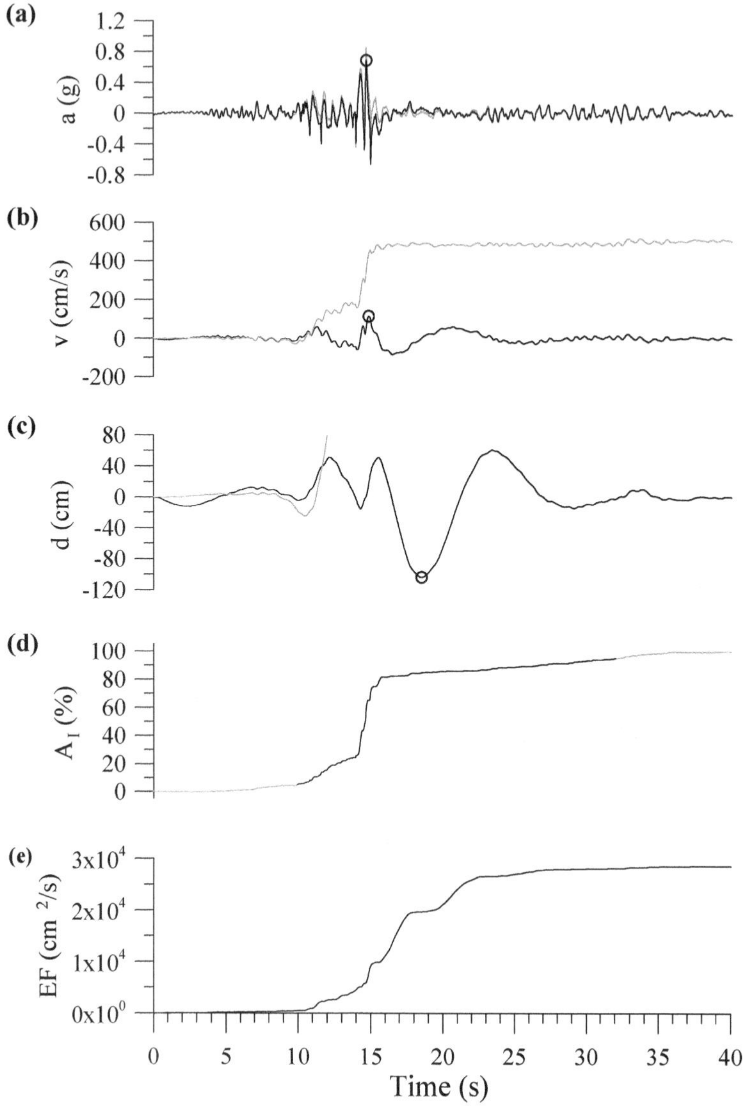

Fig. 2.14 Processing of a strong-motion record taken during the M 7.6, 1999 Chi-Chi earthquake in Taiwan. **a** Original (gray) and baseline corrected and bandpass (0.1–25 Hz) filtered accelerogram (black); **b** uncorrected (gray) and corrected ground motion velocity; **c** uncorrected (gray) and corrected (black) displacement; **d** Arias intensity (gray) the black interval indicates the significant time between 5 and 95% of A_I; **e** built up of the *SED* (energy flux). The circles marc the peak amplitudes of $a(t)$, $v(t)$, and $d(t)$

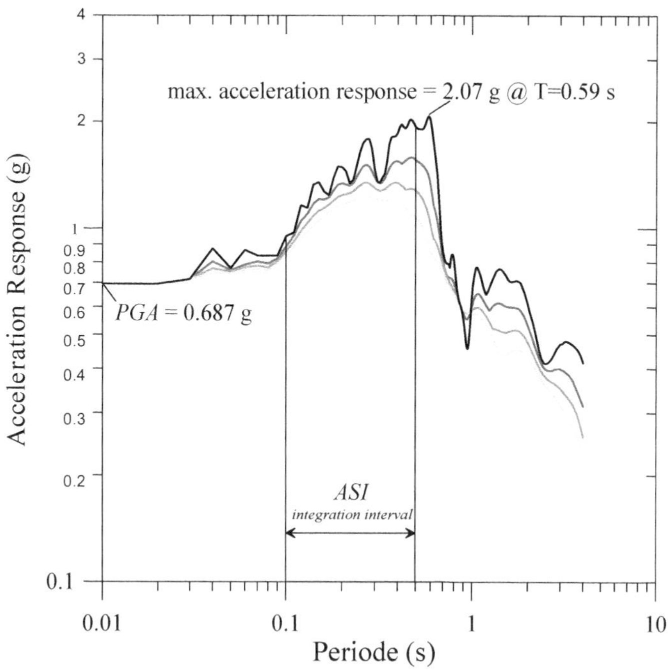

Fig. 2.15 Elastic acceleration response spectrum of the ground motion from Fig. 2.14. The black line shows the response amplitudes @ a damping of $\xi = 5\%$, The gray curves from top to bottom are for 10, 15, and 20% of critical damping, respectively. The *PGA* at the low-period end of the spectrum, maximum acceleration response and the period integration interval to determine *ASI* are indicated

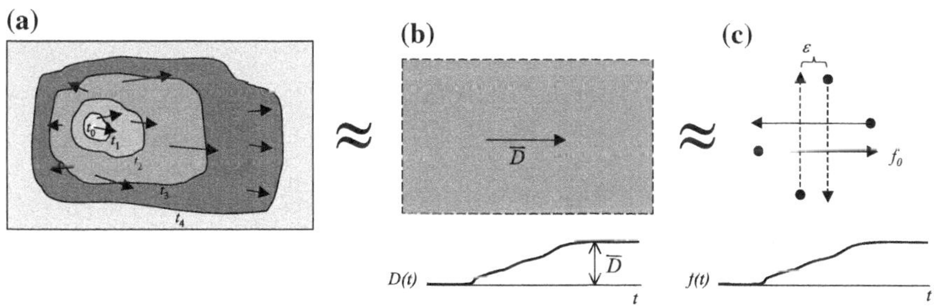

Fig. 2.16 Idealization of a shear dislocation source into an equivalent body force model. **a** Fault surface with a complex displacement history, **b** idealized dislocation model with a rectangular source area and time dependent dislocation function $D(t)$, **c** idealization of the source by an equivalent body force model with force time function $f(t)$ (after Lay and Wallace 1995)

and time, which can be approximated by a dislocation function $D(t)$. The second step of idealization includes the creation of the equivalent body force system. Instead of the dislocation function, time dependent forces are introduced, which can directly be

Table 2.1 Parameters for the record shown in Fig. 2.14

Maximum acceleration	PGA	0.687 g @ t = 14.85 s
Maximum velocity	PGV	113.45 cm/s @ t = 14.85 s
Maximum displacement	PGD	103.8 cm @ t = 18.50 s
v_{max}/a_{max}		0.168 s
Acceleration RMS	a_{rms}	0.084 g
Velocity RMS	v_{rms}	26.71 cm/s
Displacement RMS	d_{rms}	32.33 cm
Arias intensity	I_A	4.34 m/s
Specific energy density	SED	28,544 cm^2/s
Cumulative absolute velocity	CAV	1962 cm/s
Acceleration spectral intensity	ASI	0.64 gs
Velocity spectral intensity	VSI	306.5 cm
Sustained maximum acceleration	SMA	0.51 g
Sustained maximum velocity	SMV	62.4 cm/s
Effective design acceleration	EDA	0.68 g
Significant duration	D_s	22.0 s @ 5–95% I_A
Bracketed duration	D_b	35.3 s @ A_0=0.035 g
Uniform duration	D_u	18.9 s @ A_0=0.035 g
Effective duration	D_e	19.6 s @ 0.2–4.0 m/s I_A

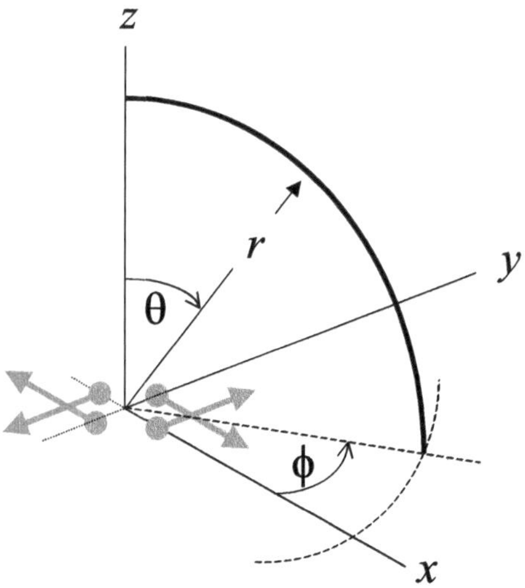

Fig. 2.17 Polar coordinate system of a double couple force model with the forces in the *xy*-plane

used in the equation of motion. The standard model for the earthquake source is that of a *double couple* of forces (Fig. 2.16c). Here, instead of the forces, the moment time function is used:

$$M_0(t) = \varepsilon \cdot f(t). \tag{2.3.1}$$

where ε is the lever between the forces (Fig. 2.16). This double couple of forces does not introduce a resultant moment into the medium, as it would be the case with a single couple of forces. However, the *radiation pattern* of seismic waves is equivalent to that of a shear dislocation point source. Using the polar coordinates shown in Fig. 2.16, the radiation pattern of seismic waves in the far field can be written as (i.e., Aki and Richards 1980; Lay and Wallace 1995; Shearer 1999):

$$\begin{aligned} u_r &= \tfrac{1}{4\pi\rho}\tfrac{1}{v_P^3}\tfrac{1}{r}\dot{M}_0\left(t - \tfrac{r}{v_P}\right)\sin^2\theta\ \sin 2\phi, \\ u_\theta &= \tfrac{1}{4\pi\rho}\tfrac{1}{v_S^3}\tfrac{1}{r}\dot{M}_0\left(t - \tfrac{r}{v_S}\right)\sin\theta\ \cos\theta\ \sin 2\phi\,, \\ u_\phi &= \tfrac{1}{4\pi\rho}\tfrac{1}{v_S^3}\tfrac{1}{r}\dot{M}_0\left(t - \tfrac{r}{v_S}\right)\sin\theta\ \cos 2\phi. \end{aligned} \tag{2.3.2}$$

The trigonometric functions in Eq. (2.3.2) describe the radiation pattern of the source in the three directions of the polar coordinates. Component u_r gives the amplitudes of the P waves, which have only a radial component of motion. The radiation pattern in Fig. 2.18 of the P waves shows quadrants with changing polarity, where a positive polarity is equivalent to a first ground motion away from the source and in quadrants with negative polarity, the first ground motion is directed towards the source. This attribute of the P wave radiation is used in the method of *fault plane solution* to determine the source mechanism of actual earthquakes. The P wave first motion direction (up or down) is determined at a large number of seismic stations distributed around the source. Mapping of the first motion directions on an imaginary sphere centered at the source allows for fixing the orientation of the nodal planes separating the quadrants. In the direction of these nodal planes the P wave amplitude is theoretically zero, and crossing a nodal plane invokes a change of polarity. However, as two equal nodal planes exist, it is not possible to decide which of the two represents the actual fault plane and which is the secondary plane.

The S wave motions result from the combination of the u_ϕ and u_θ components. In Eq. (2.2.3) both these components spread with the shear wave velocity, v_S. The xy plane for the u_θ component is a nodal plane, consequently here u_ϕ contains the total S wave motion. However, no nodal plane but only nodal points exist for the resultant S wave motion as illustrated in Fig. 2.19. In the direction of the S wave nodal points, i.e. at $\phi = 45°$ and $\theta = 90°$ the P wave radiation has its maxima. So the combination of the orientation of the fault plane and the displacement direction strongly influence the dominant sense of motion at a site, which can be important with the respect of building aspect ratios and orientations.

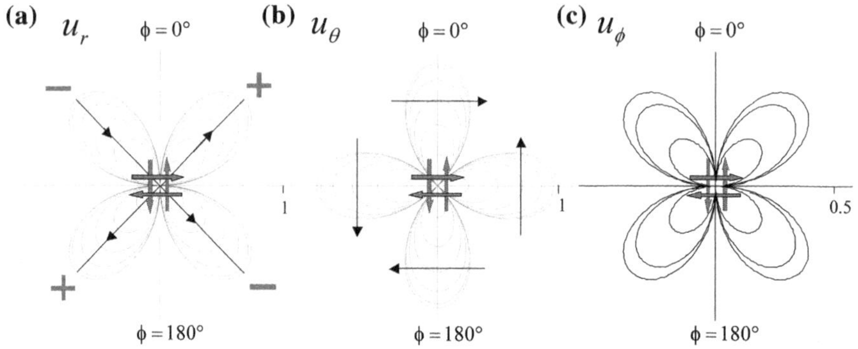

Fig. 2.18 Radiation pattern of a double couple source giving the relative amplitudes and polarities of displacement in the far filed. **a**, **b**, and **c** give the u_r, u_ϕ and u_θ, respectively. For patterns **a** and **b** the curves from outside to inside are for $\theta = 90°, 75°, 60°, 45°, 30°$, and 15°. In case of pattern **c** the radiation for $\theta = 90°$ and 0° is zero. The three curves represent $\theta = 45°, 30°$ and 15°

Fig. 2.19 Radiation pattern of the resultant S wave amplitudes of a double couple source

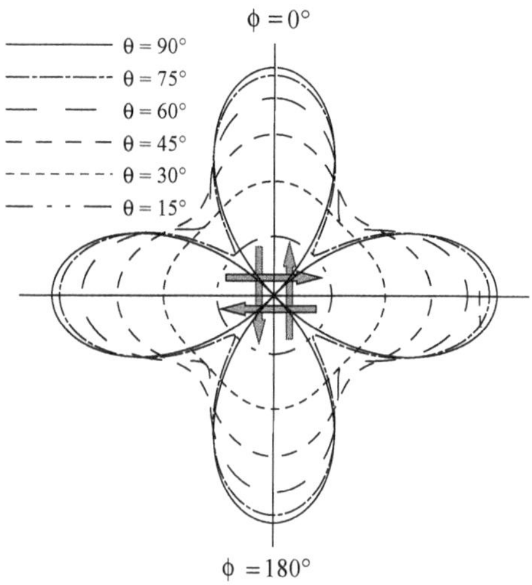

In the far field, the amplitudes of both body waves, P and S, are proportional to $1/r$. Equation 2.3.2 also indicates that the two wave types spread with different velocities, however, the form of the radiated P and S signal is alike and determined by the time derivative of the moment function, $\dot{M}(t)$. Figure 2.20 shows schematically the shape of the P and S pulse if a simple ramp function is assumed for the source.

The time independent value of M_0 corresponds to the scalar seismic moment introduced in Eq. (2.1.1).

Equation (2.3.2) and Fig. 2.18 are limited to the far field approximation of the radiation of body waves, valid only at distances much larger than the source dimension. At closer distances additional terms cannot be neglected. As shown in Eq. (2.3.2) the amplitude decay of the far field term is with $1/r$. The complete solution of the wave

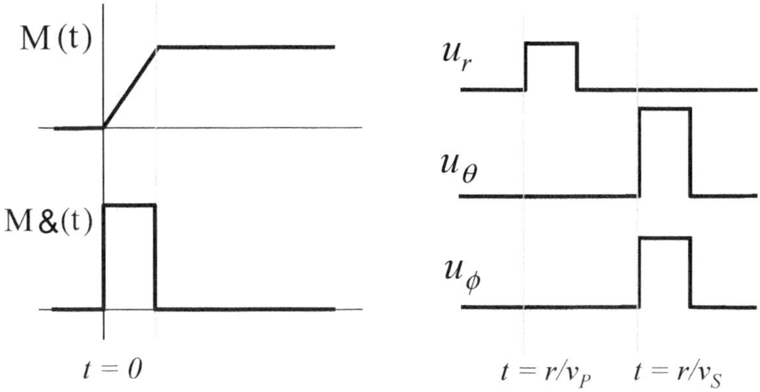

Fig. 2.20 Left: moment time function of a double couple seismic source in form of a ramp and its time derivative. Right: resulting shape of the P and S wave pulses after Lay and Wallace (1995)

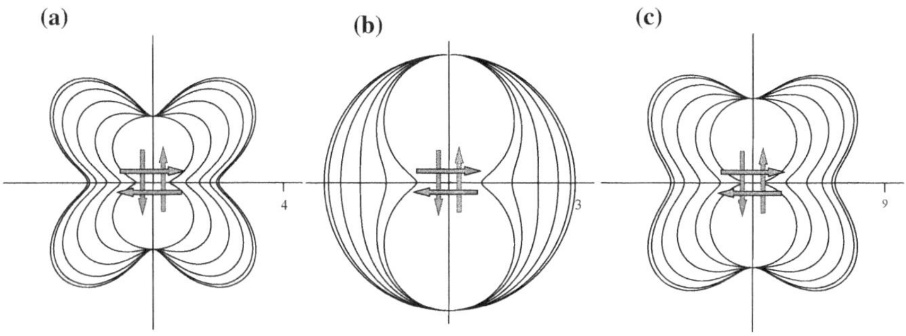

Fig. 2.21 Radiation pattern of the intermediate and near field terms of a double couple source. **a** and **b** are the intermediate term patterns which radiate with v_P and v_S, respectively. **c** Shows the pattern of the near field term

equation (i.e., Aki and Richards 1980) includes two terms with amplitude decay of $1/r^2$ and $1/r^4$, the intermediate and near field term, respectively. The intermediate field term contains two parts that travel with v_P and v_S, similar to the far field. However the form of the signal follows that of the moment time function itself and not its time derivative. In the near field term P and S waves are not separated. Figure 2.21 shows the radiation pattern for the intermediate and near filed terms. The scales in the plots differ significantly.

2.3.2 *Moment Tensor*

A more general representation of the seismic source than the double couple model allows the moment tensor (Aki and Richards 2002). It cannot only give the effects

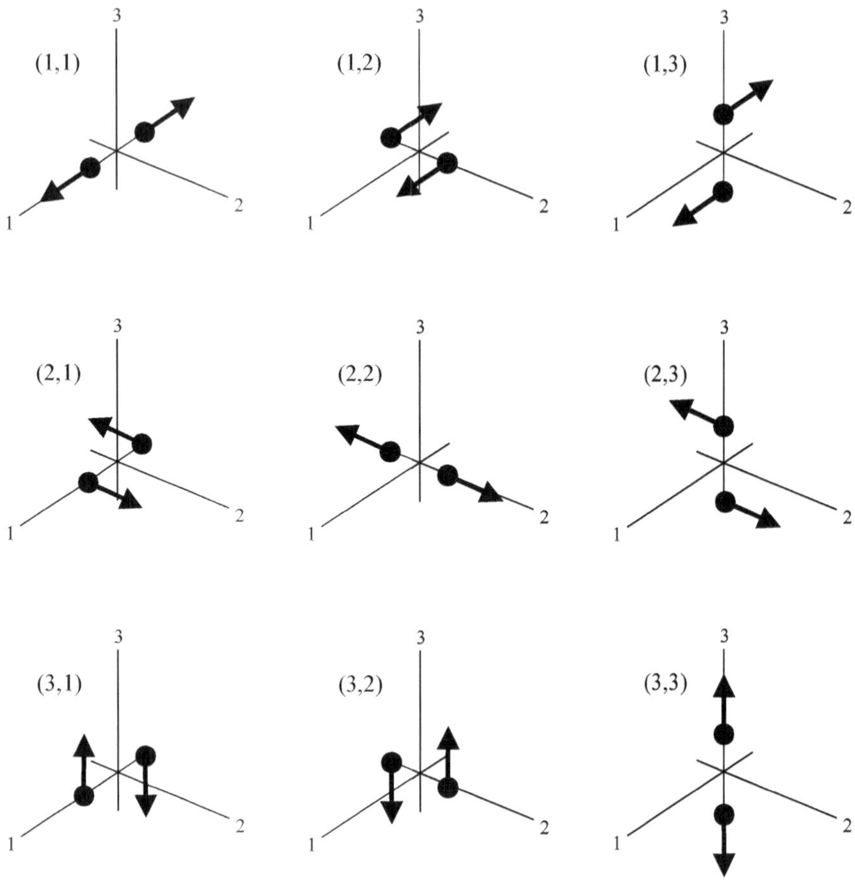

Fig. 2.22 Nine possible orientations of force couples in a Cartesian coordinate system with the directions 1, 2, and 3. The force couples give the equivalent forces, necessary to describe a common dislocation source in a medium (Aki and Richards 2002)

of an arbitrary oriented pure shear dislocation but also that of different sources, i.e. tensional fractures or explosions. The moment tensor includes all information of the seismic source, which can be deduced from records in the far field that contain wavelengths larger than the source dimension. It is a second order tensor whose elements represent the strength of force couples. Given three possible orientations of the force couples and three possible orientations of the lever of theses forces, the common tensor has nine elements (Fig. 2.22).

The moment tensor can be written as:

$$M = \begin{pmatrix} M_{11} & M_{12} & M_{13} \\ M_{21} & M_{22} & M_{23} \\ M_{31} & M_{32} & M_{33} \end{pmatrix}, \qquad (2.3.3)$$

where the first index of the tensor elements gives the orientation of the forces and the second the orientation of the lever of the couples. A pure double couple source of a shear dislocation becomes:

$$M = \begin{pmatrix} 0 & M_0 & 0 \\ M_0 & 0 & 0 \\ 0 & 0 & 0 \end{pmatrix}. \tag{2.3.4}$$

A moment tensor which contains only elements different from 0 in its diagonal represents an ideal explosion where all forces are oriented away from the source:

$$M = \begin{pmatrix} M_{11} & 0 & 0 \\ 0 & M_{22} & 0 \\ 0 & 0 & M_{33} \end{pmatrix}. \tag{2.3.5}$$

Using a time dependent formulation of the moment tensor which contains the source time function, the far field seismogram from this source can be expressed as the convolution of the moment tensor and the Green's function of the medium through which the waves travel (Jost and Herrmann 1989).

2.3.3 *Finite Source Effects*

The kinematics of the rupture growing over the fault surface of an earthquake source can have a strong influence on the radiation pattern and cause a source *directivity*, which cannot be neglected in engineering seismological problems. For strong crustal earthquakes, the source dimension can reach several hundred km. An often-used empirical relation between source dimension and magnitude was established by Wells and Coppersmith (1994). Figure 2.23 shows the relation between rupture length, rupture width, rupture surface, and average surface displacement with the magnitude.

The large range of the source dimensions of earthquakes between magnitudes 5 and 8 is obvious in Fig. 2.23. A typical M = 7.0 earthquake has surface rupture length in the order of 40 km, a rupture surface of some 700 km^2, and displacements in the range of meters.

The radiation of seismic waves from a shear dislocation with an extended rupture surface depends on the rupture dimensions, the rupture velocity, v_r, with which the rupture front expands, and the velocity with which the opposite sides of the fault move from the initial position before the earthquake to the final position at the end of the earthquake. Haskell (1964) described a simple rupture model that allows studying the influence of the rupture parameters on the radiated seismic waves. The rupture begins at one end of a rectangular rupture surface (Fig. 2.24) and spreads over the entire rupture length, L, with the constant velocity, v_r. The rupture area is arranged in segments of width, w, and length, dL. When the individual segments are treated as

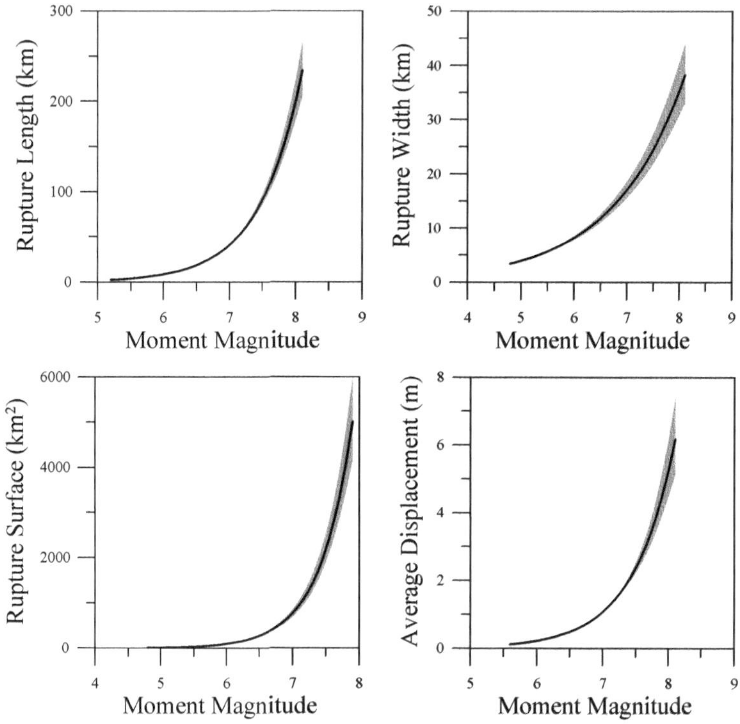

Fig. 2.23 Empirical relation between surface rupture length, rupture width in direction of the fault plane dip, rupture surface, and average surface displacement of earthquake sources and moment magnitude. The curves are shown within the magnitude interval of the data used in the regression calculations. The gray shaded ranges indicate the standard deviation (after Wells and Coppersmith 1994)

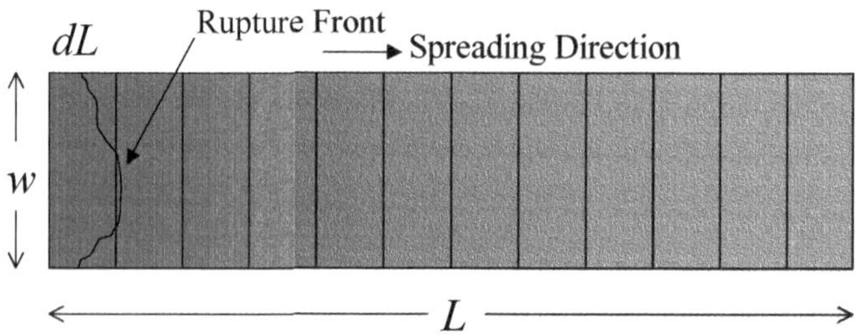

Fig. 2.24 Model of an extended seismic source with a rupture travelling from left to right (Haskell 1964). The segments of the fault break consecutively following the constant rupture velocity

point sources, the model becomes a line source. The seismograms at a station in the vicinity of the source can be calculated from the superposition of the signals from the individual segments with an appropriate time delay, determined by v_r.

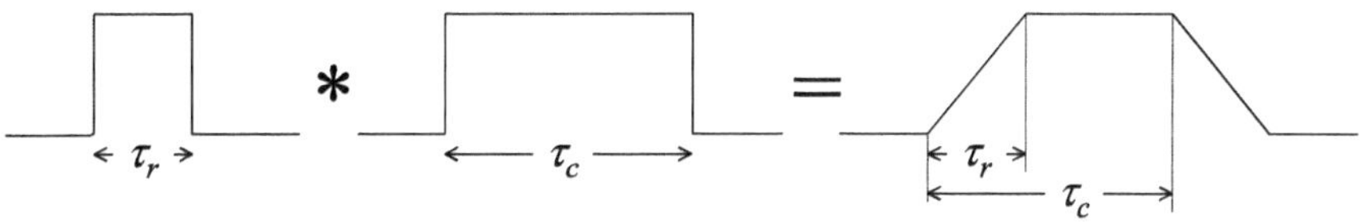

Fig. 2.25 The convolution of two boxcar functions representing the rupture rise time and the spreading time along the fault results in a trapezoidal pulse

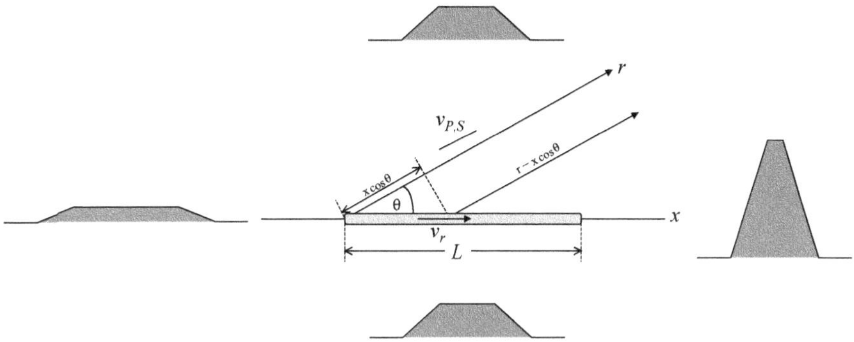

Fig. 2.26 Schematic form of body wave pulses from an extended source with the rupture travelling from left to right after Lay and Wallace (1995)

We assume that the far field displacement of a rupture segment has a boxcar form (Fig. 2.20) with the duration, τ_r, the *rise time* of the rupture, which can be regarded as the time the rupture takes to spread over the width, w. The rupture sequence can also be expressed by a boxcar function with the duration $\tau_c = L/v_r$. The displacement pulse of the extended source results on the convolution of the two boxcar functions with a trapezoidal form (Fig. 2.25).

This form applies to both, P and S waves. If the displacement pulse of a shear source has been measured and is corrected for the amplitude loss due to the spreading effects and the radiation pattern at the source, $R^{P,S}$, the integration of the pulse directly reveals the seismic moment:

$$4\pi\rho v_{P,S}^3 \frac{1}{R^{P,S}} \int_{-\infty}^{\infty} u(r,t)\,dt = M_0. \tag{2.3.6}$$

Besides the rupture dimensions and rupture velocity, the amplitude and shape of the body wave pulse also depends on the azimuth at which a site is located with respect to the spreading rupture. For most earthquakes the rupture velocity is smaller than the shear wave velocity. In this case, the wave energy from a segment that is just breaking will always arrive later at the site than the energy from the previously broken segment. Following the rupture geometry shown in Fig. 2.26, the time τ_c becomes:

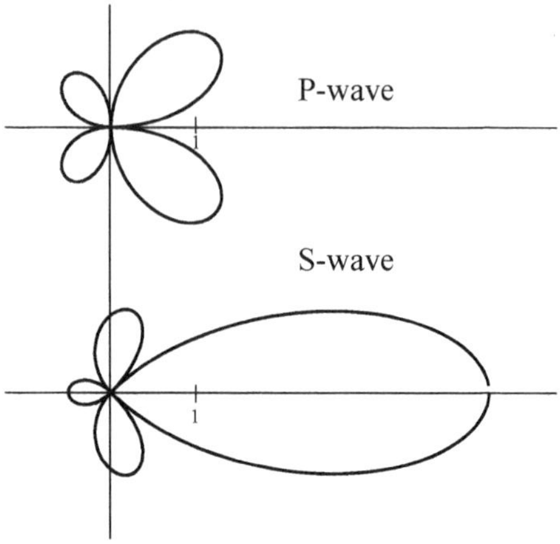

Fig. 2.27 Combined effect of radiation pattern and directivity for a rupture linear spreading from left to right with $v_r = 0.8\ v_S$

$$\tau_c = \frac{L}{v_r} - \left(\frac{L\cos\theta}{v_{P,S}}\right). \tag{2.3.7}$$

The effect of the unilateral spreading rupture is shown schematically in Fig. 2.26. The amplitude variation with azimuth, the *directivity*, leads to a relative short, large impulse in the direction of the spreading rupture. In the opposite direction, the impulse is elongated in time, but smaller in amplitude. The area under the impulses is always proportional to M_0, which is independent of the azimuth.

The directivity effect is stronger in S waves than in P waves, because v_S is closer to v_r than it is the case for v_P. The directivity is superimposed on the radiation pattern. Figure 2.27 shows the combination of radiation pattern and directivity for P and S waves from a unilateral rupture where $v_r = 0.8\ v_S$. The strong influence on the amplitudes, especially towards the rupture direction points out the importance of directivity in engineering applications.

Observations have shown, that these effects are not always symmetric with respect to that fault i.e., Sommerville and Abrahamson (1995) and Abrahamson and Silva (1997) observed larger ground motion amplitudes on the hanging wall than on the footwall of a fault. The hanging wall is the block above an inclined fault plane and the footwall the part below it (Fig. 2.4). Sites located on the hanging wall have larger amplitudes than average. This effect is probably a combination of radiation, directivity and wave spreading. The fault might act as a 'wave barrier' and seismic energy gets trapped in the footwall block. These observations have been included in complex ground motion prediction models (Abrahamson and Silva 1997, Campbell and Bozorgnia 2003).

The tectonic environment, classified in Sect. 2.2.1, also has a significant influence on ground motion amplitudes and especially ground motion attenuation. In general, the frequent large earthquakes in active tectonic environments tend to lower stress drop values and anelastic attenuation is higher than average in these regions. On the other hand in stable tectonic environments large earthquakes are infrequent, however, stress drop tends to higher values and the anelastic attenuation is smaller than average. This again shows, that care has to be taken that ground motion prediction relations are based on data which suit the tectonic and seismic features of the area for which the prediction has to be made. Naturally, these relations should also not be extended beyond the distance and magnitude limits of the data they are based in.

2.3.4 Seismic Source Spectrum

Figure 2.25 illustrated that the form of the seismic impulse and its duration depending on the dimensions of the extended seismic source. The convolution between the impulses representing the rise time and the rupture duration, τ_r and τ_c, respectively, corresponds to a multiplication of the Fourier spectra of the two boxcar signals in frequency domain, which is given by the sin c-function:

$$\sin c(x) = \frac{\sin(x)}{x}. \tag{2.3.8}$$

The far field displacement in frequency domain is the product of the two sin c-functions:

$$\hat{u}(\omega) = M_0 \left| \frac{\sin(\omega\tau_r/2)}{\omega\tau_r/2} \right| \left| \frac{\sin(\omega\tau_c/2)}{\omega\tau_c/2} \right|. \tag{2.3.9}$$

Example 2.4
Assume a rupture rise time of $\tau_r = 0.4$ s and a rupture duration $\tau_c = 5$ s. With a rupture velocity of 80% of the shear wave velocity at $v_S = 3.0$ km/s, this corresponds to a rupture length of 12 km and a magnitude of ca. 6. Figure 2.28 shows the *sinc*-functions of the two boxcars and their combination into the source spectrum. The spectral amplitudes decay with increasing frequency. At frequencies smaller than $2/\tau_r$ the spectra show a plateau; at higher frequencies the decay rate is $1/\omega$. The frequency at which the extrapolated plateau and the $1/\omega$ asymptote subtend is called the *corner frequency* of the spectrum. The product of the two spectra (Fig. 2.28b) shows three sections, the plateau at low frequencies, the $1/\omega$ amplitude decay in the middle and a $1/\omega^2$ decay at the high-frequency end of the spectrum. The value of the plateau is a measure for the seismic moment M_0 and the spectral amplitudes can be approximated by:

$$\hat{u}(\omega) = \begin{cases} M_0 & \omega < \frac{2}{\tau_c} \\ \frac{M_0}{\omega\tau_c/2} & \frac{2}{\tau_c} < \omega < \frac{2}{\tau_r} \\ \frac{M_0}{\omega^2(\tau_r\tau_c/4)} & \omega > \frac{2}{\tau_r} \end{cases}. \tag{2.3.10}$$

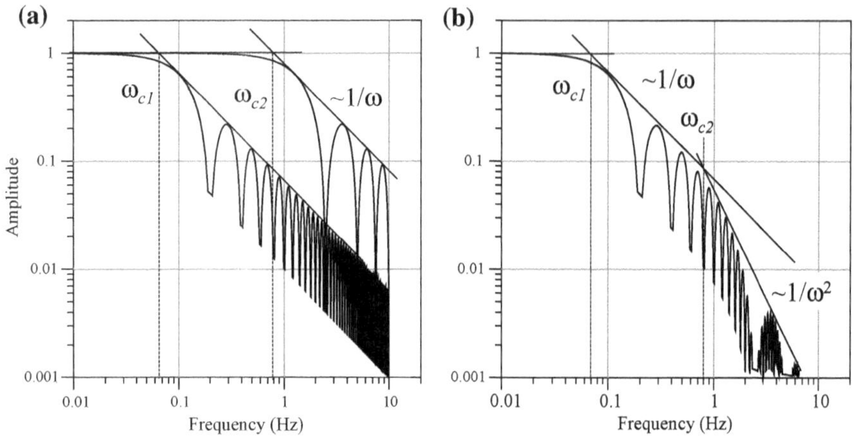

Fig. 2.28 **a** Fourier spectrum of two boxcar functions with durations of $\tau_r = 0.4$ s and $\tau_c = 5$ s, representing rise time and rupture duration of an earthquake source. **b** Product of the two spectra

The plateau is expected at periods which are larger than the rupture duration. At periods between the rupture duration and the rise time, a spectral decay of $1/\omega$ is expected and at periods smaller than the rise time the spectral amplitude decay is $1/\omega^2$. Corresponding to this frequency decay, the model is termed ω^2-model.

In practice, often only the corner frequency between the plateau and the $1/\omega^2$ asymptote is obvious in a measured spectrum. Figure 2.29 shows the Fourier amplitude spectrum of a ground displacement from a small magnitude 3.8 earthquake measured at 95 km epicentral distance. The ground motion of the S wave was corrected for wave spreading effects and amplitude loss due to anelastic attenuation. From the plateau amplitude and the corner frequency of 2.7 Hz a source dimension of 1 km and a seismic moment of $M_0 = 2.5 \times 10^{14}$ Nm can be deduced.

Example 2.5

In Sect. 2.3 the complex interaction of radiation, directivity, and tectonic regime on the ground motion was shown. However, a rough estimate of ground motion amplitudes can be made with an engineering approach given by Schneider (1980) if only a few source parameters are known or can be estimated. If the effective stress, the difference between the shear stress before the earthquake and the frictional stress on the source plane, shear modulus, and shear wave velocity can be estimated, the maximum ground velocity in the vicinity of the source is (Schneider 1980):

$$v_{\max} = \frac{\sigma_{eff}}{\mu} v_S. \tag{2.3.11}$$

With $\sigma_{eff} = 1 \times 10^7$ Pa, $\mu = 3 \times 10^{10}$ Pa and $v_S = 3.0$ km/s, the maximum velocity is ca. 1 m/s.

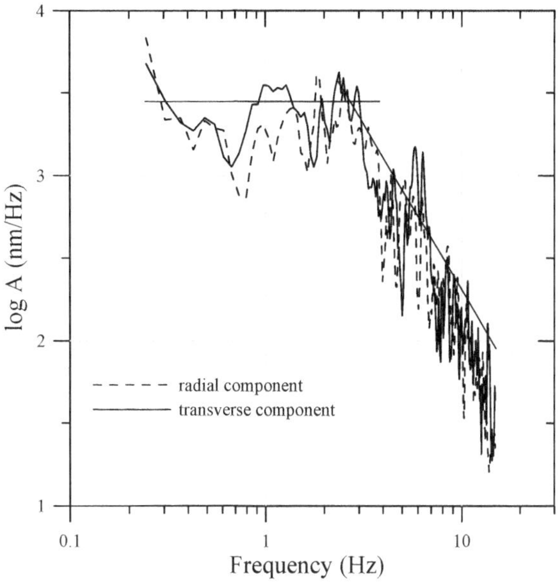

Fig. 2.29 Measured source spectrum of a local earthquake with magnitude 3.8 at 95 km epicentral distance

A second simple estimate is based on the average dislocation, the S wave velocity, rupture velocity, and source dimension:

$$v_{\max} = \frac{2}{\pi}\overline{D}\frac{v_S}{v_r}\frac{v_S}{L}. \qquad (2.3.12)$$

Assuming $\overline{D} = 0.2$ m, $v_S = 3.0$ km/s and a v_r of 80% of v_S and a source dimension of $L = 4.0$ km, the maximum velocity is 0.12 m/s.

2.4 Site Effects

Besides the source effects discussed in Sect. 2.3, the local site conditions are the most influential parameters for the ground motion during an earthquake (Fig. 2.30).

The two seismograms in Fig. 2.17 were recorded during the same earthquake of $M_L = 5$, both at a distance of ca. 30 km. The large differences in amplitude and duration of the ground motion are mainly due to differences in the subsurface under the stations. While one station is located on hard rock, the second stands on layered soft rock of a sedimentary basin. On the sediments, amplitudes a larger by a factor of 3 and the duration is more than double (Fig. 2.31).

Reverberations and resonance effects in soft sediments are the cause of the larger amplitudes and the longer duration of strong ground motions. Within a sedimentary basin, seismic energy can be 'trapped'. At internal boundaries with changes of the

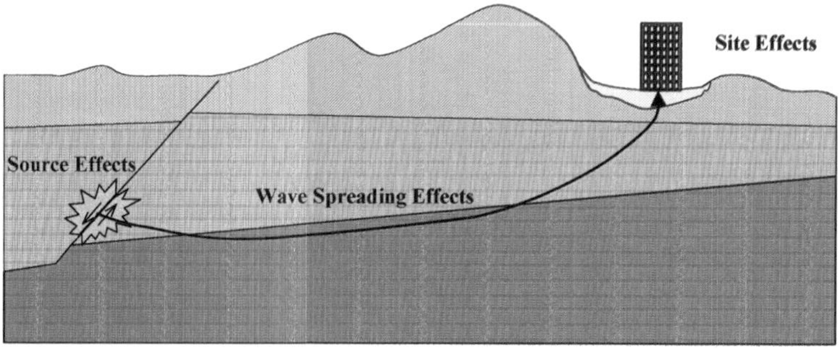

Fig. 2.30 Schematic travel path of seismic waves, structured into the source region, the geologic subsurface, and the site

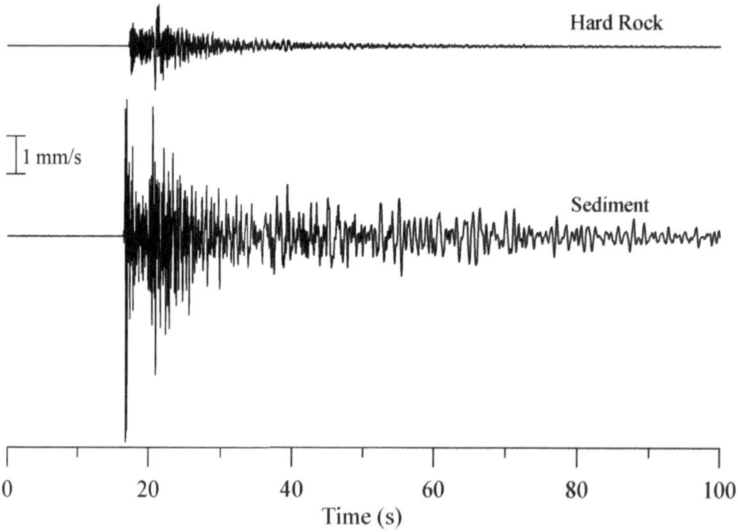

Fig. 2.31 Vertical component of ground velocity measured during a M_L 5.0 earthquake at 30 km distance. Maximum ground velocity is 1.2 mm/s on hard rock and 4.0 mm/s on sediment

seismic velocities, the seismic waves are forced to change their direction, following Snell's law. Close to the Earth's surface often layers with highly reduced P and S wave velocities exist. These can be softrock layers of sedimentary basins or loose fill at valley floors. Even when rock is outcropping at Earth's surface, the upper few meters are often weathered and show reduced wave speeds with respect to the competent rock at depth. The often-sharp velocity reduction close to the surface bends the wavefront of seismic waves towards a direction parallel to the Earth's surface. Therefore one-dimensional models are often regarded sufficient to study the amplification effects of sedimentary cover. Figure 2.32 shows the terms used to address the different locations of ground motion.

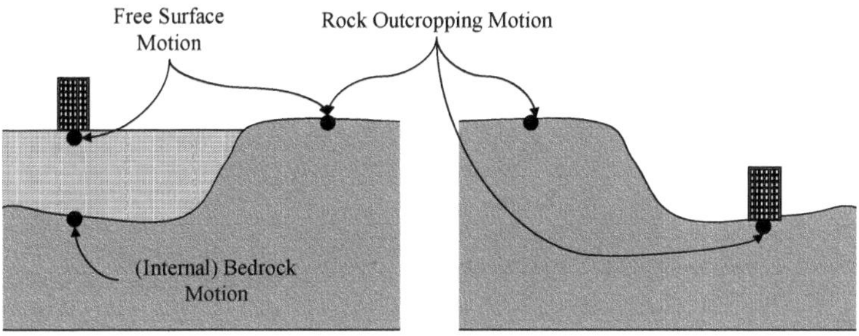

Fig. 2.32 Terms describing the ground motion at and below earth's surface

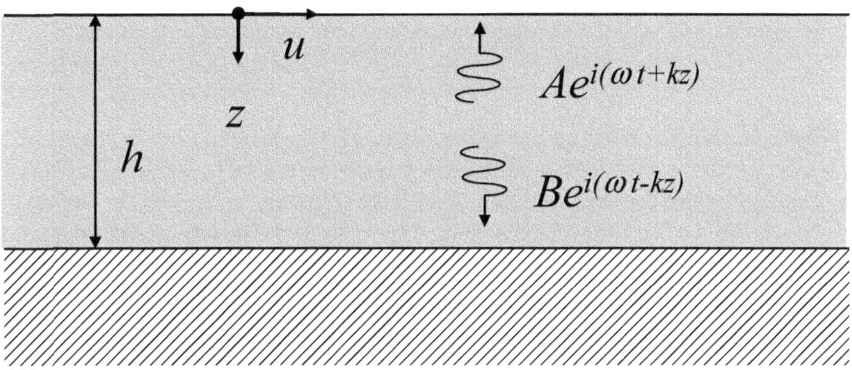

Fig. 2.33 Sediment layer of thickness h on top of a stiff half space

Transfer functions are used to quantify the influence of layering in the subsurface. These transfer functions can be established for ground displacement, velocity, and acceleration, as well as shear stresses and shear deformation. These measures can be related for example to the *internal bedrock acceleration*.

2.4.1 Single Layer Without Damping

We start with the model of a single sedimentary layer over a stiff half space (Fig. 2.33) and ignore possible damping effects in the sediments. A harmonic horizontal oscillation of the half space will generate a vertically propagating shear wave in the layer.

The horizontal displacement dependent on depth, z, and time can be written as:

$$u(z,t) = Ae^{i(\omega t+kz)} + Be^{i(\omega t-kz)}. \tag{2.4.1}$$

Where $k = \omega/v_S$ is the wavenumber and A and B are the amplitudes of the upwards and downwards travelling waves. At the free surface ($z=0$) the shear stress is zero:

$$\tau(0,t) = \mu \frac{\partial u(0,t)}{\partial z} = 0. \tag{2.4.2}$$

With this condition and differentiation we get:

$$\mu ik\left(Ae^{ik(0)} - Be^{-ik(0)}\right)e^{i\omega t} = \mu ik(A-B)e^{i\omega t} = 0, \tag{2.4.3}$$

A non trivial solution of Eq. (2.4.3) in case $A=B$ can be written as:

$$u(z,t) = 2A\frac{e^{ikz} + e^{-ikz}}{2}e^{i\omega t} = 2A\,\cos kz e^{i\omega t}. \tag{2.4.4}$$

Equation (2.4.4) describes a standing wave of amplitude $2A \cos kz$ that results from the interference of the up- and down-going waves. Due to the stiff half space, the energy is trapped in the sediment layer. This equation can be used to formulate the transfer function of the displacement amplitudes between any two points in the layer (Kramer 1996). If the two points are chosen at the basise of the layer and at the Earth's surface, the transfer function is (Kramer 1996):

$$F(\omega) = \frac{u_{\max}(0,t)}{u_{\max}(h,t)} = \frac{2Ae^{i\omega t}}{2A\cos khe^{i\omega t}} = \frac{1}{\cos kh} = \frac{1}{\cos(\omega h/v_S)}. \tag{2.4.5}$$

The modulus of the complex transfer function describes the amplitude amplification within the layer:

$$|F(\omega)| = \sqrt{\{\mathrm{Re}(F(\omega))\}^2 + \{\mathrm{Im}(F(\omega))\}^2} = \frac{1}{|\cos(\omega h/v_S)|}. \tag{2.4.6}$$

As the denominator of Eq. (2.4.6) cannot be larger than 1, the displacement at the surface does not get smaller than that at the base of the layer. However, at certain frequencies the surface displacement gets much larger than the excitation motion at the base of the layer. Due to the stiff half space no wave energy gets back across the interface. If the ratio $\omega h/v_S$ reaches uneven multiples of $\pi/2$, the denominator of (2.2.6) goes to zero and the amplification to infinity; resonance occurs at certain frequencies, which depend on the layer thickness and the shearer wave velocity (Fig. 2.34).

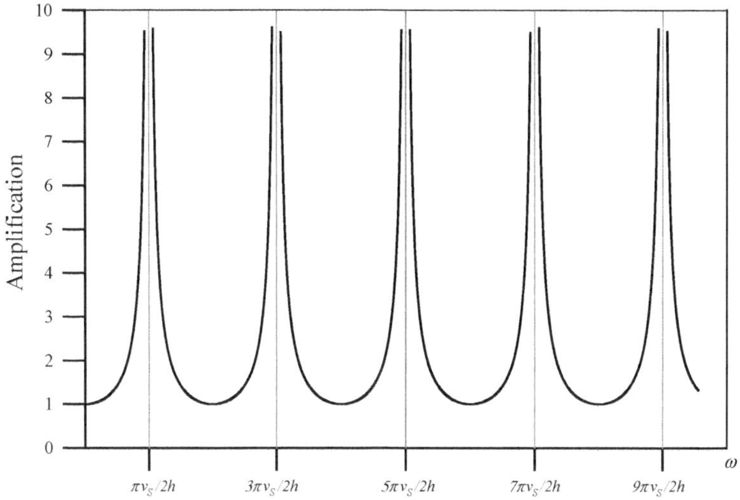

Fig. 2.34 Transfer function of a sediment layer without damping on top of a stiff half space for a vertically propagation shear wave

2.4.2 *Single Layer with Damping*

The infinite amplifications at the resonance frequencies in Sect. 2.4.1 are only theoretical; in a more realistic model which includes damping within the layer, the amplification stays finite. Introducing damping can be formulated by a complex shear modulus:

$$\mu^* = \mu(1 + i2\xi) \tag{2.4.7}$$

where ξ is the material damping. In case of a complex shear modulus, v_S also has to be complex:

$$v_S^* = \sqrt{\frac{\mu^*}{\rho}} \approx v_S(1 + i\xi) \text{ and } k^* = \frac{\omega}{v_S^*} \approx k(1 - i\xi). \tag{2.4.8}$$

In this case the transfer function is (Kramer 1996):

$$|F(\omega)| \approx \frac{1}{\sqrt{\cos^2 kh + (\xi kh)^2}} = \frac{1}{\sqrt{\cos^2(\omega h/v_S) + [\xi(\omega h/v_S)]^2}}. \tag{2.4.9}$$

As $\xi > 0$ the denominator of the transfer function cannot become zero and the amplification stays finite as shown in Fig. 2.35.

Maxima of the amplification are found at the frequencies:

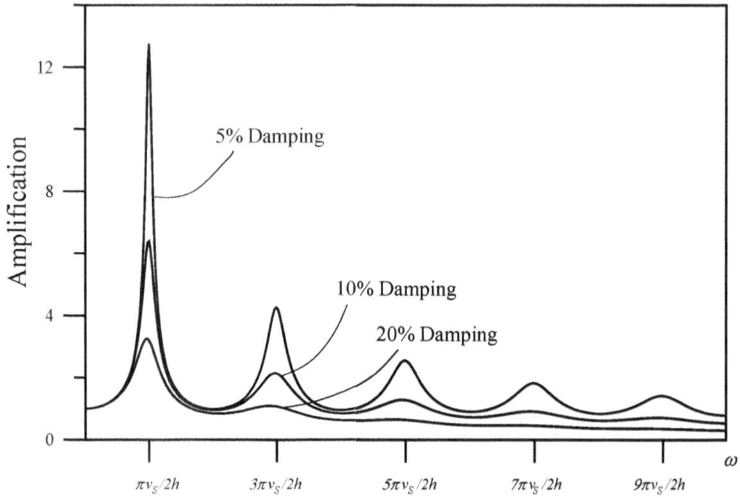

Fig. 2.35 Transfer function of a sediment layer with damping on top of a stiff half space for a vertically propagation shear wave

$$\omega_n \approx \frac{v_S}{h}\left(\frac{\pi}{2} + n\pi\right) \quad n = 0, 1, 2, \ldots, \infty. \tag{2.4.10}$$

The damping effect increases with increasing frequency, because at higher frequencies more wavelengths fit into the layer. The largest amplification occurs at the fundamental mode at a period of:

$$T_0 = \frac{2\pi}{\omega_0} = \frac{4h}{v_S}. \tag{2.4.11}$$

This *characteristic site period* can be used as a rule of thumb to estimate the resonance characteristic of a sediment layer: assuming an average $v_S = 2000$ m/s in the sediments; resonance frequencies of 1, 5, and 10 Hz correspond to sediment thicknesses of 500, 100 and 50 m, respectively.

Even the most competent bedrock is not ideally stiff as assumed in the previous models. If the half space is modelled as an elastic medium, part of the energy of the down going wave will be transmitted back to the half space (Fig. 2.36). Still, if the layer were without damping, the energy transmission back to the half space would prevent infinite amplification. Using the index *Se* for the sediment layer and *Ro* for the rock half space, the corresponding displacement amplitudes can be written as:

$$\begin{aligned} u_{Se}(z_{Se}t) &= A_{Se}e^{i\left(\omega t + k^*_{Se}z_{Se}\right)} + B_{Se}e^{i\left(\omega t - k^*_{Se}z_{Se}\right)} \\ u_{Ro}(z_{Fe}t) &= A_{R0}e^{i\left(\omega t + k^*_{Ro}z_{Ro}\right)} + B_{Ro}e^{i\left(\omega t - k^*_{Ro}z_{Ro}\right)} \end{aligned} \tag{2.4.12}$$

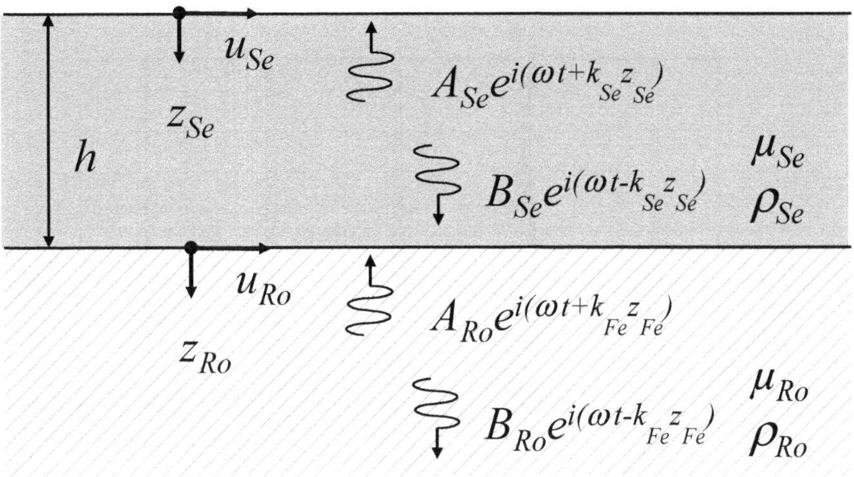

Fig. 2.36 Up and down going waves in the model of an elastic layer over an elastic half space

A and *B* are again the amplitudes of the up, and down going waves, respectively. The boundary condition between sediment and rock requires continuity of displacements and stress.

The transfer function between the free surface motion and the internal rock motion (Fig. 2.32) is (Kramer 1996):

$$F(\omega) = \frac{1}{\cos\left(\omega h / v^*_{SSe}\right) + iI^* \sin\left(\omega h / v^*_{SSe}\right)}, \tag{2.4.13}$$

where I^* is the complex impedance ratio between the sediment rock material, defined by the densities, ρ_{Se}, ρ_{Ro} and the complex velocities, $v_{S\,Se}$, $v_{S\,Ro}$:

$$I^* = \frac{\rho_{Se} v^*_{S\,Se}}{\rho_{Ro} v^*_{S\,Ro}} = \frac{\mu_{Se} k^*_{Se}}{\mu_{Ro} k^*_{Ro}}. \tag{2.4.14}$$

The transfer function cannot be straightforwardly expressed in terms of the damping. However, in the special case of $\xi = 0$, the complex velocities in Eq. (2.4.13) become real and the transfer function is:

$$|F(\omega, \xi = 0)| = \frac{1}{\sqrt{\cos^2_{k_{Se}h + I^2 k_{Se} h}}}. \tag{2.4.15}$$

The larger the impedance ratio becomes, the more energy goes back to the half space and reduces the resonance at the fundamental mode as well as at the higher modes as shown in Fig. 2.37.

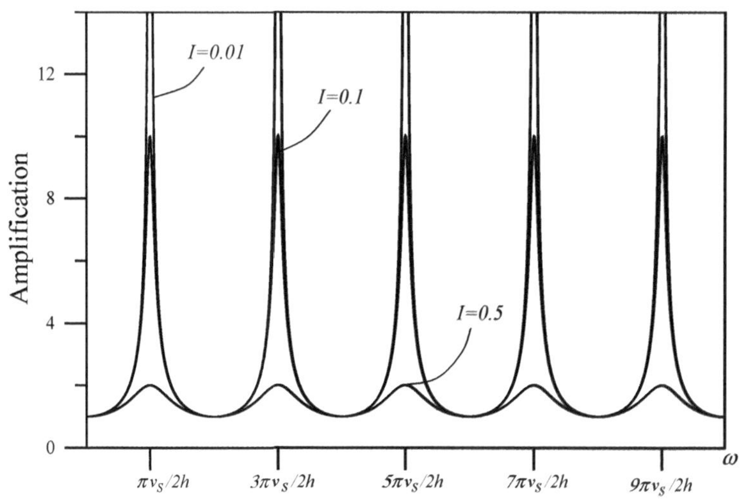

Fig. 2.37 Transfer function of a sediment layer with damping $\xi = 0$ on top of an elastic half space for a vertically propagation shear wave for three different impedance ratios, I

2.4.3 *Multiple Layers with Damping*

While the total sediment thickness and an average v_S can be used to roughly estimate the period of the fundamental mode of a structured sediment package (Eq. 2.4.11), the local geology can make it necessary to study the amplification in more detail with a model of N layers over a half space. Especially in case of strong impedance contrasts within the sediments, such models are essential as the largest amplification might occur at higher modes and not at the fundamental mode. The equation of motion still has the form of Eq. (2.4.1) and the boundary condition of continuous displacement and stress applies to all internal boundaries. If the complex impedance ratio between layers m and $m+1$ is I_m, the amplitudes in layer $m+1$ can be written as (Fig. 2.38):

$$\begin{aligned} A_{m+1} &= \tfrac{1}{2}A_m\left(1 + I_m^*\right)e^{ik_m^* h_m} + \tfrac{1}{2}B_m\left(1 - I_m^*\right)e^{-ik_m^* h_m} \\ B_{m+1} &= \tfrac{1}{2}A_m\left(1 - I_m^*\right)e^{ik_m^* h_m} + \tfrac{1}{2}B_m\left(1 + I_m^*\right)e^{-ik_m^* h_m} \end{aligned}. \quad (2.4.16)$$

Amplitudes in layer m are a function of amplitudes in layer $m = 1$ following:

$$\begin{aligned} A_m &= a_m(\omega)A_1 \\ B_m &= b_m(\omega)B_1 \end{aligned}. \quad (2.4.17)$$

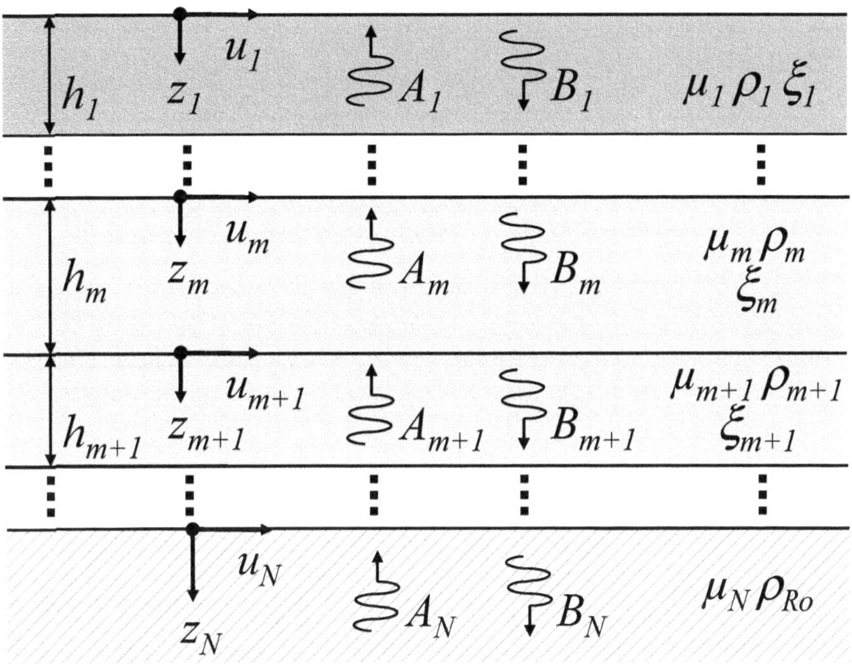

Fig. 2.38 Package of $N - 1$ sedimentary layers on to of a half space

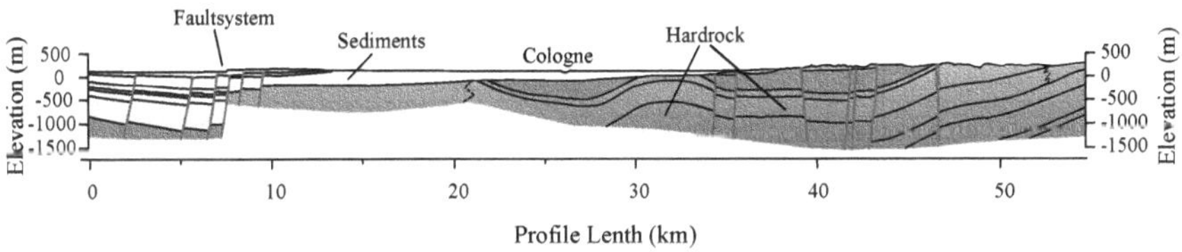

Fig. 2.39 Simplified geologic cut through the eastern part of the Lower Rhine Embayment. Dark shaded layers are Devonian rock and light collored layers are Tertiary and Quaternary sediments

Equation (2.4.17) gives two recursion formulas which can be used to calculate the displacement in an arbitrary layer, if the displacement in one other layer is known. The transfer function of the displacement amplitudes in layer i with respect to those in layer j becomes:

$$F_{ij}(\omega) = \frac{|u_i|}{|u_j|} = \frac{a_i(\omega) + b_i(\omega)}{a_j(\omega) + b_j(\omega)}. \tag{2.4.18}$$

Example 2.6

Especially at the edges of sedimentary basins, strong and rapidly changing ground motion amplification can occur. Figure 2.40 shows a simplified cross section through

Table 2.2 Sediment thickness, frequency of the fundamental mode and maximum amplification for the 4 models from Fig. 2.41

Pofile	A	B	C	D
Sediment thickness (m)	1000	250	25	7
Frequency fundamental mode (Hz)	0.17	0.67	3.61	17.3
Max. amplification	4.15	4.0	3.6	4.5

the eastern part of the Lower Rhine Embayment, a young sedimentary basin. At the eastern end of the profile Devonian rock reaches the surface and is only overlain by thin sedimentary or layers of weathered rock. Westward, the sediment thickness increases steadily towards the centre of the basin. In the range of Cologne City, the sediment thickness is 250–300 m. At the western end of the profile (Fig. 2.39) a significant fault system offsets the base rock. West of this fault the sediment thickness exceeds 1000 m.

Figure 2.40 shows in its bottom parts the depth distribution of the shear wave velocity and the density for four selected sites on the profile. Profile (A) is west of the fault system with 1000 m total thickness of sediments, interbedded strata of sand, gravel, clay and lignite. The latter forms the low velocity channel at a depth of 450 m. Profile (B) with ca. 250 m of sediments represents a section within the city limits of Cologne. Profile (C) with 25 m of sediments is a location east of the Rhine River and the profile (D) has only 7 m of sediments. The Program SIMUL (Scherbaum 1994) has been used to calculate the one-dimensional transfer functions of the sediment packages for the four models. The corresponding amplification functions are shown in Fig. 2.40. For all models it was assumed, that in the transition between had rock and sediments a 40 m thick weathered rock layer is located. This lowers the impedance contrast between sediments and half space. For all profiles, an earthquake with M_W 5.0 at an epicentral distance of 5 and 10 km depth was assumed. The change of the frequency of the fundamental mode from low to high frequencies when lowering the sediment thickness is obvious. Table 2.2 summarizes these frequencies and the amplification.

Figure 2.41 shows example seismograms for the 4 profiles. The delayed arrival of the ground motion with increasing sediment thickness is obvious. This is caused by the longer travel paths in the slow sediments. With decreasing depth of the hard rock surface, the frequency of the seismograms increases significantly as well as the maximum accelerations which reach 80% of g in case of the 7 m layer; however in a frequency range higher than the resonance frequencies of most buildings.

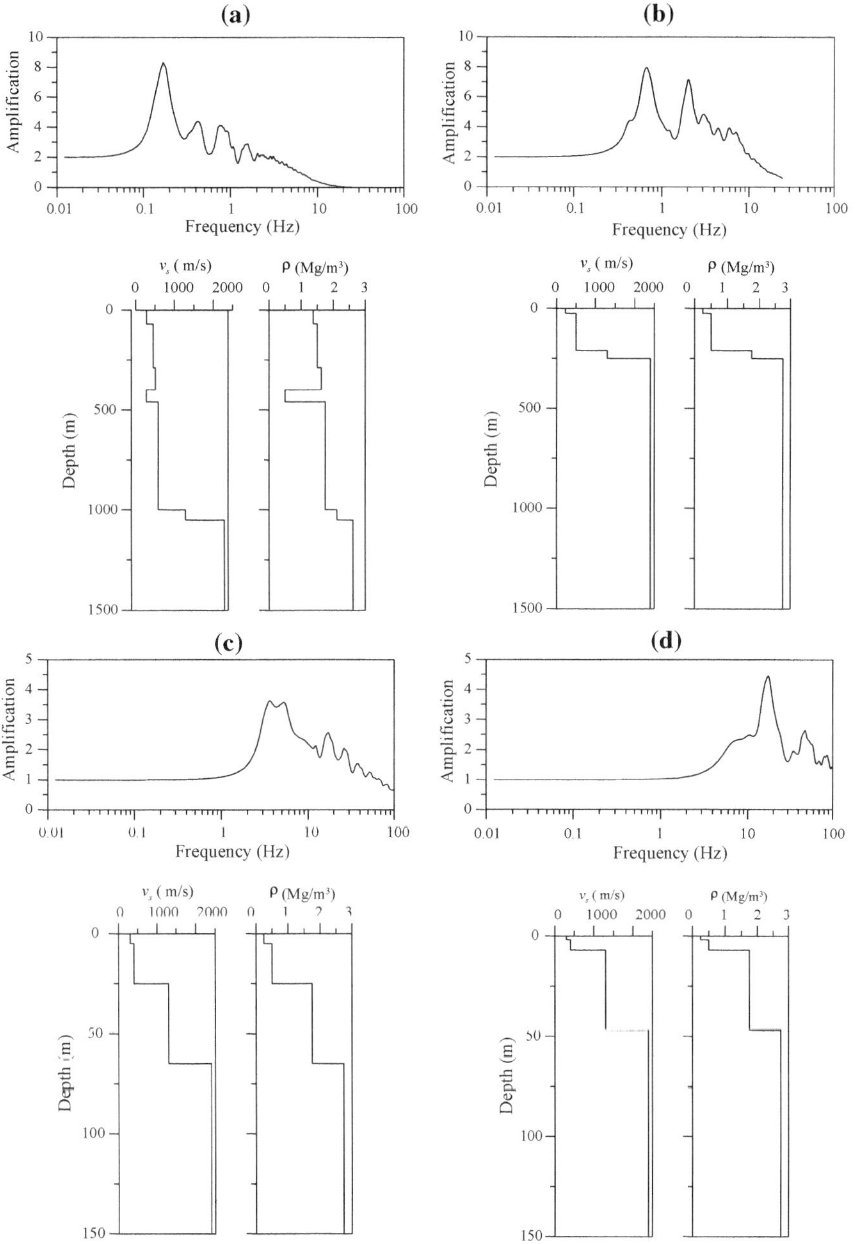

Fig. 2.40 Four examples of sedimentary structures from the profile in Fig. 2.39. The bottom part of each plot shows the depth distribution of the seismic velocity, the density and the Q value. The top graphs show the amplification of vertically propagating S waves

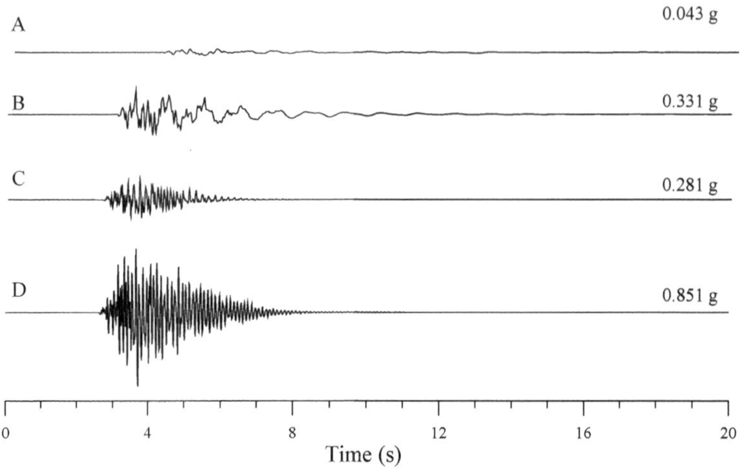

Fig. 2.41 Synthetic acceleration seismograms of horizontal ground motion for the four different sediment profiles from Fig. 2.40. The source is in all cases an earthquake with magnitude 5 in 5 km distance at a depth of 10 km. Numbers at the and of the equally scaled seismograms are *PGA* in units of g

2.5 Design Ground Motions

The most important bridge between seismology and the engineering environment is the response spectrum introduced in Chap. 1. Shape and amplitude levels of response spectra are determined by seismological models and tools, either using a deterministic or a probabilistic seismic hazard analysis, abbreviated as DSHA and PSHA, respectively. In both approaches it is common practice to start with the definition of appropriate earthquake sources. Depending on the knowledge about the target region and its seismotectonic setting, theses sources can be modelled as point-, line-, surface-, or volume-sources. Earthquake catalogs of recent and historic earthquakes and geologic maps are used to define the sources. If available also paleoseismic and archaeoseismic data should be taken into account to cover a time window as large as possible.

Independent of the approach, all sources which are capable of generating significant ground motions at the site for which the seismic load has to be determined, have to be taken into account. Often a radius of 200 km from the site is regarded as sufficient. In addition to the definition of the source geometry and strength that is necessary in both approaches, in the PSHA the probability distribution of earthquakes within each source have to be estimated. In a DSHA usually a probability of 1 is given for each source to the shortest distance between the source and the site.

As buildings are designed for ground motions, and not for earthquakes, appropriate ground motion parameters have to be estimated for the construction site for all earthquakes. Ground motion prediction equations introduced in Sect. 2.2 are used to estimate these ground motions. In the DSHA, single values of magnitude, distance and number of standard deviations of the ground motion prediction are used for each source.

In PSHA all possible deterministic earthquake scenarios, i.e. all magnitudes and location combinations, are considered and combined with ground motion probabilities of the model (Fig. 2.42). The number of individual earthquake scenarios for which the site-specific ground motion is calculated can become rather large in PSHA. The rate at which the ground motion is equal or larger than a specified level is the hazard, expressed in hazard curves (step 4 in Fig. 2.42); the number of times a certain ground motion level is exceed is plotted versus the ground motion parameter itself.

While in DSHA the largest ground motions from the relative few scenarios can be directly selected for the design, in PSHA first a hazard level has to be chosen for which the design ground motion parameters are determined.

Authorities sometimes try to leave it to the scientific community to suggest such a hazard level; however, setting such a level is not a scientific but a political issue and the smaller the probability of occurrence, the larger is the resulting ground motion parameter. For standard buildings it is common regulation practice to use a 10% probability of exceedance in 50 years; however, is not really clear why this level was once chosen. In case of constructions with a significant hazard in case of failure as dams, or chemical plants, higher safety levels are required such as 2% probability of exceedance in 50 years. For nuclear facilities, 'worst-case' ground motions are required for design. Though it sounds easy to select the scenario with the largest ground motion values, these values will strongly depend on the number of standard deviations regarded throughout the whole hazard model.

PSHA has four major advantages when compared with the DSHA. (1) The method recognizes that by far more small earthquakes happen than large ones. As we are looking for the design ground motion and not a design earthquake, PSHA takes into account, that (2) smaller earthquakes in rare cases can produce exceptionally large ground motions, a fact introduced into the model by the number of standard deviations of the GMPEs regarded. In contrast to DSHA in the probabilistic model (3) activity rates, the average number of earthquakes per year can vary from source to source, and (4) increased hazard from multiple faults or sources is covered.

If the design process includes testing with time series, spectrum compatible seismograms have to be determined. In order to use appropriate scenarios for the magnitude and distance of the simulated time series in a disaggregation of the results of a PSHA it can be determined which magnitude-distance combinations have the largest share to the ground motion hazard. For these combinations time series with appropriate envelopes and duration can then be calculated.

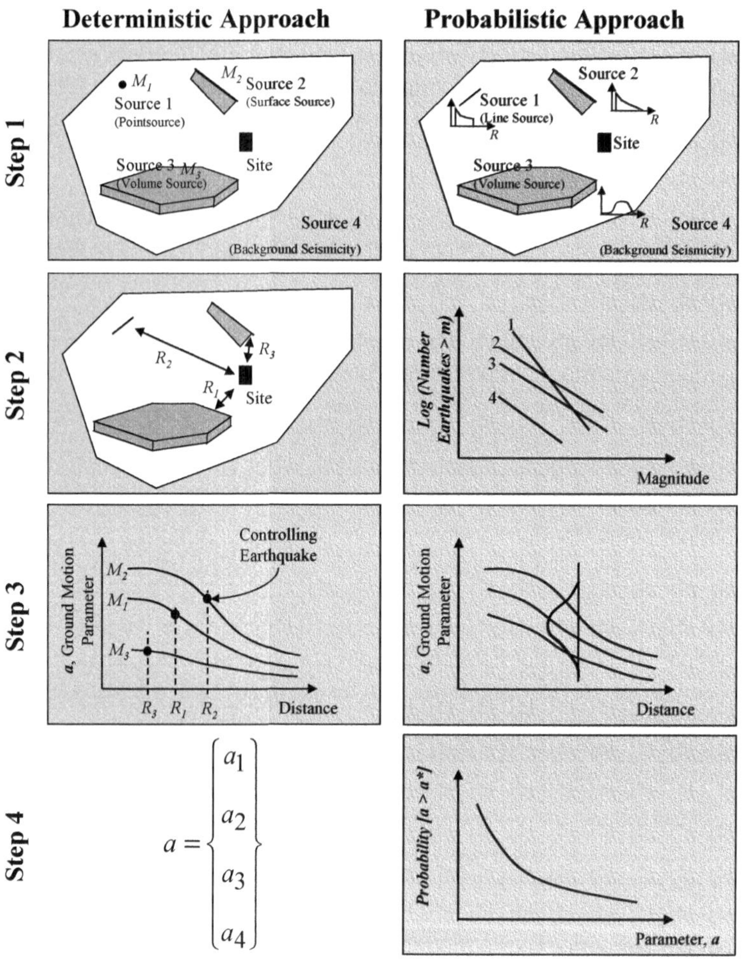

Fig. 2.42 Basic steps of a deterministic (left) and probabilistic (right) seismic hazard analysis (after Reiter 1990; Kramer 1996)

2.6 Examples of Application

For the civil engineer, the response spectrum is the basis of earthquake safe design of a construction, no matter whether a frequency domain analysis is made directly with the acceleration response values at the main eigenfrequencies of the construction, or whether a full time domain analysis is needed. Anyway, detailed seismological aspects of certain earthquakes are not the focus of interest during this step. However, it is important also for the engineer to have a feeling of how seismic source parameters influence the response spectrum and particularly the time series. In this chapter we use synthetic seismograms and the corresponding response spectra to highlight some of these influences.

Synthetic ground motion signals were derived using a stochastic approach (Wang 1999). The source is represented by a rectangular fault plane on which circular subsources of randomly varying size are distributed. The ground motion of each subsource at selected measuring points is calculated based on the Green's function of the subsurface. The effects from all subsources are superposed to get the complete ground motion. The number and size of subsources is selected so that the total seismic moment is in agreement with the target magnitude of the earthquake. The place and time of each subsource mimics the rupture process on the fault plane.

Example 2.7

Synthetic ground motions were calculated for an earthquake source with normal faulting mechanism on a vertical fault plane of a length of 45 km, corresponding to a moment magnitude of M_W 7.0. Figure 2.43 shows a map of the assumed geometry of observational points. A total of 20 stations are placed at a distance of 20 km from the surface projection of the fault plane. Measuring points are labelled MP followed by the azimuth with respect to north as seen from the source. Five measuring points are placed at a constant distance of 20 km along both the east and western side of the fault line.

The rupture of the normal fault was assumed to originate in the middle of the fault and spread to north and south with a constant rupture velocity of 2.0 km/s.

The calculated three-component time histories of ground acceleration are shown in Fig. 2.44 within a 60 s time window and with the same amplitude scale for each plot. The time windows start at the moment of rupture initiation so that the travel time from the source to the station can be read directly from the seismograms. As the rupture starts in the middle of the fault, the travel time toward the stations at azimuth of 90° and 270° is slightly shorter than that to the other stations, though the distance between the fault and all stations is constant. The largest acceleration amplitudes are observed also at azimuths of 90° and 720° in the vertical component. Due to the radiation pattern north and south of the fault line the maximum amplitudes are observed in the EW component.

The corresponding acceleration response spectra in Fig. 2.45 show the large differenced in response amplitudes with respect to azimuth of the observation point and among the tree ground motion components.

Figure 2.46 shows the ground acceleration for the measuring point distribution shown in Fig. 2.43 for the M_W 5.0 Earthquake. Besides the magnitude and the length of the fault, which is in this example 0.9 km, all other parameters are the same as in case of the M_W 7.0 simulation. The smaller source dimension not only limits the radiated seismic energy, also the duration of the rupture process is significantly smaller. This leads to the much shorter durations of the significant ground motions.

The seismograms in Fig. 2.47 were calculated for the same geometry as shown in Fig. 2.43 for the M_W 7.0 earthquake. The difference to the example shown in Figs. 2.44 and 2.45 is the strike slip source mechanism and the rupture direction. Here it was assumed that the rupture starts at the southern end of the fault and

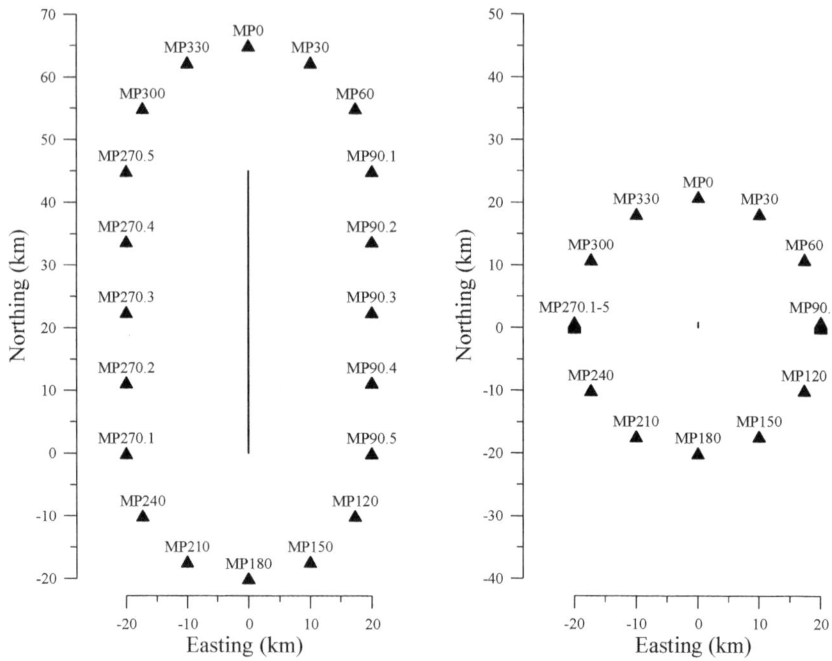

Fig. 2.43 Distribution of assumed measuring points (triangles with labels) surrounding a 45 km long north-south striking and 90° dipping fault of a magnitude 7 earthquake (left) and for a magnitude 5 earthquake with an 0.9 km long fault (right)

propagates again with a rupture velocity of 2.0 km/s towards north. The changed source mechanism and the different rupture process results in ground accelerations with different character; the similar scales of Figs. 2.44 and 2.47 allow a direct comparison. The increasing travel times from stations in the south to those in the north are obvious. This is the effect of the northward travelling rupture. At a rupture length of 45 km, the total rupture duration for the strike slip earthquake is 21.0 s. Therefore the duration of the significant ground motion is larger than in case of the normal faulting example. Largest acceleration amplitudes are observed in the NS components at 90° and 270° azimuth.

The acceleration response spectra of the strike slip simulation are compared to the normal faulting simulation in Fig. 2.48. The switch between NS and Z as the components with the largest accelerations at 90° and 270° azimuth between the two simulations is obvious. However, the significantly larger duration of the signal is lost when only response spectra are used. This underlines the necessity to make adequate assumptions about the envelope when spectrum compatible ground motions are derived from response spectra.

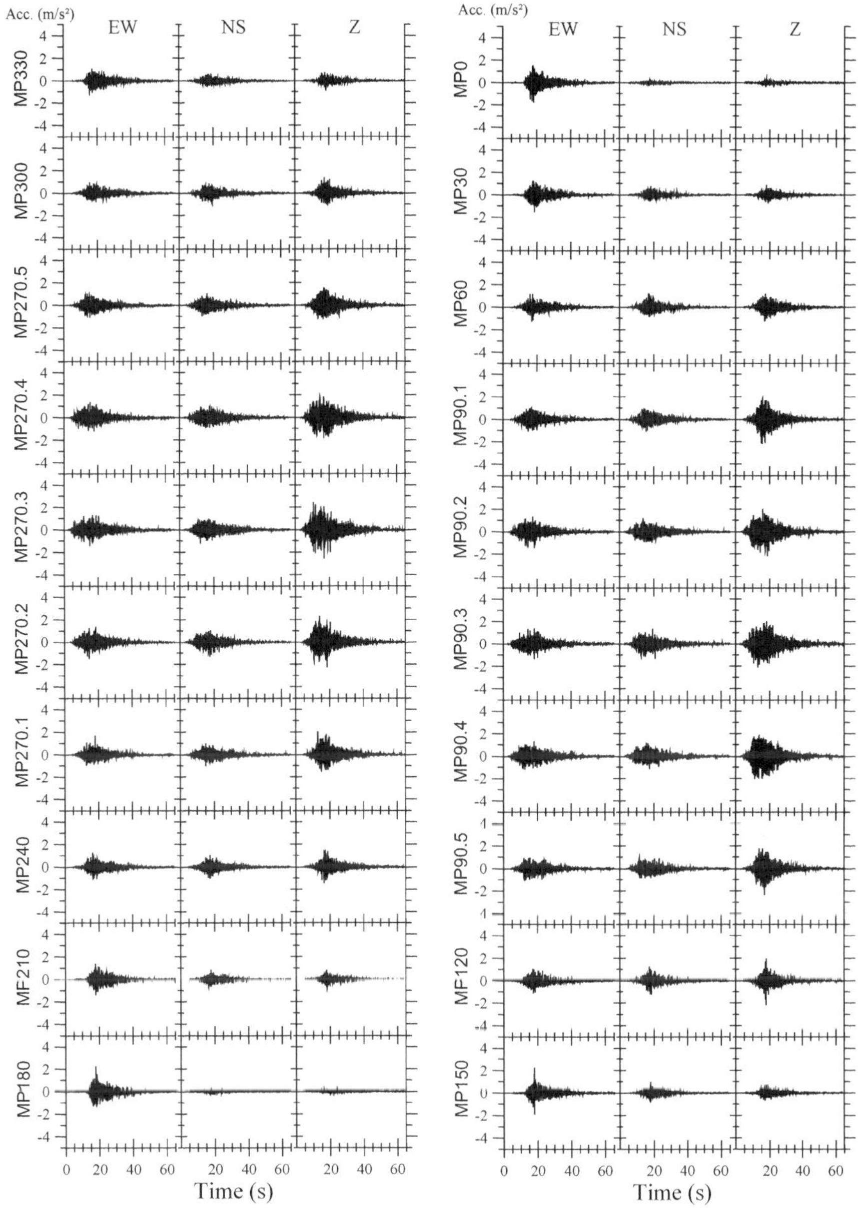

Fig. 2.44 Synthetic acceleration seismograms of the east-west (EW), north-south (NS), and vertical (z) component of ground motion for a M_W 7.0 earthquake of normal faulting mechanism. The station distribution is shown in Fig. 2.43

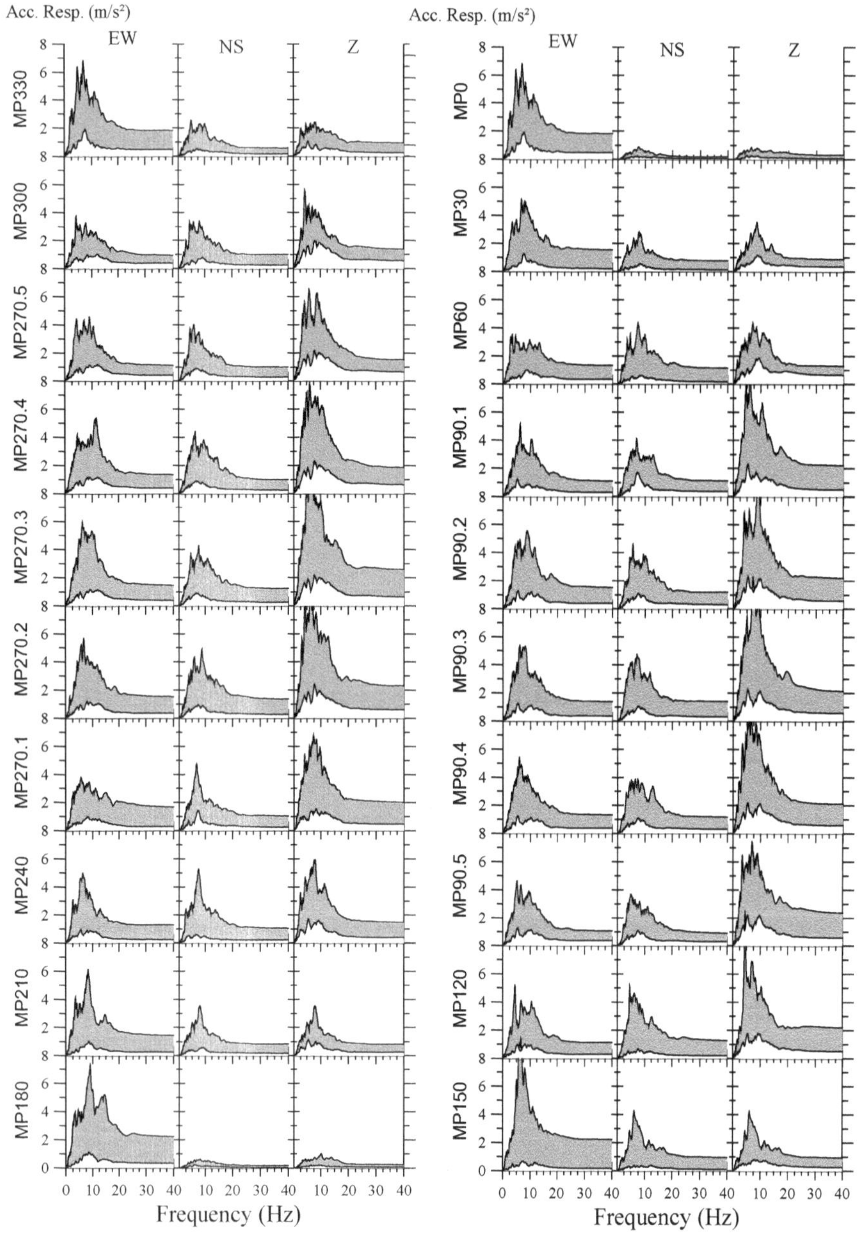

Fig. 2.45 Acceleration response spectra of the seismograms shown in Fig. 2.44 (grey fill) and Fig. 2.46 (white fill)

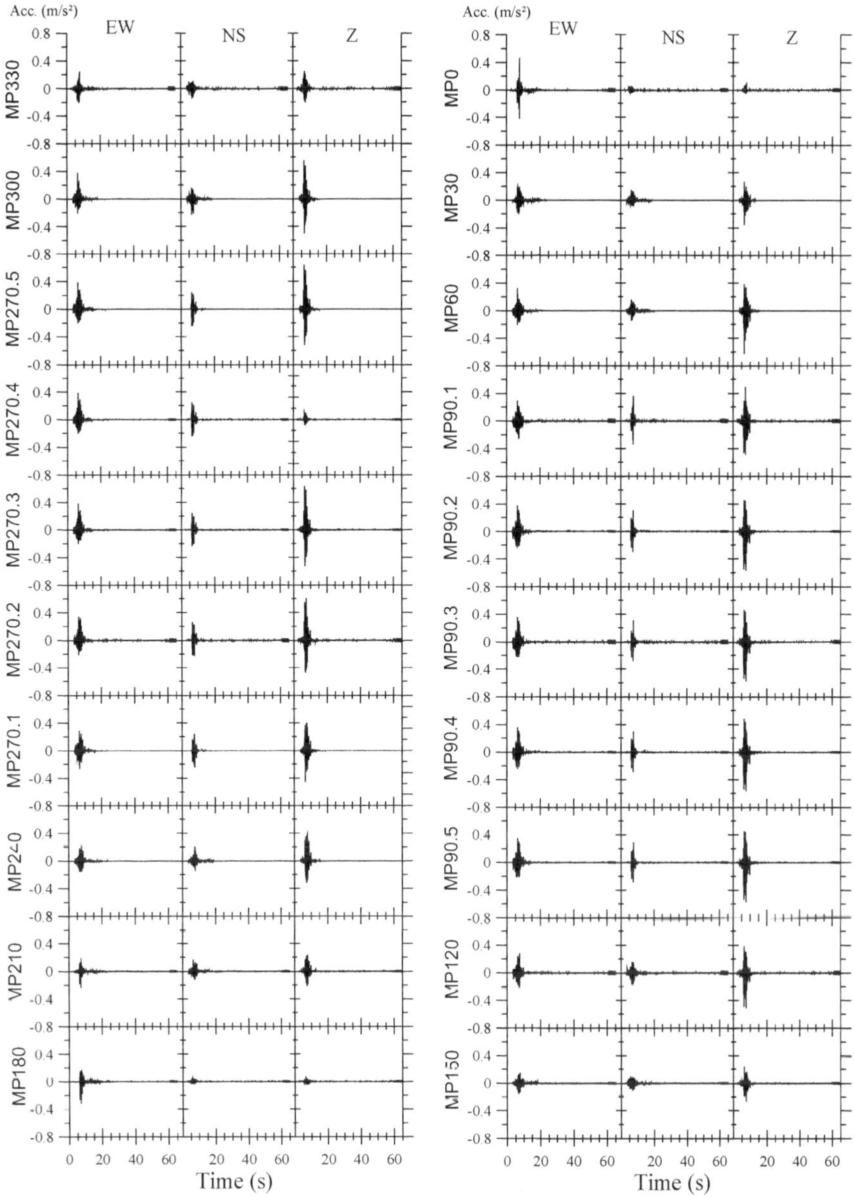

Fig. 2.46 Synthetic acceleration seismograms of the east-west (EW), north-south (NS), and vertical (z) component of ground motion for a M_W 5.0 earthquake of normal faulting mechanism. The station distribution is shown in Fig. 2.43

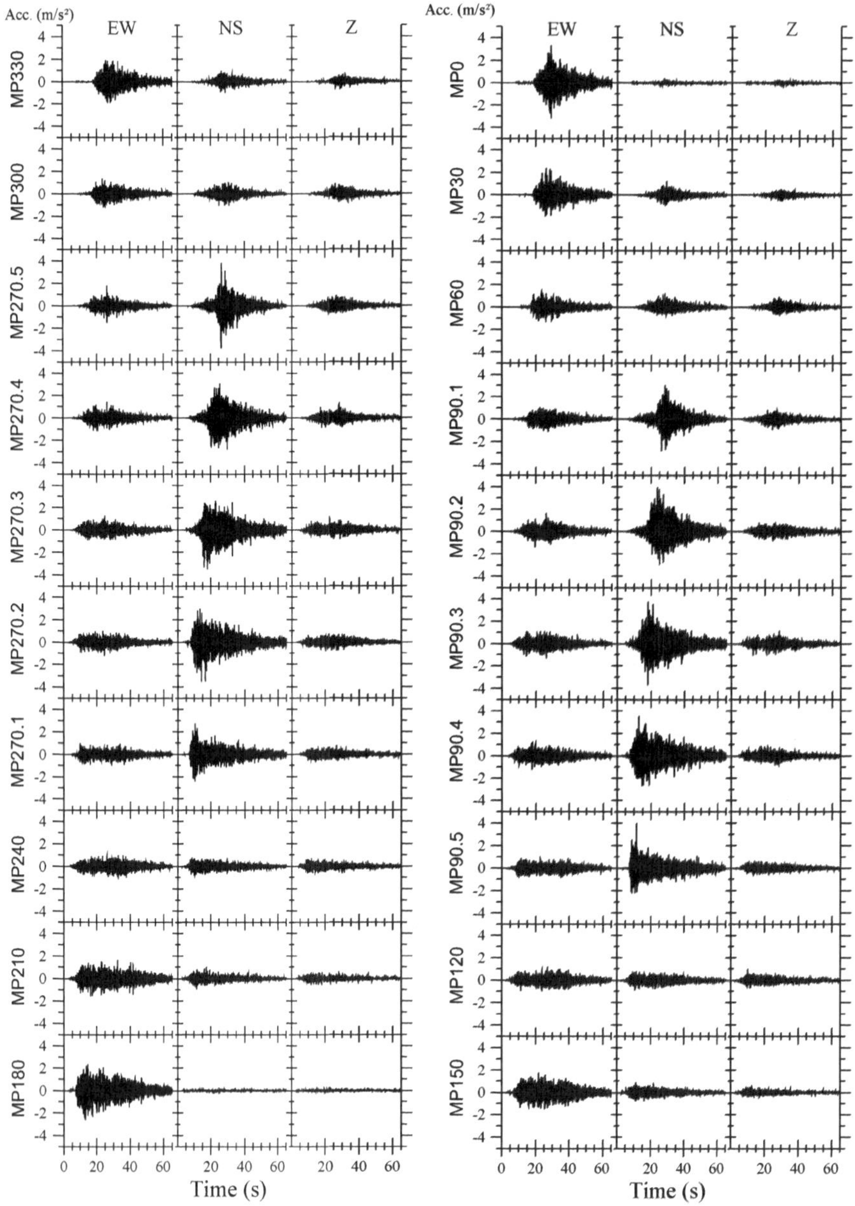

Fig. 2.47 Synthetic acceleration seismograms of the east-west (EW), north-south (NS), and vertical (z) component of ground motion for a MW 7.0 earthquake of strike-slip mechanism. The station distribution is shown in Fig. 2.43

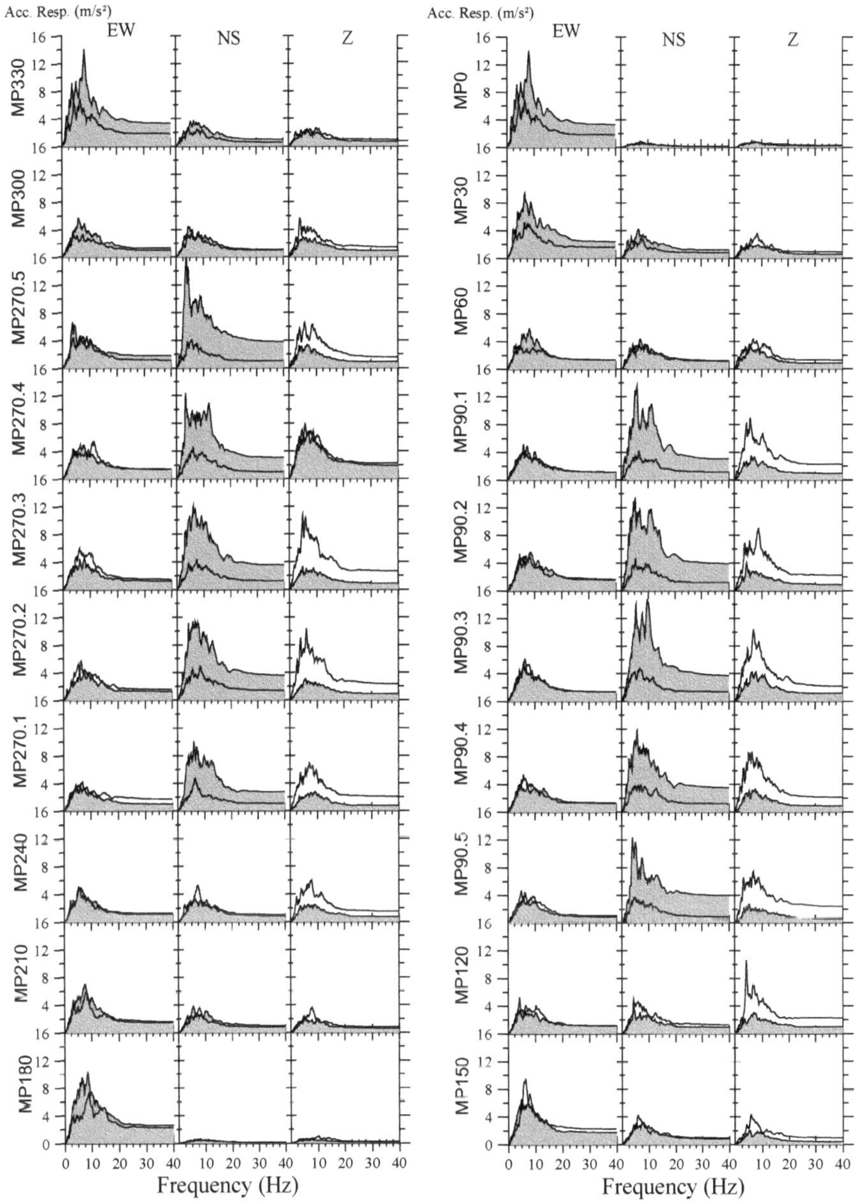

Fig. 2.48 Acceleration response spectra of the seismograms shown in Fig. 2.47 (grey fill) and for comparison those from Fig. 2.44 (no fill)

References

Abrahamson, N.A., Shedlock, K.M.: Overview. Seismol. Res. Lett. **68**, 9–23 (1997)
Abrahamson, N.A., Silva, W.J.: Empirical response spectral attenuation relations for shallow crustal earthquakes. Seismol. Res. Lett. **68**, 94–127 (1997)
Aki, K.: Generation and propagation of G waves from the Niigata earthquake of June 16, 1964. 2. Estimation of earthquake moment, released energy, and stress-strain drop from G wave spectrum. Bull. Earthq. Res. Inst. **44**, 23–88 (1966)
Aki, K.: Scaling law of earthquake time-function. Geophys. J. Roy. Astron. Soc. **31**, 3–25 (1972)
Aki, K., Richards, P.: Quantitative Seismology: Theory and Methods, vol. 1 and 2. W.H. Freeman, San Francisco, California (1980)
Aki, K., Richards, P.: Quantitative Seismology, 2nd edn, p. 700. University Science Books, Sausalito, California (2002)
Arias, A.: A measure of earthquake intensity. In: Hansen, R.J. (ed.) Seismic Design for Nuclear Power Plants, pp. 438–483. MIT press, Cambridge, Massachusetts (1970)
Båth, M.: Introduction to seismology. Birkhäuser, Basel und Stuttgart (1973)
Benjamin, J.R.: A criterion for determining exceedance of the operating basis earthquake. EPRI Report NP-5930, Electric Power Research Institute, Palo Alto, California (1988)
Bolt, B.A. Duration of strong motions. In: Proceedings of the 4th World Conference an Earthquake Engineering, Santiago, Chile, pp. 1304–1315 (1969)
Bommer, J.J., Martinez-Pereira, A.: The effective duration of earthquake strong ground motion. J. Earthquake Eng. **3**, 127–172 (1999)
Boore, D.M., Joyner, W.B.: Prediction of ground motion in North America. In: Proceedings of the ATC-35 Seminar on new Developments on Earthquake Ground Motion Estimates an Implications for Engineering Design Practice, Applied Technology Council, Redwood City, pp. 1–14 (1994)
Boore, D.M., Joyner, W.B., Fum, T.E.: Estimation of response spectra and peak accelerations from western North American earthquakes: An interim report. U.S. Geological Survey Open-File Report 95–509, pp. 72 (1993)
Campbell, K.W.: Near-source attenuation of peak horizontal acceleration. Bull. Seismol. Soc. Am. **71**, 2039–2270 (1981)
Campbell, K.W.: Predicting strong ground motion in Utah. In: Gori, P.L., Hays, W.W. (eds) Assessment of Regional Earthquake Hazards and Risk Along the Wasatch Front, Utah, Vol. II, U.S. Geological Survey, Open-File Report 87-585, L1-L90 (1987)
Campbell, K.W.: Engineering models of strong ground motion. In: Chen, W.-F., Scawthorn, C. (eds.) Earthquake Engineering Handbook. CRS Press, Boca Raton, FA (2003)
Campbell, K.W., Bozorgnia, Y.: Updated near-source ground motion (attenuation) relations for the horizontal and vertical components of peak ground acceleration and acceleration response spectra. Bull. Seismol. Soc. Am. **93**, 314–331 (2003)
Davis, E.F.: The Marvion strong motion seismograph. Bull. Sesimol. Soc. Am. **3**, 195–202 (1913)
Draper, N.R., Smith, H.: Applied regression analysis, 2nd edn. Wiley, New York (1981)
Erdik, M., Durkal, E.: Simulation modeling of strong ground motions. In: Chen, W.-F., Scawthorn, C. (eds.) Earthquake Engineering Handbook. CRS Press, Boca Raton, FA (2003)
Haskell, N.A.: Total energy and energy spectral density of elastic waves from propagating faults. Bull. Seismol. Soc. Am. **54**, 1811–1841 (1964)
Hinzen, K.-G., Fleischer, C.: A strong-motion network in the lower rhine embayment (SeFoNiB), Germany. Seismol. Res. Lett. **8**, 502–511 (2007)
Husid, R.L.: Analisis de terremotos. Analisis General, Revista del IDIEM, 8, Santiago, Chile, pp. 21–42 (1969)
Jost, M.O., Herrmann, R.B.: A student's guide to and review of moment tensors. Seismol. Res. Lett. **60**, 37–57 (1989)
Joyner, W.J., Boore, D.M.: Peak horizontal acceleration and velocity from strong ground motion recordings including records from the 1979 Imperial Valley, California earthquake. Bull. Seismol. Soc. Am. **71**, 2011–2038 (1981)

Kanamori, H.: The energy release in great earthquakes. J. Geophys. Res. **82**, 2981–2987 (1977)
Kanamori, H., Anderson, D.L.: Theoretical basis of some empirical relations in seismology. Bull. Seismol. Soc. Am. **65**, 561–590 (1975)
Keilis Borok, V.I.: On the estimation of the displacement in an earthquake source and of source dimensions. Ann. Geofis. **12**, 205–214 (1957)
Kramer, S.L.: Geotechnical Earthquake Engineering. Prentice Hall, Upper Saddle River, N.J, pp. 205–214 (1996)
Lawson, A.C.: The California earthquake of April 18, 1906: Report of the State Earthquake Investigation Commission: Carnegie Institution of Washington Publication 87, 2 vols (1908)
Lay, T., Wallace, T.C.: Modern Global Seismology, p. 517. Academic Press, San Diego, California (1995)
Newmark, N.M., Hall, W.J.: Earthquake spectra and design. Earthquake Engineering Research Institute, Berkeley, California (1982)
Nuttli, O.W.: The relation of sustained maximum ground acceleration and velocity to earthquake intensity and magnitude, State-of-the-Art for Assessing Earthquake Hazards in the United States, Report 16, Misc. Paper S-73-1, U.S. Army Engineer Waterways Experiment Station, Vicksburg, Mississippi (1979)
Rathje, E.M., Abrahamson, N.A., Bray, J.D.: Simplified frequency content estimates of earthquake ground motions. J. Geotech. Eng. Div. ASCE **1998**(124), 150–159 (1998)
Reid, H.F.: The California earthquake of April 18, 1906. Publication 87, 21, Carnegie Institute of Washington, Washington, D.C (1910)
Reiter, L.: Earthquake Hazard Analysis—Issues and Insights, p. 254. Columbia University Press, New York (1990)
Richter, C.F.: An instrumental earthquake scale. Bull. Seismol. Soc. Am. **25**, 1–32 (1935)
Richter, C.F.: Elementary Seismology. W.H. Freeman, San Francisco (1958)
Scherbaum, F.: Modelling the Roermond Earthquake of April 13, 1992 by stochastic simulation of its high frequency strong ground motion. Geophys. J. Int. **119**, 31–43 (1994)
Schnabel, P.B., Bolton Seed, H.: Accelerations in rock for earthquakes in the western United States. Bull. Seismol. Soc. Am. **63**, 510–516 (1973)
Schneider, G.: Naturkatastrophen, p. 364. Enke Verklag, Stuttgart (1980)
Shearer, P.: Introduction to Seismology, p. 272. Cambridge University Press (1999)
Somerville, P.G., Abrahamson, N.A.: Ground motion prediction for thrust earthquakes. In: Proceedings of SMIP95 Seminar, California Division of Mines and Geology, San Francisco, California, pp. 11–23 (1995)
Trifunac, M.D., Brady, A.G.: A study of the duration of strong earthquake ground motion. Bull. Seismol. Soc. Am. **65**, 581–626 (1975)
von Thun, J.L., Rochim, L.H., Scott, G.A., Wilson, J.A.: Earthquake ground motions for design and analysis of dams. In: Earthquake Engineering and Soil Dynamics II—Recent Advances in Ground-Motion Evaluation (GSP 20), pp. 463–481. ASCE, New York (1988)
Wang, R.: A simple orthonormalization method for stable and efficient computation of Green's functions. Bull. Seismol. Soc. Am. **89**, 733–741 (1999)
Wells, D.L., Coppersmith, K.J.: New empirical relationships among magnitude, rupture length, rupture width, rupture area, and surface displacement. Bull. Seismol. Soc. Am. **84**, 974–1002 (1994)
Wilson, R.C.: Relation of Arias intensity to magnitude and distance in California. Open File Report 93-556, U.S. Geological Survey, Reston, Virginia, p. 42 (1993)

Chapter 3
Stochasticity of Wind Processes and Spectral Analysis of Structural Gust Response

Rüdiger Höffer and Ana Cvetkovic

Abstract Wind loads have great impact on many engineering structures. Wind storms often cause irreparable damage to the buildings which are exposed to it. Along with the earthquakes, wind represents one of the most common environmental load on structures and is relevant for limit state design. Modern wind codes indicate calculation procedures allowing engineers to deal with structural systems, which are susceptible to conduct wind-excited oscillations. In the codes approximate formulas for wind buffeting are specified which relate the dynamic problem to rather abstract parameter functions. The complete theory behind is not visible in order to simplify the applicability of the procedures. This chapter derives the underlying basic relations of the spectral method for wind buffeting and explains the main important applications of it in order to elucidate part of the theoretical background of computations after the new codes. The stochasticity of the wind processes is addressed, and the analysis of analytical as well as measurement based power spectra is outlined. Short MATLAB codes are added to the Appendix 3 which carry out the computation of a single sided auto-spectrum from a statistically stationary, discrete stochastic process. Two examples are presented.

Keywords Wind turbulence · Alan G. Davenport wind loading chain · Gust wind response · Spectral analysis

In this section, the most important basics of the analysis of wind buffeting in the frequency domain are introduced. The random character of both, loading and structural response, is considered. The section serves also as an introduction to section 6 where the application of the concepts to line—like vertical structures is shown.

Electronic supplementary material The online version of this chapter (https://doi.org/10.1007/978-3-662-57550-5_3) contains supplementary material, which is available to authorized users.

K. Meskouris et al., *Structural Dynamics with Applications in Earthquake and Wind Engineering*, https://doi.org/10.1007/978-3-662-57550-5_3

3.1 Short Review of the Stochasticity of Wind Processes

3.1.1 General

The lower layer of the Earth's atmosphere is the planetary or atmospheric boundary layer (ABL). Earth's atmosphere is, on average, more than 100 km thick; the ABL is only a thin part of it with the height of up to 1 km during day on average. Its thickness depends on terrain roughness, wind intensity and other factors. The dynamic wind loads on structures depend directly on the wind properties which can be diverse for different wind events. In Europe, extreme winds due to non-tropical cyclones as the main wind type in this region are most commonly considered for a wind-resistant design of structures. Other types of storms, such as tornadoes, thunderstorms or fall winds can indeed cause significant damage to the structures. However, an event of the latter type is local in character and its influence as well as its energy is limited to smaller areas of the built environment. Therefore, the following sections put the focus on dynamic wind effects due to the impact of non-tropical cyclones.
Wind-resistant design of structures deals at first with dynamic wind effects which can lead to a forced fracture or to an overload breakage of structural components. The following sections focus on this wind loading case. However, not the rare extreme loading events alone give a picture of a wind load scenario. The complete ensemble of load effects during the life-time of the structure is central for the evolution of structural degradation. It is of equal importance as a further matter of design to have models at hand which adequately reflect the experienced time histories of impacts. Information about such aspects are available in Höffer (2009).
There are two basic strategies when solving the dynamic problem of wind processes: the deterministic and the non-deterministic approach. The first one is applied if the process is fully known at any point of time and at any required space point. If this is not the case—as for wind related quantities in principle—but the processes can be described in statistical sense, then a non-deterministic approach of random wind processes is applied. The random wind processes are analysed by the application of statistic operators and of the procedures of Digital Signal Processing (DSP). Since wind speed and load can be measured only at discrete time points, it is treated as a time discrete signal and not as a continuous function. Therefore, in the following, the focus is put on discrete, spatially distributed signals.

3.1.2 Stochastical Description of the Turbulent Wind

3.1.2.1 Statistical Collective

Processes with non-regularly fluctuating amplitudes are mostly considered to be random or stochastic. A stochastic process is a statistical collective consisting of individual realisations of random functions instead of discrete events. A deterministic function, e.g. defined by $x(t) = \sin(\omega t)$, describes a well-defined value x which is assigned to the time t_i. The function is reproducible, and x is predictable. A random (e.g. time dependent) function has a random value at the time t_i. The function is not reproducible, and x is not predictable. The value of the function at time t, $x(t)$, is replaced by a probabilistic statement, e.g.

$$P_{x \leq a}(a, t) = \int_{-\infty}^{a} f_x(\xi, t)\, d\xi \tag{3.1}$$

$P_{x \leq a}(a, t)$ is the probability that a realisation with a value $x \leq a$ at the time t occurs, and $f_x(\xi, t)$ is the probability density function of the stochastic variable X at level $x = \xi$ and at time t.

3.1.2.2 Statistical Parameters

A structure is subjected to a single random excitation, e.g. due to wind. For the description of such loading, statistical estimators are applied. Important estimators are:

$$\text{arithmetical mean} : \overline{x} = \lim_{T \to \infty} \frac{1}{T} \int_{-T/2}^{T/2} x(t)\, dt \tag{3.2}$$

$$\text{square mean} : \overline{x^2} = \lim_{T \to \infty} \frac{1}{T} \int_{-T/2}^{T/2} x^2(t)\, dt \tag{3.3}$$

$$\text{variance of scattering} : \sigma_x^2 = \overline{(x - \overline{x})^2} = \lim_{x \to \infty} \frac{1}{T} \int_{-T/2}^{T/2} [x(t) - \overline{x}]^2\, dt$$

$$\text{or } \sigma_x^2 = \overline{x^2} - \overline{x}^2 \tag{3.4}$$

ctically only be obtained from limited collectives. Thus, they are only estimates of the expected, 'true' values of a collective with an infinite number of elements. In general, the probability density function is non-stationary, $f_x(\xi, t_1) \neq f_x(\xi, t_2)$, which means that the statistical properties in general are time-dependent. The essential conditions within in this section are:

1st assumption Stochastic vibrations are supposed to be stationary, which means that $f_x(\xi, t) = f_x(\xi)$ is not dependent on the time t. Stationarity causes the joint probability density function between the collectives $X_1 = [x_1(t_1), x_2(t_1), \ldots, x_N(t_1)]$ and $X_2 = [x_1(t_2), x_2(t_2), \ldots, x_N(t_2)]$ to not depend on times t_1 and t_2 but on the time difference $\tau = t_2 - t_1$ only, and $\overline{x}_E(t_1) = \overline{x}_E(t_2) = \overline{x}_E(t)$.

2nd assumption The process is ergodic; its statistical properties are completely reflected by each realisation which means that $\overline{x}_{T,1} = \overline{x}_{T,2} = \overline{x}_T$ and $\overline{x}_T = \overline{x}_E$. The simplest statistic measure is the (arithmetical) mean value $\overline{x}$. Averaging with respect to the ensemble at time t leads to Fig. 3.1:

$$\overline{x}_E(t) = \lim_{N\to\infty} \frac{1}{N} \sum_{i=1}^{N} x_i(t) \tag{3.5}$$

Averaging over the time for the ith realisation results into:

$$\overline{x}_{T,i} = \lim_{T\to\infty} \frac{1}{T} \int_{-T/2}^{T/2} x_i(t)\, dt \tag{3.6}$$

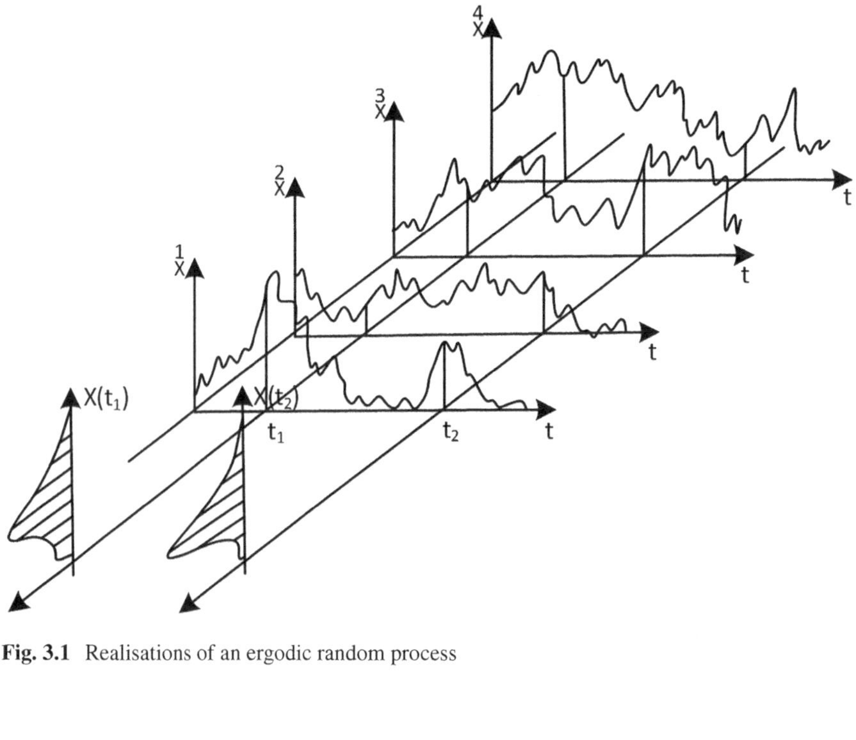

Fig. 3.1 Realisations of an ergodic random process

The load is considered to be a Gaussian process with the probability density function

$$f(x) = \frac{1}{\sqrt{2\pi} \cdot \sigma_x} \cdot e^{-\frac{(x-\overline{x})^2}{2\sigma_x^2}} \quad \text{(normal or Gaussian distribution function).} \tag{3.7}$$

The probability density function is completely described by the mean value $\overline{x}$ and the variance σ_x^2, where the latter value carries information about the probability of load amplitudes. Other representations of the scattering are:

- standard deviation: $\sigma = \sqrt{variance}$
- intensity: $I_x = \sigma/\overline{x}$

Statistical information about the properties of the process in time is required for a vibration analysis. In the time domain such information is provided by the auto-correlation function $R(\tau)$. In the frequency—domain the spectral density $S(\omega)$ is mathematically equivalent, ω denotes the circular frequency.
For Gaussian processes $R(\tau)$ allows for probability statements concerning the load amplitudes at two different time instants differing by τ can be made. The autocorrelation function R is a statistical measure for the memory regarding the sequences of signals of a process under consideration. R is expressed as function of the argument τ, which is explained in Fig. 3.2 as the time shift. A real stochastic process, e.g. the time history of velocity fluctuations in turbulent wind, is characterized by a quickly decaying autocorrelation function.

The autocorrelation function is defined by

$$R(\tau) = \lim_{T\to\infty} \frac{1}{T} \int_{T/2}^{T/2} x(t) \cdot x(t+\tau)\, d\tau \tag{3.8}$$

The function can be characterized as the following properties:

1. $R(0) = \overline{x}^2$—square mean
2. for $x = 0$ (zero-mean process) is $R(0) = \sigma^2$—variance
3. $R(-\tau) = R(\tau)$—the autocorrelation is an even function.

Expected values of various powers of one or more random variables are called *statistical moments*. For a single variable, $E[X]$ is the first moment, $E[X^2]$ is the second moment, $E[X^n]$ is the nth moment. For two (or more) random variables, $E[X_1^m X_2^n]$ is joint moment of X_1 and X_2 of the $(m+n)$th order.

Obviously the mean value μ_X is actually the first moment of X. Then $E[(X - \mu_X)^n]$ is the nth central moment of X. The second central moment $E[(X - \mu_X)^2] = E[X] - \mu_X^2 = \sigma^2$ is then equal to the variance of the random variable X.

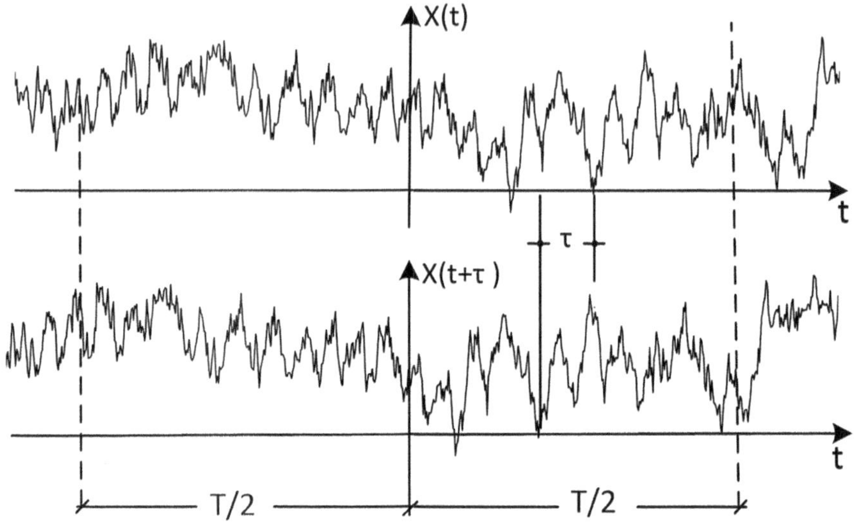

Fig. 3.2 Autocorrelation

The third moment $E[(X - \mu_X)^2]$ normalized by the variance is used to calculate the skewness of the process. The sample skewness can be estimated as

$$s_1 = \frac{\frac{1}{n}\sum_{i=1}^{n}(x_i - \overline{x})^3}{\left(\sqrt{\frac{1}{n}\sum_{i=1}^{n}(x_i - \overline{x})^2}\right)^3} \tag{3.9}$$

In case of MDOF systems, the relationship between two or more stochastic processes has to be defined. A two-dimensional process is expressed by two stochastic processes, $\eta(t)$ and $\xi(t)$. It is determined if the joint distribution of the random variables of the two named processes is defined. If the joint distributions are Gaussian (and hence the respective processes themselves), then the whole joint process is said to be a Gaussian process. The cross-correlation function for the two stochastic processes is defined by

$$R_{\eta\xi}(\tau) = E[\eta(t)\xi(t + \tau)] \tag{3.10}$$

and the cross-spectrum is defined as its Fourier transform

$$S_{\eta\xi}(\omega) = \frac{1}{2\pi}\int \lim_{-\infty}^{\infty} R_{\eta\xi}(\tau)\mathrm{e}^{i\omega\tau}\mathrm{d}\tau \tag{3.11}$$

Since the cross-correlation is not an even function of τ, the cross-spectral density will, in general, be a complex quantity. It is further shown in Lin (1967) that a n-dimensional stochastic process, consisting of stochastic processes $\eta_1(t), \eta_2(t), \ldots,$ $\eta_n(t)$, is characterized by the cross-spectral density matrix

$$S = \begin{bmatrix} S_{\eta_1\eta_1}(\omega) & S_{\eta_1\eta_2}(\omega) & \ldots & S_{\eta_1\eta_n}(\omega) \\ \vdots & \ddots & & \\ S_{\eta_n\eta_1}(\omega) & S_{\eta_n\eta_2}(\omega) & \ldots & S_{\eta_n\eta_n}(\omega) \end{bmatrix}$$

Similarly to the autocorrelation function, higher order correlations can be defined. Analogously, the third order correlation is:

$$R_{xxx}(t_1, t_2, t_3) = E[x(t_1)x(t_2)x(t_3)] \tag{3.12}$$

For stationary signal, it will depend on time lags only:

$$R_{xxx}(\tau_1, \tau_2) = E[x(t)x(t + \tau_1)x(t + \tau_2)] \tag{3.13}$$

For $\tau_1 = \tau_2$, the correlation function gives the 3rd central statistical moment of x:

$$\mu_3(x) = R_{xxx}(0, 0) \tag{3.14}$$

This correlation function is symmetric with respect to the axis $\tau_1 = \tau_2$. Similarly, the fourth order correlation is

$$R_{xxx}(\tau_1, \tau_2, \tau_3) = E[x(t)x(t + \tau_1)x(t + \tau_2)x(t + \tau_3)] \tag{3.15}$$

3.1.2.3 Stochastical Properties of Turbulent Wind

Natural wind is highly turbulent and therefore producing fluctuating loads on structures. When characterizing the turbulent wind for standard wind engineering purposes, the following assumptions are made:

- the wind is stationary in the horizontal plane
- the wind direction does not change with the change of height above ground
- for along wind, neutral thermal stability is assumed.

The wind-excited forces and responses are normally decomposed in two orthogonal directions: parallel to the mean wind velocity $U(z)$ (alongwind), and orthogonal in lateral direction (crosswind or lift). The mean wind speed depends only on the height and macro-meteorological conditions. The velocity components at time t are defined as

- in the longitudinal direction $U(z) + u(x, y, z, t)$
- in the lateral direction $v(x, y, z, t)$
- in the vertical direction $w(x, y, z, t)$

where u, v and w are zero-mean fluctuating components of the instantaneous wind field. These components are defined by their standard deviations, power spectral density functions and co-spectra.

The turbulence intensity $I_u(z)$ in direction of the flow is given by

$$I_u(z) = \frac{\sigma_u}{U(z)} \tag{3.16}$$

Similarly, the turbulence intensity is defined in other two directions. According to Dyrbye and Hansen (1997), for flat terrain it can be approximated by $I_u(z) = \frac{1}{ln(z/z_0)}$, where τ_0 is an appropriate terrain roughness length. For simplification, also an average of the three turbulence intensities is employed, see Sockel (1994):

$$I_u = \frac{\sqrt{\frac{1}{3}(\overline{u^2} + \overline{v^2} + \overline{w^2})}}{U} \tag{3.17}$$

As specified below, the flow fluctuations are caused by superposition of different eddies, each related to an appropriate frequency. Therefore the fluctuation can be defined through the power spectrum by

$$\overline{\sigma_u^2} = \int_0^\infty S_n(f)df \tag{3.18}$$

Assuming a flat terrain, the standard deviations depend only on height above the ground while their values are maximal at the zero height and decrease with the height. It was experimentally shown by Davenport and other authors later, that this decrease is very small in the atmospheric boundary layer, i.e. up to the height of ordinary structures. Up to the heights of 200 m above ground, the standard deviations are assumed to be normally distributed. However, this distribution does not always describe the wind field structure appropriately. Weibull and extreme value distributions can also be assumed.

Many random variables have a distribution which can be closely approximated by *Gaussian* (normal) probability density function:

$$p_X(x) = \frac{1}{\sqrt{2\pi}\sigma} e^{-\frac{(x-\mu)^2}{2\sigma^2}} \tag{3.19}$$

An estimate of an extreme value of a Gaussian random variable X can be constructed as $X = \mu + k \cdot \sigma$

A gaussian distribution of a wind load, as well as of other physical phenomena, is often assumed due to the central limit theorem, which can be found with its proof in Lin (1967). The variables of this distribution are closed under linear operations, meaning that the linear functions of Gaussian variables stay normally distributed.

Another assumption for wind load (wind velocities) as input signal is *the assumption of stationarity*: the statistical properties (e.g. average, variance) are invariant during the process. This is shown to be accurate enough for a relevant time of observation e.g. over a period of 10 min.

When dealing with the sampled data and assuming a continuous distribution that will fit the distribution of the data, there will in general be a difference between the two, which can be estimated as a standard error. More samples of the same signal are taken, the mean value might have different values. This possible deviation of the sample mean from the whole signal mean is called standard error of the mean:

$$SE_{\mu} = \frac{\sigma}{\sqrt{N}} \tag{3.20}$$

For large numbers of measurement points, the standard error of the standard deviation can be approximated as

$$SE_{\sigma} = \frac{\sigma}{\sqrt{2(N-1)}} \tag{3.21}$$

When analyzing a response to a wind load, one is mostly interested in the extreme response that can occur in a given time range. This is a relevant information for the design of the structure. The extreme response depends on the characteristics of the process. If $X(t)$ is a Gaussian process, then the normalized process is described by

$$Y(t) = \frac{X(t) - \mu_X}{\sigma_X} \tag{3.22}$$

The peak value distribution for a time period T is characterized with its own mean $\mu_{Y,max}$ and standard deviation, and for a Gaussian process it converges to $\sigma_{Y,max}$ (see Dyrbye and Hansen 1997 and Grigoriu 2002)

$$\mu_{Y,max} = \sqrt{2 \ln \nu T} + \frac{2}{\sqrt{2 \ln \nu T}} \tag{3.23}$$

where $\gamma = 0.5772$ is the Euler constant, and νT is the expected number of zero-upcrossings during T. The zero upcrossing rate ν is given by Grigoriu (2002)

$$\nu = \sqrt{\frac{m_2}{m_0}}$$

The spectral moments, that were introduced by Vanmarcke in 1972, m_i are defined as

$$m_i = \int_0^\infty f^i S_Y(f) df$$

Calculated in this manner, $\mu_{Y,max}$ gives the peak factor k_p for the extreme value. For the non-normalized Gaussian process $X(t)$, it follows that the mean extreme value is

$$\mu_{X,max} = \mu_X + k_p \cdot \sigma_X \tag{3.24}$$

Kwon and Kareem (2009) presented peak factors for the estimation of extreme values of a stationary non-Gaussian process, which can, with some discussed restrictions, be applied to wind pressures. Additionally to the peak factor for a Gaussian process, this peak factor takes into account the skewness and the kurtosis of the process.

3.1.3 Spectral Analysis of Wind Processes

3.1.3.1 Wiener-Khinchin Theorem and Parseval's Equality Theorem

The Wiener-Khinchin theorem plays a major role in DSP. It states that the Fourier transform of a ACF of a random signal (or generally process) corresponds to the frequency function called power spectral density (power spectrum) $S(\omega)$ of the signal:

$$S_x(\omega) = \frac{1}{2\pi} \int_{-\infty}^{\infty} R_x(\tau) e^{-i\omega\tau} d\tau \tag{3.25}$$

Its Fourier inverse is given by

$$R_x(\tau) = \int_{-\infty}^{\infty} S_x(\omega) e^{i\omega\tau} \mathrm{d}\omega \tag{3.26}$$

For $\tau = 0$, it follows:

$$R_x(\tau = 0) = E[x^2] = \int_{-\infty}^{\infty} S_x(\omega) \mathrm{d}\omega \tag{3.27}$$

This means that the area under the curve of spectral density is equal to the second statistical moment, i.e. the variance of the process in question.

For discrete time signals it follows:

$$S_x(f) = \sum_{\tau=-\infty}^{\infty} R_x(\tau) e^{-i(2\pi f)\tau} \tag{3.28}$$

Equation 3.28 formally defines the power spectrum. However, with the possibilities to utilize FFT with modern software, it is common to compute a PSD directly from the the time series, as further explained.

The instantaneous power of a signal $x(t)$ at the time t is defined as the squared magnitude of the signal. The expected power of stationary signal is

$$P = E[x^2(t)] = \lim_{T\to\infty} \frac{1}{T} \int_{-T/2}^{T/2} x^2(t)dt$$

which is, in case of sampled signal with N samples, equal to

$$P = \frac{1}{T} \sum_{n=0}^{N-1} x_n^2(t) \approx \frac{1}{NT} \int_0^{NT} x^2(t)dt$$

In this case, P is just an estimation of the power of the signal. Further it can be shown (Stearns and Hush 2011) that the power in time and frequency domains are equal. This result is known as *Parseval's equality theorem*:

$$P_{avg} = \frac{1}{N} \sum_{n=0}^{N-1} x_n^2 = \frac{1}{N^2} \sum_{m=0}^{N-1} |X_m|^2 \tag{3.29}$$

P_{avg} defines the total or average power, equal to the integral of the power spectrum. The power density in this form can be applied to stationary signals of final length, from which segments (windows or blocks) are extracted for analysis. Also, in this case, energy, as an integral of power over time, can be used for analysis. If the frequency units are changed from Hz-s (as above, corresponding to the frequency step of $1/N$) to Hz or rad/s, the power density must be accordingly scaled (Stearns and Hush 2011). Assume $P_{xx}(m)$ is constant over m and equal to P. For different frequency notations, the following scaling must be applied: $P_{xx}(f) = TP$ and $P_{xx}(\omega) = TP/(2\pi)$.

Referring to the complex representation of non-periodic signals through Fourier series, for the signals with zero-mean, the variance is

$$\sigma^2 = \sum_{-\infty}^{\infty} \frac{a_n^2 + b_n^2}{2} = \sum_{-\infty}^{\infty} |c_n|^2 = \sum_{n=1}^{\infty} \Delta\sigma_n^2 \tag{3.30}$$

where

$$\Delta\sigma_n^2 = |c_n^2| = \frac{1}{T^2} |X_n(\omega_n)|^2 \tag{3.31}$$

Consider that

$$T = \frac{2\pi}{\Delta\omega_n}$$

and for the non-periodic signals $T \to \infty$ it follows that $\Delta\omega \to 0$ and from 3.31 that $\Delta\sigma \to 0$. Consequently, the total variance is

$$d\sigma^2 = \frac{1}{4\pi} \lim_{T\to\infty} \frac{2}{T} |X(\omega)|^2 d\omega \tag{3.32}$$

For a continuous random signal x the power spectral density function is expressed as

$$S(\omega) = \lim_{T\to\infty} \frac{2}{T} |X(\omega)|^2$$

Therefore the total variance is equivalent to the integral of the PSD:

$$\sigma^2 = \frac{1}{4\pi} \int_{-\infty}^{+\infty} S(\omega) d\omega \tag{3.33}$$

3.1.3.2 Cross-spectra of Two Random Signals

The cross-spectral analysis of two stochastic processes shows their statistical relation in the frequency domain. Equation 3.28 shows that the autospectrum is defined as the Fourier transform of the ACF. Similar to the power spectrum density, the cross-spectrum density is defined as the Fourier transform of the cross-correlation function of two random signals:

$$S_{xy}(2\pi f) = \sum_{\tau=-\infty}^{\infty} R_{xy}(\tau) e^{-i(2\pi f)\tau} \tag{3.34}$$

and

$$S_{yx}(2\pi f) = \sum_{\tau=-\infty}^{\infty} R_{yx}(\tau) e^{-i(2\pi f)\tau} \tag{3.35}$$

Similar to the Eq. 3.29, the cross-spectrum can be written in terms of Fourier transform as

$$S_{xy} = \frac{1}{N^2} \sum_{k=0}^{N-1} X_k^* Y_k$$

In complex notation this is

$$S_{xy} = c_{xk} c_{yk} e^{i(\alpha_{xk} - \alpha_{yk})} = c_{xk} c_{yk} (\cos(\alpha_{xk} - \alpha_{yk}) + i \sin(\alpha_{xk} - \alpha_{yk}))$$

The real part is called coincidence spectrum and it represents the in-phase signal, while the imaginary part, quadrature spectrum, represents the out-of-phase signal.

The coincidence (coherence) spectrum describes how highly correlated are the processes for the given frequency. The squared coherence estimates the percentage of variance of x that can be estimated from y and vice verse. The quadrature (phase) spectrum defines the phase relationship between the processes for a given frequency.

The condition for the cross-spectra to exist is that $x(t)$ and $y(t+\tau)$ are uncorrelated for $\tau \to \infty$ and that none of the signals is equal to zero. It can be shown (see Newland 1993) that $S_{xy}(f)$ and $S_{yx}(f)$ are complex-conjugates.

3.1.3.3 Estimation of Power Spectra

The measurement of wind velocities in the wind tunnels (or in real scale experiments) gives a recording of a part of the total time history of velocities. From such samples, it is possible to make an estimation of the power spectrum, especially considering that wind velocities are treated as random.

The Nyquist-Shannon Sampling Theorem describes properties of the conversion of continuous into sampled signal with discrete numerical values. It states that, for the discrete signal sampling rate f_s, the frequencies that can be evaluated in the frequency domain are in a range of f_s/N (frequency resolution) to $f_s/2$ (Nyquist frequency), where N is the length of the Fourier transform.

There are two common methods to obtain the power spectrum of a random signal. The first is the Blackman-Tuckey method. The second one, which uses FFT, is the Cooley-Tuckey method. This method obtains the power spectrum directly from the FFT of the time series. It was further improved by Welch 1967.

The power spectral density computed by the mentioned methods always leads to a statistical error. It represents only the estimation of the true power spectral density, which is a continuous integral. It is shown that the standard error of PSD is governed by a chi-square distribution. To obtain estimates with an acceptable statistical error, further averaging is necessary. This averaging can be done over frequencies or over parts of time series called windows (or blocks). In the first case, the PSD is averaged over m adjacent frequency intervals $\Delta f = f_r$ to obtain the average over frequency bandwidth $m \cdot f_r$. This averaging leads to a lower frequency resolution, which can be disadvantageous.

The second case is known as *ensemble averaging*. The signal (time series) is divided into smaller segments (windows) that may or may not overlap. Assume the standard error for one power spectrum of a signal to be equal to one. Then for an ensemble average of n windows, the error is is $\frac{1}{\sqrt{n}}$. This method is common and also known as PSD estimation of an averaged periodogram. Moreover, the power spectrum can be defined as a periodogram that has been smoothed by any smoothing functions (windows) to reduce the sampling error (Warner 1998). However, if the signal of a specific length can be divided into a greater number of windows, usually these windows are shorter, which leads to a more reliable PSD estimation, but with lower frequency resolution (Fig. 3.3). These two demands are in conflict, so the

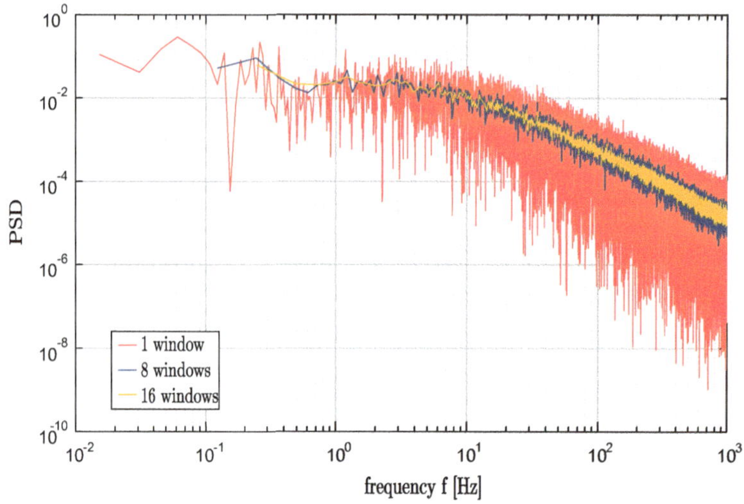

Fig. 3.3 Scatter reduction through ensemble averaging of PSD for different number of windows

balance should be found for every individual application. The choice of a specific window is a matter of a subjective appraisement.

The overlapping of the windows improves the power spectrum estimate. The use of windows results in a smoother PSD and it is especially recommended for stationary random signals with a broad spectrum, such as wind. There are a few main points for using overlapping. Firstly, it is an efficient way of averaging the signal, and it actually significantly reduces the averaging waiting time. Other than that, for window functions (other than rectangular), it reduces the error cause by the end shaping effects (Silva 1999). With overlapping, the computation power is more efficiently used and the statistical error is also reduced (Fig. 3.4).

Another known error of FFT is *leakage*. Leakage represents a distortion of the power spectrum where the power of one frequency seems to slip into adjacent frequency intervals. This error in the frequency domain happens when the signal in the time domain is divided (i.e, multiplied) into windows of finite length. The error that arises from the difference of a true and a windowed signal $X(t) - X(t) * W$ is called truncation error. This occurs as a FFT is computed over a finite frequency range (0 to $F_s/2$) and the series data is not periodic in the windowed time range. It causes undesired side lobes in the PSD which alters the shape of the spectrum. Leakage can be minimized by using windowing functions, such as Hann, Hamming, Blackman, Welch instead of simple rectangular ones. This is in detail discussed in the literature (i.e. Stearns and Hush 2011; Newland 1993).

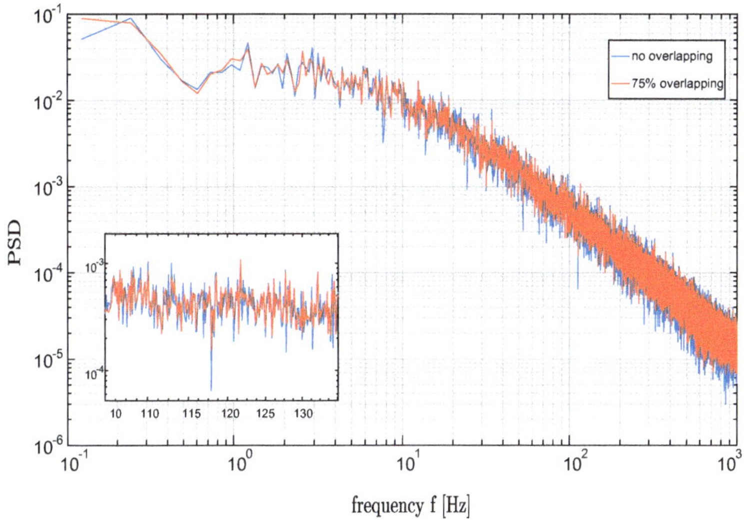

Fig. 3.4 PSD calculated with overlapping and not overlapping windows

3.1.3.4 Wind Velocity Spectra

As mentioned, with the assumption of a horizontally homogeneous flow, the standard deviations σ_x, σ_y and σ_z depend only on the height z.
The frequency distribution of u is described by the normalized non-dimensional power spectrum density function R_f, dependent on frequency $f[Hz]$ and height $z[m]$, of general form:

$$R_f(z, f) = \frac{f S_u(z, f)}{\sigma_u^2(z)} \tag{3.36}$$

In small scale physical simulations the wind power spectrum $S_u(z, f)$ is determined through wind velocities measurements in wind tunnel tests. Many different experimental relations are established for describing $S_u(z, f)$ of the longitudinal wind component, and often do not depend on the height z. The spectra of the other two turbulence components v and w can be determined in an equivalent manner. Davenport was one of the first to develop such an expression, which is independent of the height:

$$\frac{f S_u(f)}{u_*^2} = 4 \frac{f_1^{*2}}{(1 + f_1^{*2})^{4/3}} \tag{3.37}$$

where f_1^* is a non-dimensional frequency given by $f_1^* = \frac{1200 f}{\overline{u}(10)}$, and u_*^2 is a non-dimensional velocity u.

Kaimal (Dyrbye and Hansen 1997) proposed the relation:

$$\frac{\frac{2}{3}\lambda f_z}{(1+\lambda f_z)^{5/3}} \qquad \lambda = 50 \qquad f_z = \frac{2}{3\lambda} \tag{3.38}$$

A common and, according to Holmes (2001), rather precise relation was developed by von Karman and adopted by Harris as follows:

$$\frac{f S_u(f)}{\sigma_u^2} = \frac{4\frac{L_{ux} f}{U}}{\left[1 + 70.8(\frac{f L_{ux}}{U})^2\right]^{5/6}} \tag{3.39}$$

The Eurocode gives the following expression for a normalized spectrum:

$$\frac{f S_u(f)}{\sigma_u^2} = \frac{6.8 f_L}{(1+10.2 f_L)^{5/3}} \tag{3.40}$$

In the high frequency range, the large eddies are broken down into smaller ones, thus transferring energy to the latter through inertia. This frequency range is known as the inertial subrange. Here, the Kolmogorow hypothesis is applied:

$$\frac{f S_u(f)}{u_*^2} = 0.26 f^{*-2/3} \tag{3.41}$$

All the spectra normalized by the variance σ_u^2 and with an appropriate frequency normalization f_0 show a good accordance with the $f^{-2/3}$ line. In this case, all the spectra have the general shape

$$\frac{f S_u(f)}{\sigma_u^2} = \frac{0.16(f/f_0)}{1+0.16(f/f_0)^{5/3}} \tag{3.42}$$

where f_0 can be estimated as $1/4 f_{max}$ (see Sockel 1994).

For the MATLAB routines in the scope of this work, a Fichtl-McVehil spectrum (Eq. 3.43) was adopted for comparison to the values of a wind power spectrum obtained from measurements in wind tunnel tests:

$$\frac{f S_u(f)}{\sigma_u^2} = \frac{4 f T_{ux}}{(1+b(f T_{ux})^r)^{5/3r}} \tag{3.43}$$

Here, r, a and b are constants which are obtained empirically.

$$a = 1.5^{\frac{1}{r}} \cdot r \cdot \frac{\Gamma(\frac{5}{3}r)}{\Gamma(\frac{2}{3}r) \cdot \Gamma(\frac{1}{r})} \tag{3.44}$$

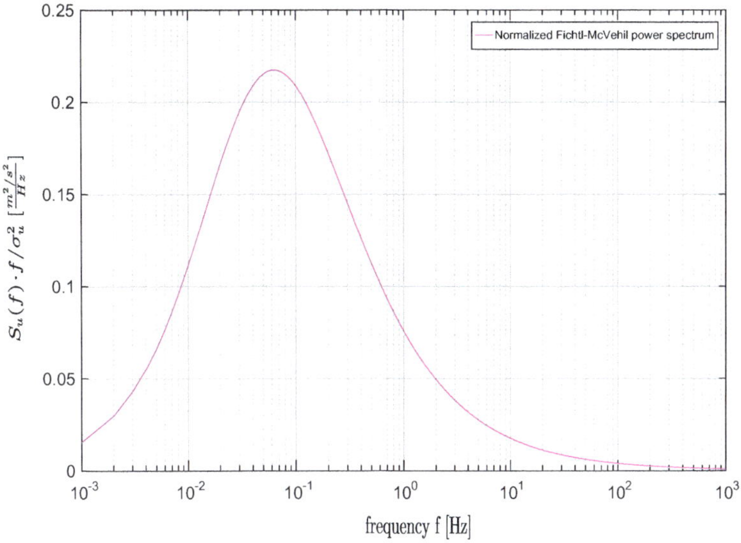

Fig. 3.5 Normalized Ficht-Mc Vehil spectrum, $r = 1$, $T_{ux} = 4$

$$b = 1,5(\frac{4}{a})^r \tag{3.45}$$

$$T_{ux} = \frac{L_{ux}}{U} \tag{3.46}$$

L_{ux} is the integral length of turbulent wind, and Γ is the Gamma-function. Depending on the value of the parameter r, it is possible to get different PSD curves: for $r = 2$, a von Karman curve is obtained; for $r = 1$, a Kaimal curve, and for $r = 0.845$, a Fichtl-McVehil curve. However, in order to obtain the most precise results, it is recommended by Schrader (1995) to use the value of $r = 1$ (Fig. 3.5).

3.2 Spectral Analysis of the Structural Gust Response

The spectral analysis of the dynamic responses of structures to along-wind excitation is normally conducted through Davenport's procedure, developed in the early 1960s. This procedure makes use of the quasi-static theory. It is assumed that the load fluctuations occur only due to the fluctuation in the turbulent wind, while the possible fluctuations induced by the structure are neglected. This is true for very stiff structures with very small displacements. For a flexible structure, where a dynamic response will be evoked, this approach is modified to a quasi-static approach in the frequency domain, resulting in Davenport's procedure. The turbulent wind is related to the

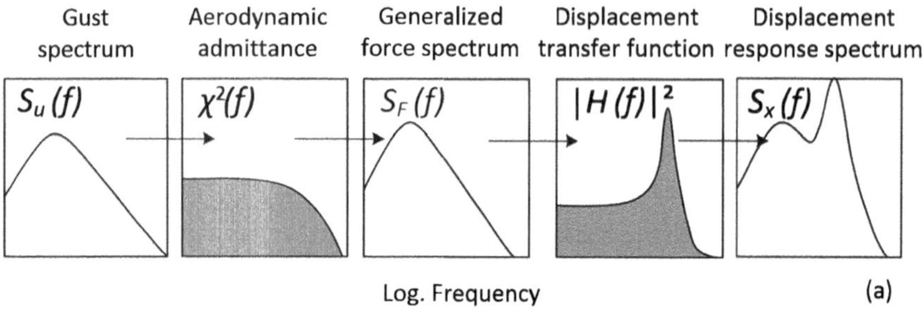

Fig. 3.6 Alan G. Davenport wind loading chain and spectral analysis of along-wind effects on structures. [Zhou and Kareem (2003)]

load by means of the aerodynamic admittance function. The load and the structural response are linked via the transfer (mechanical impedance) functions. Although the load is still assumed to be induced only by wind, the structure does not respond to all load frequencies evenly. The procedure is illustrated in the Fig. 3.6.

3.2.1 Quasi-stationary Load Models

The wind load on a structure can generally be divided into a time-constant mean part of static nature F_q, and into the turbulent null-mean fluctuating part F_t. The largest (characteristic) wind load at a point of a structure that can occur in a certain time period is

$$F_{max} = F_q + k_p\sigma_F \tag{3.47}$$

where k_p is a peak factor (most usual value 3–5) and σ_F is the standard deviation of a point load. The characteristic structural response can be expressed in an equivalent manner.

The use of the Davenport model (1962) of turbulent wind load on a structure in an atmospheric boundary layer is associated with the aerodynamic admittance functions to obtain the wind load on the structure. In this model, the total load F_{tot} on a point of a structure is given by

$$F_{tot} = C_D \cdot A\frac{1}{2}\rho\, u_{tot}^2 = F_q + F_t \tag{3.48}$$

where the drag coefficient C_D depends on the Reynolds number. The total velocity vector u_{tot} is of the intensity

$$u_{tot}^2 = (U + u)^2 + v^2 + w^2 = U^2 + 2\,Uu + u^2 + v^2 + w^2 \tag{3.49}$$

Since the mean wind velocity is usually significantly larger than the turbulent components, the squared fluctuation terms in the previous equation contribute only with a few percentage to the total load and can be neglected, leading to the linearisations:

$$u_{tot}^2 = U^2 + 2\,U u \tag{3.50}$$

$$F_q = C_D A \frac{1}{2} \rho U^2 \tag{3.51}$$

$$F_t = C_D A \rho U u \tag{3.52}$$

In case of a Gaussian distribution of the wind speed fluctuations, the load distribution will as well be Gaussian. Different authors have investigated the influence of these terms to the total loading, where the distribution becomes non-Gaussian, see Gusella and Materazzi (1998), Holmes (1981), Kareem and Tognarelli (1998).

3.2.2 *Aerodynamic Admittance Function*

Except for structures with rather small wind exposed surfaces, the extreme wind load can not be calculated by a simple summation of wind pressures at different position of the structures. If the relevant dimension of the structure is greater than the size of the eddies in the oncoming turbulent wind, which is mostly the case, the pressures on the surfaces are not fully correlated. The proportion of the gust, that generates the surface pressure on the structure, depends on the ratio of the size of eddies and the dimension of structure. The capability of the structure to "accept" the wind load is in the frequency domain denoted as the aerodynamic admittance function χ^2 as introduced by Davenport and further investigated by Vickery (Tamura and Kareem 2013). It can also be defined as the ratio of the pressure coefficient at a given frequency to the one at zero frequency.

Let the ratio

$$\lambda = \frac{U}{f}$$

be denoted as the gust wavelength. Let $\sqrt{A}$ be the characteristic dimension of the structure, where A is the wind influenced area of the structure. Then, in general, the aerodynamic admittance function depends on the ratio $\frac{\lambda}{\sqrt{A}}$. Consequently, the dynamic response of the structure is affected.

Davenport proposed a relation that takes into account the dimensions of the structure, as well as the integral length scale (Stengel 2015):

$$|\chi(f)|^2 = \frac{1}{\left(1 + \frac{C_{uz}}{3}\frac{h \cdot f}{U}\right)\left(1 + \frac{C_{uy}}{2}\frac{b \cdot f}{U}\right)} \tag{3.53}$$

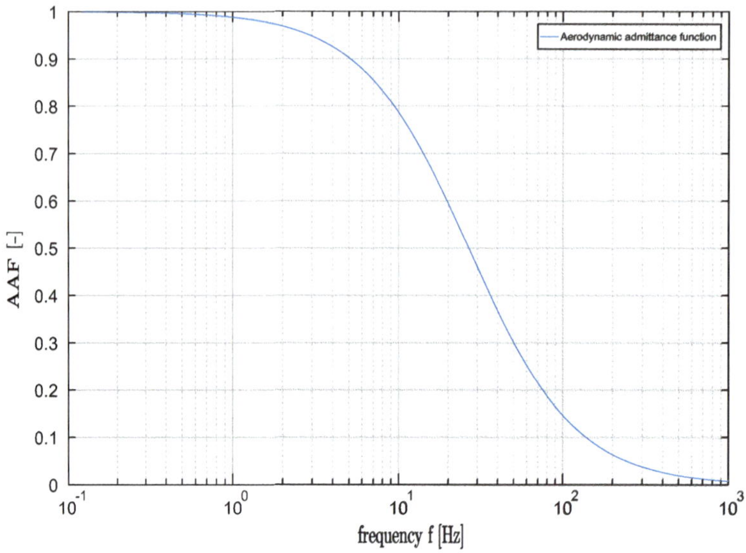

Fig. 3.7 Aerodynamic admittance function (after Vickery)

with $C_{uy} = 20$ and $C_{uz} = 8$.

Vickery introduced the lattice-plate theory which also models the lack of correlation of wind pressures, Eq. 3.54. He introduced the aerodynamic admittance function, though it is limited to the sharp-edged prismatic objects with the area A, and is therefore not generally applicable:

$$|\chi(f)|^2 = \left(1 + \frac{2f\sqrt{A}}{U}\right)^{-2} \tag{3.54}$$

He improved his model with an empirical formula that is applicable to different shapes of bluff bodies and depends of the dimension of the body D, which can be nominally taken as $D = \sqrt{A}$:

$$|\chi(f)|^2 = \left(1 + (\frac{2fD}{U})^{4/3}\right)^{-2} \tag{3.55}$$

where it is assumed that the pressures are fully correlated in horizontal direction (Fig. 3.7).

Velozzi and Cohen introduced a model that accounts for the correlation in all three directions:

$$|\chi(f)|^2 = \left(\frac{1}{\zeta} - \frac{1}{2\zeta^2}(1 - e^{-2\zeta})\right)\left(\frac{1}{\gamma} - \frac{1}{2\gamma^2}(1 - e^{-2\gamma})\right)\left(\frac{1}{\mu} - \frac{1}{2\mu^2}(1 - e^{-2\mu})\right) \tag{3.56}$$

The correlation coefficients ζ, γ and μ are given by

$$\zeta = \frac{3.85 f \Delta x}{U*}$$

$$\gamma = \frac{11.5 f \Delta y}{U*}$$

$$\mu = \frac{3.85 f \Delta z}{U*}$$

$$U^* = \frac{1}{H}\int_0^H U(z)dz$$

Hölscher proposed a model based on wind tunnel experiments on circular cylinders:

$$|\chi(f)|^2 = \frac{\beta}{\left(1 + (\gamma \frac{fD}{U})^2\right)^{\delta/6}} \tag{3.57}$$

where β, γ and δ are modifiable parameters (see also Aldasoro 2014).

By accounting the aerodynamic admittance function, the power spectrum of wind load becomes (Fig. 3.8):

$$S_F(f) = \frac{4F_q^2}{U^2} \cdot \chi^2 S_u(f) \tag{3.58}$$

3.2.3 Transfer Functions

The response of the SDOF in the frequency domain is obtained as

$$x(t) = \frac{1}{2\pi}\int_{-\infty}^{\infty} H(i\omega)c(i\omega)\mathrm{e}^{i\omega t}\mathrm{d}\omega \tag{3.59}$$

where $H(i\omega)$ is the complex transfer or frequency-response function, and $c(i\omega)$ is the Fourier transform of forcing function $p(t)$:

$$c(i\omega) = \int_{-\infty}^{\infty} p(t)\mathrm{e}^{-i\omega t}\mathrm{d}t$$

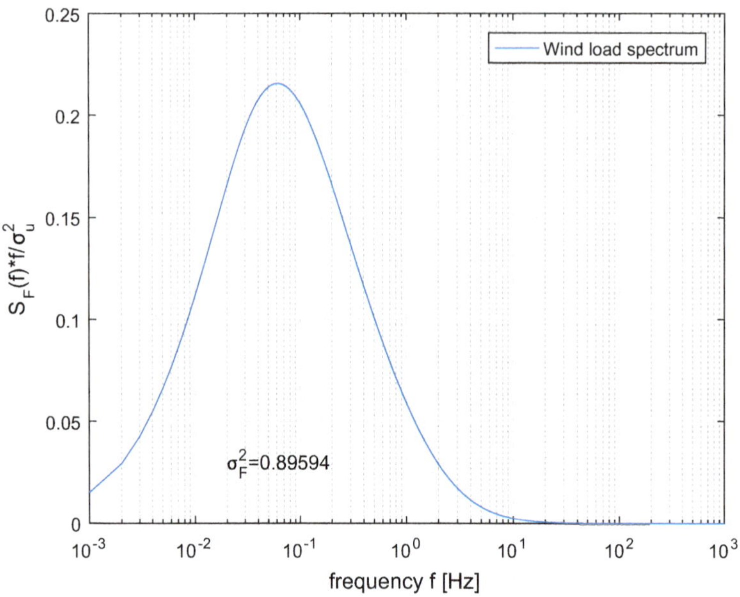

Fig. 3.8 Load spectrum derived from a Ficht-McVehil velocity spectrum

The inverse Fourier transform is

$$p(t) = \frac{1}{2\pi} \int_{-\infty}^{\infty} c(i\omega) \mathrm{e}^{i\omega t} \mathrm{d}\omega \tag{3.60}$$

Since $p(t)$ consists of harmonics over the whole time range, it follows from Eq. 3.59

$$x(t) = H(i\omega) c(i\omega) \mathrm{e}^{i\omega t} \tag{3.61}$$

One obtains the transfer function as

$$\begin{aligned} H(\omega) &= \frac{1}{-\omega^2 m + i\omega c + k} \\ &= \frac{1}{k} \frac{1}{(1 + 2i\xi\beta - \beta^2)} \end{aligned} \tag{3.62}$$

where $\beta = \omega/\omega_n$. It depends on the load frequency and structural properties. Its magnitude is

$$|H(\omega)| = \frac{1}{k\left((1-\beta^2)^2 + (2\xi\beta)^2\right)}$$

For frequencies $f = \omega/(2\pi)$, the transfer function becomes

$$H(f) = \frac{1}{4\pi^2 k} \cdot \frac{1}{1 - (f/f_n)^2 - 2i\xi(f/f_n)}$$

with the magnitude

$$|H(f)| = \frac{1}{4\pi^2 k} \cdot \frac{1}{(1 - (f/f_n)^2)^2 + (2\xi(f/f_n))^2} = \frac{1}{4\pi^2}|H(\omega)|$$

If the periodic loading would be rewritten for discrete frequencies ω_n, the result would be

$$p(t) = \sum_{n=-\infty}^{\infty} c_n \mathrm{e}^{i\omega_n t} \tag{3.63}$$

where

$$c_n = \frac{1}{T_p} \int_0^{T_p} p(t)\mathrm{e}^{-i\omega_n t}\mathrm{d}t$$

It can be seen that $H(\omega)$ has a complex conjugate in $H(-\omega)$. Therefore, for discrete frequencies ω_n, it is possible to obtain the steady state response by means of superposition as

$$x(t) = \sum_{n=-\infty}^{\infty} H(\omega_n)c_n \mathrm{e}^{i\omega_n t} \tag{3.64}$$

The transfer function applies to any quantity that depends linearly on the load:

$$x_{dyn} = |H(f)| \cdot x_{st}$$

3.2.4 Response Spectrum

The dynamic response of the structure is divided into two part: the mean and the instantaneous component

$$x = X + x'$$

where X is obtained as the response to a quasi-static mean wind load, while the fluctuating component x' is calculated by means of spectral analysis. The response spectrum is computed as

$$S_x = |H(f)|^2 S_F(f) \tag{3.65}$$

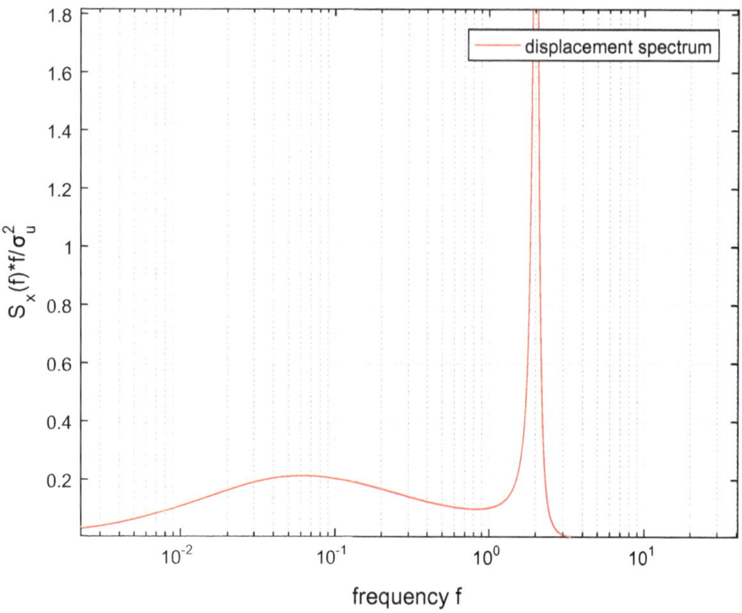

Fig. 3.9 Response spectrum of a single degree of freedom oscillator exposed to modal loading according to a Fichtl-McVehil spectrum

An example of the response spectrum is given in Fig. 3.9 with the following parameters: $\xi = 0.05,\ f_1 = 2Hz, U = 16\frac{m}{s}$

By integrating the previous equation, one obtains

$$\sigma_x^2 = \int_0^\infty S_x(f)\mathrm{d}f = \int_0^\infty \frac{4X^2}{U^2}|\chi(f)|^2|H(f)|^2 S_{uu}(f)\mathrm{d}f \tag{3.66}$$

This mean square fluctuating response can be approximated by background and resonance components 3.10:

$$\sigma_x^2 = \frac{4X^2}{U^2}\int_0^\infty |\chi(f)|^2|H(f)|^2 S_{uu}(f)\mathrm{d}f = \frac{4X^2}{U^2}(\sigma_B^2 + \sigma_R^2) = \frac{4X^2}{U^2}[B + R] \tag{3.67}$$

The background component B represents the response due to low frequent turbulent loads, which are assumed to not inducing any resonant response.

$$B = \int_0^\infty |\chi(f)|^2 \frac{S_u(f)}{\sigma_u^2}\mathrm{d}f$$

$$R = |\chi(f)|^2 \frac{S_u(f)}{\sigma_u^2}\int_0^\infty |H(f)|^2\mathrm{d}f$$

The integral $\int_0^\infty |H(f)|^2 \mathrm{d}f$ is shown to be equal to $\frac{\pi f_1}{4\xi}$ (Holmes 2001). The approximation of the spectrum given by (3.67) is used in different codes for the evaluation of the structural response.

3.2.5 Implementation of Spectral Analysis in the Eurocode Procedure

The wind code as part of Eurocode 1 (EN 1991-1-4, 2010) defines the wind actions on structures relevant for the design practice in Europe. It belongs to a part of *Eurocode 1* which defines general actions on structures. The structural design is carried out with respect to the serviceability limit state (SLS) and the ultimate limit state (ULS). EN 1991-1-4 provides a calculational procedure which covers the structural response to the along-wind turbulence, where only the first mode (with constant signs) is considered. It refers to structures of heights up to 200 m, as well as bridges with spans up to the same length. It does not address the problem of torsional vibrations, bridge deck vibrations from transverse wind turbulence, or cable supported bridges, nor the aeroelastic responses. Guyed masts and lattice towers are treated separately, namely in Eurocode 1993-3-1.

After the Eurocode, a wind force is calculated through vector summation of wind pressures over the respected area, or by applying the wind force coefficients described in the Eurocode. It has been noticed that the latter approach will give more accurate results (Vrouwenvelder and Steenbergen 2005). Whichever approach is applied, surface pressures should always be calculated for the design of structural elements (such as cladding or its supporting elements).

The Eurocode uses the aerodynamic admittance function based on the one proposed by Velozzi and Cohen (Eq. 3.56):

$$|\chi(f)|^2 = R_y R_z = \left(\frac{1}{\eta_B} - \frac{1}{2\eta_B^2}(1 - e^{-2\eta_B})\right)\left(\frac{1}{\eta_H} - \frac{1}{2\eta_H^2}(1 - e^{-2\eta_H})\right)$$

where the non-dimensional coefficients η_B and η_H of the correlation in across-wind and vertical direction respectively are calculated at the reference height, taken as 60% of the full height of the structure:

$$\eta_B = \frac{K_y C_y f B}{U(z_s)} = \frac{0.4 \cdot 11.5 \cdot f \cdot B}{U(z_s)}$$

$$\eta_H = \frac{K_z C_z f H}{U(z_s)} = \frac{0.4 \cdot 11.5 \cdot f \cdot H}{U(z_s)}$$

The formulation of the aerodynamic admittance function neglects the correlation in along-wind direction. Vickery showed in his work that this assumption is acceptable.

The peak velocity pressure q_p depends on the mean wind velocity q_m and short-time velocity fluctuation as follows:

$$q_p(z, z_0) = \frac{1}{2}\rho v_p^2 = (1 + 7I_v(z, z_0))\frac{1}{2}\rho v_m^2$$

$$q_p(z, z_0) = (1 + 7I_v(z, z_0)) \cdot q_m(z, z_0)$$

where z_0 represents the terrain roughness, and I_v is the turbulence intensity.

The resonance peak is accounted for through the dynamic factor c_d, and the decay of correlation with distance through the size factor c_s, which are combined into the structural factor $c_d c_s$. Wind forces are determined with the aerodynamic force coefficient c_f.

$$F_w = c_s c_d c_f q_p(z_e) A_{ref}$$

where A_{ref} is the reference area, z_e is the reference height for the calculation of the aerodynamic force coefficient.

The structural factor $c_s c_d$, discussed in Sect. 6.1, accounts for "the effects of wind actions from the n on-simultaneous occurrence of peak wind pressures on the surface (c_s) together with the effect of vibration of structures due to turbulence (c_d)" (EN 1991-1-4, 2010). It might be taken as equal to 1 for special cases (buildings with height less than 15 m, facade and roof elements with natural frequency greater than 5 Hz etc.). It is explicitly defined in an Annex for the multistory steel and concrete buildings and chimneys without liners, see Sect. 6.2.

In case that only the first (fundamental) mode vibrations are significant, the structural factor is calculated according to the detailed procedure described in EN 1991-1-4. In this case, it may be decomposed into a size factor c_s and a dynamic factor c_d. Its calculation is based on Davenport's procedure and the gust response factor G. The gust response refers to the mean velocity pressure q_m, and the EC factor $c_d c_s$ refers to the peak velocity pressure, which must yield the same result, therefore:

$$c_d c_s q_p = G \cdot q_m$$

The gust factor contains background and resonant excitation contributions to the structural response (B and R), which are illustrated in Fig. 3.10. Its value is assumed as

$$G = 1 + 2k_p I_v \sqrt{B^2 + R^2}$$

where k_p is a peak factor. It follows

$$c_d c_s = \frac{1 + 2k_p I_v \sqrt{B^2 + R^2}}{1 + 7I_v} \tag{3.68}$$

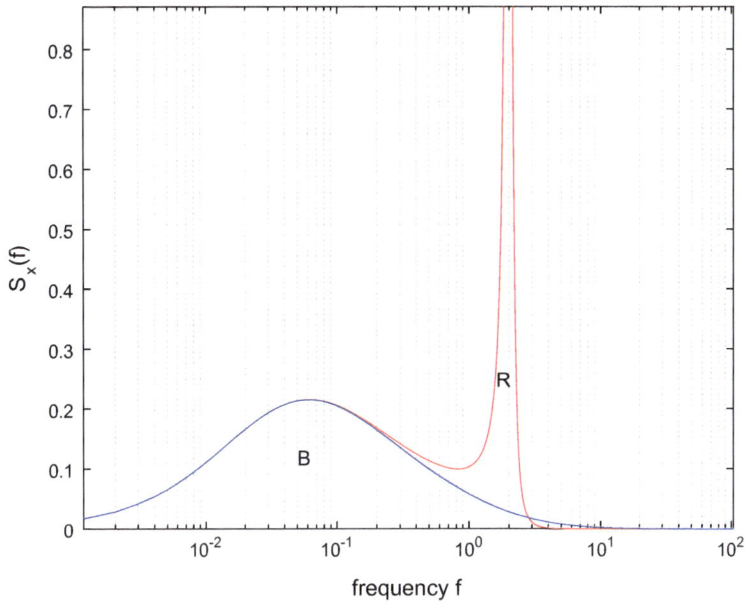

Fig. 3.10 Background and resonant part of a response spectrum of a single degree of freedom oscillator exposed to modal loading according to a Ficht-McVehil spectrum

The size factor c_s takes into consideration the reduction of the wind load due to the non-simultaneous occurrence of peak pressures at different positions.

$$c_s = \frac{1 + 7I_v\sqrt{B^2}}{1 + 7I_v}$$

The dynamic factor c_d takes into consideration the increased vibration amplitudes in case of resonance (due to turbulence).

$$c_d = \frac{1 + 2k_p I_v(z_s)\sqrt{B^2 + R^2}}{1 + 7I_v\sqrt{B^2}}$$

Eq. 3.68 can be applied if specific conditions are met, namely that only the first mode is significant (while the higher modes are neglected) and that it has a constant sign. Otherwise, a detailed procedure must be applied.

3.2.6 *Modal Analysis for Structural Response Due to the Wind Loading*

3.2.6.1 Linear N-degree of Freedom Systems

Consider a linear N-degree of freedom system, which is simultaneously excited with stochastic loads $p_n(t)$ at each degree of freedom, i.e. a lumped mass system. Since the system is linear, the total response will be a superposition of the responses to the single loads. The behavior of the system in different modes is defined by different transfer functions. These are in general defined as the ratio of the Fourier transforms of the response and load amplitudes for different frequencies ω at the same DOF:

$$H_n(\omega) = \frac{F_{xn}(\omega)}{F_{pn}(\omega)} = \frac{|F_{xn}(\omega)|}{|F_{pn}(\omega)|}\mathrm{e}^{-i\phi_n(\omega)} \tag{3.69}$$

where $\phi_n(\omega)$ is a frequency-dependent phase between excitation and response.

The Fourier transform of the response is given by:

$$F_x(\omega) = \int_0^T (x_1(t) + \cdots + x_N(t))\mathrm{e}^{-i\omega t}\mathrm{d}t = \sum_{n=1}^{N} H_n(\omega)F_p(\omega) \tag{3.70}$$

In many cases, it is hard to find the transfer function of the continuous system. For this reason, an approximate method for its calculation has been derived, based on the experimental investigations that have shown that, in many cases, only a limited number of natural modes contributes to the response of the system to the random excitation.

Rayleigh was the first to show in his work that undamped vibrating linear systems can undergo modal motions, and the same concept is extended to damped systems, with the introduction of Rayleigh damping. This means that the motion is simply a superposition of the harmonic oscillations at a specific frequency ω. Each of these harmonic components form a displacement pattern called a normal mode, and the corresponding frequency of occurrence is called the modal frequency ω_j. As it is already known, these properties are obtained from the matrix eigenvalue problem of the system:

$$det\left(K - \omega_j^2 M\right) = 0 \tag{3.71}$$

where K and M are stiffness and mass matrix of the system, respectively. Solving this equation for eigenfrequencies gives eigenvectors ϕ_j, describing the displacement pattern of nodes 1 to N for the mode ***j***

$$\phi_j = \left[\phi_j(1) \ldots \phi_j(N)\right]^T \tag{3.72}$$

which are orthogonal. The eigenfrequencies are determined as

$$\omega_j = k_j / m_j \tag{3.73}$$

where k_j and m_j are modal stiffness and modal mass, respectively. Since the modes are here considered orthogonal and uncoupled, the transfer function for mode $\boldsymbol{j}$ depends on modal properties of that mode only:

$$H_j(\omega) = \frac{1}{m_j\left(\omega_j^2 - \omega^2 + 2i\xi_j\omega_j\omega\right)} \tag{3.74}$$

This transfer function controls the contribution of the mode **j** to the total response at given point. Any of the modal transfer functions consists of a real and an imaginary part, that can be depicted as:

$$H_j = |H_j|(\cos\alpha_j - i\sin\alpha_j) \tag{3.75}$$

where α_j is a phase shift between the excitation and the response, equal to

$$\alpha_j = \frac{2\xi_j\omega\omega_j}{\omega_j^2 - \omega^2} \tag{3.76}$$

The magnitude of the modal transfer function is given by

$$|H_j| = \frac{1}{\omega_j^2 m_j\sqrt{(1 - (\omega/\omega_j)^2)^2 + 4\xi_j^2(\omega/\omega_j)^2}} \tag{3.77}$$

With the transfer functions defined, the spectrum of the response can be estimated. Assume that the system is subjected to a stochastic load acting at each node i. The generalized load vector for mode n becomes

$$P_n(t) = \sum_{i=1}^{N}\phi_n(i)P(i,t) = \phi_n^T P(t) \tag{3.78}$$

When the load $P_n(t)$ is a Gaussian process at each point, the modal cross-spectrum matrix for $P_m(t)$ and $P_n(t)$ becomes

$$S_{p_m p_n}(\omega) = \sum_{i=1}^{N}\sum_{k=1}^{N}\phi_m(i)\phi_n(k)S_{p_i p_k}(\omega) = \phi_m^T S_{pp}\phi_n \tag{3.79}$$

The response spectrum S_x', which is a function of the frequency ω will then be found as

$$S'_x = \sum_m \sum_n S'_{x_m x_n} \tag{3.80}$$

where $S_{x_m x_n}$ is the cross-spectrum for modal responses x_m and x_n.

$$S'_{x_m x_n} = \phi_m \phi_n H^*_m H_n S_{p_m p_n} \tag{3.81}$$

$$S'_x = \sum_n \sum_m H^*_m H_n S_{P_m P_n} \tag{3.82}$$

Observe the product of the complex conjugate of one transfer function with the other one. If the transfer functions are written in complex form as in 3.75, then it follows

$$\begin{aligned} H^*_m H_n &= |H_m|(\cos(\alpha_m) + i\,\sin(\alpha_n)) \cdot |H_n|(\cos(\alpha_n) + i\,\sin(\alpha_n)) \\ &= |H_m||H_n|(\cos(\alpha_m - \alpha_n) + i\,\sin(\alpha_m - \alpha_n)) \end{aligned} \tag{3.83}$$

$$S'_x = \sum_m \sum_n \phi_m H^*_m \phi_n H_n \sum_j \sum_k \phi_m(j)\phi_n(k) S_{jk} \tag{3.84}$$

The load cross-spectrum can be as well divided into the real (coincidence) Co and the imaginary (quadrature) spectrum Qu

$$S_{p_j p_k} = Co_{jk} - i\,Qu_{jk} \tag{3.85}$$

where for $j = k$ an autospectrum is obtained for which the imaginary part vanishes. Inserting Eqs. 3.83 and 3.85 into the Eqs. 3.84, it yields

$$\begin{aligned} S'_x = \sum_m \sum_n \phi_m \phi_n |H_m||H_n| \sum_j \sum_k \phi_m(j)\phi_n(k)\big(\cos(\alpha_m - \alpha_n) \cdot Co_{jk} + \\ + \sin(\alpha_m - \alpha_n) \cdot Qu_{jk} + i(\sin(\alpha_m - \alpha_n) \cdot Co_{jk} - \cos(\alpha_m - \alpha_n) \cdot Qu_{jk})\big) \end{aligned} \tag{3.86}$$

The real part of the response spectrum is then

$$\begin{aligned} S_x = \sum_m \sum_n \phi_m \phi_n |H_m||H_n| \sum_j \sum_k \phi_m(j)\phi_n(k)\big(\cos(\alpha_m - \alpha_n) \cdot Co_{jk} + \\ + sin(\alpha_m - \alpha_n) \cdot Qu_{jk}\big) \end{aligned} \tag{3.87}$$

It can be noticed that both, the real and the imaginary part of the excitation spectra, contribute to the real response spectrum. It can be shown that in the case of small damping and clearly separated eigenfrequencies, the cross terms of the transfer

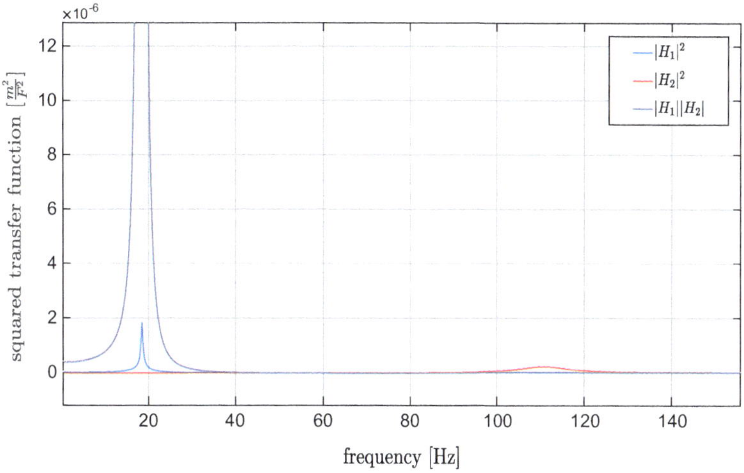

Fig. 3.11 Cross H component and modal components of the transfer function

functions have a significantly smaller influence than the one of the resonant mode for the given eigenfrequency ω_m:

$$|H_m(\omega_m)||H_n(\omega_m)| << |H_m(\omega_m)|^2 \tag{3.88}$$

This is illustrated in Fig. 3.11. Further, as illustrated in Fig. 3.12, the difference in phase shifts for 2 different modes is zero apart from resonance. Since the quadrature spectrum of the excitation is weighted with the sinus function of this difference, its contribution can be neglected.

Introducing these two approximations, the spectrum of the response is obtained as:

$$S_{rr}(\omega) = \sum_m \phi_m^2 |H_m|^2 \sum_j \sum_k \phi_m(j)\phi_m(k) \cdot Co_{jk} \tag{3.89}$$

This Equation shows that the modal responses are independent and that the total response is a superposition of separate modal responses.

With the assumption of Rayleigh damping, the MDOF system can be analyzed by means of modal analysis analogously to the undamped system. However, it was shown in different studies that this assumption is often incorrect, leading to inaccurate results. The real systems often show non-Rayleigh (non-classical) damping, and some or all of the modal equations of motions are coupled. This leads to complex, and not real eigenfrequencies and eigenvectors, which are no longer orthonormal as defined for classical damping.

When carrying out modal analysis, it is assumed that the mode shapes are ideal, which is not always the case, especially systems exposed to wind load. Only a few

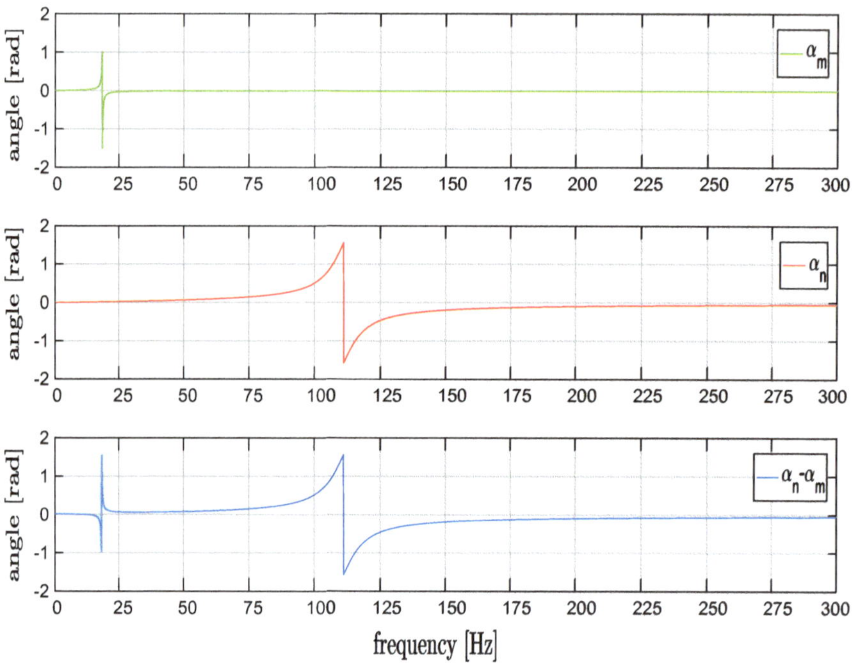

Fig. 3.12 Phase shifts of two transfer functions and differences

studies have been done to account for this non-ideal modes. One of the required correction procedures is presented by Zhou and Kareem (2002).

3.2.6.2 Application Scheme for the Spectral Analysis of a Cantilever

The wind force $P'(t) = c_F(f) \cdot A \cdot q'(t)$ in a turbulent flow depends on

- the area A and the aerodynamic shape of the application area of the load,
- the actual dynamic pressure,
- the frequency f of the fluctuations of the velocities,

$c_F(f)$ is the frequency dependant force coefficient. The normalization of the equation with reference to the mean load results into

$$\frac{P'(t)}{\overline{P}} = \frac{c_F(f)}{c_F(0)} \cdot \frac{q'(t)}{\overline{q}}, \quad or \quad \frac{P'(t)}{\overline{P}} = 2 \cdot \chi_F \cdot \frac{u'}{\overline{u}} \tag{3.90}$$

if the definition of the aerodynamic transfer function

$$\chi_F = \frac{c_F(f)}{c_F(0)}$$

and the linearisation

$$\frac{q'}{\overline{q}} = 2 \cdot \frac{u'}{\overline{u}}$$

are used. The transformation in the frequency domain results into the following SDF:

$$\frac{S_P}{\overline{p}^2} = 4 \cdot \chi_F^2 \cdot \frac{S_u}{\overline{u}^2}, \quad \text{or} \quad \frac{S_P \cdot f}{\sigma_P^2} = \left(\frac{2 \cdot I_u}{I_P}\right)^2 \cdot \chi_F^2 \cdot \frac{S_u \cdot f}{\sigma_u^2} \tag{3.91}$$

if the following normalizations are used:

$$I_u = \frac{\sigma_u}{\overline{u}}, \; I_P = \frac{\sigma_P}{\overline{P}}.$$

The aerodynamic transfer function χ_F has the same meaning for the dynamic excitation as the aerodynamic coefficient has for the static wind load. In general, it has to be determined by experiments. χ_F depends on the position of the considered point and on the ratio $\frac{L}{\lambda}$. $\lambda = \frac{\overline{u}}{f}$ is the wave length, and L_A is a characterizing length of the loaded area A, $L_A = \sqrt{A}$. For $\frac{f \cdot \sqrt{A}}{\overline{u}} \to 0$ is $\chi \to 1$ and for $\frac{f \cdot \sqrt{A}}{\overline{u}} \to \infty$ is $\chi \to 0$. The following estimation was empirically determined for sharp-edged bodies (Fig. 3.13):

$$\chi_F = \frac{1}{1 + \left(2 \cdot \frac{f \cdot \sqrt{A}}{\overline{u}}\right)^{\frac{4}{3}}} \tag{3.92}$$

Setting $\chi_F = 1$ as an estimation on the safe side it follows that $\frac{2 \cdot I_u}{I_P} = 1$, which results into:

$$\frac{S_P \cdot f}{\sigma_P^2} = \frac{S_u \cdot f}{\sigma_u^2}. \tag{3.93}$$

The cross-SDF of two wind forces P_1 and P_2 is complex. It can be estimated from the cross-SDF of the wind velocities u_1 and u_2.

$$\frac{S_{P12}(if)}{\overline{P}_1 \cdot \overline{P}_2} = 2 \cdot \chi_{F1}(f) \cdot 2 \cdot \chi_{F2}(f) \cdot \frac{S_{u12}(if)}{\overline{u}_1 \cdot \overline{u}_2}$$

The square contribution of a cross-SDF is real:

$$|S_{12}|^2 = S_{12} \cdot S_{12}^* = Co_{12}^2 \cdot Qu_{12}^2$$

so that

$$S_{12}(if) = |S_{12}| \cdot \mathrm{e}^{i \cdot \Phi_{12}},$$

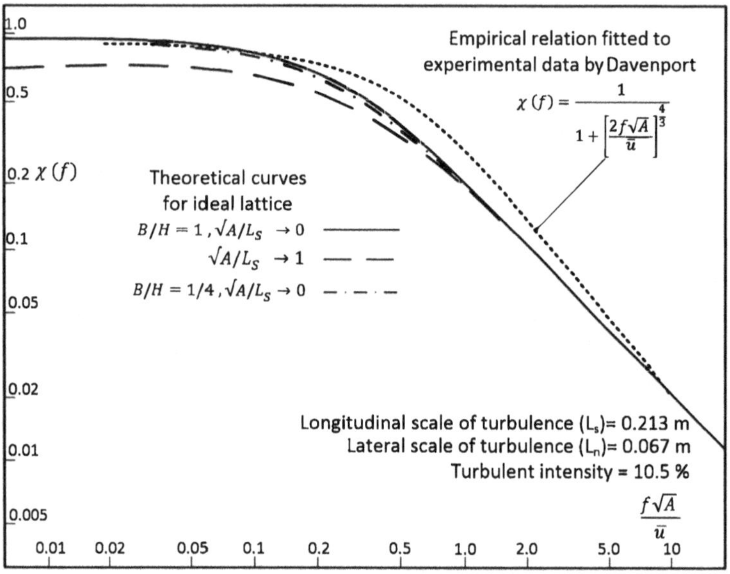

Fig. 3.13 Empirical aerodynamic transfer function by Vickery based on experiments conducted by (Vickery, 1966) and (Davenport, 1962)

where the phase is a function of the frequency:

$$\Phi_{12} = \Phi_{12}(f)$$

The standardisation of the modulus yields the coherence:

$$\gamma_{12}(f) = \frac{|S_{12}|}{\sqrt{S_1 \cdot S_2}}$$

The cross-SDF of the velocity is expressed by modulus and phase:

$$S_{u12} = \gamma_{u12} \cdot \sqrt{S_{u1} \cdot S_{u2}} \cdot \mathrm{e}^{i \cdot \Phi_{12}}$$

Neglecting the height dependency of S_u: $S_{u1} = S_{u2} = S_u$, $S_{u12} = \gamma_{u12} \cdot \mathrm{e}^{i \cdot \Phi} \cdot S_u$ and applying the approximation: $\chi_{F1} = \chi_{F2} = 1$ and $\sigma_{u1} = \sigma_{u2} = \sigma_u$ yields:

$$\frac{S_{P12} \cdot f}{\sigma_{P1} \cdot \sigma_{P2}} = \gamma_{u12} \cdot \mathrm{e}^{i \cdot \Phi_{12}} \cdot \frac{S_u \cdot f}{\sigma_u^2}$$

The coherence and the phase between point 1 and 2 in Fig. 3.14 can be approximated by

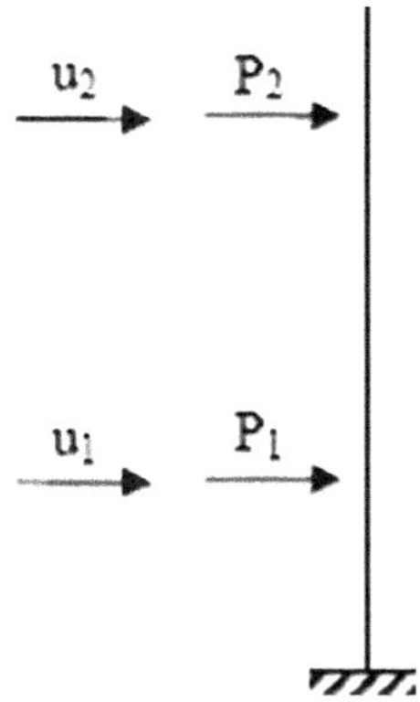

Fig. 3.14 Wind fluctuations and force fluctuations

$$\gamma_{u_j u_k} = \mathrm{e}^{-c_z \cdot \frac{f \cdot |\Delta z|}{\overline{u}_m}},$$
$$\Phi_{12}(f) = -c_z' \cdot \Delta z \cdot \frac{f}{\overline{u}_m}.$$

with

$$\Delta z = z_j - z_k \tag{3.94}$$

and the average velocity

$$\overline{u}_m = \frac{1}{2}\left(\overline{u}_j + \overline{u}_k\right) \tag{3.95}$$

j, k being points of observation with vertical distance
c_z: empirical constant depending on terrain, e.g. open terrain $c_z = 7$. The sign of the phase can be positive or negative which depends on whether points above or below are related to the point of reference. Hereby, the groundswell due to an arriving gust and the tailing off after of the gust body can be taken in consideration. The empirical constant c_z' depends also on the terrain. A value of $c_z' = 2.6$ can be assumed for open terrain (Fig. 3.15).

3.2.6.3 Example—Frequency Domain Analysis of a Cantilever Beam Model

A cantilever beam model of a high-rise building is prepared for a wind tunnel test. The model is of cross-section 4 × 8 cm and of height 78.5 cm. The measured stiffness in direction of interest is K = 1444.83 $\frac{\mathrm{N}}{m}$. The eigenfrequency, modal mass and the critical damping ratio for the first mode are $f_1 = 19.07$ Hz, $m_1 = 0.1036$ kg and $\xi = 0.0039$, respectively. Measures are undertaken to ensure the cantilever behavior of the model. Wind velocities time histories are measured at the top and middle of

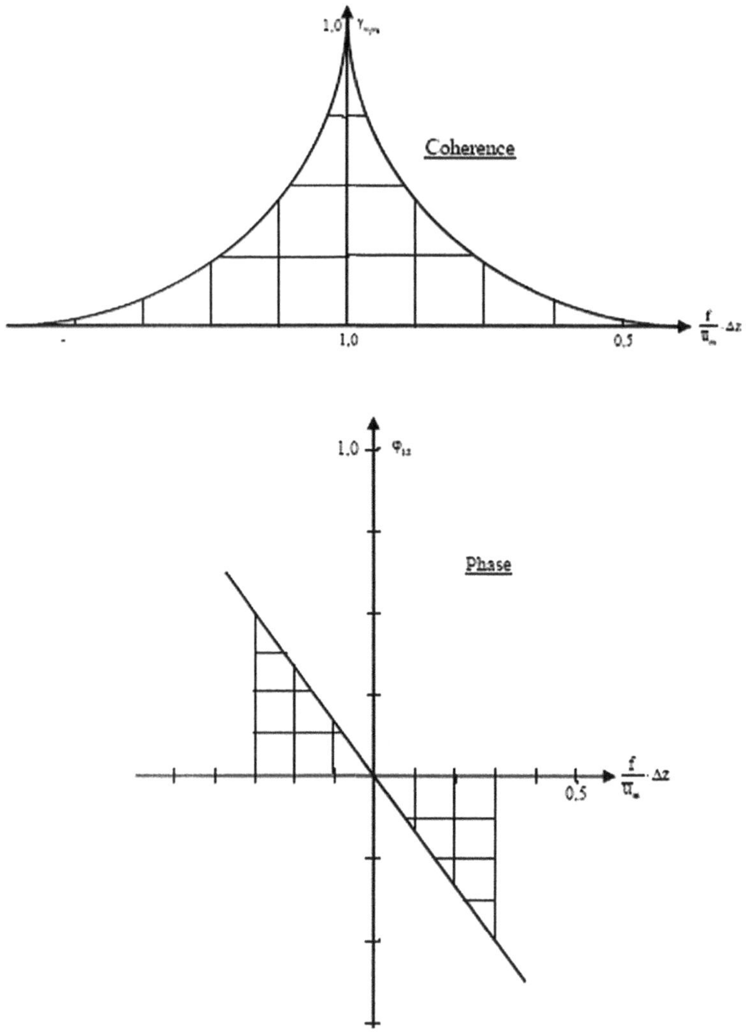

Fig. 3.15 Semi-empirical coherence and phase functions of velocity fluctuations in turbulent wind

the model, in total $N = 2^{17}$ data points. These time histories are transformed into the power spectrum using the provided MATLAB function *AutoSpectrumFunction*. In order to get a more reliable estimation, the data is divided into 15 windows of length $wl = N/8$ with 50% overlapping. Since the sampling frequency of the used sensors is 2000 Hz, frequencies of up to the Nyquist frequency f_n = 1000 Hz with the frequency resolution f_r = 0.1221 Hz can be considered. Each of the windows consists of zero-mean fluctuating data representing wind turbulence.

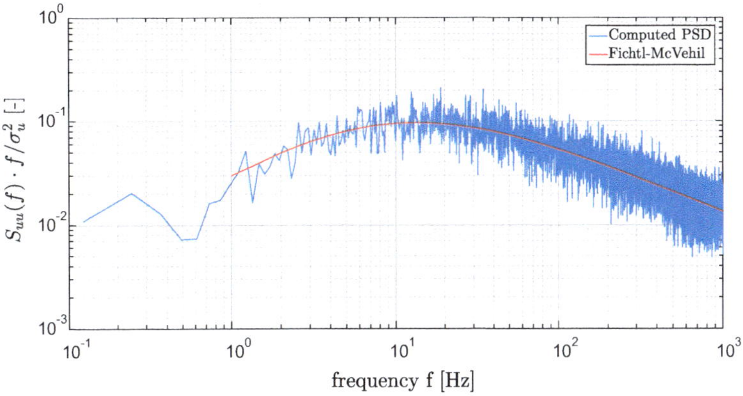

Fig. 3.16 Normalized power spectrum of the wind velocities at the top point

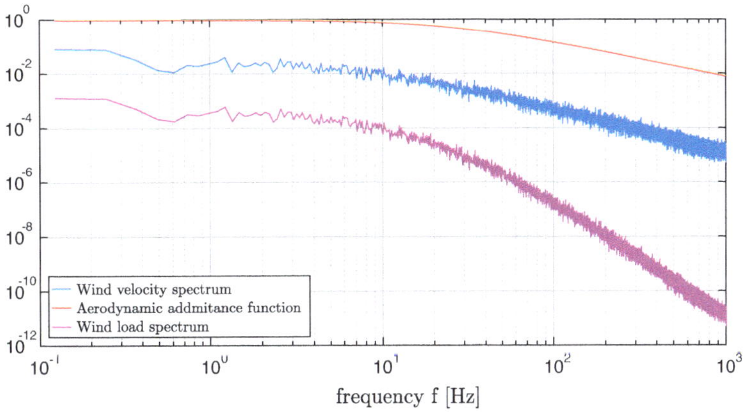

Fig. 3.17 Comparison of wind velocity and wind load spectra

Single Degree of Freedom Model

The wind spectrum at the top point, normalized with the variance of the whole signal, is presented in Fig. 3.16. The spectrum obviously follows a Fichtl-McVehil curve with $r = 1$ and consequently $b = 6$, which was fitted to the data with the *Curve Fitting Tool* of MATLAB. The integral time scale estimated from the curve is $T_{ux}(FMV) = 0.018$ s.

Figure 3.17 illustrates the influence of the aerodynamic admittance on the values of the wind load power spectrum, which is calculated according to Eq. 3.58:

$$S_{FF}(f) = 4(\frac{1}{2}C_D A\rho)^2 \chi^2 S_{uu}(f) \tag{3.96}$$

The variance of the response (displacement) is calculated by integrating the response spectrum. For a single degree of freedom, this turns out to be

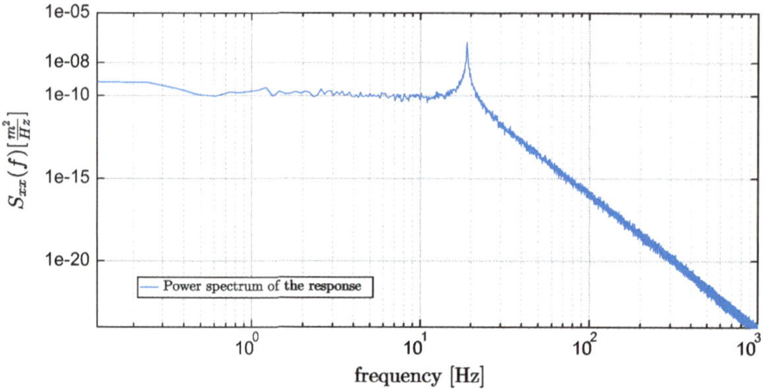

Fig. 3.18 Response spectrum of a single-degree of freedom model

$\sigma_x^2 = 5.3926 \times 10^{-8}\,\text{m}^2$, and the standard deviation of the response is

$$\sigma_x = 2.3382 \times 10^{-4}\,\text{m}$$

The mean response is calculated assuming the mean force F_q as the static load:

$$\overline{x} = \frac{F_q}{k} = \frac{0.4116\text{N}}{1449.42\text{N/m}} = 2.8429 \times 10^{-4}\,\text{m}$$

The total response power spectrum is shown in Fig. 3.18.

It can be noticed that, for very low frequencies, the power of the response is without trend until the frequencies approach the eigenfrequency with the distinct peak in the spectrum. The significant drop of the power magnitude for high frequencies happens because of the effects of inertia and aerodynamic admittance.

The peak factor for the Gaussian process is calculated as

$$k_p = 3.8785$$

which is in the expected range of values of 3–5. The mean peak value of the response is then

$$x_{max} = \overline{x} + k_p\sigma_x = (2.8396 + 3.8785 \times 2.3382) \times 10^{-}4 = 0.0012\,\text{m}$$

Multi-degree of Freedom Model

Since there have been done two simultaneous measurements of wind velocities in the wind tunnel, a two degree-of-freedom model is also used. The analysis was done separately for two modes, and the results are then superimposed. The second

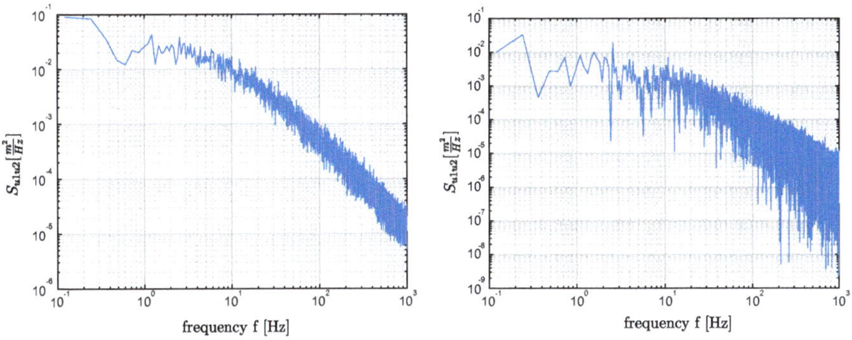

Fig. 3.19 Auto-spectra of the wind velocity processes at the middle point and cross-spectrum

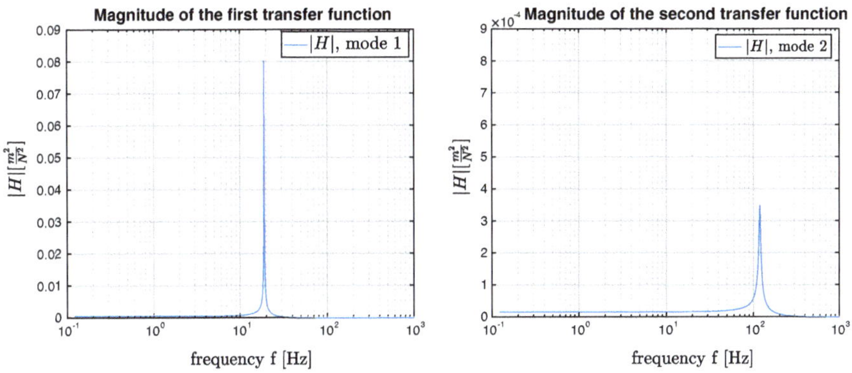

Fig. 3.20 Modal transfer functions of a two degree-of-freedom model

eigenfrequency is $f_2 = 119.51$ Hz, and the Rayleigh critical damping ratio for the second mode is $\xi_2 = 0.0244$.

The power spectrum of the wind velocities at the middle point as well as the cross-spectrum of these two processes are shown in Fig. 3.19.

The magnitudes of the transfer function for both modes can be seen in Fig. 3.20. Because of the difference of the damping ratios ($\xi_1 = 0.0039$ and $\xi_2 = 0.0244$) and natural frequencies (in effect causing different stiffnesses), the peaks are of different order of the magnitude (pay attention to the units size).

The spectra of the response (displacements) are computed as the superposition of results for both modes. For example, the total spectrum at the top point is:

$$S_{x_2x_2}(f) = S_{1,x_1x_2} + S_{1,x_2x_2} + S_{2,x_1x_2} + S_{2,x_2x_2}(f) \tag{3.97}$$

The standard deviation of the displacement of the top point yields

$$\sigma_{x2} = 1.4313 \times 10^{-4}\,\text{m},$$

and the standard deviation of the displacement of the middle point is

$$\sigma_{x1} = 3.4983 \times 10^{-5}\,\mathrm{m}$$

Appendix 3: MATLAB-Codes

Function: autoSpectrumFunction.m

```
%% autoSpectrumFunction computes one-sided auto-spectrum ...
    of the stationary input signal x.
%
% INPUT: required: signal x; extra arguments (in ...
    following order): sampling frequency - Fs; number of
% windows the signal will be divided into - nw; percent ...
    of the window that
% is overlapping - shift;
%
%OUTPUT: 2-column matrix. First column: frequency domain; ...
    Second column:
%values of auto-power-spectrum for the given frequency ...
    domain. Spectrum is
%not normalized
%
%Used window function: rectangular
%Ana Cvetkovic, last updated 05.05.2017

function [ autoSpectrum ] = autoSpectrumFunction( x, ...
    varargin )
%%Extra arguments: sampling frequency, number of windows, ...
    overlapping
%in case that the extra arguments are not defined, they ...
    will take the
%default values: sampling frequency 2000 Hz; number of ...
    windows 8, making
%the window length 1/8 of the length of the input signal; ...
    overlapping of
%50 percent
if isempty(cell2mat(varargin(1)))
    Fs=2000;
    else
    Fs=varargin(1);
end
if isempty(cell2mat(varargin(2)))
    nw=8;
    else
    nw=varargin(2);
end
if isempty(cell2mat(varargin(3)))
    shift=0.5;
    else
    shift=1-varargin(3);
end
%% Window
```

```
% Defines windows according to the previous parameters; N ...
    is length of the
% input signal; wl is length of one window; shift gives ...
    the 'displacement'
% of the beggining of the next window relative to the ...
    previous one. Third
%extra paramter must be strictly smaller than 1 (full ...
    overlapping; doesn't
%make sense) and greater or equal to 0(no overlappng, ...
    shift equal to window
% length). In there is overlapping, the number of used ...
    windows will
% increase.
    N=length(x);
    wl=floor(N/nw);
    shift=shift*wl;
    df=Fs/wl;
if shift-wl≠0.0000
        nw=floor((N-wl)/shift+1);
end
% Split signal into matrix of (wl)x(nw). Each column ...
    represents one window
% that is used independently for estimation of ...
    power-spectrum. A mean of
% each window (column of x_matrix) is substracted and ...
    spectrum is estimated
% from the fluctuating part of the signal. By definition, ...
    a two-sided
% spectrum with negative frequencies included is computed
    x_matrix=zeros(wl,nw);                %initialization ...
        of the matrix before the loop
    for i=1:nw
        x_matrix(:,i)=x((i-1)*shift+1:(i-1)*shift+wl);
    end
    xw=bsxfun(@minus,x_matrix,mean(x_matrix));
    psd2=zeros(wl,nw);
    for w=1:nw
        psd2(:,w)=abs(fft(xw(:,w))).^2;
    end
%Two-sided spectrum psd_vec is estimated as the average ...
    of all the previously computed spectra
    psd_vec=mean(psd2,2);
% scaling with the time step (inverse of the sampling ...
    frequency) to change
% from circular to temporal frequency
    psd_vec=psd_vec/wl/Fs;
%One sided spectrum psd1 for positive frequencies with ...
    the conserved power of
%the signal
    psd1=psd_vec(1:wl/2+1);
    psd1(2:end-1)=2*psd1(2:end-1);
%Define frequency domain for positive frequencies
    freqDomain=df*(0:wl/2);
    freqDomain=freqDomain';
autoSpectrum=[freqDomain,psd1];
end
```

Function: SpectralAnalysisSDOF.m

```
%% SDOF oscillator (cantilever beam) excited by wind load
%Calculates dynamic response of single-degree of freedom ...
    oscillator to wind
%load that is reduced to a load at the DOF. Calculates ...
    extreme value of
%the response. Structural properties and input velocities ...
    from experimental data
%from the wind tunnel experiment.
%
% INPUT: required: signal (wind velocities time history) ...
    u for calculation
% of auto-spectrum with possible extra arguements (see
% autoSpectrumFunction); drag force coefficient; air ...
    density; structural properties
%
% Used window function: rectangular
% Ana Cvetkovic, last updated 08.05.2017

%% Input parameters (here from the WT experiment)
%Area of the along-wind-exposed surface of the structure
area=0.785*0.04/2;
%Drag force coefficient and air density
Cd=1.0;            %[-]
rho=1.18;          %[kg/m^3]
%Structural properties: first natural frequency, critical ...
    damping ratio and
%stiffness
f_nat=19.071;      %[Hz]
ksi=0.0039;        %[-]
k=1449.42;         %[N]
%% Input wind velocties
%Get the frequency domain and power-spectral density of ...
    the wind velocities
%from the autoSpectrumFunction. Assume x is already saved ...
    in workspace
[f,spectrum]=autoSpectrumFunction(u);

%Possible normalization with the variance of the input ...
    signal to check the
%PSD estimation. The ratio of the area under the spectrum ...
    and the variance
%should be approximately equal to 1.
variance=var(x);
psdArea=trapz(f,spectrum);
ratio=psdArea/variance;
%% Aerodynamic addmitance function
% mean wind velocity
Umean=mean(u);
% Aerodynamic addmitance function by Vickery
aaf=1./(1+(2*sqrt(area)/umean*f).^(4/3));

%% Wind load power spectrum
meanForce = 1/2*Cd*area*rho*Umean^2;
spectrumOfForce=(2*Cd*area*rho)*aaf.^2.*spectrum;

%% Displacement spectrum
% Define transfer function for SDOF
```

```
tf=power(1-(f./f_nat).^2+1i*2*ksi*f./f_nat,-1);
absTF=abs(tf);

% Calculate displacement spectrum
sx_f=k^(-2)*absTF.^2.*spectrumOfForce;
displacementVariance=trapz(f,sx_f);

%% Peak factor for a Gaussian process
% Standard deviation of the response
sigmaX=sqrt(displacementVariance);
% Mean force for the mean (static) response
meanResponse=meanForce/k;
% First frequency must be set as different that zero because
% of the following division
f(1)=f(2)/2;
% Spectral moments for peak factor
m0=trapz(f,sx_f./f);
m2=trapz(f,sx_f.*f);
f(1)=0;

%Calculating peak factor and extreme response
nu=sqrt(m2/m0);
peakFactor=sqrt(2*log(nu*(N*T)))+0.577/sqrt(2*log(nu*(N*T)));
maxResponse=meanResponse+peakFactor*sigmaX;     %[m]

%% Plot results
% Plot wind velocities spectrum, aerodynamic addmitance ...
    function and wind
% load psectrum
figure;
plot(f(2:end),psd1(2:end)); hold on;
plot(f(2:end),aaf(2:end)); hold on;
plot(f(2:end),spectrumOfForce(2:end),'Color','m');
grid on;
xlabel('frequency f [Hz]');
legend('Wind velocity spectrum','Aerodynamic addmitance ...
    function','Wind load spectrum');
legend('Location','southwest')
set(gca,'xscale','log');
set(gca,'yscale','log');

%Plot transfer function
figure;
plot(f,absTF);
xlabel('Frequency');
ylabel('Transfer function magnitude [-]');
grid on;
set(gca,'xscale','log');
% set(gca,'yscale','log');

%Plot displacement spectrum
figure;
plot(f,sx_f,'Color','r');
grid on;
xlabel('frequency f');
ylabel('S_x(f)*f/\sigma_u^2');
legend('Displacement spectrum');
set(gca,'xscale','log');
% set(gca,'yscale','log');
```

References

Aldasoro, H.A.: On gust buffeting design of slender chimneys—building interference and fatigue. PhD thesis, Universität zu Braunschweig (2014)

Davenport, A.G.: The application of statistical concepts to the wind loading of structures. In: Proceedings of the Institute of Civil Engineering, No. 6480 19 (1962)

Dyrbye, C., Hansen, C.O.: Wind Load on Structures. Wiley (1997)

EN 1991-1-4 (2010). Actions on structures—Part 1-4: General actions—Wind actions

Grigoriu, M.: Stochastic Calculus: Application in Science and Engineering. Birkhäuser Boston (2002)

Gusella, V., Materazzi, A.L.: Non-Gaussian along-wind response analysis in time and frequency domains. J. Eng, Struct (1998)

Höffer, R.: Damage-oriented actions and environmental impact on materials and structures. In: Stangenberg, F., Breitenbücher, R., Bruhns, O.T., Hartmann, D., Höffer, R., Kuhl, D., Meschke, G. (eds.) Lifetime-Oriented Structural Design Concepts. Springer (2009)

Holmes, J.D.: Non-Gaussian characteristics of wind pressure fluctuations. J. Wind Eng, Ind (1981)

Holmes, J.D.: Wind Loading of Structures. Spon Press (2001)

Kareem, A., Tognarelli, M.: Modeling and analysis of the quadratic term in the wind effects on structures. J. Wind Eng. Ind, Aerodyn (1998)

Kwon, D.-K., Kareem, A.: Peak factor for a non-Gaussian process revisited. In: The seventh Asia-Pacific conference on Wind engineering, Taipei (2009)

Lin, Y.K.: Probabilistic Theory of Structural Dynamics. McGRAW-HILL Book Company (1967)

Newland, D.E.: An Introduction to Random Vibrations. Spectral and Wavelet Analysis, Pearson Education Limited (1993)

Schrader, P.: Die statistische Stabilität gemessener integraler Längenmaße und anderer Windparameter. Ruhr-Universität Bochum (1995)

de Silva, C.: Vibration: Fundamental and Practice. CRC Press (1999)

Sockel, H.: Wind-Excited Vibrations of Structures. Springer, Technical University of Vienna (1994)

Stearns, S.D., Hush, D.R.: Digital Signal Processing with Examples in MATLAB. CRC Press (2011)

Stengel, D.: Naturmessungen und numerische Simulationen zur Untersuchung von Freileitungsseilen im natürlichen Wind. Universität zu Braunschweig (2015)

Tamura, Y., Kareem, A.: Advanced Structural Wind Engineering. Tokyo Polytechnic University, Springer Japan (2013)

Vickery, B.J.: Fluctuating lift and drag on a long cylinder of square crosssection in a smooth and turbulent stream. J. Fluid Mech. **25**, 481–494 (1966)

Vrouwenvelder, T., Steenbergen, H.: Implementation of Eurocodes: Handbook 3—Actions effects for buildings-Chapter 3: Wind actions. Leonardo Da Vinci Pilot Project (2005)

Warner, R.: Spectral Analysis of Time-Series Data. The Guilford Press (1998)

Welch, P.: The use of fast fourier transform for the estimation of power spectra: a method based on time averaging over short, modified periodograms. IEEE Trans. Audio Electroacoust. **15** (1967)

Zhou, Y., Kareem, A.: Mode shape corrections for wind load effects. J. Eng. Mech. **128** (2002)

Zhou, Y., Kareem, A.: Aerodynamic Admittance Functions of Tall Buildings. Technical Report. University of Notre Dame, USA (2003)

Chapter 4
Earthquake Resistant Design of Structures According to Eurocode 8

Linda Giresini and Christoph Butenweg

Abstract The chapter initially provides a summary of the contents of Eurocode 8, its aim being to offer both to the students and to practising engineers an easy introduction into the calculation and dimensioning procedures of this earthquake code. Specifically, the general rules for earthquake-resistant structures, the definition of design response spectra taking behaviour and importance factors into account, the application of linear and non-linear calculation methods and the structural safety verifications at the serviceability and ultimate limit state are presented. The application of linear and non-linear calculation methods and corresponding seismic design rules is demonstrated on practical examples for reinforced concrete, steel and masonry buildings. Furthermore, the seismic assessment of existing buildings is discussed and illustrated on the example of a typical historical masonry building in Italy. The examples are worked out in detail and each step of the design process, from the preliminary analysis to the final design, is explained in detail.

Keywords Seismic design · Eurocode 8 · Design examples · Response spectrum
Pushover analysis

Electronic supplementary material The online version of this chapter (https://doi.org/10.1007/978-3-662-57550-5_4) contains supplementary material, which is available to authorized users.

K. Meskouris et al., *Structural Dynamics with Applications in Earthquake and Wind Engineering*, https://doi.org/10.1007/978-3-662-57550-5_4

4.1 General Introduction and Code Concept

Eurocode 8-1 (2004) deals with the seismic design and earthquake resistant construction of buildings and civil engineering works of common reinforced concrete, steel, composite steel - concrete, timber and masonry. The objective of the code is the protection of human lives, the limitation of structural damage and the ensuring of the functional efficiency of buildings and structures that are of relevance to public safety and infrastructure. An especially important aspect when using the code is compliance to the validity limits on standard buildings. This means that Part 1 of Eurocode 8-1 (2004) is not applicable for structures and parts of structures that could pose additional risks in the event of an earthquake. This refers to structures from the areas of traffic infrastructure (e.g. bridges), water supply (e.g. reservoirs, dams), energy supply (e.g. power stations) or industrial facilities (e.g. tanks, production installations). These are only a few examples for cases where a direct application of Part 1 is not permissible. The reason for this limitation is that the safety level for standard buildings cannot be transferred to these "special structures" and that Part 1 does not include any special rules for such structures. Special structures are dealt with in the other parts of Eurocode 8-1 (2004). Eurocode 8-2 (2004) deals with bridge structures, Eurocode 8-4 (2004) comprises silos, tanks and pipelines and Eurocode 8-6 (2005) provides special regulations for masts and chimneys. Eurocode 8-3 (2005) of the code deals with existing structures, in which special attention is paid to the condition of the building, the level of knowledge and the built-in materials. Eurocode 8-5 (2003) is of relevance for all types of buildings and provides rules for the seismic design of foundations.

4.1.1 Contents

A glance at the layout of Eurocode 8-1 (2004) shows that the rules for the design and execution of earthquake resistant structures are of great importance. Adhering to these rules, the structural engineer can perform the calculation and the subsequent dimensioning with simple planar calculation models, saving time and effort. In case of non-compliance, the standard prescribes more complex calculation models, since the dynamic behaviour cannot be predicted with sufficient accuracy by simple models. A cursory glance reveals furthermore that the material-specific regulations for steel, concrete, timber and masonry structures are explained in great detail. The reason for the detailed description of the material-specific rules is the possibility to reduce seismic actions by a structural behaviour factor that quantifies the structure's capability to dissipate seismic energy. In order to make use of this reduction possibility, observance of certain construction rules is required that ensure the actual presence of the required structural ductility. Another important part is the definition of the seismic action by means of design response spectra taking into account subsoil conditions. Figure 4.1 depicts the structure and the main chapters of the code, namely

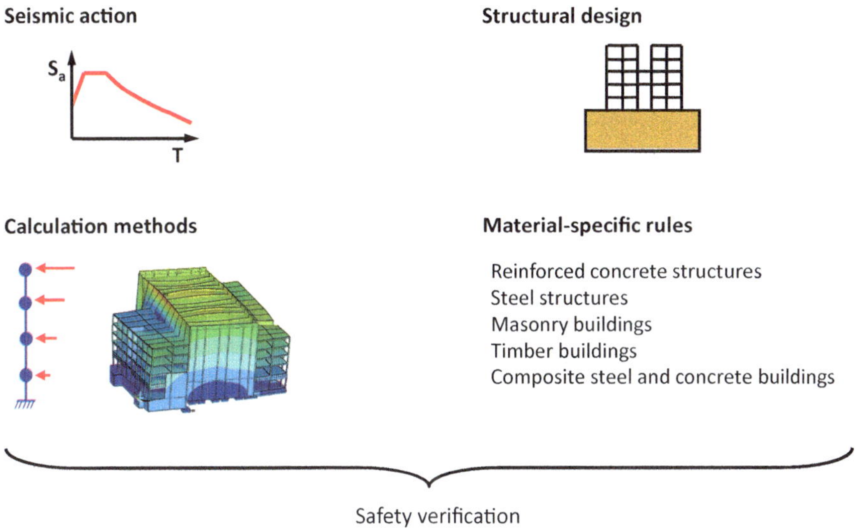

Fig. 4.1 Content and structure of Eurocode 8-1 (2004)

the definition of the seismic action, rules on earthquake-resistant design, calculation methods and material-specific rules. These topics are described in more detail in the next sections.

4.1.2 Performance and Compliance Criteria

The design of structures according to Eurocode 8-1 (2004) is carried out with the aim of fulfilling the no-collapse and damage limitation requirements of the code. The no-collapse requirement guarantees that the structure is designed and constructed to withstand the design seismic actions without local or global collapse, while even retaining a residual load-carrying capacity, after a seismic event with a probability of exceedance of 10% in 50 years and a corresponding return period of 475 years. The damage limitation requirement ensures an adequate level of reliability against damage associated with limitations of use or unacceptable repair costs in comparison to the costs of the structure itself. Eurocode 8-1 (2004) recommends for damage limitation a seismic action with a probability of exceedance of 10% in 10 years and a corresponding return period of 95 years.

4.1.3 General Rules for Earthquake Resistant Structures

Section 4 of Eurocode 8-1 (2004) presents general rules for earthquake resistant structures, paying attention to structural regularity and appropriate foundation concepts. In particular, designers should provide direct paths for the transmission of seismic forces through the structure to the ground. This goal is easier to achieve when the building is as simple as possible in terms of its load-carrying mechanisms. For instance, structures that are symmetric in-plan perform much better under earthquake loading than asymmetric ones. Symmetric structures are also much easier to analyse with simplified models, and therefore the insights based on numerical calculations are more reliable.

4.1.3.1 Regularity in Plan

A building is considered to be regular in plan when its layout is compact and the in-plan stiffness of its floors is sufficiently large compared to the lateral stiffness of its vertical structural elements. As an example, re-entrant corners or edge recesses do not enhance the in-plan stiffness of the floors, the deformation of which might influence the distribution of forces in the vertical load-carrying elements. A building is considered to be irregular in plan when the ratio between its largest and shortest in-plan dimension is greater than 4. However, irregular in-plan buildings can be transformed to regular ones by means of separation joints (Fig. 4.2).

The minimum distance between the centre of mass and the centre of stiffness is crucial for a favourable distribution of inertia forces, since torsional effects are

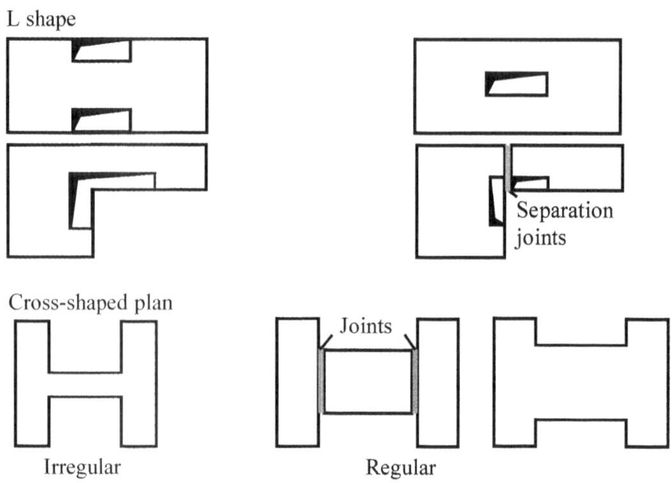

Fig. 4.2 In-plane regularity

especially small when this distance is close to zero. This objective is easily reached with uniformity in plan. According to Eurocode 8-1 (2004) uniformity in plan is given if the eccentricity between the centre of mass and the stiffness centre is limited to

$$e_{0i} \leq 0.30 \cdot r_i \tag{4.1}$$

and the radius of gyration l_s of the floor mass in plan satisfies the following condition:

$$l_s \leq r_i \tag{4.2}$$

where

e_{0i} Distance between the centre of stiffness and the centre of mass, measured along the direction perpendicular to the direction of analysis considered.

r_i Square root of the ratio of the torsional stiffness to the lateral stiffness in the direction of loading ("torsional radius").

l_s Radius of gyration of the floor mass in plan (square root of the ratio of (a) the polar moment of inertia of the floor mass in plan with respect to the centre of mass of the floor to (b) the floor mass).

4.1.3.2 Regularity in Elevation

The seismic behaviour of buildings is also influenced by the regularity of the building in elevation. Table 4.1 shows examples of unfavourable and favourable configurations of building in elevation. In general, all lateral load resisting systems, such as cores, structural walls, or frames should run without interruption from their foundations to the top of the building.

Both the lateral stiffness and the mass of the individual storeys must remain essentially constant, without abrupt changes, from the base to the top of the building. If abrupt changes of the lateral stiffness lead to a soft storey mechanism, the building becomes highly susceptible to a total building collapse. Figure 4.3 shows favourable and unfavourable lateral stiffness distributions in elevation and two typical soft storey mechanisms.

If setbacks are present, the conditions according to Sect. 4.2.3.3 of Eurocode 8-1 (2004) must be taken into account. In addition, the connections between two adjacent structures should allow relative movements to avoid damage caused by out-of-phase vibrations. Adjacent buildings with vertical offsets of the floor slabs or parts of buildings with different vibration behaviour should be separated by joints to prevent dangerous pounding effects. Finally, it must be noted that tall and slender buildings as well as buildings with large masses at great heights lead to extreme seismic action effects on the foundations.

Table 4.1 Regularity criteria in elevation

	Unfavourable	Favourable
High slender buildings	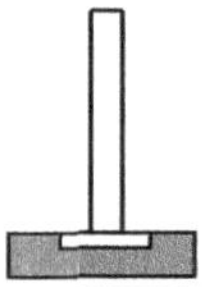	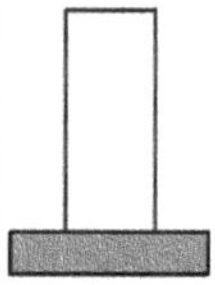
Concentrated masses in upper stories	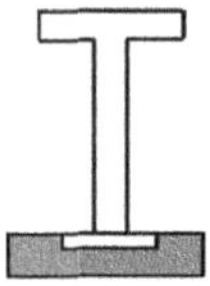	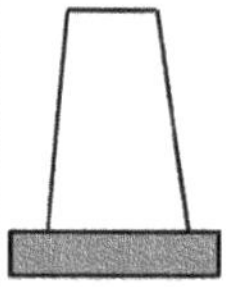
Horizontal offsets of columns	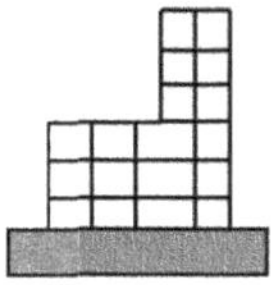	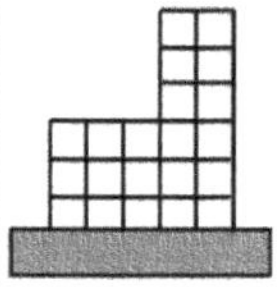
Vertical offset of floors	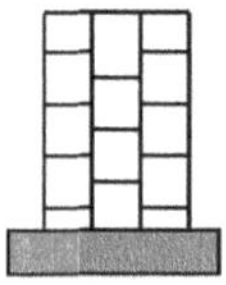	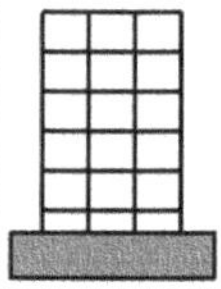
Connection of building parts	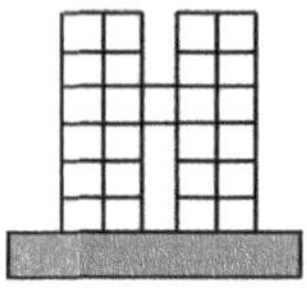	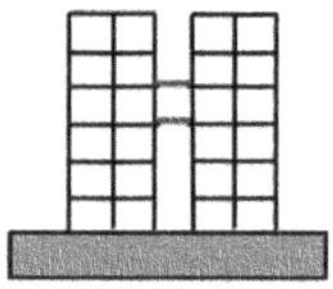
Building with differences in elevation	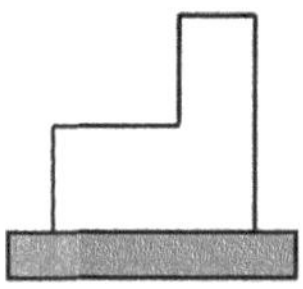	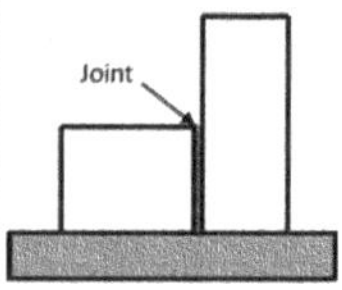

4.1.3.3 Redundancy

Redundancy is also a key concept for earthquake resistant structures (Bertero and Bertero 1999). Redundancy is obtained through a targeted arrangement of structural members, aiming at a more favourable seismic response through the redistribution of inertia forces and increased energy dissipation. A redundant structure clearly cannot be statically determinate, which ensures that its overall safety is not compromised in the case of failure of a single, perhaps not very important, structural element. Anyway, such structural elements should be secondary ones, so that their failure does not cause global collapse and the affected areas are not overly large.

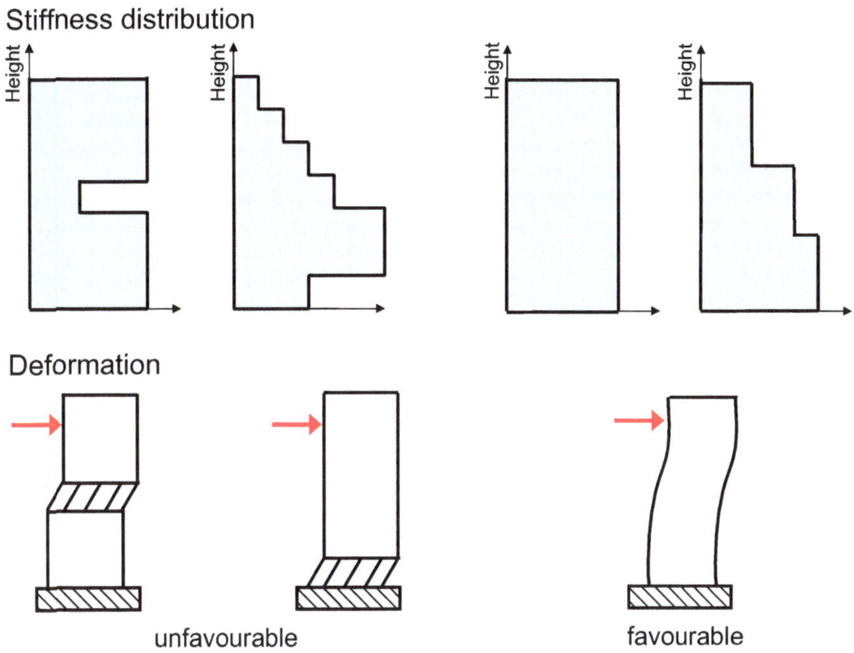

Fig. 4.3 Favourable and unfavourable lateral stiffness distributions and soft storey mechanisms (Bachmann 2002)

4.1.3.4 Seismic Resistant Foundations

The foundations of structures must be designed in such a way that the entire structure undergoes a uniform displacement due to seismic ground motions. As a simple practical construction rule it is always recommended to connect single or strip foundations by coupling beams, thus preventing differential displacements. If the foundation is not continuous, the structural system must be designed considering possible differential displacements as a result of seismic ground motions. In addition, all foundation elements should lie on subsoil with similar geotechnical properties in order to avoid differential settlements due to increased soil pressure during an earthquake.

4.1.4 Seismic Actions

The horizontal seismic action is described by two orthogonal components assumed as being independent and represented by the same response spectrum. The design response spectrum is defined piecewise between four control periods T_A, T_B, T_C and T_D as a function of the structural period (Fig. 4.4).

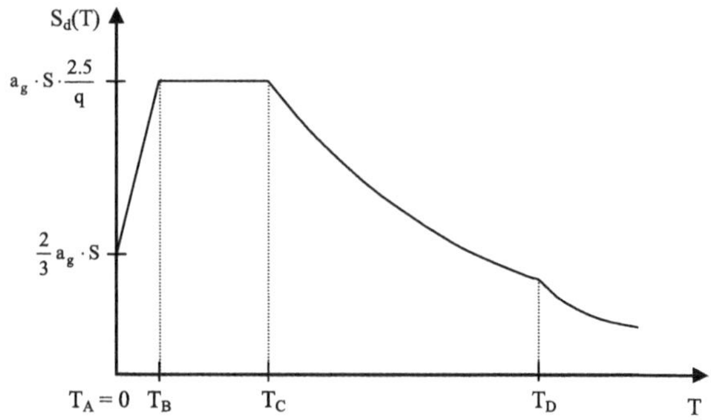

Fig. 4.4 Shape of the design response spectrum with control periods T_A, T_B, T_C and T_D

Table 4.2 Importance classes and corresponding return periods (RP)

Class	Building	γ_I	Return period (years)
I	Buildings of minor importance for public safety, e.g. agricultural buildings, etc.	0.8	225
II	Ordinary buildings, not belonging in the other categories	1.0	475
III	Buildings whose seismic resistance is of importance in view of the consequences associated with a collapse, e.g. schools, assembly halls, cultural institutions etc.	1.2	825
IV	Buildings whose integrity during earthquakes is of vital importance for civil protection, e.g. hospitals, fire stations, power plants, etc.	1.4	1225

The horizontal design spectrum $S_d(T)$ is given by the following expressions:

$$T_A \leq T \leq T_B:\ \ S_d(T) = a_g \cdot S \cdot \left[\frac{2}{3} + \frac{T}{T_B} \cdot \left(\frac{2.5}{q} - \frac{2}{3}\right)\right] \tag{4.3}$$

$$T_B \leq T \leq T_C:\ \ S_d(T) = a_g \cdot S \cdot \frac{2.5}{q} \tag{4.4}$$

$$T_C \leq T \leq T_D:\ \ S_d(T) = a_g \cdot S \cdot \frac{2.5}{q} \cdot \frac{T_C}{T} \geq \beta \cdot a_g \tag{4.5}$$

$$T_D \leq T:\ \ S_d(T) = a_g \cdot S \cdot \frac{2.5}{q} \cdot \frac{T_C T_D}{T^2} \geq \beta \cdot a_g \tag{4.6}$$

where:

$S_d(T)$ Spectral ordinate of the design spectrum
a_g Design ground acceleration on type A ground ($a_g = \gamma_I \cdot a_{gR}$)
$T_{A\text{-}D}$ Control periods of the spectrum
S Soil factor
γ_I Importance factor according to Table 4.2
β Lower bound factor.

The vertical seismic design spectrum is given by expressions (4.3)–(4.6) in which the design ground acceleration a_g is replaced by the vertical design ground acceleration a_{vg} and S is taken equal to 1.0. The reference peak ground acceleration a_{gR} must be determined by using seismic hazard maps, which in European countries are provided on a national level. The control periods and soil factors are tabulated in Eurocode 8-1 (2004), Sect. 3.2.2.2. The horizontal and vertical spectra are assigned to a reference return period of 475 years with a probability of exceedance of 10% in 50 years. The importance of structures is taken into account by scaling the spectra with an importance factor γ_I, whereby the reference return period of 475 years corresponds to an importance factor of 1.0 for standard residential buildings. The scaling is carried out by means of four importance classes with importance factors ranging between 4.8 and 1.4 (Table 4.2). A modification of the importance factor can be interpreted as a change of the return period or the probability of exceedance, which is described by a Poisson distribution:

$$P_N = 1 - e^{-N/T_L}. \tag{4.7}$$

N is the number of years for which the probability of exceedance is sought and T_L is the desired return period. The relationship to the importance factor γ_I can be described via the reference return period T_{LR}:

$$\gamma_I = \frac{1}{(T_{LR}/T_L)^{1/k}}. \tag{4.8}$$

The exponent k depends on the local seismicity and can be estimated to be in the order of 3. Alternatively, the importance factor can also be determined with the probabilities of exceedance P_{LR} and P_L corresponding to T_{LR} and T_L as

$$\gamma_I = \frac{1}{(P_L/P_{LR})^{1/k}}. \tag{4.9}$$

Table 4.2 specifies the return periods calculated using the above relationships for the individual importance classes.

It should be noted that the importance factor only scales the ordinate of the response spectrum and, thus, the seismic input of the relevant seismic zone. The range of seismic zones is not changed with this approach although stronger earthquakes, i.e. earthquakes with a lower probability of exceedance or a higher return period resp., generally affect larger geographical regions. The use of scaling factors, however, is a common approach of international codes. Alternatively, the design ground acceleration can be taken directly from seismic hazard maps of the relevant return period or probability of exceedance in case such maps are available.

The behaviour factor q takes the global structure ductility into account and transforms the elastic response spectrum to an inelastic design spectrum. This procedure is a radical simplification as the tacit assumption is made that a global ductility results from the combination of the existing local dissipative reserves. Furthermore,

Table 4.3 Limits of behaviour factors for different building materials

Behaviour factor	Reinforced concrete	Steel	Timber	Masonry
min q	1.5	1.5	1.5	1.5
max q	3	8	4	2.5

the use of inelastic spectra for elastic calculations on the basis of the response spectra procedure with a subsequent superposition of the calculated quantities is strictly speaking not correct as the superposition principle is no longer valid. Nevertheless, the use of behaviour factors is a pragmatic approach in building practice, providing an adequate estimation of the real non-linear dynamic behaviour of buildings under seismic actions. The behaviour factor q is defined as being the quotient between the reacting force in the elastic system R_{el} and the force R_{nl} in the non-linear system:

$$q = \frac{R_{el}}{R_{nl}}. \tag{4.10}$$

Eurocode 8-1 (2004) gives values of the behaviour factor q as a function of the building material. Table 4.3 shows minimum and maximum behaviour factors for the different building materials.

The given behaviour factors are the result of comprehensive experimental and theoretical research activities (Chopra 2001; Nensel 1986; Tomazevic 2004). The range of the values shows that the factors are only approximations describing the complex non-linear behaviour of structures in a simplified manner with just one global factor.

4.1.5 Analysis Methods

Analysis methods that are able to simulate the building response under earthquake loading can be either linear or non-linear. Linear analysis approaches include the lateral force analysis method and the response spectrum analysis as introduced in Sect. 1.2.5. Non-linear methods comprise non-linear static (pushover) analyses and non-linear transient analyses. If particular conditions of regularity are satisfied, linear methods can be applied for the sake of simplicity. For more complex configurations and/or notable non-linear material behaviour, non-linear analyses generally are more suitable. The level of complexity increases from the lateral force method to non-linear transient analyses. The main advantage of linear analyses is that results are easy to obtain and to interpret. Nevertheless, the pushover analysis is often used for many types of building materials whose non-linear behaviour cannot be neglected.

4.1.5.1 Lateral Force Method of Analysis

The lateral force method of analysis is a linear elastic method. As all linear methods, it can be performed by using two planar models, one for each main direction of the building, if the in-plan regularity criteria explained in Sect. 4.1.3 are met. Even if the structure does not fulfil the in-plan regularity criteria, two planar models can still be used, provided that the special regularity conditions given in Eurocode 8-1 (2004), Sect. 4.3.3.1 (8) are satisfied. As only the seismic horizontal force corresponding to the fundamental period of vibration of the structure is used, the lateral force method can be applied only to buildings whose response is not significantly affected by contributions of higher modes of vibration. Here, the base shear force is equal to:

$$F_b = S_d(T_1) \cdot m \cdot \lambda, \tag{4.11}$$

where $S_d(T_1)$ is the ordinate of the design spectrum at the fundamental period of vibration T_1 in the considered direction and m is the total mass of the building. λ is a correction factor that reduces the force to be applied to the building to 85% ($\lambda = 0.85$) in cases where the building has more than two storeys and $T_1 \leq 2T_c$. In other cases, λ is set to $\lambda = 1.0$. In other words, the factor λ accounts for the fact that the effective modal mass of the fundamental mode of buildings with at least three storeys is on average 85% of the total building mass.

The distribution of the total earthquake force F_b on the structure is assumed to be affine to the fundamental modal shape or, simplified, linear over the structure height. The horizontal seismic forces F_i are applied at the height of the floor slabs. The seismic horizontal forces F_i on each floor level can be calculated as follows:

$$F_i = F_b \cdot \frac{s_i m_i}{\sum s_j m_j}. \tag{4.12}$$

Here, m_i, m_j are the storey masses and s_i, s_j are the corresponding displacements of the masses in the fundamental mode. If a linear approximation of the fundamental mode is used, s_i, s_j correspond to the heights z_i, z_j of the masses m_i, m_j above the foundation level. The distributions of the total earthquake force F_b using the fundamental modal shape and the triangular form are shown in Fig. 4.5.

The determination of the fundamental period of vibration can be carried out using a simplified empirical relationship. Thus, T_1 can be estimated as

$$T_1 = C_t \cdot H^{3/4} \tag{4.13}$$

for buildings with heights of up to 40 m. Here, H is the height of the structure in (m), and C_t is a factor considering the structural type. C_t is 0.085 for moment resisting steel frames, 0.075 for moment resisting reinforced concrete frames and eccentrically braced steel frames and 0.050 for all other structures. Alternatively, the fundamental period of vibration can be calculated by using expressions based on standard methods of structural dynamics (e.g. Rayleigh method).

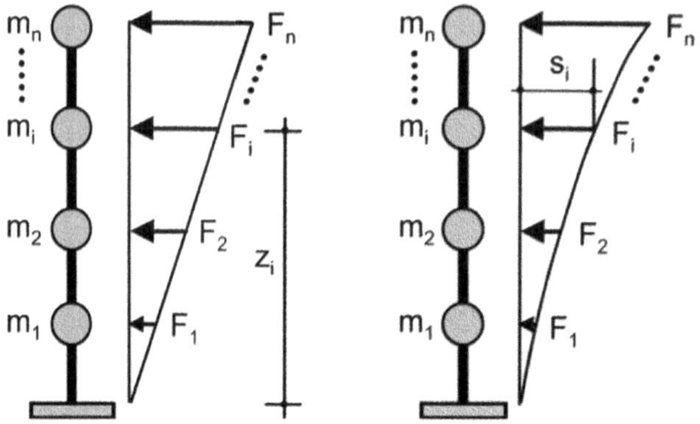

Fig. 4.5 Distribution of the total earthquake force F_b over the building height

4.1.5.2 Modal Response Spectrum Analysis

The modal response spectrum analysis is a linear method, one of the most used standard methods applicable to all types of buildings. The modal response spectrum approach considers the dynamic response of the building but it is not computationally as expensive as a non-linear analysis. The response of all modes of vibration significantly contributing to the global response has to be taken into account. This means that only modes with effective modal masses greater than 5% of the total mass must be considered. The sum of the effective modal masses for the modes taken into account must amount to at least 90% of the total mass of the structure. This percentage can be reduced in some cases, for which it is difficult to reach 90% due to the presence of higher modes (e.g. in very rigid structures). One central problem in the modal response spectrum analysis is the realistic combination of the modal responses in each direction of the considered seismic actions. The responses in two vibration modes may be considered to be independent from each other if their periods T_i and T_{i+1} satisfy the following condition:

$$T_i \geq 0.90 \cdot T_{i+1} \tag{4.14}$$

with $T_i > T_{i+1}$. In this case, the maximum value E_E of a seismic action effect (force, displacement, etc.) may be considered as:

$$E_E = \sqrt{\sum\nolimits_i E_{Ei}^2} \tag{4.15}$$

where E_{Ei} is the value of the seismic action effect due to the vibration mode i. This combination procedure is known as the "Square Root of the Sum of Squares" (SRSS) method. SRSS turns out to give poor results when the vibration modes cannot be considered independent of each other. In this case, a more accurate method must be adopted. In most cases the "Complete Quadratic Combination" (CQC) is used (Wilson et al. 1981). According to it, the maximum value E_E of a seismic action effect may be taken as:

$$E_E = \sqrt{\sum_i \sum_j \rho_{ij} \cdot E_i \cdot E_j} \tag{4.16}$$

where E_i and E_j are the seismic effects of the modes i and j and ρ_{ij} is the correlation coefficient between the modes i and j (NTC 2008):

$$\rho_{ij} = \frac{8\xi^2(1+\beta_{ij})\beta_{ij}^{\frac{3}{2}}}{\left(1-\beta_{ij}^2\right)^2 + 4\xi^2\beta_{ij}(1+\beta_{ij})^2}. \tag{4.17}$$

The term ξ refers to the viscous damping ratio of modes i and j. The coefficient β_{ij} is equal to the ratio of the periods of the two modes i and j:

$$\rho_{ij} = \frac{T_j}{T_i}. \tag{4.18}$$

The greater the difference between periods, the lower ρ_{ij}. In other words:

$\rho_{ij} = 1$ → completely correlated modes
$0 < \rho_{ij} < 1$ → partially correlated modes
$\rho_{ij} = 0$ → uncorrelated modes.

The structural response to each horizontal seismic component is separately evaluated, using the combination rules for modal responses. The resulting maximum value of an action effect can be estimated by the square root of the sum of the squared values of the action effects due to each horizontal seismic component. This superposition rule generally leads to a conservative estimate. Alternatively, the action effects due to the combination of the horizontal components of the seismic action can be computed using the following combinations:

$$\begin{aligned} &1.00 \cdot E_{Edx} \oplus 0.30 \cdot E_{Edy} \\ &0.30 \cdot E_{Edx} \oplus 1.00 \cdot E_{Edy} \end{aligned} \tag{4.19}$$

where

$\oplus$ To be combined with alternating signs
E_{Edx} Action effects due to the seismic action in x-direction
E_{Edy} Action effects due to the same seismic action in y-direction.

If the vertical design ground acceleration a_{vg} is greater than 0.25 g the vertical seismic effects E_{Edz} must be taken into account for horizontal or nearly horizontal structural members spanning 20 m or more, for horizontal or nearly horizontal cantilever components longer than 5 m, for horizontal or nearly horizontal pre-stressed components, for beams supporting columns and in base-isolated structures. The effects of the vertical seismic component need to be taken into account only for these elements and their directly associated supporting elements or substructures. If the horizontal components of the seismic action are also relevant for these elements, the combinations (4.19) can be used for the computation of the action effects, extended to three components:

$$\begin{aligned}&1.00 \cdot E_{Edx} \oplus 0.30 \cdot E_{Edy} \oplus 0.30 \cdot E_{Edz}\\&0.30 \cdot E_{Edx} \oplus 1.00 \cdot E_{Edy} \oplus 0.30 \cdot E_{Edz}\\&0.30 \cdot E_{Edx} \oplus 0.30 \cdot E_{Edy} \oplus 1.00 \cdot E_{Edz}.\end{aligned} \tag{4.20}$$

4.1.5.3 Non-linear Static (Pushover) Analysis

A non-linear static analysis allows a more accurate estimation of the inelastic structural response than linear methods using behaviour factors, since the formation of plastic effects and the redistribution of forces are considered. A pushover analysis can be regarded as a compromise between a simple linear static analysis and a time-consuming non-linear transient analysis. The non-linear static analysis estimates the overall building load-carrying capacity by means of a non-linear load-displacement curve determined under monotonously increasing horizontal loads while the vertical loads are kept constant. Such an investigation is commonly called "pushover analysis". The resulting non-linear load-displacement curve is shortly denoted as pushover curve. Eurocode 8-1 (2004) and numerous international standards and guidelines ATC-40 (1996), FEMA 273 (1997), FEMA 274 (1997), FEMA 356 (2000) propose the pushover analysis as one of the standard non-linear calculation methods. Figure 4.6 depicts the pushover curve of a two-storey frame representing the total base shear F_b as a function of the roof displacement Δ_{top}. The pushover curve is calculated under permanent vertical loads and a monotonic displacement-controlled lateral load pattern that continuously increases through the elastic and the inelastic range until the ultimate displacement is attained. The lateral load can be assumed proportional to the distribution of mass along the building height, affine to the fundamental modal shape considering reduced stiffness and damage effects or simply as linear along the building height. In the latter case at least one further investigation should be carried out with a rectangular load pattern to consider both the initial behaviour and the behaviour close to failure. Furthermore, second-order effects must be considered as the displacements in the non-linear range are continuously increasing, which magnifies moments and axial forces in the horizontal load-bearing elements.

As an example, Fig. 4.7 shows a two-storey frame with the following characteristics: The bending stiffness of the columns is 120,000 kNm^2 and their fully plastic moments are equal to 700 kNm, whereas the bending stiffness of the beams is 70,000 kNm^2 and their plastic moments 300 kNm. With the floor masses given in Fig. 4.7, a fundamental period of $T_1 = 0.318$ s is obtained. For an associated spectral acceleration of $S_a = 1.0$ m/s^2, storey forces of $F_1 = 17.41$ kN and $F_2 = 23.59$ kN

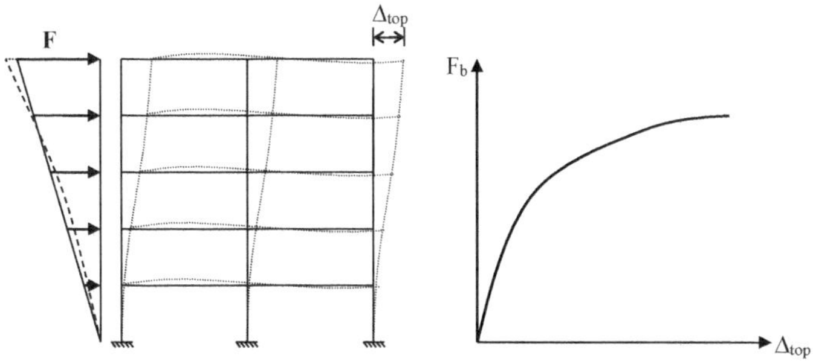

Fig. 4.6 Determination of the non-linear load-displacement curve (pushover curve)

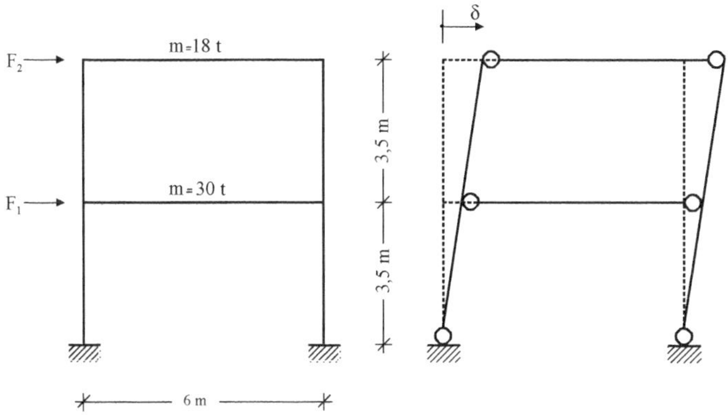

Fig. 4.7 Two-storey frame and deformed shape with plastic hinges

are calculated. If these lateral loads are continuously increased, the capacity curve depicted in Fig. 4.8 results. The load carrying capacity curve represents the total base shear as a function of the displacement at roof level, whereby the formation of the plastic hinges becomes noticeable in the form of abrupt stiffness changes and an overall decreasing stiffness until failure occurs after the formation of six plastic hinges.

A complex calculation of the entire building can be avoided if the overall load carrying capacity curve of the overall building is determined by the superposition of the stiffening elements capacities in the regarded direction of seismic action. This approach is mostly applied for buildings stiffened by reinforced concrete or masonry shear walls. As an example, the non-linear load carrying capacity curve of a single reinforced concrete shear wall with 8 m length, 5 m height and a thickness of 30 cm is depicted in Fig. 4.10. The concrete strength class of the wall is C25/30 and the wall is reinforced with 1.9 cm^2/m mesh reinforcement on both sides. The ductility of the wall is ensured by additional high ductility longitudinal reinforcement bars of 14.6 cm^2 of type B according to Eurocode 2-1-1 (2004) at each wall end. The wall is part of a two-story reinforced concrete building with storey heights of 4 m and was

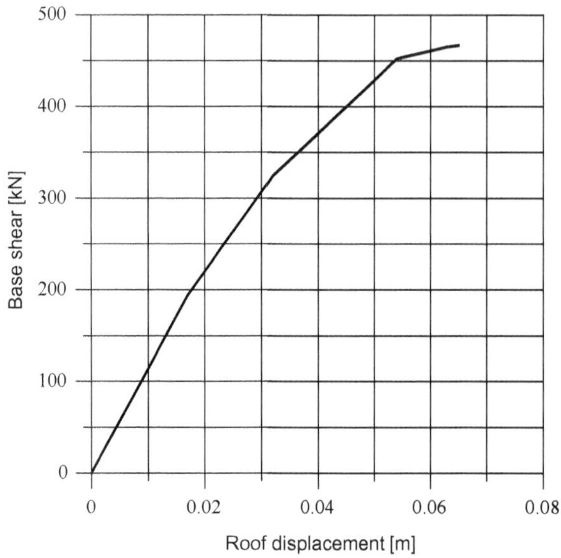

Fig. 4.8 Pushover curve of the two storey frame shown in Fig. 4.7

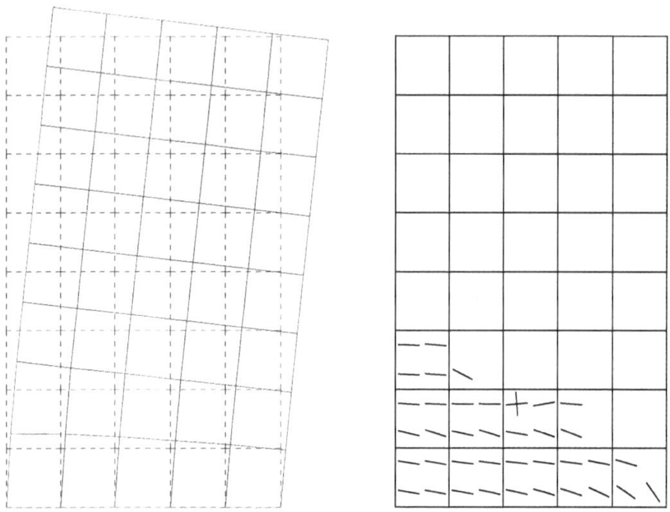

Fig. 4.9 Deformation shape and crack pattern of the wall close to failure

analysed in detail by Noh (2001). Figure 4.9 shows the resulting pushover curve of the wall in terms of the total base shear as a function of the horizontal displacement at the top of the wall.

In addition, pushover analyses can be used to calculate the multiplication factor α_u/α_1 as a measure of ductility of the whole structure. The parameter α_1 corresponds

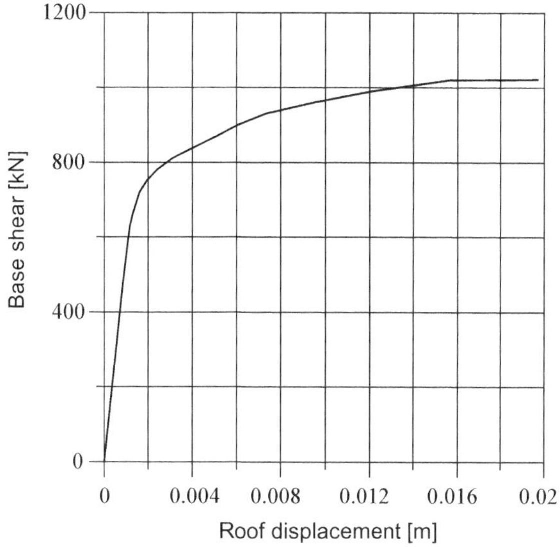

Fig. 4.10 Pushover curve of the reinforced concrete wall shown in Fig. 4.9

to the multiplier of the horizontal design seismic action at the formation of the first plastic hinge in the system, and the parameter α_u denotes the multiplier of the horizontal seismic design action at the system collapse level with the formation of a global plastic mechanism.

The pushover analysis is carried out either with two- or with three-dimensional models, depending on the structural regularity, explained in Sect. 4.1.3. Two independent pushover analyses with lateral loads applied in each direction of the seismic action may be performed when the criteria of structural regularity are satisfied.

4.1.5.4 Capacity Spectrum Method

Among the displacement-based approaches of non-linear static pushover analyses, the capacity spectrum method (CSM) developed by Freeman et al. (1998) is one of the most popular and widely used non-linear static methods. The capacity spectrum method has been developed within the scope of a pilot project for the assessment of the seismic vulnerability of buildings in the Puget Sound Naval Shipyard for the U.S. Navy (Freeman et al. 1975). In this method the seismic action is described by means of a response spectrum and the load-carrying capacity of the building by an inelastic static load-deformation curve. Both curves are transformed into a spectral acceleration-spectral displacement diagram (Fig. 4.11). The point of intersection of both curves (Performance Point) corresponds to the displacement demand. The secant stiffness of the capacity spectrum at the origin of the coordinate system corresponds to the square of circular natural frequency ω^2.

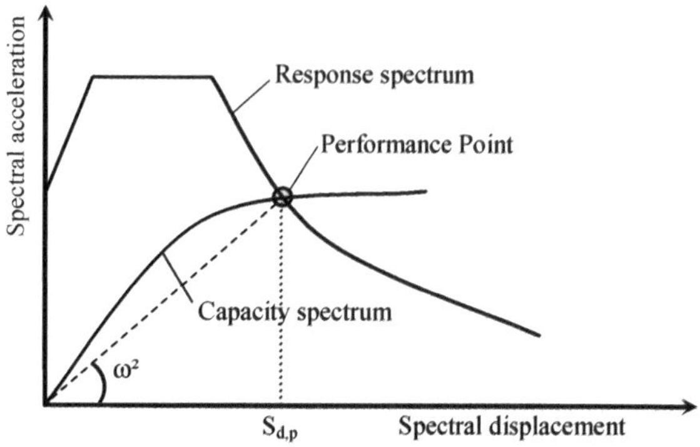

Fig. 4.11 Superposition of the capacity and response spectrum to obtain the performance point

According to the basic idea of the capacity spectrum method the pushover curve is transformed into the S_a-S_d-diagram using an equivalent single-degree-of-freedom (SDOF) oscillator and the fundamental mode shape of the system. Each point *i* of the capacity curve ($F_{b,i}$, $\Delta_{Roof,i}$) is transformed by the following expressions:

$$S_{d,i} = \frac{\Delta_{Roof,i}}{\beta_1 \cdot \phi_{1,Roof}} \tag{4.21}$$

and

$$S_{a,i} = \frac{F_{b,i}}{M_{Tot,eff} \cdot \alpha_1} \tag{4.22}$$

where

$\Delta_{Roof,i}$	Horizontal displacement of point *i* at roof level
$F_{b,i}$	Total base shear of point *i*
$\phi_{1,Roof}$	Amplitude of the fundamental mode at roof level
β_1	Modal participation factor for the fundamental mode of the system
$M_{Tot,eff}$	Effective total building mass
α_1	Modal mass coefficient for the fundamental mode, calculated as the ratio of the effective modal mass M_{eff} to the total mass $M_{Tot,eff}$.

The determination of the modal participation factor β_1 and the modal mass coefficient α_1 requires the calculation of the fundamental natural frequency of the building. Simplified, the building can be idealized as a multi-degree-of-freedom system with horizontal degrees-of-freedom and concentrated masses at each floor level. It is recommended to recalculate the fundamental natural period for each point of the capacity curve by using an updated secant stiffness to consider the decrease of the period caused by non-linear effects. The transformation of the response spectrum into the S_a-S_d-diagram is carried out for each point *i* of the response spectrum using the following formula:

$$S_{d,i} = \frac{T_i^2}{4\pi^2} \cdot S_{a,i}. \tag{4.23}$$

The influence of energy dissipation is considered by a reduction of the linear elastic response spectrum by means of an effective viscous damping ξ_{eff}. The development of a damped response spectrum allows a direct determination of the "Performance Point" without carrying out a time-consuming iterative solution.

In Eurocode 8-1 (2004) the influence of viscous damping is characterised by a viscous damping correction factor η with a reference value of $\eta = 1$ for 5% viscous damping. The value of the damping correction factor is calculated as:

$$\eta = \sqrt{\frac{10}{5+\xi}}, \tag{4.24}$$

where ξ is the viscous damping ratio of the structure in percent. The influence of energy dissipation within the non-linear range of the capacity curve is considered by an equivalent viscous damping ξ_{eq}. This damping part represents the hysteretic material behaviour and has to be recalculated for each point of the capacity spectrum. The equivalent viscous damping ξ_{eq} can be calculated according to Chopra (2001):

$$\xi_{eq} = \frac{1}{4\pi} \frac{E_D}{E_{So}}. \tag{4.25}$$

Here, E_{So} is the maximum strain energy and E_D specifies the hysteretic energy, corresponding to the area of the hysteresis loop (Fig. 4.12).

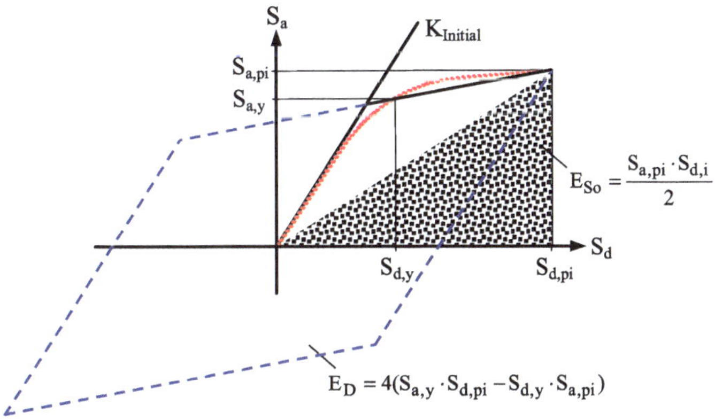

Fig. 4.12 Strain energy E_d and hysteretic energy E_{SO}

Table 4.4 Values of damping modification factor κ according to ATC-40 (1996)

Hysteretic behaviour	ξ_{eq} (%)	κ	Function plot
A Stable hysteresis loops with high energy dissipation	≤ 16.25	1.0	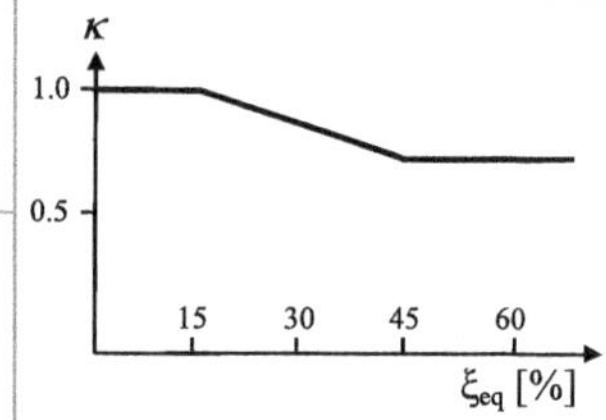
	> 16.25	$1.13 - 0.51 \cdot \frac{S_{a,y} \cdot S_{d,pi} - S_{d,y} \cdot S_{a,pi}}{S_{a,pi} \cdot S_{d,pi}}$	
B Hysteresis loops with moderate reduction of the hysteresis loop areas	≤ 25	0.67	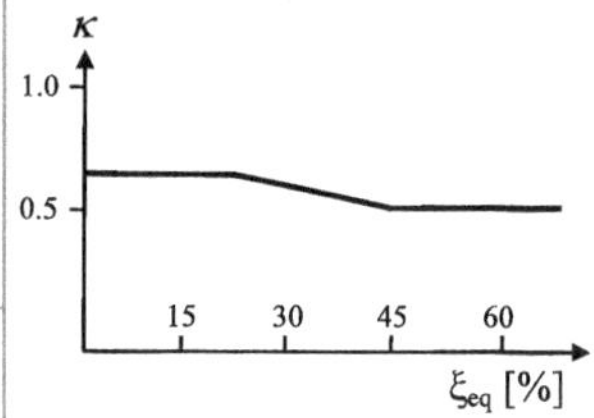
	> 25	$0.845 - 0.446 \cdot \frac{S_{a,y} \cdot S_{d,pi} - S_{d,y} \cdot S_{a,pi}}{S_{a,pi} \cdot S_{d,pi}}$	
C Severely pinched hysteresis loops with low energy dissipation		0.33	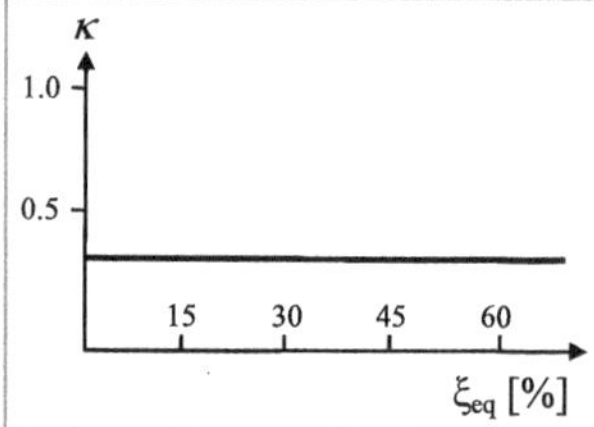

If the load displacement-curve can be idealized as a bilinear curve, the equivalent viscous damping ξ_{eq} can be calculated from Fig. 4.12 as follows (ATC-40 1996):

$$\xi_{eq} = 0.637 \frac{S_{a,y} \cdot S_{d,pi} - S_{d,y} \cdot S_{a,pi}}{S_{a,pi} \cdot S_{d,pi}}. \tag{4.26}$$

The effective total damping ξ_{eff} is determined as the sum of the viscous damping and the equivalent viscous damping ξ_{eq}:

$$\xi_{eff} = \xi_0 + \xi_{eq}. \tag{4.27}$$

Since the real hysteresis curve does not quite look like the parallelogram shown in Fig. 4.12, ATC 40 (1996) recommends a reduction of the equivalent viscous damping by the damping modification factor κ (Table 4.4):

$$\xi_{eff} = \xi_0 + \kappa \cdot \xi_{eq}. \tag{4.28}$$

The values for the damping modification factor are determined from the ratio of the area enclosed by the actual hysteresis loop to the area resulting from the bilinear approach. ATC-40 (1996) proposes damping modification factor for practical

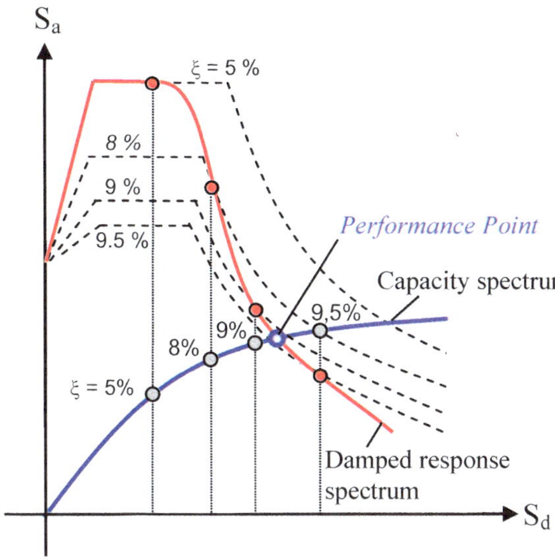

Fig. 4.13 Determination of the "Performance Point" using a damped response spectrum

application as listed in Table 4.4. These factors depend on the type of hysteretic behaviour of the structure.

The calculation of the damped response spectrum allows a direct determination of the "Performance Point" in the common S_a-S_d-diagram (Fig. 4.13). The simple idea of pre-calculating the damped response spectrum avoids the application of time-consuming iterative procedures as included in ATC 40 (1996).

4.1.5.5 Non-linear Pushover Analysis Based on the N2-Method

Among the different approaches proposed in the literature, the N2 method, developed by Fajfar (1989, 1998) is described and recommended in Eurocode 8-1 (2004) for evaluating the seismic response of newly designed and existing buildings. This version of the method combines the advantages of the visual representation of the capacity spectrum method, developed by Freeman et al. (1998), with the physical basis of inelastic demand spectra. Based on the statistical evaluation of the inelastic spectra of Vidic et al. (1994), Fajfar (1999) developed an even simpler approach for applying the method to code spectra. For this purpose, he simplified the response spectra obtained with non-linear SDOF oscillators and replaced the period T_0 by the control period T_C of the design response spectrum (Sect. 4.1.4). By using this substitution the reduction function R_μ proposed by Vidic et al. (1994) simplifies to:

$$\begin{aligned} R_\mu &= (\mu - 1) \cdot \tfrac{T}{T_c} + 1 && \text{for } T \leq T_c \\ R_\mu &= \mu && \text{for } T > T_c. \end{aligned} \tag{4.29}$$

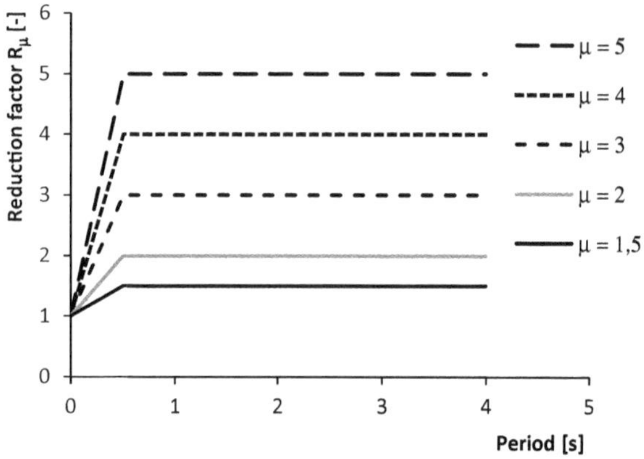

Fig. 4.14 Reduction factor R_μ as a function of the period for $\mu = 1.5$ to $\mu = 5$

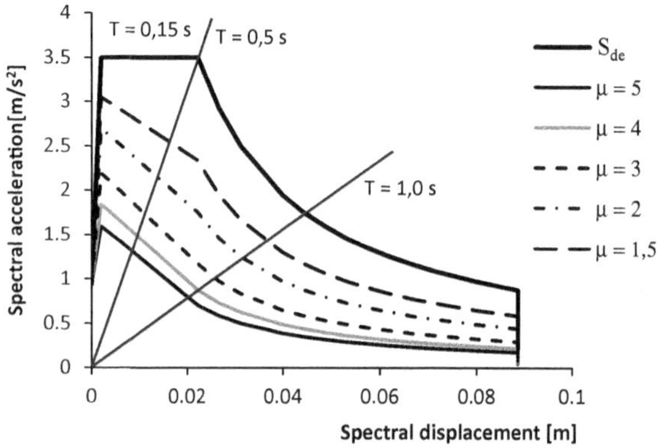

Fig. 4.15 Elastic response spectrum S_{de} with reduced spectra for $\mu = 1.5$ to $\mu = 5$

The reduction function considers a ductility factor μ calculated as the ratio of the yield displacement d_y to the ultimate displacement d_{max}:

$$\mu = \frac{d_{\max}}{d_y}. \tag{4.30}$$

The reduction function R_μ is the basis for the N2 method implemented in the Informative Annex B of Eurocode 8-1 (2004). Figure 4.14 shows the reduction function R_μ for ductility factors μ ranging from 1.5 to 5. Figure 4.15 depicts the reduced inelastic spectra for an elastic spectrum according to Eurocode 8-1 (2004) assuming a spectrum type I and subsoil class E. It is evident that higher ductility factors significantly reduce the response spectra.

4.1.5.6 Application of the N2-Method

The N2-method is a simplified displacement-based verifications concept for determining the "Performance Point" as the intersection point of the response spectrum and the building capacity curve. Unlike the capacity spectrum method, the N2-method does not require a graphical representation of the curves in a common S_a-S_d-diagram. The N2-method combines the pushover analysis of a multi-degree-of-freedom (MDOF) model with the response spectrum analysis of an equivalent SDOF oscillator and is formulated in the S_a-S_d format. The safety check is based on the comparison of the calculated displacement demand and the maximum allowable displacement of the structure. In the following the steps of the non-linear static analysis and safety verification using the N2-method are presented.

Step 1: Determine the elastic response spectrum
This step consists of the choice of an elastic response spectrum according to the employed design code. As no energy dissipation effects have to be considered in this step, a behaviour factor $q = 1$ is applied.

Step 2: Set-up of the dynamic system
Regular structures are idealized as multi-degree of freedom (MDOF) systems with concentrated masses at each storey level. The required control node of the MDOF system is usually selected at the roof level of the building. The lateral load can be assumed to be proportional to the distribution of mass along the building height or affine to the fundamental mode shape, whereas the displacements are normalized to the displacement of the control node. Figure 4.16 shows an example of a four-storey frame idealized as a MDOF system. The node of the mass m_n at roof level is selected as the control node so that the normalized displacement for this node is $\Phi_n = 1$.

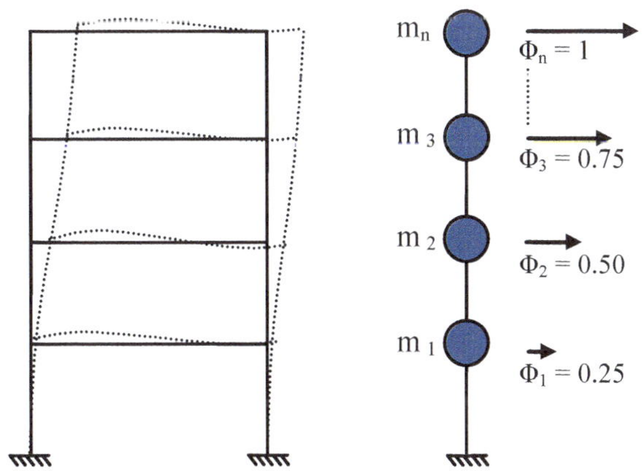

Fig. 4.16 MDOF system with normalized displacement vector

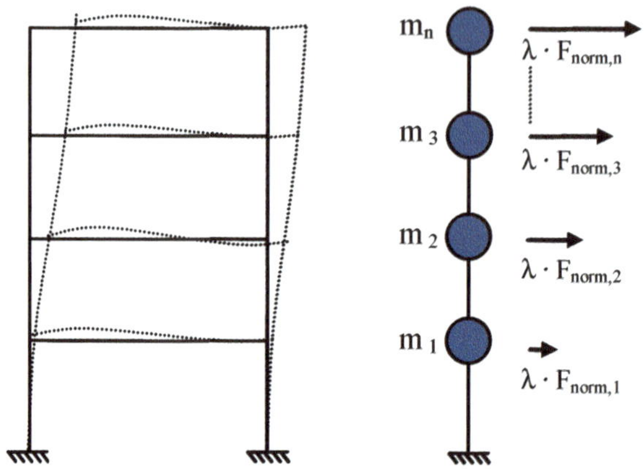

Fig. 4.17 MDOF system with normalized force distribution

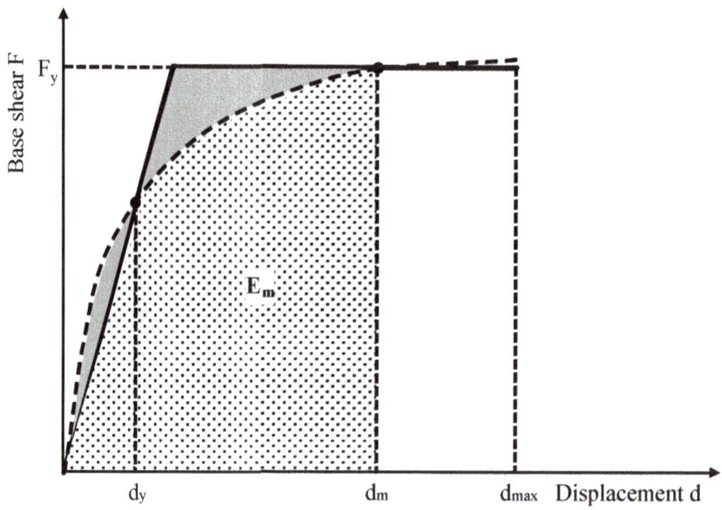

Fig. 4.18 Non-linear pushover curve and bilinear approximation based on the principle of equal energy

Step 3: Determine the pushover curve

The pushover curve is determined by gradually increasing the earthquake loading acting at each floor level as a function of the displacement vector Φ and the corresponding floor mass. The lateral load pattern is normalized with respect to the control node and continuously increased with the load factor λ through the elastic and inelastic range until the ultimate displacement is reached (Fig. 4.17):

$$\underline{F}_{norm} = \lambda \cdot \frac{\underline{\underline{M}} \cdot \underline{\Phi}}{m_n \cdot \Phi_n}. \tag{4.31}$$

Eurocode 8-1 (2004) requires the use of at least two distributions of the lateral loads and a simplified consideration of torsional effects according to Eurocode 8-1 (2004), Sect. 4.3.3.4.2.7. The two distributions are (a) a uniform pattern with lateral forces proportional to the distribution of mass along building height and (b) a modal pattern affine to the fundamental mode shape.

The result of the pushover analysis is a non-linear displacement curve, which is the basis for the definition of the yield displacement d_m, the maximum yield force F_y and the maximum displacement d_{max} as shown in Fig. 4.18. These parameters are further used to define the bilinear approximation of the non-linear displacement curve. From the principle of equal energy of the areas under the actual and the idealized load-displacement curve, the initial stiffness of the bilinear approximation is determined. Thereafter, the yield displacement d_y is determined, which defines the transition between the elastic and the plastic region. The yield displacement d_y follows from the deformation energy E_m under the actual load-deformation curve:

$$d_y = 2\left(d_m - \frac{E_m}{F_y}\right). \tag{4.32}$$

Step 4: Determine the pushover curve for the equivalent mass oscillator
The bilinear pushover curve (Fig. 4.19) of the equivalent SDOF oscillator is calculated by scaling with the modal participation factor Γ:

$$F^* = \frac{F}{\Gamma} \tag{4.33}$$

and

$$d^* = \frac{d}{\Gamma}. \tag{4.34}$$

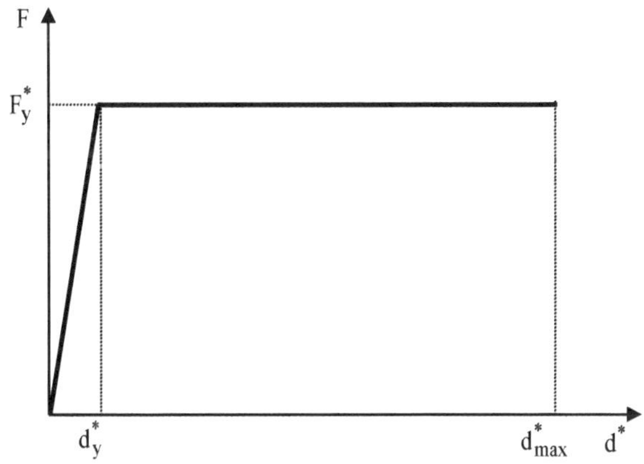

Fig. 4.19 Idealized pushover curve of the equivalent single-mass oscillator

The modal participation factor Γ is expressed as:

$$\Gamma = \frac{\sum_{i=1}^{n} m_i \cdot \Phi_i}{\sum_{i=1}^{n} m_i \cdot \Phi_i^2} \tag{4.35}$$

where

n Number of stories
m_i Mass of floor i
ϕ_i Ordinates of the fundamental mode shape normalized to the control node

The equivalent mass $m*$ of the equivalent SDOF oscillator is calculated as:

$$m^* = \sum_{i=1}^{n} m_i \cdot \Phi_i. \tag{4.36}$$

The simple bilinear curve of the equivalent SDOF oscillator is defined by the elastic limit force F_y^*, the yield displacement d_y^*, the elastic stiffness $k^* = F_y^*/d_y^*$ and the maximum displacement capacity d_{max}^*.

Step 5: Determine the period for the equivalent mass oscillator
The period T^* of the equivalent SDOF oscillator is calculated as:

$$T^* = 2\pi\sqrt{\frac{m^*}{k^*}} = 2\pi\sqrt{\frac{m^* d_y^*}{F_y^*}}. \tag{4.37}$$

Step 6: Determine the target displacement of the equivalent SDOF oscillator
The elastic displacement d_{et}^* of the equivalent SDOF oscillator with period T^* and unlimited linear elastic behaviour is given by:

$$d_{et}^* = S_e(T^*)\left[\frac{T^*}{2\pi}\right]^2. \tag{4.38}$$

Herein, $S_e(T^*)$ is the ordinate of the elastic response spectrum constructed in step 1 for the period T^*. The target displacement d_t^* of the equivalent SDOF oscillator must be determined from d_{et}^* as a function of the period range and the material behaviour. A total of three cases must be considered for the calculation of the target displacement:

Case 1: $T^* < T_C$ and $\frac{F_y^*}{m^*} \geq S_e(T^*)$
This case refers to structures with linear behaviour in the short period range. The target displacement d_t^* is equal to the elastic displacement d_{et}^* assuming an unlimited linear material behaviour:

$$d_t^* = d_{et}^*. \tag{4.39}$$

The elastic displacement d_{et}^* of the equivalent SDOF oscillator is calculated for the period $T = T^*$ with (4.38).

Case 2: $T^* < T_C$ and $\frac{F_y^*}{m^*} < S_e(T^*)$
This case refers to structures with non-linear behaviour in the short period range. The target displacement d_t^* is calculated from the elastic displacement d_{et}^* of the equivalent SDOF oscillator taking into account the required ductility:

$$d_t^* = \frac{d_{et}^*}{q^*}\left[1 + \left(q^* - 1\right)\frac{T_c}{T^*}\right] \geq d_{et}^*. \tag{4.40}$$

Here, q^* is the ratio of the acceleration in the structure with unlimited elastic behaviour $S_e(T^*)$ and in the structure with limited strength F_y^*/m^*:

$$q^* = \frac{S_e(T^*)m^*}{F_y^*}. \tag{4.41}$$

Case 3: $T^* \geq T_C$
This case refers to structures with periods in the range of medium and long fundamental periods. The target displacement d_t^* is equal to the elastic displacement d_{et}^* of the equivalent SDOF oscillator assuming an unlimited linear material behaviour:

$$d_t^* = d_{et}^*. \tag{4.42}$$

Step 7: Calculation of the target displacement of the MDOF oscillator
The target displacement of the MDOF oscillator corresponding to the control node is calculated with the modal participation factor Γ:

$$d_t = \Gamma \cdot d_t^*. \tag{4.43}$$

Step 8: Verification of the target displacement
The verification according to Sect. 4.3.3.4.2.3 of Eurocode 8-1 (2004) is satisfied if the determined target displacement d_t does not exceed the maximum displacement d_{max} reduced by the safety factor of 1.5 to limit the utilization of the plastic reserves:

$$d_t \leq \frac{d_{max}}{1.5}. \tag{4.44}$$

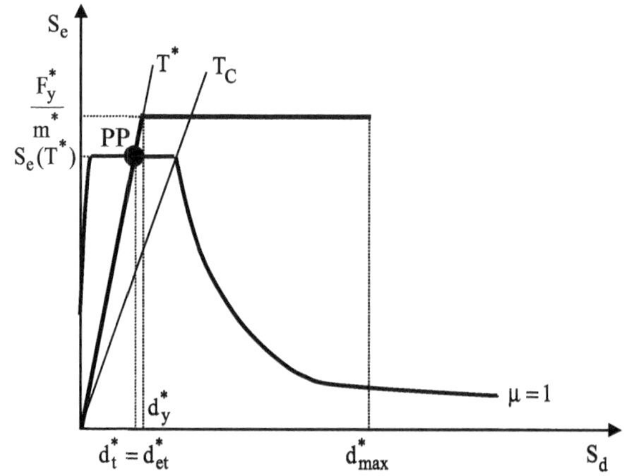

Fig. 4.20 Case 1: $T^* < T_C$, linear elastic material behaviour

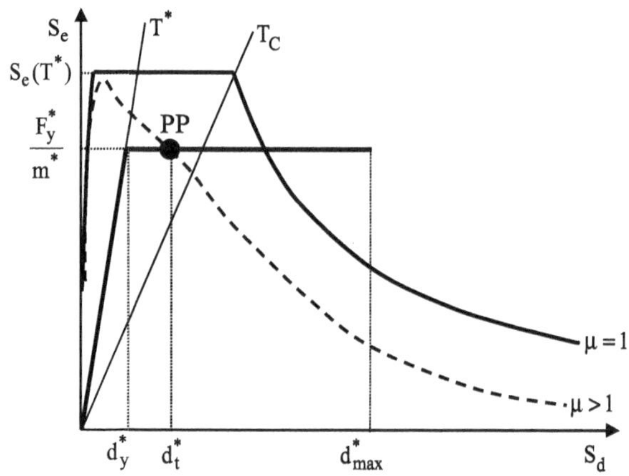

Fig. 4.21 Case 2: $T^* < T_C$, non-linear material behaviour

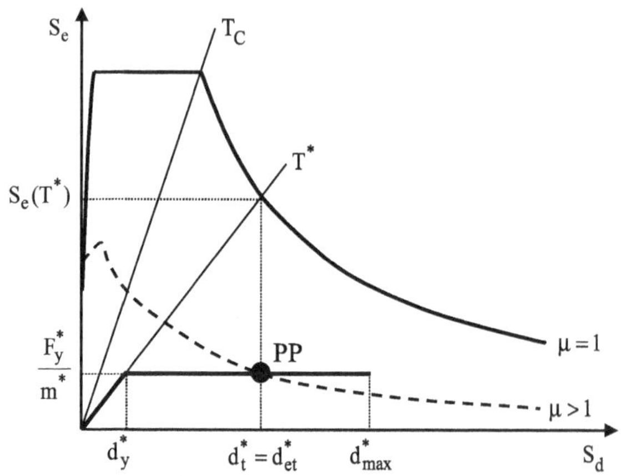

Fig. 4.22 Case 3: $T^* \geq T_C$, non-linear material behaviour

Step 9: Determination of the structural ductility
The structural ductility demand corresponds to the ratio between the target displacement d_t and the yield displacement d_y:

$$\mu = \frac{d_t}{d_y}. \tag{4.45}$$

Step 10: Graphical representation in the S_a-S_d diagram
A graphical representation of the target displacement allows a better understanding and control of the calculation results. For this purpose the reduced response spectrum and the pushover curve of the SDOF oscillator are superposed in the S_a-S_d diagram. The reduction of the spectra is carried out by means of the reduction factor R_μ introduced in Sect. 4.1.5.5. The point of intersection of both curves corresponds to the target displacement d_t^* of the SDOF oscillator. In Figs. 4.20, 4.21 and 4.22 the graphical representation of the three cases described in step 6 are shown qualitatively. The performance point (PP), as in the capacity spectrum method, is the point of intersection of the two curves and it yields the identical target displacement d_t^*.

4.1.5.7 Non-linear Dynamic Analysis

Non-linear dynamic analyses of MDOF oscillators are procedures for evaluating the dynamic response of structures over time. Such analyses are the most accurate calculation methods in that they simulate the transient behaviour of structures taking into account their non-linear material behaviour. Indeed, the practical application of non-linear dynamic analyses is limited as they are computationally expensive and the huge amount of the produced time-dependent results is not easy to use for the subsequent dimensioning of the structural elements. However, in some cases a detailed non-linear time-history analysis can be reasonable. The structural response can be obtained through a direct numerical integration of the differential equations of motion. The number of accelerograms to be used as inputs must be at least three and Eurocode 8-1 (2004) proposes in Sect. 3.2.3.1.2 the following three types:

- Artificial accelerograms: generated to match the elastic site specific response spectra for 5% viscous damping. The minimum duration should be 10 s.
- Recorded accelerograms: real seismic records recorded by stations can be used provided that they are adequately qualified with respect to the seismogenetic features of the sources and to the soil conditions appropriate to the site. The records must be scaled to the value of $a_g S$ for the earthquake zone under consideration (Fig. 4.23).
- Simulated accelerograms: accelerograms generated by site-specific hazard analysis considering parameters such as seismic sources, rupture types, site characteristics and the travel path mechanism (Plevris et al. 2017).

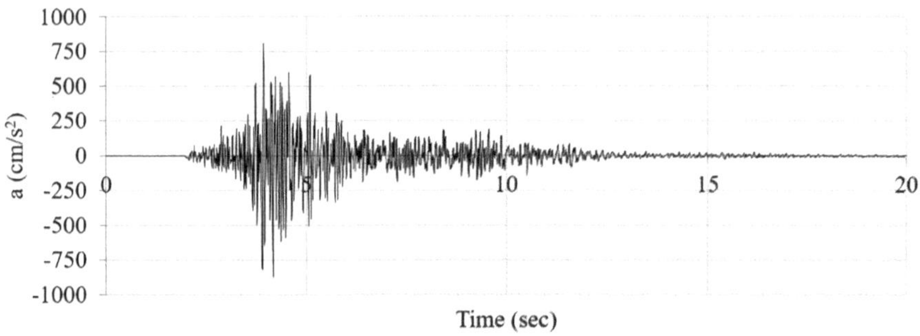

Fig. 4.23 Recorded acceleration time-history (Mirandola, vertical, Accelerograms Database of seismic shocks on May, 29th 2012, Emilia Romagna region (Italy) (Department of Italian Civil Protection 2018)

If a non-linear material is assumed, the mechanical behaviour under post-elastic unloading and reloading cycles must be defined for all primary seismic load-carrying elements. The material models must reflect the energy dissipation capacity over the range of displacement amplitudes expected in the seismic design situation. This condition is generally difficult to satisfy since the correct mechanical cyclic behaviour can be realistically determined only with access to experimental results. The design value of the action effect (4.57) can be calculated according to Sect. 4.3.3.4.3 (3) of Eurocode 8-1 (2004):

- As the average of the response quantities, if at least seven independent time histories of the ground motion are considered.
- As the most unfavourable value of response quantities, if between three and seven accelerograms are considered.

When using non-linear time-history analyses and employing a spatial model of the structure, simultaneously acting accelerograms must be applied in the three main directions.

4.1.6 Torsional Effects

4.1.6.1 Buildings with Symmetrical Distribution of Stiffness and Mass

For structures with symmetrical distribution of lateral stiffness and mass, the accidental torsional effects can be taken into account by increasing the calculated stress resultants in the individual load resisting elements by a factor δ defined as:

$$\delta = 1 + 0.6\frac{x}{L_e}. \tag{4.46}$$

Here, x is the distance between the regarded structural element and the centre of mass of the structure, measured perpendicular to the direction of the applied earthquake action, and L_e is the distance between the two outermost lateral structural elements, measured perpendicular to the direction of the applied seismic action.

4.1.6.2 Buildings with Asymmetrical Distribution of Stiffness and Mass

If buildings with an asymmetrical distribution of lateral stiffness and mass are analysed with three-dimensional calculation models, only the accidental torsional effects must be considered for each of the storeys. The torsional effects are represented by accidental eccentricities of the mass centre, which must be applied on all storey levels in both directions of the seismic actions with changing sign:

$$e_{1i} = \pm 0.05 \cdot L_{1i}, \tag{4.47}$$

Here, L_i is the dimension of the floor i perpendicular to the seismic direction. The accidental eccentricity e_{1i} can be taken into account in the calculation model by applying additional torsional moments as equivalent pairs of forces along the boundaries of the floor slabs. The torsional moments have to be applied with changing signs, but in the same direction for all storeys. As an alternative, buildings with an asymmetrical distribution of lateral stiffness and mass can be analysed with two planar systems in the main directions of the building as long as the following requirements according to Sect. 4.3.3.1 of Eurocode 8-1 (2004) are met:

(a) The building has well-distributed and relatively rigid cladding and partition walls.
(b) The building is not higher than 10 m and has well-distributed rigid outer and inner walls.
(c) The in-plane stiffness of the floors allows the assumption of a rigid diaphragm.
(d) The centres of lateral stiffness and mass are approximately on a vertical line and the following conditions are satisfied in the two horizontal directions of analysis: $r_x^2 > l_s^2 + e_{0x}^2$ and $r_y^2 > l_s^2 + e_{0y}^2$.

Herein, l_s^2 is the square of the radius of gyration, which corresponds to the quotient of the mass moment of inertia of the regarded storey for rotations around the vertical axis through its centre of gravity and the storey mass. The square of the radius of gyration l_s^2 for a rectangular plan with dimensions L and B and uniformly distributed mass is equal to:

$$l_s^2 = (L^2 + B^2)/12. \tag{4.48}$$

The square of the torsional radius corresponds to the ratio of torsional and lateral stiffness of the storey in the direction of the seismic action. The torsional radii r_i^2, r_j^2 for the directions *i, j* of the horizontal seismic actions are calculated as:

$$r_i^2 = \frac{K_T}{K_j} = \frac{\sum_{i=1}^{n} k_i \cdot r_{di}^2 + \sum_{j=1}^{l} k_j \cdot r_{dj}^2}{\sum_{j=1}^{n} k_j}, \quad r_j^2 = \frac{K_T}{K_i} = \frac{\sum_{i=1}^{n} k_i \cdot r_{di}^2 + \sum_{j=1}^{l} k_j \cdot r_{dj}^2}{\sum_{i=1}^{n} k_i} \tag{4.49}$$

where:

K_T Total torsional stiffness of the considered storey
K_i, K_j Total translational stiffness parallel and perpendicular to the direction of seismic action of the considered storey
k_i, k_j Translational stiffnesses of stiffening elements parallel and perpendicular to the direction of seismic action
r_{di}, r_{dj} Distance of the stiffening elements to the centre of stiffness

If the aforementioned conditions (a)–(d) are fully met, the torsional effects can be calculated simply by increasing the calculated stress resultants in the individual load-resisting elements by a factor δ defined as:

$$\delta = 1 + 1.2\frac{x}{L_e}, \tag{4.50}$$

or by doubling the accidental eccentricity e_1 of expression (4.47).

$$e_{1i} = \pm 0.1 \cdot L_i. \tag{4.51}$$

Alternatively, a more accurate calculation approach can be used if the centres of stiffness and masses of the individual stories are approximately on a vertical line and if the conditions $r_x^2 > l_s^2 + e_{ox}^2$ and $r_y^2 > l_s^2 + e_{oy}^2$ are satisfied. The approach considers torsional effects in each direction of the seismic action by means of the structural eccentricity e_o, the additional eccentricity e_2 (consideration of dynamic effects due to simultaneous translational and torsional vibrations) and the accidental eccentricity e_1. The additional eccentricity e_2 results as a minimum from the following expressions:

$$\text{Min}\begin{cases} e_2 = 0.1 \cdot (L+B) \cdot \sqrt{\frac{10 \cdot e_0}{L}} \leq 0.1 \cdot (L+B) \\ e_2 = \frac{1}{2e_o}\left[l_s^2 - e_0^2 - r^2 + \sqrt{\left(l_s^2 + e_0^2 - r^2\right)^2 + 4 \cdot e_0^2 \cdot r^2}\right] \end{cases}. \tag{4.52}$$

The maximum and minimum eccentricities for each storey are calculated from the eccentricities e_0, e_1 and e_2 (Fig. 4.24):

$$\begin{aligned} e_{\text{max}} &= e_0 + e_1 + e_2 \\ e_{\text{min}} &= 0.5\, e_0 - e_1. \end{aligned} \tag{4.53}$$

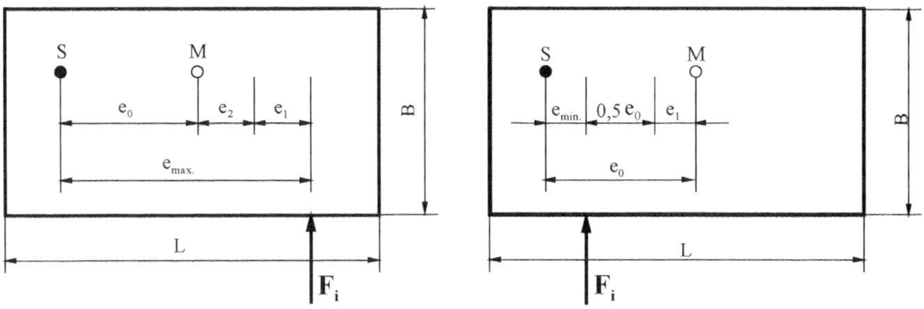

Fig. 4.24 Maximum and minimum eccentricities for planar models

Thereafter, the resulting seismic forces in the stiffening elements parallel (index i) and perpendicular (index j) to the direction of the seismic action are calculated with the maximum and minimum eccentricities e_{min} and e_{max}. For this purpose, load distribution factors s_i, s_j for the stiffening elements in each storey are calculated parallel and perpendicular to the direction of the seismic action. The distribution factors correspond to a percentage of the horizontal seismic forces acting on the floor levels and can be calculated as follows:

$$s_i = \frac{k_i}{K_i}\left(1 \pm \frac{K \cdot r_{di} \cdot e}{K_T}\right), \; s_j = \frac{k_j \cdot r_{dj} \cdot e}{K_T}. \tag{4.54}$$

If only conditions (a)–(c) of this section are satisfied, torsional effects can be considered using the presented simplified or more accurate approach; however, all seismic action effects resulting from the analysis must be multiplied by 1.25 in this case. If none of the conditions (a)–(d) is fulfilled, a three-dimensional calculation taking into account torsional effects must be carried out. The accidental eccentricities can be considered by additional torsional moments M_{1i} to be applied on each floor level i:

$$M_{1i} = e_{1i} \cdot F_i. \tag{4.55}$$

Here, F_i is the horizontal seismic force acting on floor level i and e_{1i} is the eccentricity of the storey mass i as defined in (4.47).

4.1.6.3 Combination of Actions for the Seismic Design Situation

The design value E_{dAE} of the effects of actions in the seismic design situation is determined in accordance with Eurocode 0 (2002):

$$E_{dAE} = E\left\{\sum_{j\geq 1} G_{k,j} \oplus P_k \oplus \gamma_1 \cdot A_{Ed} \oplus \sum_{i\geq 1} \psi_{2,i} \cdot Q_{k,i}\right\} \tag{4.56}$$

where:

E	Effect of actions
$\oplus$	"in combination with"
$G_{k,j}$	Characteristic value of a permanent action j
A_{Ed}	Design value of seismic action
$Q_{k,i}$	Characteristic value of a single variable action i
$\psi_{2,i}$	Combination coefficient for quasi-permanent value of a variable action i
P_k	Characteristic value of prestressing actions
γ_1	Weighting factor for seismic actions, ($\gamma_1 = 1.0$)

The design value of the seismic action A_{Ed} is determined taking into account permanently acting vertical loads:

$$A_{Ed} = A\left\{\sum G_{k,j} \oplus \sum \psi_{E,i} Q_{k,i}\right\} = A\left\{\sum G_{k,j} \oplus \sum \varphi\, \psi_{2,i}\, Q_{k,i}\right\}. \tag{4.57}$$

The combination coefficients $\psi_{E,i} = \varphi \cdot \psi_{2,i}$ take into account the likelihood of the loads $Q_{k,i}$ not being present over the entire structure during the earthquake. The factor φ is defined in Table 4.2 of Eurocode 8-1 (2004) and the combination coefficients $\psi_{2,i}$ are given in Eurocode 0 (2002), Table A1.1. The safety verification is fulfilled, if the design value E_{dAE} of the effects of actions in the seismic design situation does not exceed the corresponding design resistance R_d for all structural elements including connections and relevant non-structural elements:

$$E_{dAE} < R_d = R\left\{\frac{f_k}{\gamma_m}\right\}. \tag{4.58}$$

The design resistance is calculated in accordance with the material-specific rules in terms of the characteristic value of material properties f_k and partial factors γ_M.

Second-order effects need not be taken into account if the interstorey drift sensitivity coefficient θ fulfils the following condition in all storeys:

$$\theta \approx \frac{P_{tot} \cdot d_e}{V_{tot} \cdot h} \leq 0.10. \tag{4.59}$$

Here, P_{tot} is the total gravity load at and above the storey, d_e is the elastic design interstorey drift, h is the storey height and V_{tot} is the total seismic storey shear. Values of the coefficient θ must not exceed 0.3. If θ lies in the range $0.1 < \theta \leq 0.2$, second-order effects can be taken into account by simply multiplying the seismic action effects by a factor of $1/(1 - \theta)$.

4.2 Design and Specific Rules for Different Materials

4.2.1 Design of Reinforced Concrete Structures

Reinforced concrete structures can be designed in accordance with Eurocode 8-1 (2004) for low-dissipative or dissipative structural behaviour. Three structural ductility classes are proposed by the standard:

DCL Low dissipative structural behaviour
DCM Medium dissipative structural behaviour
DCH High dissipative structural behaviour.

In ductility class DCL, a linear elastic structural analysis is carried out with a behaviour factor of 1.5. The subsequent design is executed according to Eurocode 2-1-1 (2004) without considering any further earthquake specific requirements. In ductility classes DCM and DCH, it is allowed to use higher behaviour factors, if the material-specific design requirements are considered. Eurocode 8-1 (2004) provides these requirements in Sect. 5.4 for ductility class DCM and in Sect. 5.5 for ductility class DCH. The upper limit of the behaviour factor for dissipative reinforced concrete buildings regular in elevation is defined according to Eurocode 8-1 (2004), Sect. 5.2.2.2 as:

$$q = q_0 \cdot k_W \geq 1.5. \tag{4.60}$$

Here, q_0 is the basic value of the behaviour factor according to Table 4.5 and the factor k_w reflects the prevailing failure mode in structural systems with walls. In case of frame or equivalent dual systems the factor takes the value 1.0. In all other cases the factor can be calculated as:

$$k_W = (1 + \alpha_0)/3 \leq 1, \text{ but } \geq 0.5. \tag{4.61}$$

The prevailing aspect ratio α_0 can be determined from the following expression, if the individual aspect ratios do not significantly differ:

$$\alpha_0 = \frac{\sum H_{wi}}{\sum l_{wi}}, \tag{4.62}$$

Table 4.5 Basic value of the behaviour factor q_o for systems regular in elevation, Eurocode 8-1 (2004)

Structural type	DCM	DCH
Frame system, dual system, coupled wall system	3.0 α_u/α_1	4.5 α_u/α_1
Uncoupled wall system	3.0	4.0 α_u/α_1
Torsionally flexible system	2.0	3.0
Inverted pendulum system	1.5	2.0

where H_{Wi} is the total height and l_{wi} is the length of the wall i. For buildings not regular in elevation, the value of q_0 must be reduced by 20%.

The factor α_u/α_1 can be evaluated by an explicit non-linear calculation. Alternatively, the following approximate values of α_u/α_1 for buildings which are regular in plan can be applied according to Eurocode 8-1 (2004), Sect. 5.2.2.2:

Frames or frame-equivalent dual systems:

- One-storey buildings: $\alpha_u/\alpha_1 = 1.1$
- Multistorey, one-bay frames: $\alpha_u/\alpha_1 = 1.2$
- Multistorey, multi-bay frames or frame-equivalent dual structures: $\alpha_u/\alpha_1 = 1.3$

Wall- or wall-equivalent dual systems:

- Wall systems with only two uncoupled walls per horizontal direction: $\alpha_u/\alpha_1 = 1.0$
- Other uncoupled wall systems: $\alpha_u/\alpha_1 = 1.1$
- Wall-equivalent dual, or coupled wall systems $\alpha_u/\alpha_1 = 1.2$.

4.2.1.1 Calculation Example: 4-Storey Reinforced Concrete Building

The following calculation example presents a four-storey reinforced concrete building stiffened in both main directions with shear walls. The seismic calculation is carried out by means of a three-dimensional model using multimodal analysis and the design process is shown for the most stressed wall for ductility classes DCM and DCH.

Structure Description

The investigated structure is a 4-storey reinforced concrete building stiffened by five 30 cm thick reinforced concrete shear walls arranged asymmetrically. Columns with a diameter of 30 cm and the shear walls are arranged in plan in a grid pattern of 5 m × 6 m and are continuous from foundation to roof. The columns are considered to be attached to the slabs by hinges, so that they only carry vertical loads. The rigid reinforced concrete floor slabs are 20 cm thick. Figure 4.25 depicts the plan of the building and Fig. 4.26 shows its elevation. The longitudinal direction of the building corresponds to the x-coordinate and the transverse direction to the y-coordinate. The seismic action is determined based on the German National Annex DIN EN 1998-1/NA (2011), as the building location is close to Aachen in Germany.

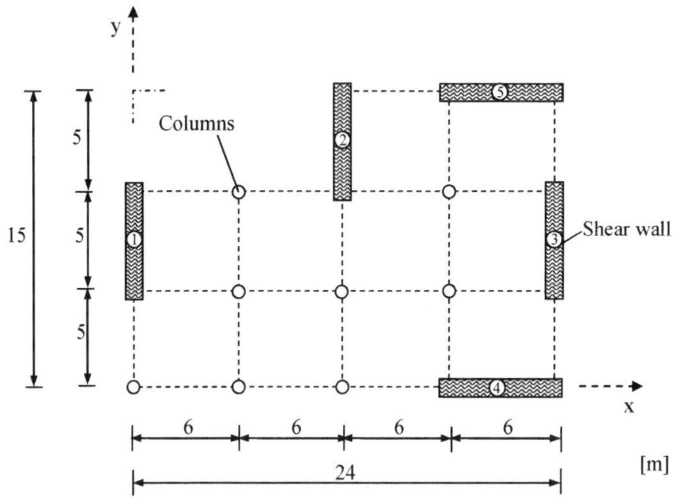

Fig. 4.25 Ground plan configuration of the reinforced concrete building

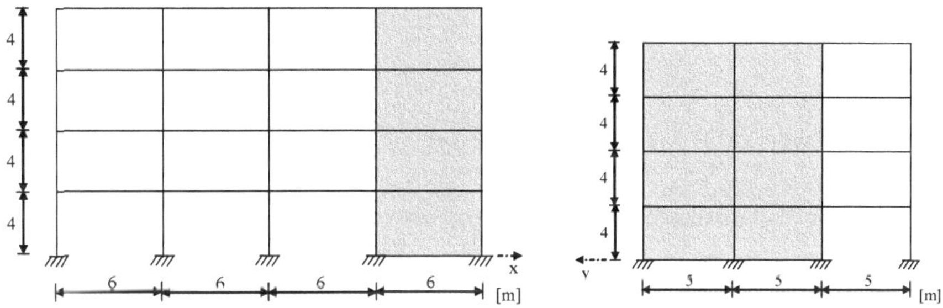

Fig. 4.26 Elevation of the building in longitudinal (x) and transverse direction (y)

The following input data are used for the seismic calculation and design:

Construction material:	Concrete C20/25 Reinforcement: BSt 500 M (A), BSt 500 S (B)
Building location:	Aachen, Germany, seismic zone 3 according to DIN EN 1998-1/NA (2011)
Building type:	Office building, Importance category II => $\gamma_I = 1.0$ The storeys are occupied independently of each other

Actions and Seismic Design Situation

The actions according to Eurocode 1-1-1 (2004) are summarized below:
Dead load
Floor slabs:

Concrete, d = 20 cm	25 kN/m^3
Plaster (d = 1.5 cm)	21 kN/m^3
Floor screed (d = 5 cm)	22 kN/m^3
Sum	6.42 kN/m^2

Roof:

Concrete (d = 20 cm)	25 kN/m^3
Plaster (d = 1.5 cm)	21 kN/m^3
Roof	1.2 kN/m^2
Sum	6.52 kN/m^2

Live loads
Floor slabs:

Live load on slabs for offices	2 kN/m^2
Surcharge for light partition walls	1.25 kN/m^2
Sum	3.25 kN/m^2

Roof:

Live loads on roof	0.75 kN/m^2
Snow load	0.75 kN/m^2

The building is designed for the seismic design situation according to (4.56). The corresponding partial safety factors and combination coefficients ψ for the individual load actions according to Eurocode 2-1-1 (2004) are summarized in Table 4.6 and Table 4.7 respectively.

Horizontal Design Spectrum

The design response spectrum is constructed according to DIN EN 1998-1/NA (2011) as the building location is close to Aachen in Germany. The location lies in seismic

Table 4.6 Partial safety factors for reinforced concrete according to Eurocode 2-1-1 (2004)

	γ_G for permanent actions G_k	γ_Q for variable actions Q_k
Favourable effect	1.00	0.00
Unfavourable effect	1.35	1.50

Table 4.7 Combination coefficients according to Eurocode 0 (2002), Table A.1.1

	Combination coefficients		
	ψ_0	ψ_1	ψ_2
Live loads on slabs in offices	0.7	0.5	0.3
Snow loads	0.5	0.2	0 (0.5)[a]

[a]The value of ψ_2 for snow loads is 0.5 according to DIN EN 1998-1/NA (2011)

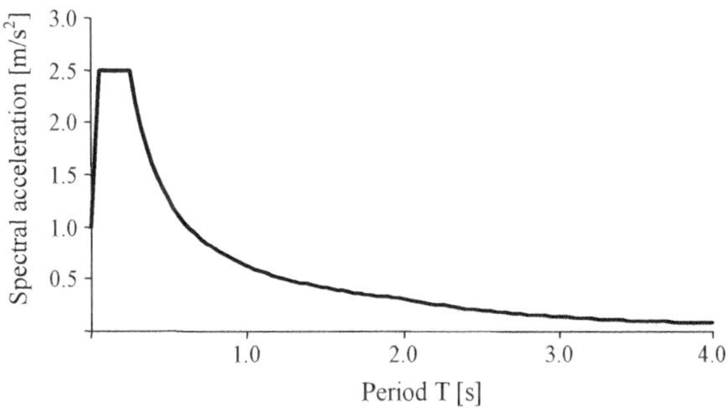

Fig. 4.27 Horizontal design spectrum (q = 1.0) according to DIN EN 1998-1/NA (2011)

zone 3 with a design ground acceleration of a_g = 0.8 m/s^2. The soil factor S and the control periods of the horizontal design spectrum are determined for the deep geology class R in combination with soil class B:

$$S = 1.25,\ T_B = 0.05\ \text{s},\ T_C = 0.25\ \text{s and}\ T_D = 2.0\ \text{s}.$$

The office building is classified in importance category II with an importance factor of γ_I = 1.0. The resulting design response spectrum assuming a behaviour factor of q = 1.0 is depicted in Fig. 4.27.

Vertical Component of the Seismic Action

According to Eurocode 8-1 (2004), Sect. 4.3.3.5.2, the vertical component of a seismic action must only be considered in the design of beams that carry columns if the vertical acceleration exceeds 0.25 g. As the horizontal acceleration is just a_{vg} = 0.04 g (50% of a_g) and the vertical load transfer of the reinforced concrete structure is continuous from roof to foundation, the vertical seismic action is not taken into account.

Behaviour Factor

The behaviour factors are determined for ductility classes DCM and DCH following the explanations and remarks in Sect. 4.2.1.

Ductility class DCM
For the investigated reinforced concrete building the basic value of the behaviour factor q_0 is determined for "uncoupled wall systems" after Table 4.5 with $q_0 = 3.0$. The resulting behaviour factor q is calculated with respect to the prevailing failure mode in structural wall systems considering the factor k_W:

$$q = q_0 \cdot k_w \geq 1.5.$$

The factor k_W reflects the prevailing failure mode in structural systems with walls, as described in Sect. 4.2.1, and is calculated to:

$$k_w = (1 + \alpha_0)/3 = 1.32 \leq 1.0 \quad => k_w = 1.0,$$

where:

$$\alpha_0 = \frac{\sum H_{wi}}{\sum l_{wi}} = \frac{5 \cdot 16m}{2 \cdot 6m + 3 \cdot 5m} = 2.96.$$

The resulting behaviour factor $q = 3.0 \cdot 1.0 = 3.0$ can be applied in both horizontal building directions.

Vertical Loads to Be Considered in the Seismic Design Situation

For the ultimate limit state verification, the design value A_{Ed} of the seismic action must be calculated taking into account all permanently acting vertical loads according to (4.57). The combination coefficients $\psi_{E,i} = \varphi \cdot \psi_{2,i}$ consider the likelihood of the loads $Q_{k,i}$ not being present over the entire structure during the earthquake. Thus, Eurocode 8-1 (2004) differentiates between independently occupied storeys and storeys with correlated occupancies reflected by a factor φ varying with respect to the load categories according to Eurocode 1-1-1 (2004). Corresponding to the use as office building, each storey can be regarded as independent and the factor φ is applied with 1.0 for the roof and with 0.5 for all other storey levels. The factor ψ_2 is applied with 0.3 for variable loads and with 0.5 for snow loads. Accordingly, the following vertical loads are considered for the calculation of the seismic active masses for the computation of the dynamic structural response (Table 4.8):

$$G_k \oplus 0.3 \cdot Q_{\text{Live Load}} \oplus 0.5 \cdot Q_{\text{Snow Load}}$$

$$G_k \oplus 0.15 \cdot Q_{\text{Snow Load}} \cdot$$

Table 4.8 Values of φ for calculating $\psi_{E,i}$ according to Eurocode 8-1 (2004)

Type of variable action	Storey	φ
Categories A–C	Roof	1.0
	Storeys with correlated occupancies	0.8
	Independently occupied storeys	0.5
Categories D–F and archives		1.0

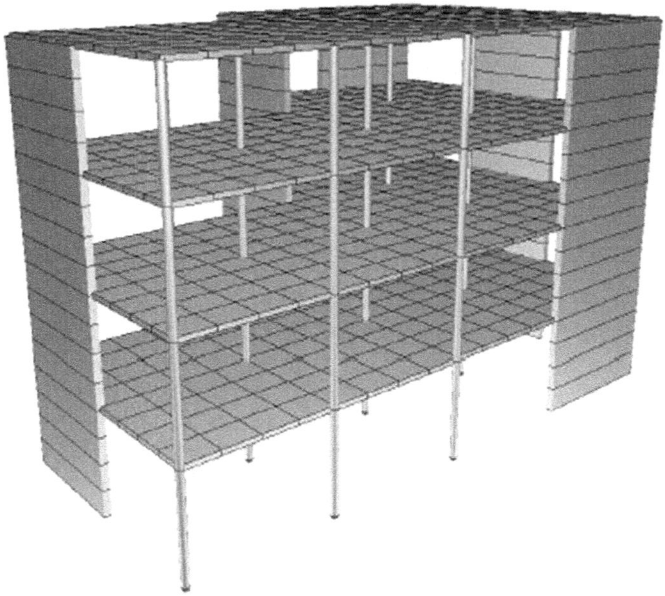

Fig. 4.28 Three-dimensional finite-element model of the office building

Modelling

The building meets the conditions of regularity in elevation, but the ground plan is irregular. Because of the irregularity in plan a three-dimensional model is used to calculate the stress resultants considering all relevant modes of vibration. The slabs and walls are idealized by shell elements and the hinged columns are modelled using truss elements. The overall building model is shown in Fig. 4.28.

Modal Analysis and Vibration Modes

The first three natural modes of the structure with natural periods of 0.30, 0.28 and 0.20 s are shown in Figs. 4.29, 4.30 and 4.31.

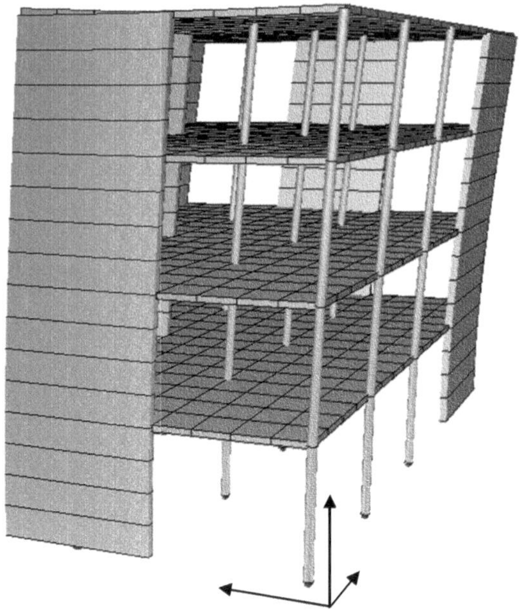

Fig. 4.29 First natural mode, $T_1 = 0.30$ s, (translational vibration mode)

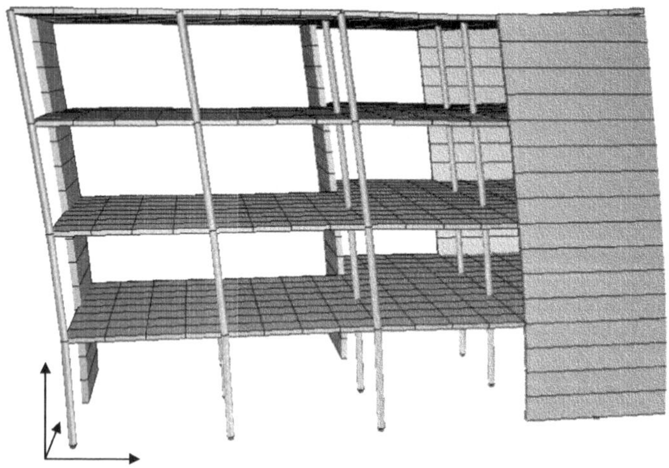

Fig. 4.30 Second natural mode, $T_2 = 0.29$ s (translational vibration mode)

If the first three vibration modes are considered, about 70% of the building mass is already activated in each main direction of the building. However, in order to achieve the required 90% according to Eurocode 8-1 (2004) a sufficient number of higher vibration modes must be considered. It should be noted that the higher vibration

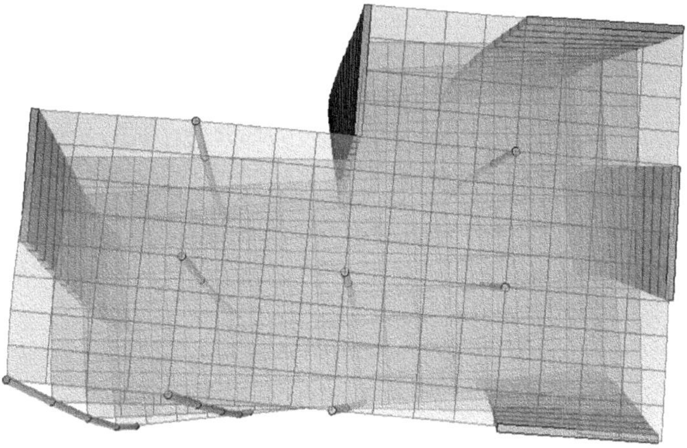

Fig. 4.31 Third natural mode, $T_3 = 0.20$ s (torsional vibration mode)

modes will also include local vibration modes of the slabs, which do not provide a significant contribution to the total activated mass.

Torsional Effects

As described in Sect. 4.1.6.2, accidental torsional effects are considered in order to account for uncertainties in the location of masses and in the spatial variation of the seismic actions. The eccentricity to be applied for this purpose is 5% of the building dimension perpendicular to the earthquake direction. For each floor level, this eccentricity is to be multiplied with the equivalent horizontal seismic forces. The torsional moments determined in this way must be considered as additional loads, and their effect must be investigated considering varying signs. For the present building, the resulting torsional moments on each storey level are summarized in Table 4.9. The basis for the calculation of the torsional moments are the total earthquake forces in the main building directions, distributed proportional to mass and height of each storey level. With a total mass of the structure of 1140 t, spectral accelerations of 1.465 m/s^2 in x-direction and 1.383 m/s^2 in y-direction lead to overall seismic forces of $F_{b,x} = 1420$ kN and $F_{b,y} = 1340$ kN. The accidental eccentricities are calculated to be $\pm\ 0.05 \cdot 24$ m $= \pm\ 1.2$ m in x-direction and $\pm\ 0.05 \cdot 15$ m $= \pm\ 0.75$ m in y-direction. The torsional moments were replaced by equivalent force pairs applied along the boundaries of the slabs on each floor.

Table 4.9 Torsional moments due to accidental eccentricities

Storey	Height (m)	Weight W (t)	Horizontal-force $F_{x,i}$ (kN)	M (earthquake x-direction) (kNm)	Horizontal-force $F_{i,y}$ (kN)	M (earthquake y-direction) (kNm)
1	4	293.7	149	111.8	141	168.8
2	8	293.7	298	223.5	281	337.7
3	12	293.7	447	335.2	422	506.5
4	16	258.9	525	394.0	496	595.3

Second-Order Effects (P-ΔEffects)

Second-order effects need not be taken into account as the interstorey drift sensitivity coefficient θ is way below 10% for all storeys of the building.

Calculation Procedure

The following calculation steps are carried out for seismic actions applied in x-direction:

- Calculation of the stress resultants for all relevant mode shapes.
- Calculation of the stress resultants $E_{Modal,x}$ by superposition of all mode shape contributions using the SRSS rule.
- Calculation of the stress resultants E_{Mx} due to torsional effects.
- Calculation of the stress resultants E_{Edx} by summing $E_{Modal,x}$ and E_{Mx}.

The calculation of the stress resultant for a seismic action in y-direction is carried out by the same procedure, substituting "x" for "y". The decisive stress resultants are obtained by combining the seismic action effects E_{Edx} and E_{Edy} in x- and y-direction:

$$E_{Edx} \oplus 0.30 \cdot E_{Edy}$$
$$0.30 \cdot E_{Edx} \oplus E_{Edy}.$$

The stress resultants are superimposed with varying positive and negative signs to identify the decisive combinations. The result of the combinations are the design values of seismic actions A_{Ed} (4.57), which are subsequently superimposed with permanent loads, proportional variable loads and proportional snow loads according to (4.56). The flow chart for the calculation of the design values of the effects of actions E_{dEA} is presented in Fig. 4.32.

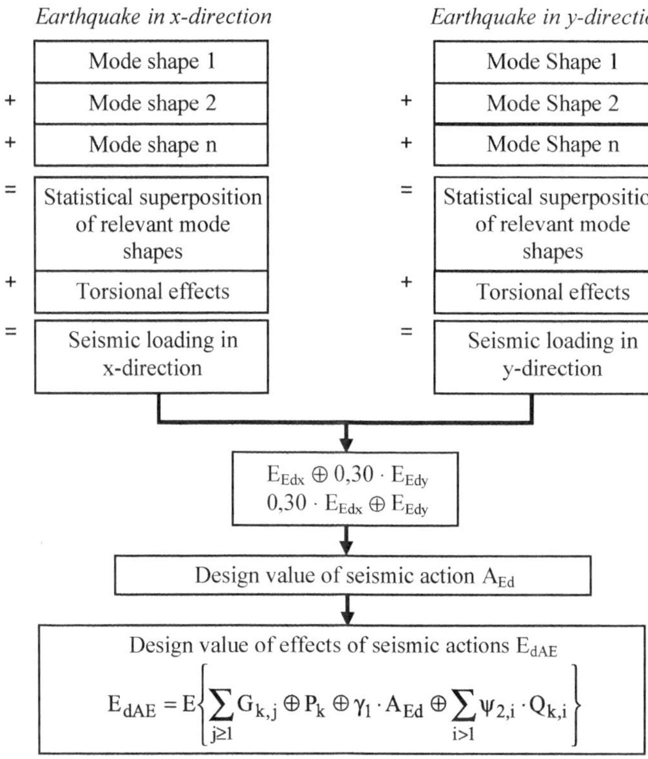

Fig. 4.32 Flow chart for the calculation of the design value of the effect of actions E_{dEA}

Table 4.10 Design stress resultants for ductility class DCM

Ductility class DCM	Wall 1	Wall 2	Wall 3	Wall 4	Wall 5
min N_{Ed} (kN)	−1139	−1334	−1334	**−1247**	−1286
(±) M_{Ed} (kNm)	2803	3067	3360	**4855**	4692
Eccentricity (M_d/N_d) (m)	2.46	2.30	2.52	**3.90**	3.65
(±) Q_{Ed} (kN)	214	178	305	**386**	347

Calculation Results

The design stress resultants for all walls based on the assumptions of ductility class DCM (q = 3) are given in Table 4.10. Since the walls run continuously from foundation to roof the critical wall section for design is the wall base.

Design and Constructive Detailing for DCM

General requirements
The design and constructive detailing is shown exemplarily for the critical wall no. 4. The basic design follows Eurocode 2-1-1 (2004) and at the same time the additional rules for ductility in accordance with Eurocode 8-1 (2004) must be considered. A minimum concrete strength class of C16/20 and high ductile reinforcement of type B according to Eurocode 2-1-1 (2004), Table C.1 for primary seismic elements are required. The concrete cover of reinforcement is assumed to be 5 cm for all sides. The design forces for q = 3 at the base of wall no. 4 according to Table 4.10 are:

Design stress resultants

$$N_{Ed} = -1247\,\text{kN}$$
$$V_{Ed} = 386\,\text{kN}$$
$$M_{Ed} = 4855\,\text{kNm}.$$

Design for shear
The design value of the shear force V_{Ed} must be increased by a factor of 1.5 following Eurocode 8-1 (2004), Sect. 5.4.2.4:

$$V_{Ed} = 1.5 \cdot 386\,\text{kN} = 579\,\text{kN}.$$

The maximum permissible shear force $V_{Rd,max}$ of the concrete compression strut is calculated assuming a strut angle of $\vartheta = 40°$:

$$V_{Rd,max} = \frac{b_w \cdot z \cdot \alpha_c \cdot f_{cd}}{\cot\theta + \tan\theta} = \frac{\left(0.3 \cdot (0.9 \cdot 5.9) \cdot 0.75 \cdot \left(20 \cdot \frac{0.85}{1.5}\right)\right)}{1.2 + \frac{1.0}{1.2}} = 6.659\,\text{MN} > 0.579\,\text{MN}.$$

This leads to required stirrup reinforcement for $\vartheta = 40°$ of:

$$erf\ \alpha_{sw} = \frac{V_{Ed}}{f_{yd} \cdot z \cdot \cot\vartheta} = \frac{0.579}{\frac{500}{1.15} \cdot (0.9 \cdot 5.9) \cdot 1.2} = 2.09\,\text{cm}^2/\text{m}.$$

To cover the required shear reinforcement, a mesh reinforcement Q 257 (A) is arranged on both sides. The provided reinforcement of $a_{sw,vorh} = 5.14\ \text{cm}^2/\text{m}$ covers the required stirrup reinforcement. It is important to note that the use of high ductile reinforcement is not necessary for the stirrup reinforcement as the shear force is completely transferred in the linear elastic range.

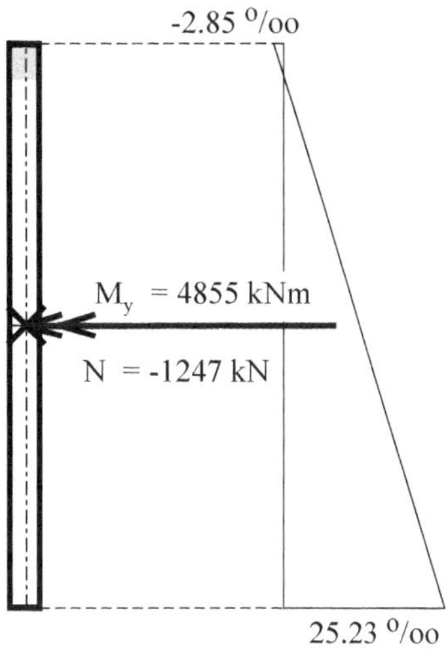

Fig. 4.33 Strain distribution for the bending design (DCM)

Design for bending and normal force

The bending design with symmetrical arrangement of reinforcement is carried out using the design software FriLo (2018). The design leads to a required bending reinforcement of 7.92 cm^2 at both ends of the wall. The design is based on the strain distribution shown in Fig. 4.33. In order to ensure a ductile bending behaviour high ductile reinforcement of 10 Ø 12 mm is selected. The contribution of the mesh reinforcement Q 257 (A) was not taken into account in the calculation of the required longitudinal reinforcement.

Special provisions for ductile walls

The wall no. 4 can be classified as a slender wall according to Eurocode 8-1 (2004), Sect. 5.4.2.4

$$\frac{H_w}{l_w} = \frac{16}{6} = 2.7 > 2.$$

Thus, additional requirements on the design and constructive detailing of the wall are given in Eurocode 8-1 (2004), Sect. 5.4.3.4.2. At first, the height of the critical zone h_{cr} above the base of the wall is estimated as:

$$h_{cr} = \max\left[l_w, \frac{H_w}{6}\right] = 6\,\text{m} \quad \leq \quad \left[\begin{matrix} 2l_w \\ h_s \text{ for } n \leq 6 \text{ stories} \end{matrix}\right] = 4 - 0.2 = 3.8\,\text{m}.$$

The design bending moment diagram along the height of the wall is defined by an envelope of the bending moment diagram from the analysis with a tension shift of the calculated height of the critical zone h_{cr} (Fig. 4.34). The envelope diagram covers uncertainties due to the distribution of the bending moment along the wall

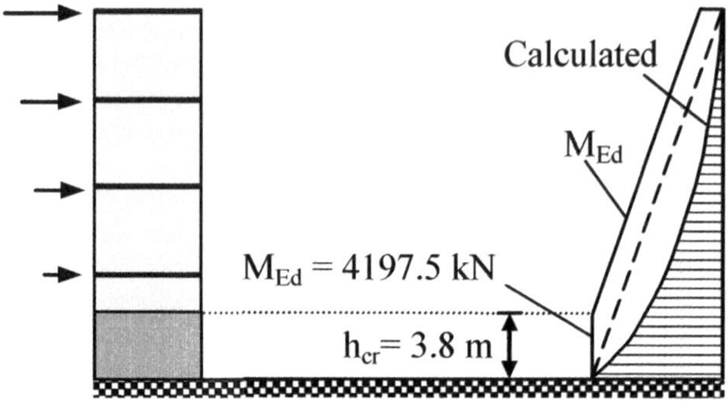

Fig. 4.34 Design envelope for the bending moment M_{sd} with vertical offset h_{cr}

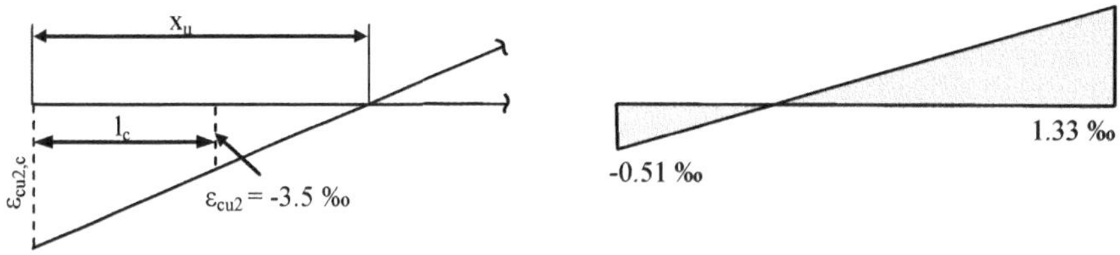

Fig. 4.35 Length l_c for strains at ultimate curvature and calculated strain distribution

height, the effects of shear forces on the flexural tension force and the effect of the decreasing normal force along the height due to the vertical offset h_{cr}.

The length l_c of the critical zone in the direction of the wall is defined as the distance between the edge compression strain and the concrete compression strain of $-3.5‰$ at which spalling is expected. The resulting strain distribution in the wall assuming the given longitudinal reinforcement and mesh reinforcement is shown in Fig. 4.35. Since the maximum edge compression strain is just $-0.47‰$ the threshold value of Eurocode 8-1 (2004), Sect. 5.4.3.4.2 becomes decisive:

$$l_c \geq 0.15 \cdot l_w = 0.9\,\text{m and } 1.50 \cdot b_w = 0.45\,\text{m} \quad \Rightarrow l_c = 0.45\,\text{m}.$$

Within the critical zone a minimum percentage of vertical reinforcement of $\rho_v =$ 0.005 is required. Thus, the minimum vertical reinforcement is equal to

$$A_{sv,min} = 0.005 \cdot 45 \cdot 30 = 6.75\,\text{cm}^2.$$

This minimum reinforcement is covered by 10 Ø 12 mm reinforcement bars distributed along the edges of the critical zones. The longitudinal bars provide a reinforcement of 11.3 cm^2/m > 6.75 cm^2. The mechanical percentage of reinforcement ω_v is:

$$\omega_v = \rho_v \cdot \frac{f_{yd,v}}{f_{cd}} = \frac{11.3}{45 \cdot 30} \cdot \frac{500/1.15}{0.85 \cdot 20/1.5} = 0.32.$$

An additional requirement must be satisfied regarding the mechanical volumetric ratio of confining stirrups within the critical region ω_{wd}:

$$\alpha \cdot \omega_{wd} \geq 30\,\mu_\Phi \cdot (\nu_d + \omega_v) \cdot \varepsilon_{sy,d} \cdot b_w/b_0 - 0.035 = 0.244$$

where

$$\mu_\Phi = 1.5 \cdot (2 \cdot q_0 - 1) = 1.5 \cdot (2 \cdot 3.0 - 1) = 7.5 \text{ for } T_1 = 0.29\,s \geq T_C = 0.25\,s$$

$$\nu_d = \frac{N_{Ed}}{A_c \cdot f_{cd}} = \frac{N_{Ed}}{A_c \cdot \alpha \cdot f_{ck}/1.5} = \frac{1.247}{6.0 \cdot 0.3 \cdot 0.85 \cdot 20/1.5} = 0.061$$

$$\omega_v = 0.32$$

$$\varepsilon_{sy,d} = f_{yk}/(E_s \cdot 1.15) = 500/(200000 \cdot 1.15) = 0.00217$$

$$b_w = 0.30\,m$$

$$b_0 = 0.20\,m.$$

The value of the confinement effectiveness factor α is determined with respect to Eurocode 8-1 (2004), Sect. 5.4.3.4.2:

$$\alpha = \alpha_n \cdot \alpha_s = 0.63 \cdot 0.67 = 0.42$$

where:

$$\alpha_n = 1 - \sum_n b_i^2/6b_0 \cdot d_0 = 1 - (8 \cdot (15^2/6 \cdot 20 \cdot 45) + 2 \cdot (10^2/6 \cdot 20 \cdot 45)) = 0.63$$

$$\alpha_s = (1 - s/2b_0)(1 - s/2d_0) = (1 - 10/(2 \cdot 20)) \cdot (1 - 10/(2 \cdot 45)) = 0.67$$

$$s = \min\left\{\frac{b_0}{2};\ 20\,cm;\ 9\,d_{sL}\right\} = 10\,cm.$$

The mechanical volumetric ratio of the confining stirrups within the critical regions is obtained with the confinement effectiveness factor α as:

$$\omega_{wd} \geq \frac{0.244}{0.42} = 0.58.$$

Thus, the required stirrup reinforcement is determined by

$$V_{\text{Stirrups}} = \omega_{wd} \cdot V_{\text{Concrete Core}} \cdot \frac{f_{cd}}{f_{yd}} = 0.54 \cdot 20 \cdot 45 \cdot 100 \cdot \frac{0.85 \cdot 20/1.5}{500/1.15}$$

$$= 1266.8\,\text{cm}^3/\text{m}.$$

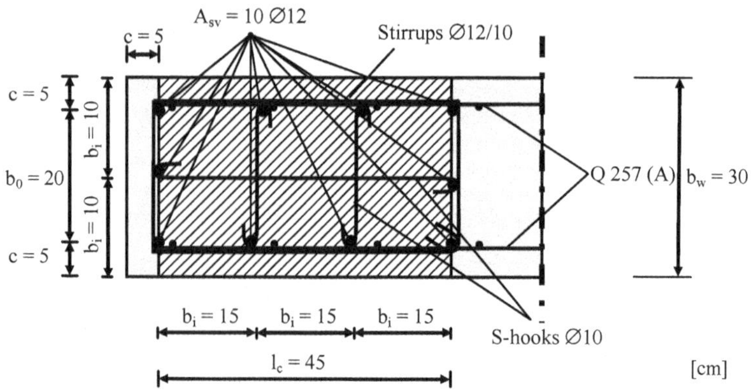

Fig. 4.36 Constructive detailing of the confined critical zone

The spacing of the stirrups is

$$s = \min\left(\frac{b_0}{2}; 200; 9 \cdot d_{sL}\right) = \min\left(\frac{200}{2}; 200; 9 \cdot 12\right) = 100\,\text{mm}.$$

The volume of one stirrup Ø12 is equal to:

$$V_{\text{Stirrups},\varnothing 12} = 1.13 \cdot (2 \cdot 20 + 2 \cdot 45) = 146.9\,\text{cm}^3.$$

This results in the following volume per meter value for a spacing of the stirrups of 10 cm:

$$V_{\text{Stirrups},\varnothing 12/10} = 146.9 \cdot 10 = 1469.0\,\text{cm}^3/\text{m}.$$

The stirrups must prevent the local buckling of the longitudinal reinforcement bars. Therefore they must be closed and at their ends have hooks bending by 135° to the inside, the length of which must be 10 times the stirrup's diameter ($10\,d_{bw}$). Additionally, the central longitudinal bars are stabilized by S-hooks of diameter Ø10. The reinforcement arrangement of the confined critical zone is depicted in Fig. 4.36.

Design and Constructive Detailing for DCH

The design of the wall for ductility class DCH demands the consideration of additional rules. The minimum quality of concrete is now C20/25, instead of C16/20. This requirement is satisfied for the office building. Furthermore, the design shear forces are magnified by the factor ε to avoid a brittle shear failure at the base of the wall in the higher deformation range:

$$V_{Ed} = \varepsilon \cdot V'_{Ed}$$

where:

$$\varepsilon = q \cdot \sqrt{\left(\frac{\gamma_{Rd}}{q} \cdot \frac{M_{Rd}}{M_{Ed}}\right)^2 + 0.1 \cdot \left(\frac{S_e(T_C)}{S_e(T_1)}\right)^2} \begin{cases} \leq q \\ \geq 1.5 \end{cases}.$$

Assuming full utilization of the section ($M_{Rd} = M_{Ed}$) the coefficient ε is obtained as:

$$\varepsilon = 4.4 \cdot \sqrt{\left(\frac{1.2}{4.4} \cdot 1\right)^2 + 0.1 \cdot \left(\frac{2.5}{2.083}\right)^2} = 2.06 \begin{cases} \leq 4.4 \\ \geq 1.5 \end{cases}$$

where

M_{Rd} Design flexural resistance at the base of the wall.
γ_{Rd} Factor to account for overstrength due to steel strain-hardening. The factor can be assumed as 1.2 in the absence of more precise data.
T_1 Fundamental period of vibration.
T_C Control period according to (4.1.4).
$S_e(T)$ Ordinate of the elastic response spectrum for T_1.

In ductility class DCM the shear force was increased by a factor of 1.5. As the bearing capacity for shear forces was significantly higher than the design shear force, a repeated design and constructive detailing is not carried out and the reader is referred to the design for DCM. The corresponding normalized axial force is significantly below the threshold value $\nu = 0.35$. Additionally, some further requirements on the constructive detailing for DCH in Eurocode 8-1 (2004), Sect. 5.5.3.4 must be considered, which are not explicitly stated here.

4.2.2 Design of Steel Structures

4.2.2.1 General Introduction: Design Concept and Specific Rules

The seismic design of steel buildings is regulated in Sect. 6 of Eurocode 8-1 (2004). The regulations have to be considered in addition to the rules given for the regular design of steel structures in Eurocode 3-1-1 (2005). Steel constructions are built in earthquake prone areas all over the world due their excellent seismic performance, as steel is a homogeneous material with high ductility and tension strength. The utilization with this high ductility property requires a detailed capacity design to guarantee the local and global deformation capacities of the structure during the successive formation of plastic hinges. Eurocode 8-1 (2004) provides three ductility classes for the design of steel structures, which differ with respect to their ductility requirements,

Table 4.11 Design concepts and associated values of behaviour factors (Eurocode 8-1 (2004), Table 6.1)

Design concept	Ductility class	Range of behaviour factor
Low dissipative structural behaviour	DCL (Low)	$\leq 1.5 - 2.0$
Dissipative structural behaviour	DCM (Medium)	≤ 4.0 Also limited by the values of Eurocode 8-1 (2004), Table 6.2
	DCH (High)	Only limited by the values of Eurocode 8-1 (2004), Table 6.2

the necessary design effort for the structural engineer and the extent of quality controls of the building materials. Furthermore, the consideration of buckling effects under cyclic loading conditions plays an important role in the verification process. In the next sections the design and ductility concepts are summarily presented and the application of the concepts is demonstrated by two real steel structures.

Design and Ductility Concepts

The choice of a certain level of structural ductility is the fundamental step in the seismic design of steel structures. Ductility is the ability of the structure to undergo significant plastic deformation beyond the yield stress without system collapse. In other words, ductility is defined as the ratio between ultimate displacement capacity and yield displacement d_y, which defines the transition between the elastic and the plastic region. Eurocode 8-1 (2004) proposes two design concepts in terms of ductility: (i) low dissipative and (ii) dissipative structural behaviour.

Table 4.11 shows the ductility classes and the range of associated behaviour factors. The higher the dissipative structural behaviour, the higher the behaviour factor q that reduces the design seismic action.

More specific behaviour factors for medium and high ductility classes depending on the structural type are given in Table 4.12. If the steel structure is non-regular in elevation, the upper limit of reference values of behaviour factors of Table 4.13 must be reduced by 20%. The largest behaviour factors are those related to the high ductility class and the following structural types: moment resisting frames without and with infills isolated from the moment frame and frames with eccentric bracings. In these cases, the behaviour factor is $q = 5\ \alpha_u/\alpha_1$ and depends on the factor α_u/α_1.

The parameter α_1 is the multiplier of the horizontal design seismic action at the formation of the first plastic hinge in the system, and the parameter α_u is the multiplier of the horizontal seismic design action at the formation of a global plastic mechanism. The value α_u/α_1 is limited to 1.6 and can be calculated from a non-linear static (pushover) analysis of the whole structure, applying a monotonically increasing lateral load pattern.

If it is not possible to perform a pushover analysis, recommended values of the multiplication factor α_u/α_1 for in-plan regular buildings can be applied. The recommended values depending on the structural type are summarized in Table 4.13. For buildings not regular in plan, an average between 1.0 and the factors given in Table 4.13 should be adopted. However, greater values of multiplication factor α_u/α_1 may be used, provided that these are confirmed by means of pushover analyses.

Table 4.12 Upper limit of reference values of behaviour factors for buildings regular in elevation according to Eurocode 8-1 (2004), Table 6.2

Structural type	Structural ductility class	
	DCM (Medium)	DCH (High)
(a) Moment resisting frames	4	$5\alpha_u/\alpha_1$
(b) Frame with concentric bracings		
Diagonal bracings	4	4
V-bracings	2	2.5
(c) Frame with eccentric bracings	4	$5\alpha_u/\alpha_1$
(d) Inverted pendulum	2	$2\alpha_u/\alpha_1$
(e) Structures with concrete cores or concrete walls	Eurocode 8-1 (2004), Sect. 5	
(f) Moment resisting frame with concentric bracing	4	$4\alpha_u/\alpha_1$
(g) Moment resisting frames with infills		
Unconnected concrete or masonry infills, in contact with the frame Connected reinforced concrete infills Infills isolated from moment frame	2	2
	Eurocode 8-1 (2004), Sect. 7	
	4	$5\alpha_u/\alpha_1$

Table 4.13 Values of ductility factors for in-plan regular buildings (Eurocode 8-1 2004)

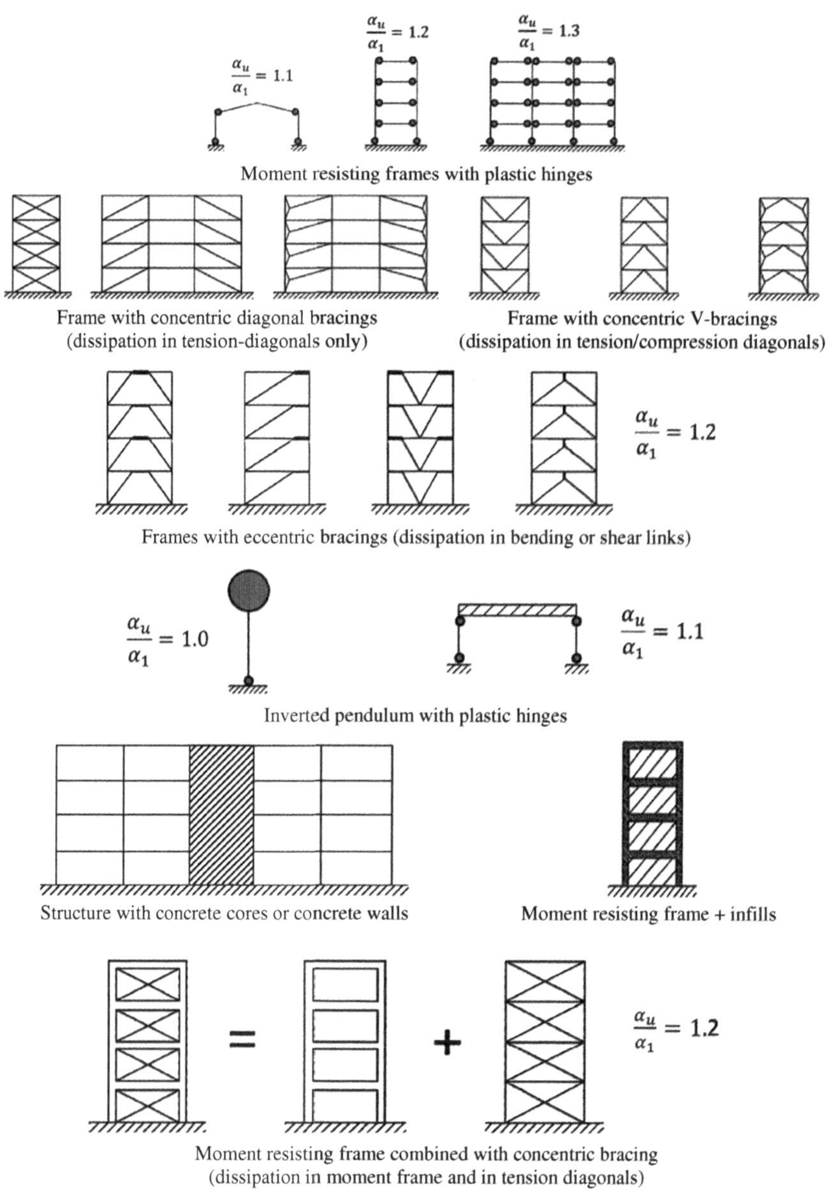

4.2.2.2 Calculation Example: Two-Story Steel Structure

The following example deals with the existing, symmetrical two-story steel structure with rigid floor slabs shown in Fig. 4.38. The seismic calculation is carried out using planar models in the transverse (x) and in the longitudinal direction (y), as the structural system fulfils the regularity criteria according to Sect. 4.1.3 in plan and elevation (Figs. 4.37, 4.38 and 4.39).

The horizontal load bearing system of the structure consists of moment resisting frames with clamped supports in x-direction and concentric diagonal bracings in the two outer planes in y-direction.

The diagonal bracing system consists of circular profiles on both storey levels. The tension-only diagonals have a diameter of 16 mm on the ground floor and 10 mm on the first floor. The moment resisting frame consists of columns with HEB 280 profiles and a horizontal beam with a HEB 180 profile. The material of all structural elements is S235 according to Eurocode 3-1-1 (2005). The design response spectrum

Table 4.14 Distributed loads due to dead and live loads

	First storey	Roof
Dead load (kN/m^2)	6	4
Live load (kN/m^2)	5	2

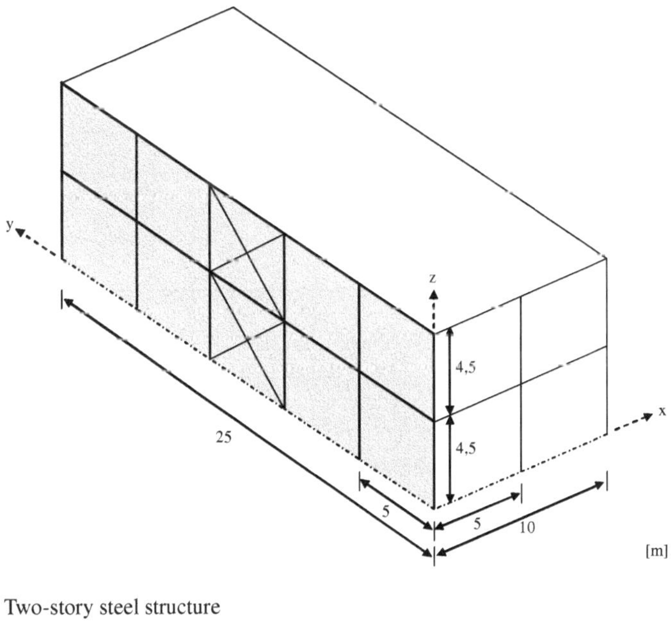

Fig. 4.37 Two-story steel structure

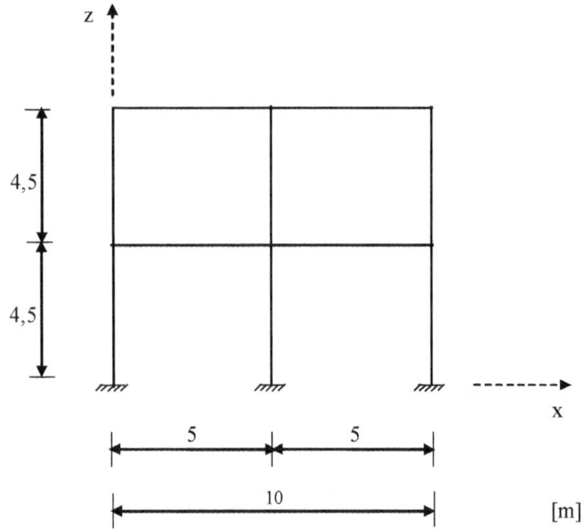

Fig. 4.38 Moment resisting frame in x-direction

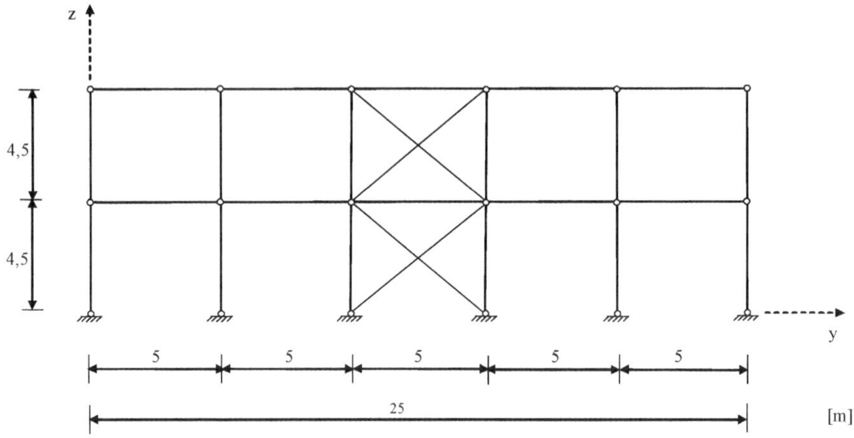

Fig. 4.39 Concentric diagonal bracing system in y-direction

is constructed according to DIN EN 1998-1/NA (2011) as the building location is close to Freiburg in Germany. The building is located in seismic zone 3 with a design ground acceleration of $a_g = 0.8$ m/s^2. The soil factor S and control periods of the horizontal design spectrum are determined for the deep geology class T in combination with soil class C:

$$S = 1.25,\ T_B = 0.1\,\text{s},\ T_C = 0.40\,\text{s and}\ T_D = 2.0\,\text{s}.$$

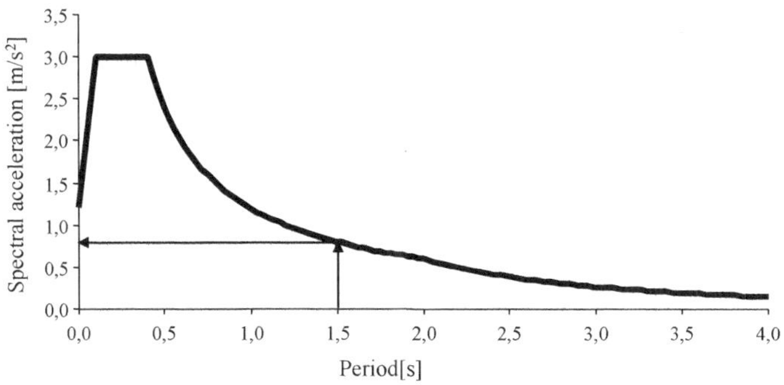

Fig. 4.40 Horizontal design spectrum according to DIN EN 1998-1/NA (2011), q = 1

The building belongs to importance category II with an importance factor of $\gamma_I = 1.2$. The resulting design response spectrum assuming a behaviour factor of q = 1.0 is depicted in Fig. 4.40.

Firstly, the total seismic design forces in the main building directions, calculated with a behaviour factor q = 1.0, are compared with the design forces due to wind actions. The design forces for wind actions in x- and y-direction are calculated according to Eurocode 1-1-4 (2004) for wind zone 3:

$$W_x = \gamma_Q \cdot c_p \cdot q \cdot A_{wx} = 1.5 \cdot 1.3 \cdot 0.8 \cdot 9 \cdot 25 = 351\,\text{kN}$$
$$W_y = \gamma_Q \cdot c_p \cdot q \cdot A_{wy} = 1.5 \cdot 1.3 \cdot 0.8 \cdot 9 \cdot 10 = 140.4\,\text{kN}.$$

Here, γ_Q is the partial safety factor for leading variable actions, c_p is the aerodynamic force coefficient, q the velocity pressure and A_w is the wind-exposed area according to DIN EN 1991-1-4/NA (2010). The design values of the seismic actions A_{Edx}, A_{Edy} in the x- and y-directions are calculated taking into account the combination of vertical loads according to Sect. 4.1.6.3. Corresponding to the intended use, each storey can be regarded as independent and the factor φ is applied with 1.0 for the roof and with 0.5 for the first storey (Table 4.7). The factor ψ_2 is applied with 0.3 for variable loads. Accordingly, the following vertical load combinations are considered for calculating the seismic storey masses (4.57):

$$\text{First storey: } G_k \oplus 0.15 \cdot Q_{\text{Live Load}}$$
$$\text{Roof: } G_k \oplus 0.3 \cdot Q_{\text{Live Load}}.$$

The application of these combinations using the distributed loads due to the dead and live loads given in Table 4.14 results to storey masses of 172.0 t (first storey) and 117.2 t (roof). The resulting mode shapes in x- and y-direction are shown in Fig. 4.41.

The steel structure complies with the regularity criteria in plan and elevation and fulfils the condition $T_1 \leq 4T_c$ in both building directions. Furthermore, the

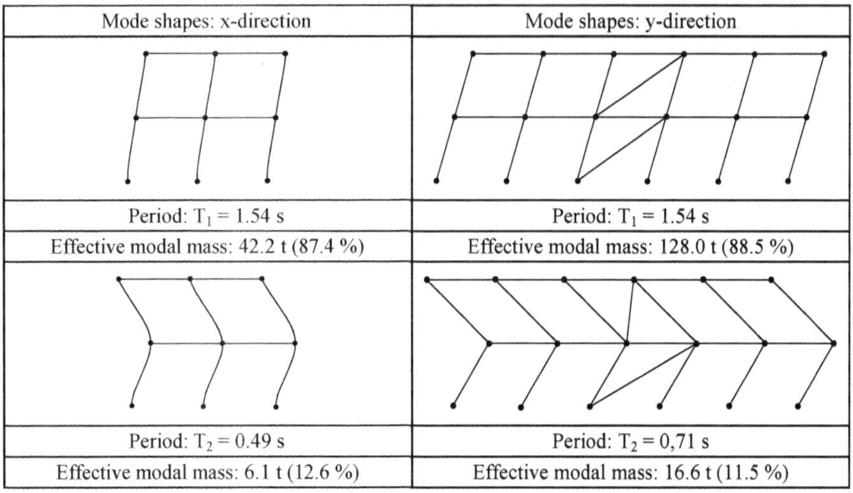

Fig. 4.41 First and second mode shapes of the steel structure in x- and y-direction

fundamental periods in the main building directions satisfy the condition T_{1x}, T_{1y} ≤ 2 s according to Eurocode 8-1 (2004), Sect. 4.3.3.2. Therefore, the lateral force method can be applied, which considers only the fundamental periods of vibration in each building direction. The total base shears F_{bx}, F_{by} in x- and y-direction are calculated according to (4.11) with a behaviour factor $q = 1.0$ and a correction factor $\lambda = 1.0$:

$$F_{bx} = S_d(T_1) \cdot M \cdot \lambda = 0.779\,\text{m/s}^2 \cdot 289.2\,t \cdot 1.0 = 225.4\,\text{kN}$$
$$F_{by} = S_d(T_1) \cdot M \cdot \lambda = 0.779\,\text{m/s}^2 \cdot 289.2\,t \cdot 1.0 = 225.4\,\text{kN}.$$

The comparison of the wind and earthquake forces shows that the earthquake forces are only critical in the y-direction, so that the following seismic design is carried out in this direction. The lateral forces are applied proportional to the modal ordinates of the fundamental mode in y-direction. This leads to lateral seismic forces of 130.73 kN on the roof level and 94.67 kN on the first storey level. The resulting normal forces due to the lateral seismic forces are shown in Fig. 4.42. Figures 4.43, 4.44 and 4.45 depict the stress resultants due to combined dead and proportional live loads for the seismic design situation.

Verification in Ductility Class DCL

In the following the seismic safety of the existing steel structure is checked for ductility class DCL. This ductility class assumes low structural dissipation capacity represented by a behaviour factor of $q = 1.5$. The cross section design is carried out

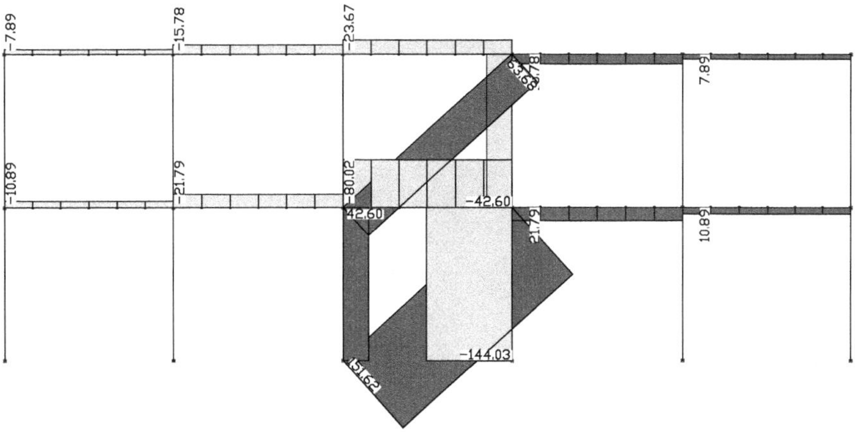

Fig. 4.42 Normal forces due to lateral seismic forces (kN)

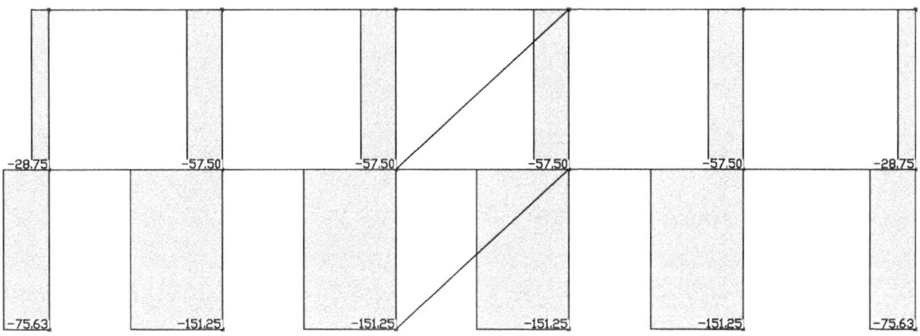

Fig. 4.43 Normal forces of the columns due to dead and proportional live loads (kN)

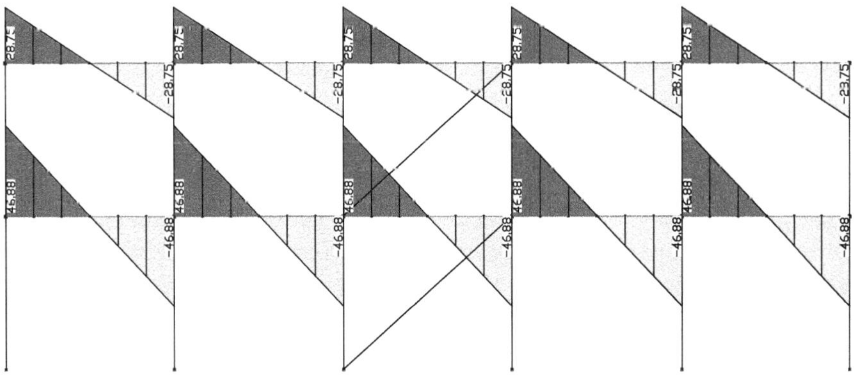

Fig. 4.44 Shear forces of the horizontal beam due to dead and proportional live loads (kN)

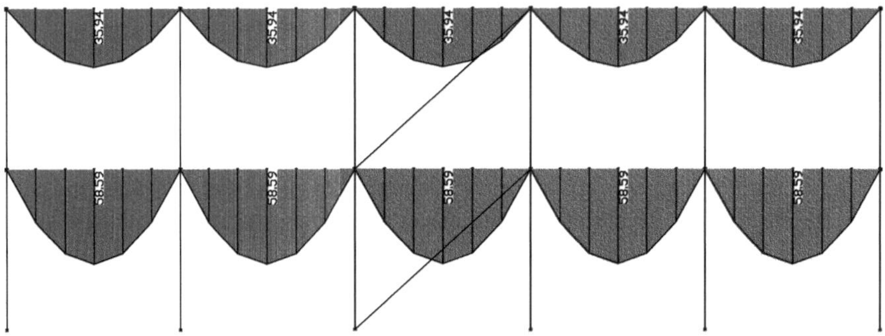

Fig. 4.45 Bending moments of the horizontal beam due to dead and proportional live loads (kNm)

according to Eurocode 3-1-1 (2005) without taking any specific seismic rules into account. The normal forces yield the following results for the diagonals:

Diagonals in the first floor (∅10):

$$\frac{N_{Ed}}{N_{Rd}} = \frac{63.7/1.5}{A \cdot f_{yk}/\gamma_M} = \frac{42.47}{0.79 \cdot 23.5} = 2.29 > 1 \Rightarrow \text{Diagonal } \varnothing 10 \text{ is not sufficient!}$$

Diagonals in the ground floor (∅16):

$$\frac{N_{Ed}}{N_{Rd}} = \frac{151.6/1.5}{A \cdot f_{yk}/\gamma_M} = \frac{101.07}{2.01 \cdot 23.5} = 2.14 > 1 \Rightarrow \text{Diagonal } \varnothing 16 \text{ is not sufficient!}$$

The seismic safety of the actual stiffening system in y-direction cannot be successfully verified in ductility class DCL for the given cross sections. Therefore, a safety check in ductility class DCM is carried out in the following section in order to try to avoid measures for strengthening the stiffening system.

Verification in Ductility Class DCM

In ductility class DCM the tension-only diagonals are designed as dissipative elements with a maximum behaviour factor of $q = 4$ for frames with concentric diagonal bracings. The choice of the required behaviour factor q depends on the utilization of the diagonals:

$$\text{Diagonals in the first floor } (\varnothing 10)\text{: } \frac{N_{Ed}}{N_{Rd}} = \frac{N_{Ed}}{A \cdot f_y} = \frac{63.7}{0.79 \cdot 23.5} = 3.43$$

$$\text{Diagonals in the ground floor } (\varnothing 16)\text{: } \frac{N_{Ed}}{N_{Rd}} = \frac{N_{Ed}}{A \cdot f_y} = \frac{151.6}{2.01 \cdot 23.5} = 3.21$$

Based on the utilization levels the seismic design is performed with a behaviour factor $q = 3.5$. The ratios Ω yield with $q = 3.5$:

$$\text{Diagonals of the first floor } (\varnothing 10)\text{:} \quad \Omega = \frac{N_{Rd}}{N_{Ed}} = \frac{3.5 \cdot 0.79 \cdot 23.5}{63.7} = 1.02$$

$$\text{Diagonals of the ground floor } (\varnothing 16)\text{:} \quad \Omega = \frac{N_{Rd}}{N_{Ed}} = \frac{3.5 \cdot 2.01 \cdot 23{,}5}{151.6} = 1.09$$

The slenderness ratio $\overline{\lambda}$ of the critical diagonals ($\varnothing 16$) according to Eurocode 3-1-1 (2005) is calculated as:

$$\overline{\lambda}_k = \frac{\lambda_k}{\lambda_a} = \frac{s_k/i}{\pi \cdot \sqrt{21.000/24}} = \frac{673/0.40}{\pi \cdot \sqrt{21.000/24}} = \frac{2692}{92.93} = 18.11$$

$$\text{with: } i = \sqrt{\frac{\pi/4.r^4}{\pi \cdot r^2}} = 0.40\,\text{cm.}$$

The slenderness ratio $\overline{\lambda}$ fulfils the condition $\overline{\lambda}_k \geq 1.5$ according to the German seismic code DIN 4149 (2005). The requirements of Sect. 6.7.3 of Eurocode 8-1 (2004) are not completely taken into account, as the structure is subjected to relatively low seismic actions.

The columns are designed according to Eurocode 8-1 (2004), Sect. 6.7.4 for the maximum design value of the axial force N_{sd}:

$$N_{sd} = N_{SG} \pm 1.1 \cdot \gamma_{ov} \cdot \Omega \cdot N_{SE} = -151.25 - 1.1 \cdot 1.25 \cdot 1.02 \cdot 144.03/3.5 = -208.96\,\text{kN.}$$

The maximum design value of axial force N_{sd} does not exceed the design resistance to normal forces of the cross-section for uniform compression according to Eurocode 3-1-1 (2005):

$$N_{Rd} = \chi \cdot N_{pl,d} = 0.69 \cdot 1945 = 1342\,\text{kN}$$

where:

$$\chi = \frac{1}{\Phi + \sqrt{\Phi^2 - \overline{\lambda}^2}} = 0.69$$

$$\Phi = 0.5 \cdot \left[1 + \alpha\left(\overline{\lambda} - 0.2\right) + \overline{\lambda}^2\right] = 0.98$$

$$\overline{\lambda} = \sqrt{\frac{A \cdot f_y}{N_{cr}}} = \sqrt{\frac{91.0 \cdot 23.5}{2906.8}} = 0.86.$$

Imperfection factor of buckling curve b

$$\alpha = 0.34.$$

Elastic critical force for the relevant buckling mode:

$$N_{cr} = \frac{EI_z \pi^2}{s_{k,z}^2} = \frac{21,000 \cdot 2840 \cdot \pi^2}{450^2} = 2906.8\,\text{kN}.$$

The buckling safety of the columns is verified.

Furthermore the horizontal beams must be checked for shear and combined bending and axial loading. The design values of the critical stress resultants obtained from the seismic design situation are calculated as:

$$N_{Ed} = \pm 80.2/3.5 = \pm 22.91\ \text{kN},\ V_{Ed} = \pm 46.88\ \text{kN},\ M_{Ed} = 58.59\ \text{kNm}.$$

The verification of the horizontal beam in the centre of the girder is carried out for combined bending and axial force. It must be checked that the design bending moment M_{Ed} does not exceed the reduced design value of the resistance to bending moments under presence of axial forces $M_{N,Rd}$:

$$M_{Ed} \leq M_{N,Rd}.$$

The resistance to bending moments $M_{N,Rd}$ does not require a reduction due to the presence of axial forces, as the conditions according to Eurocode 3-1-1 (2005), Sect. 6.2.9 are adhered to:

$$N_{Ed} = 22.91\,\text{kN} < 0.25 \cdot N_{pl,Rd} = 0.25 \cdot A \cdot f_y = 0.25 \cdot 65.3 \cdot 23.5 = 383.64\,\text{kN}$$

and

$$N_{Ed} = 22.91\,kN < \frac{0.5 \cdot h_w \cdot t_w \cdot f_y}{\gamma_{M0}} = \frac{0.5 \cdot 12.2 \cdot 0.85 \cdot 23.5}{1.0} = 121.85\,\text{kN}.$$

The design bending moment M_{Ed} does not exceed the plastic moment resistance $M_{pl,y,Rd}$:

$$\frac{M_{Ed}}{M_{pl,y,Rd}} = \frac{58.59}{113.13} = 0.52 \leq 1$$

where:

$$M_{pl,y,Rd} = \frac{W_{pl,y} \cdot f_y}{\gamma_{M0}} = \frac{481.4 \cdot 23.5}{1.0} = 113.13\,\text{kNm}.$$

Additionally the shear capacity must be verified in the critical section close to the columns:

$$\frac{V_{Ed}}{V_{pl,Rd}} = \frac{46.88}{274.8} = 0.17 \leq 1$$

where:

$$V_{pl,Rd} = \frac{A_v \cdot (f_y/\sqrt{3})}{\gamma_{M0}} = \frac{0.3102 \cdot 65.3 \cdot (23.5/\sqrt{3})}{1.0} = 274.8\,\text{kN}.$$

A uniform deformation behaviour can be assumed if the ratio of the maximum and minimum ratios Ω_i does not exceed the threshold value of 1.25 according to Eurocode 8-1 (2004):

$$\frac{\Omega\text{max}}{\Omega\text{min}} = \frac{1.09}{1.02} = 1.07.$$

This condition is fulfilled, as the maximum percentage of 7% is far below the limit value of 25%.

Finally, the capacity design requires an adequate design of the connections considering overstrength effects, to guarantee the formation of the plastic hinges in the dissipative diagonals. The connections of the diagonals must provide the following design values of resistances according to Eurocode 8-1 (2004):

Diagonals in the first floor (∅10):

$$R_{dy} = 1.1 \cdot \gamma_{ov} \cdot N_{pl} = 1.1 \cdot 1.25 \cdot 0.79 \cdot 23.5 = 25.53\,\text{kN}.$$

Diagonals in the ground floor (∅16):

$$R_{dy} = 1.1 \cdot \gamma_{ov} \cdot N_{pl} = 1.1 \cdot 1.25 \cdot 2.01 \cdot 23.5 = 64.95\text{ kN}.$$

As an example, the screws at each end of the diagonals are designed according to the principle of capacity design. If a connecting joint plate of 10 mm is assumed, screws of M12 and M16 with strength 10.9 can be installed:

Screwed connections:

Diagonals in the first floor:

M12, 10.9, SLV, shaft:	$V_{a,R,d}$ = 56.5 kN > 25.53 kN
Hole bearing, joint plate: 10 mm, e = 30:	$V_{l,R,d}$ = 45.1 kN > 5.53 kN
$V_{a,R,d}/V_{l,R,d}$ = 56.5/45.1 = 1.25 > 1.2	

Diagonals in the ground floor:

M16, 10.9, SLV, shaft	$V_{a,R,d}$ = 101.0 kN > 64.95 kN
Hole bearing, joint plate: 10 mm, e = 45:	$V_{l,R,d}$ = 72.9 kN > 64.95 kN
$V_{a,R,d}/V_{l,R,d}$ = 101.0/ 72.9 = 1.39 > 1.2	

Fig. 4.46 **a** Front and **b** side view of the steel building

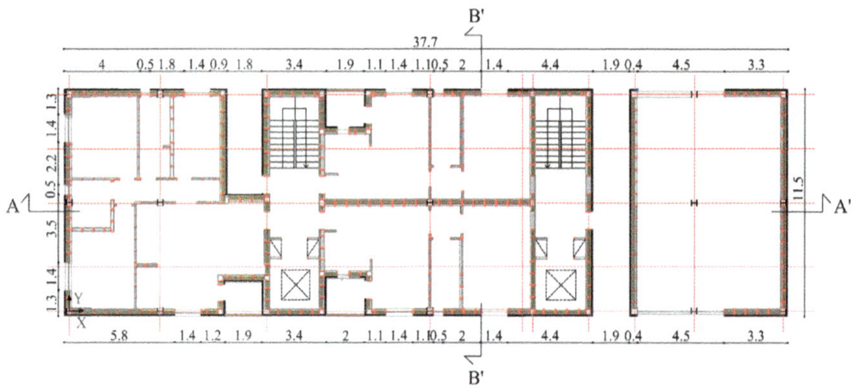

Fig. 4.47 Ground plan configuration of the steel building with dimensions (m)

4.2.2.3 Calculation Example: Seven-Story Steel Frame Building

The system considered is a seven-story residential building with dimensions of 37 × 11 square meters located in Italy. Figure 4.46 shows the front and side views of the steel building and Fig. 4.47 the ground plan configuration. Steel is adopted as structural material for columns, primary and secondary beams and bracing systems. The floors are made of composite slabs with reinforced concrete and encased steel beams. Their design is outside the scope of the following calculation and therefore will not be illustrated.

Due to architectonical demands, the plan is not perfectly symmetrical with respect to the principle directions of the building. Figure 4.48 shows the sections in the principal directions of the structure. The columns are placed at the intersection points of the vertical axes (1–9) and horizontal axes (A–C) (Fig. 4.49). Each storey counts 9 × 3 = 27 columns with an individual height of 3.1 m. Frames are arranged along

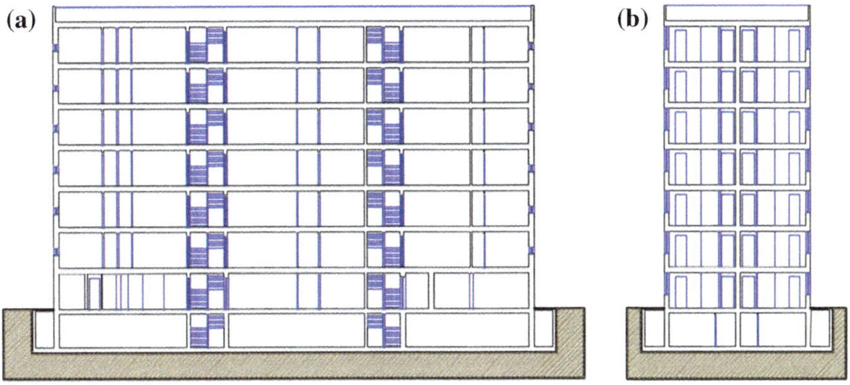

Fig. 4.48 Section A-A′ (**a**) and section B-B′ (**b**) with reference to Fig. 4.47

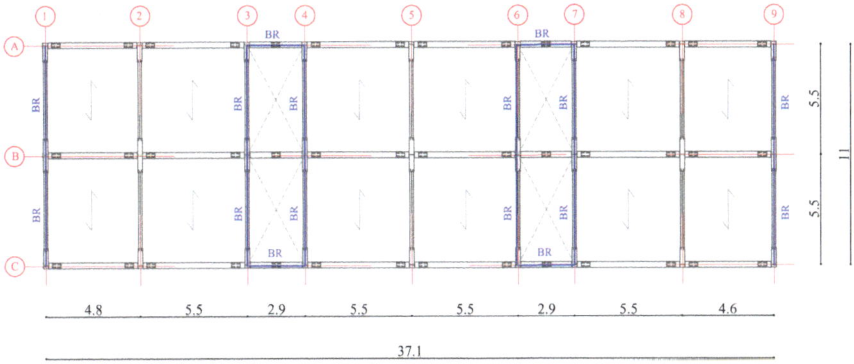

Fig. 4.49 Main horizontal and vertical building axes and positions of the bracings (BR)

the horizontal axes A–C and the vertical axes 1–9. Figure 4.49 also displays the roof framework between axes 3, 4 and 6, 7 and the position of the concentric diagonal bracing systems, four in X direction and four in Y direction. A detailed description of the bracings will be given in Section "Preliminary Design of the Bracings".

As required for earthquake resistant buildings, this structure is planned considering the following guiding principles: structural simplicity, uniformity, symmetry and redundancy. In addition, the rigidity of the floors is guaranteed and the foundation system is designed earthquake resistant, but these details are not presented here.

Structural Ductility

The design is carried out for ductility class DCM with medium dissipative structural behaviour. According to Table 4.12, the upper limit of the reference value of the behaviour factor is 4. As such a high ductility is not in principle required for the seven-storey steel structure, the behaviour factor is chosen as $q = 2.0$. The assump-

Table 4.15 Parameters for the determination of the seismic response spectra according to NTC (2008)

Soil type and behaviour factors	
Soil category	D
Topographic category	T1
Soil coefficient S	1.8
Horizontal behaviour factor	2
Vertical behaviour factor	1.5
Design ground acceleration and further parameter	
a_g (g)	0.118
F_0 (–)	2.404
T_C (s)	0.279
C_C (–)	2.365
Damping and control periods	
η	0.5
T_A (s)	0
T_B (s)	0.22
T_C (s)	0.661
T_D (s)	2.073

tion of ductility class DCM allows a linear elastic global analysis using the response spectrum reduced by the behaviour factor q. The resistances of the structural elements and the connections are designed in accordance with Eurocode 3-1-1 (2005) considering further specific seismic rules given in Sect. 6.5 of Eurocode 8-1 (2004).

Design Response Spectrum

The horizontal elastic and design response spectrum is constructed with the expressions given in Sect. 4.1.4. The determination of the response spectrum for the steel structure located in Italy also takes into account the regulations of the Italian code NTC (2008). The input parameters of the seismic action, including the design ground acceleration a_g on type D ground, the soil factor S and the control periods $T_{A\text{–}C}$ are listed in Table 4.15.

The horizontal elastic response spectrum (RS), the horizontal design response spectrum (HRS) and the vertical design response spectrum (VRS) are shown in Fig. 4.50. The vertical seismic action can be neglected here due to the reasons already explained in Sect. 4.1.5.2. It follows that the vertical design response spectrum is not further required in the following steps of calculation and design.

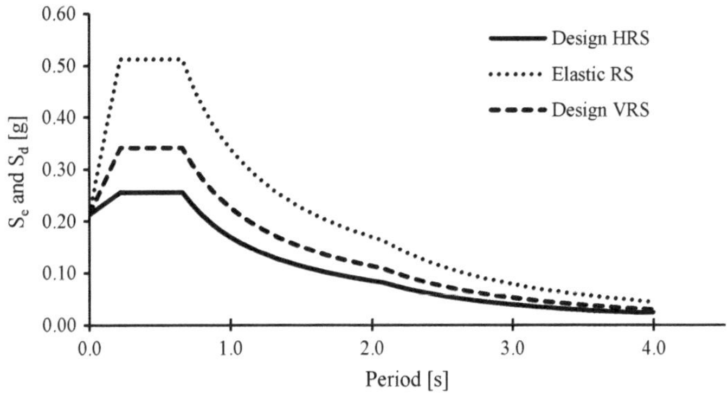

Fig. 4.50 Horizontal elastic response spectrum and horizontal and vertical design response spectra

Vertical Actions and Seismic Masses

The actions considered are determined in accordance with Eurocode 1-1-1 (2004), including dead loads of the frame structure, floors, partitions, external walls and proportional live loads. As the steel structure is a residential building, the category of use is A. For this category Eurocode 1-1-1, Table 6.2 (2002) specifies a characteristic value of uniformly distributed live loads of $q_k = 2.0$ kN/m^2, which must be applied on all floors and on the accessible roof. Table 4.16 summarizes unit weights, uniformly distributed loads and geometrical dimensions required to determine the seismic masses. The design value of the seismic action A_{Ed} is determined taking into account permanent loads $G_{k,j}$ and proportional vertical live loads $Q_{k,i}$ as introduced in Sect. 4.1.6.3. The combination coefficients $\psi_{2,i}$ are taken equal to 0.3 for variable actions and the factor φ is chosen conservatively with 1.0 for the roof and all other floors. It is important to point out that the permanent loads of columns and beams are computed with conservative unit weights, because they are usually not known accurately beforehand for preliminary evaluations. Afterwards, these assumptions can be checked by the designer and possibly modified. In addition, the last column "Design load" provides the design values obtained by multiplying the characteristic values in the "Load" column with the safety factors $\gamma_G = 1.35$ and $\gamma_Q = 1.50$. These safety factors are given in Table A2.4 (B) of Eurocode 0 (2002) for the permanent and temporary design situation.

Estimation of the Total Seismic Base Shear

The sum of the vertical loads in Table 4.16 calculated by means of the combination rule (4.57) leads to the total seismic weight of the structure:

$$W = 23457\,\text{kN}.$$

Table 4.16 Permanent and live loads of the steel building (g_1 = permanent structural loads, g_2 = permanent non-structural loads, q_k = characteristic live loads)

	Unitary weight (kN/m)	Uniformly distributed load (kN/m^2)	Length (m)	Area (m^2)	Number (−)	Vertical load (kN)	Design load (kN)
Columns (g_1)	1.03	–	21.7	–	81	1810	2444
Beams (g_1)	0.88	–	37	–	3	98	132
Secondary beams (g_1)	0.76	–	11	–	3	25	34
Floors (g_1+g_2)	–	4.50	–	433.6	7	13,657	18,437
Live loads (q_k)	–	2.00	–	433.6	7	6070	9105
Internal partitions (g_1)	–	1.60	–	433.6	7	4856	6555
Outer panels (g_1)	12.4	–	96	–	1	1190	1607

Assuming that the maximum horizontal acceleration corresponds to the plateau value of the horizontal design response spectrum (HRS) presented in Fig. 4.50, a spectral acceleration of $S_{ad} = 0.256$ g is applied to calculate the maximum base shear F_b:

$$F_b = 0.256 \cdot 23,457 = 6005\,\text{kN}.$$

Naturally, this is an upper limit of the actual total base shear, as the maximum spectral acceleration was applied. Probably, due to the fact that the structure is slender, the first period of vibration will exceed the corner period T_C and the resulting spectral acceleration will be less than 0.256 g. However, it is always important to get a first conservative estimation of the total seismic base shear in order to later be able to check the results obtained with more sophisticated numerical calculations, presented in the following.

Preliminary Design of Primary Structural Members

A preliminary design of the steel profiles for columns and beams under vertical loading conditions is carried out first. For all structural elements steel grade S275 is assumed. The design value of the yield strength f_{yd} is calculated to:

$$f_{yd} = \frac{f_{yk}}{\gamma_{M0}} = \frac{275}{1.05} = 261.9\,\text{N/mm}^2.$$

The partial safety factor γ_{M0} is applied conservatively according to NTC (2008), although Eurocode 3-1-1 (2005), Sect. 6.1 would allow a partial safety factor of $\gamma_{M0} = 1.0$.

The maximum design normal force $N_{Ed,max}$ for the internal columns is calculated with a tributary area of 5.55 × 5.55 m^2 for the permanent and temporary design situation according to Eurocode 0 (2002):

$$N_{Ed,max} = 1.35 \cdot \left[1.03 \cdot 21.7 + (4.5 + 1.6) \cdot 5.55^2 \cdot 7\right] + 1.50 \cdot 2.0 \cdot 5.55^2 \cdot 7 = 2453\,\text{kN}.$$

The minimum cross-sectional area computation based on the compression stress verification according to Eurocode 3-1-1 (2005), Sect. 6.2.4 yields:

$$A_{min} = \frac{N_{Ed,max}}{f_{yd}} = 93.7\,\text{cm}^2.$$

A steel profile HEB280 with a cross-sectional area $A = 131$ cm^2 is chosen.

The design moment M_{Ed} of the horizontal beams in axis B, oriented in the longitudinal (X) direction and subjected to a uniform load $\bar{p}$, is equal to:

$$M_{Ed} = \frac{\bar{p}L^2}{8} = [1.35 \cdot (4.5 + 1.6) + 1.5 \cdot 2.0] \cdot 5.5 \cdot \frac{5.0^2}{8} = 193\,\text{kNm}.$$

The required minimum section modulus is calculated as:

$$W_{min} = \frac{M_{Ed}}{f_{yd}} = 737.3\,\text{cm}^3.$$

A profile HEA300 with a section modulus of $W = 1259.6$ cm^3 is chosen.

The design moment M_{Ed} of the horizontal beams, oriented in the transverse (Y) direction and subjected to a uniform load $\bar{p}$ acting on the tributary width of $b = 0.25$ m, is equal to:

$$M_{Ed} = \frac{\bar{p}L^2}{8} = \{[1.35 \cdot (4.5 + 1.6) + 1.5 \cdot 2.0] \cdot b + 12.4\} \cdot \frac{5.5^2}{8} = 54\,\text{kNm}.$$

The required minimum section modulus is calculated as:

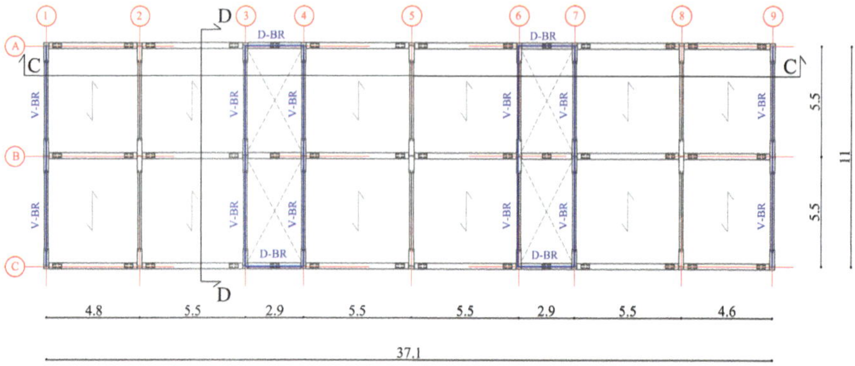

Fig. 4.51 Top view: Positions of bracings in-plan: V-BR = concentric diagonal bracings 8 (V-shaped) and D-BR: Diagonal bracings

$$W_{min} = \frac{M_{Ed}}{f_{yd}} = 205.2\,\text{cm}^3.$$

A joint profile consisting of 2 UPN260 with a section modulus of $W = 742\ \text{cm}^3$ is chosen.

The profiles are selected with sufficient load-bearing reserves in order to be able to conform to the more severe checks for buckling which are not presented here.

Preliminary Design of the Bracings

The horizontal stiffness of frame systems with concentric bracings mainly depends on the axial strength of the bracings. For the given structure diagonal concentric bracings are chosen in both building directions. This choice requires the consideration of specific seismic design rules, discussed and presented in the following. The in-plan bracing positions are shown in Fig. 4.51. In the longitudinal (X) direction separated concentric bracings are installed (Fig. 4.52), while in the transverse (Y) direction the concentric bracings are placed side-by-side in V-form (Fig. 4.53). The average lengths of the diagonals are 3.5 m in X direction and 5.2 m in Y direction.

Based on experience the steel profiles of the diagonals are staged over the building height in order to obtain a uniform deformation behaviour and a similar utilization of the diagonals. The design must guarantee that yielding of the diagonals in tension takes place before failures of the connections, beams or columns occur. This is required in Sect. 6.7 of Eurocode 8-1 (2004). An important rule is the limitation of the non-dimensional slenderness $\overline{\lambda}$, which has to be less than or equal to 2.0. The definition of the non-dimensional slenderness is given in Eurocode 3-1-1 (2005):

$$\overline{\lambda} = \sqrt{\frac{A f_y}{N_{cr}}}.$$

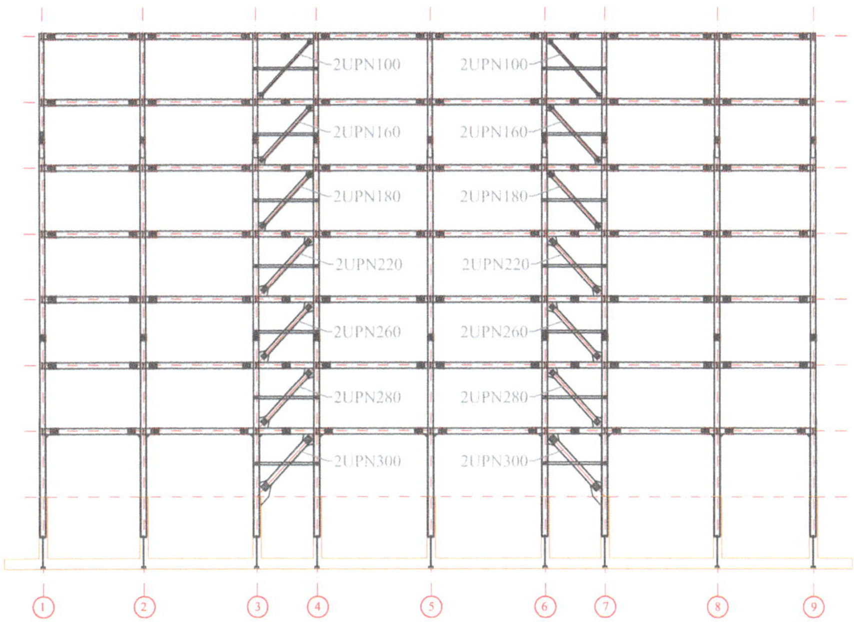

Fig. 4.52 Section C-C in Fig. 4.51 (Plan XZ): position and steel profiles of the concentric bracings

Here, A is the cross section of the profile, f_y is the yield strength and N_{cr} is the elastic critical force for the relevant buckling mode. Non-dimensional slenderness values are given for each bracing system in Tables 4.19 and 4.20 together with the overstrength ratios. In order to satisfy a homogeneous dissipative behaviour of the diagonals, the maximum overstrength should not differ from the minimum value by more than 25%.

Table 4.17 Interstorey drift sensitivity coefficient $\boldsymbol{\theta}$ (response spectrum analysis in X direction)

Storey	N (kN)	V_y (kN)	V_z (kN)	d_{ex} (m)	d_{ey} (m)	d_x (m)	d_y (m)	θ_x (−)	θ_y (−)
1	30643.9	3073.9	1185.2	0.005	0.001	0.009	0.001	0.030	0.009
2	26123.4	2906.4	1123.6	0.008	0.001	0.015	0.001	0.044	0.010
3	21724.0	2628.3	1021.0	0.014	0.001	0.028	0.003	0.076	0.018
4	17325.7	2291.2	890.9	0.018	0.002	0.036	0.003	0.087	0.020
5	12940.1	1894.0	725.9	0.025	0.002	0.049	0.004	**0.109**	0.025
6	8556.9	1419.3	519.2	0.028	0.002	0.056	0.005	**0.109**	0.026
7	4197.2	1165.7	479.1	0.033	0.003	0.067	0.006	0.078	0.017

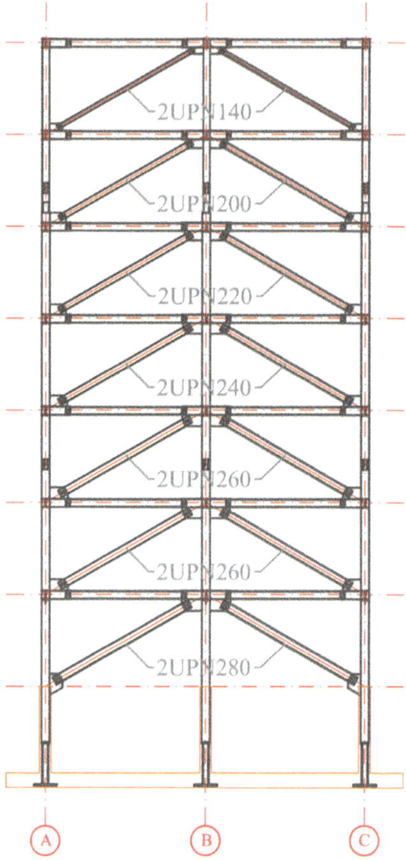

Fig. 4.53 Section D-D in Fig. 4.51, (Plan YZ): Position and steel profiles of the concentric bracings

Table 4.18 Interstorey drift sensitivity coefficient ***θ*** (response spectrum analysis in Y direction)

Storey	N (kN)	V_y (kN)	V_z (kN)	d_{ex} (m)	d_{ey} (m)	d_x (m)	d_y (m)	θ_x (−)	θ_y (−)
1	30643.6	922.7	3949.0	0.002	0.002	0.003	0.004	0.035	0.010
2	26123.1	872.5	3743.6	0.002	0.002	0.004	0.005	0.041	0.011
3	21723.7	789.0	3401.8	0.004	0.005	0.009	0.009	0.078	0.019
4	17325.5	687.8	2968.5	0.005	0.005	0.010	0.011	0.084	0.020
5	12939.9	568.6	2418.5	0.007	0.007	0.015	0.015	**0.109**	0.025
6	8556.8	426.1	1729.8	0.008	0.008	0.016	0.015	**0.106**	0.025
7	4196.8	349.7	1596.9	0.010	0.009	0.020	0.019	0.076	0.016

Analysis and Design Procedure

A linear static analysis is carried out under vertical loading due to dead and live loads and a response spectrum analysis is performed for the seismic actions in the main building directions. According to Eurocode 8-1 (2004), Sect. 6.7.2, the linear elastic analysis must be carried out without taking the bracing elements into account. Thereafter, the results of the linear static analysis and the response spectrum analyses are combined according to (4.56) to calculate the design values of the effects of actions. The stress resultants of the seismic design situation are then used for design purposes. The design of the columns and beams for bending, normal force with bending, and shear forces, is carried out with the verification concepts given in Eurocode 3-1-1 (2005). However, for the sake of brevity, these results are not presented here. The design of the bracing systems is summarized in Section "Design and Verification of the Bracing Systems, Columns and Beams".

Finite-Element Model

The frame structure with concentric bracings is modelled with finite elements in SAP2000 (2018). Beams and columns are idealized with beams and the diagonals with truss elements. The connections to the foundations are hinged. The finite-element model is shown in two- and three-dimensional views in Fig. 4.54.

Modal Analysis

A modal analysis is carried out to identify the vibration modes of the structure and the corresponding periods of vibration. The results of the modal analysis are shown in Fig. 4.55. For each of the first six mode shapes the period and the activated mass in the relevant direction of activation are given. The first and second mode shapes are translational modes in X- and Y-direction, whereas the third and the fourth are torsional shapes. The fifth and sixth modes are combined mode shapes with mass activation in both X- and Y-direction. The sum of the effective modal masses of 90% is attained, if the first and the fourth modes in X direction, and the second, third and sixth mode in Y direction are considered.

Response Spectrum Analysis

The response spectrum analysis is carried out separately for each building direction using the design response spectrum given in Section "Design Response Spectrum". The results of the single modes of vibration are superposed by means of the complete quadratic combination as described in Sect. 4.1.5.2. Selected results of the response spectrum analysis, qualitatively shown, are displayed in Fig. 4.56.

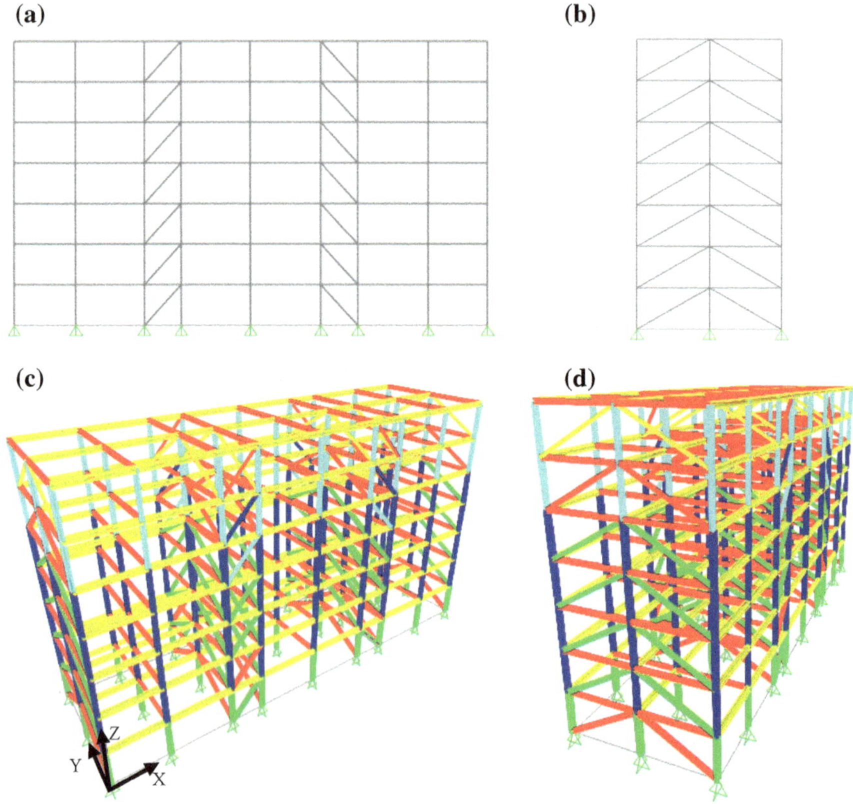

Fig. 4.54 Finite-element model: XZ plane (**a**); YZ plane (**b**); three-dimensional views (**c**–**d**)

Second Order Effects

In the analysis of steel buildings the influence of second order effects must be checked. In particular, the value of the interstorey drift sensitivity coefficient θ has to be checked as described in Sect. 4.1.6.3:

$$\theta \approx \frac{P_{tot} \cdot d_{ei}}{V_{tot} \cdot h} \leq 0.10.$$

Here, P_{tot} is the total gravity load at and above the storey, d_{ei} is the elastic design interstorey drift, h is the interstorey height and V_{tot} is the total seismic storey shear. This check has to be performed in the two main directions of the building, whilst the lateral displacements d_x, d_y in each storey are calculated with:

$$d_x = q\, d_{ex} \text{ and } d_{xy} = q\, d_{ey}$$

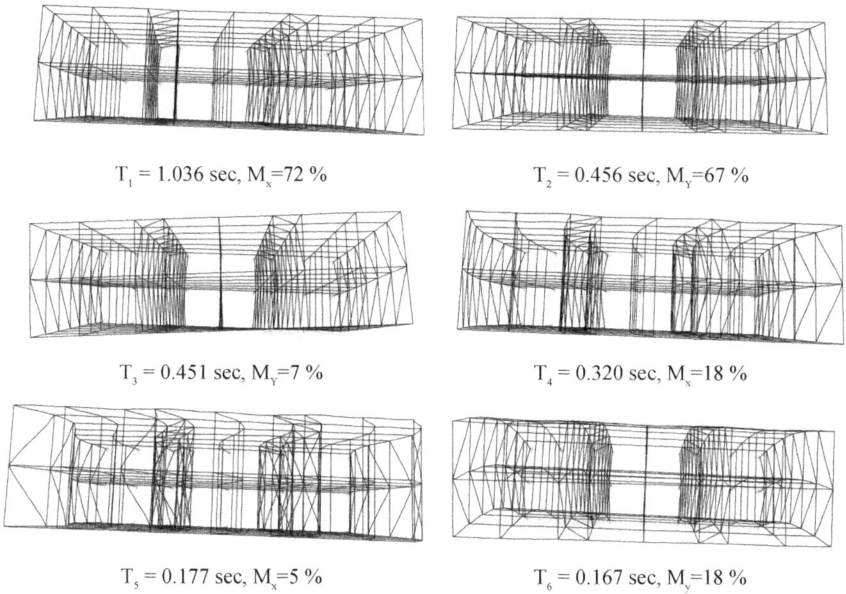

Fig. 4.55 Modal analysis results for the first six mode shapes

where q is the behaviour factor and d_{ex}, d_{ey} are the elastic displacements. The results of the interstorey drift sensitivity coefficient θ in Tables 4.17 and 4.18 show that the lower limit of 0.1 is exceeded in just four cases. Thus, second-order effects have to be considered for these cases by multiplying the effects of the seismic horizontal action by $1/(1-\theta)$.

Design and Verification of the Bracing Systems, Columns and Beams

If the overstrength factors Ω_i defined in Sect. 6.7.4 of Eurocode 8-1 (2004) do not differ from the minimum value Ω_{min} by more than 25%, a homogeneous dissipative behaviour of the diagonals can be assumed. This requirement and the limitation of non-dimensional slenderness $\overline{\lambda}$ govern the dimensioning of the diagonals. The overstrength factor is calculated for each diagonal i to:

$$\Omega_i = \frac{N_{pl,Rd,i}}{N_{Ed,i}}.$$

Here, $N_{pl,Rd,i}$ is the design plastic resistance to normal forces of diagonal i and $N_{Ed,i}$ is the design axial force of diagonal i, obtained from the response spectrum analysis. Tables 4.19 and 4.20 contain the non-dimensional slenderness $\overline{\lambda}$ around axis Y and Z and the overstrength factors for each diagonal in X and Y direction

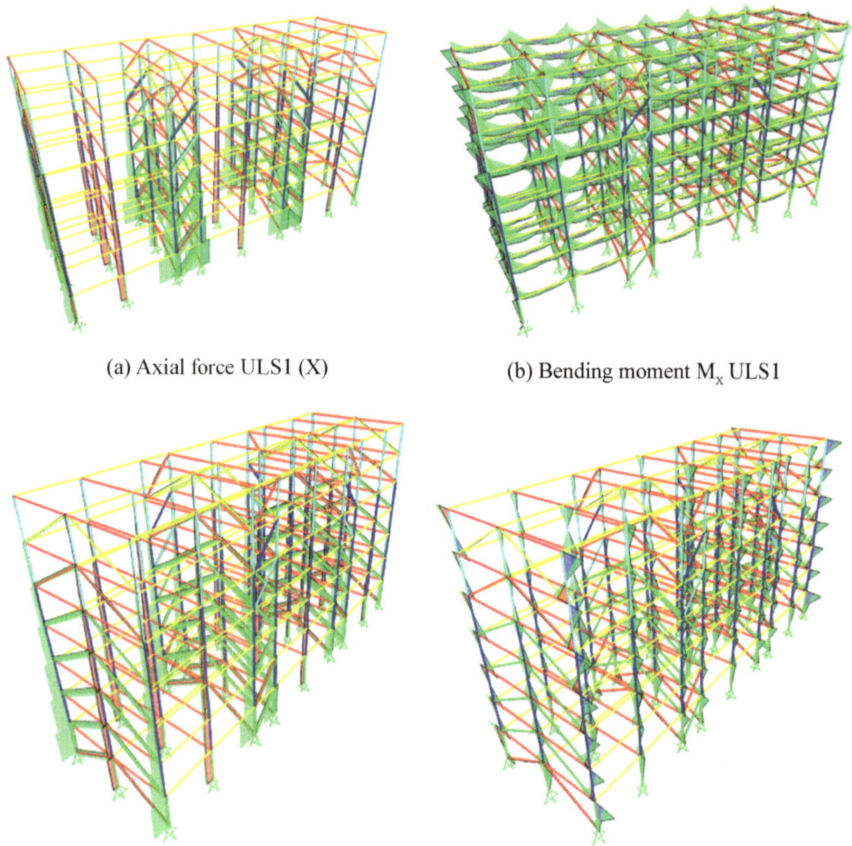

(a) Axial force ULS1 (X)

(b) Bending moment M_X ULS1

(c) Axial force ULS2 (Y)

(d) Bending moment M_Y ULS2

Fig. 4.56 Qualitative results of response spectrum analyses at ultimate limit states ULS1 and ULS2: X-direction (**a**, **b**) and in Y-direction (**c**, **d**)

respectively. The maximum overstrength differs from the minimum one by 24% in X direction and by 20% in Y direction. Furthermore the condition $\overline{\lambda} \leq 2.0$ for the non-dimensional slenderness is fulfilled for all diagonals. The successful verifications of the columns and beams in X- and Y-direction for the critical stress resultants of the seismic design situation are summarized in Tables 4.21, 4.22 and 4.23, where ULSi indicates the *i*th ultimate limit state ULS and *s* corresponds to the clear distance between UPN profiles.

Table 4.19 Non-dimensional slenderness $\overline{\lambda}$ and overstrength factors Ω_i for the bracing in X direction

Storey	Profile	$\overline{\lambda}_Y$ (−)	$\overline{\lambda}_Z$ (−)	N_{Ed} (kN)	N_{cr} (kN)	(−)
1	2UPN300 s12	0.33	0.86	1748.0	1888.0	1.080
2	2UPN280 s12	0.36	0.94	1402.0	1614.0	1.151
3	2UPN260 s12	0.04	0.11	1024.3	1375.0	1.342
4	2UPN220 s12	0.46	1.05	783.0	961.7	1.228
5	2UPN180 s12	0.66	0.66 ara>	500.9	593.1	1.184
6	2UPN160 s12	0.88	1.44	418.0	475.5	1.138
7	2UPN100 s12	1.06	1.51	166.0	207.1	1.248

Table 4.20 Non-dimensional slenderness $\overline{\lambda}$ and overstrength factors Ω_i for the bracing in Y direction

Storey	Profile	$\overline{\lambda}_Y$ (−)	$\overline{\lambda}_Z$ (−)	N_{Ed} (kN)	N_{cr} (kN)	(−)
1	2UPN280 s12	0.55	1.44	732.7	935.0	1.276
2	2UPN260 s12	0.60	1.53	726.5	772.7	1.064
3	2UPN260 s12	0.60	1.53	651.8	772.7	1.186
4	2UPN240 s12	0.65	1.56	555.7	625.3	1.125
5	2UPN220 s12	0.71	1.62	463.6	518.1	1.118
6	2UPN200 s12	0.78	1.77	359.6	405.5	1.128
7	2UPN140 s12	1.10	1.96	182.0	202.3	1.112

Table 4.21 Verification of beams HEA300 in X-direction

Cross section	Combination	N (kN)	V_y (kN)	V_z (kN)	M_T (kNm)	M_y (kNm)	M_Z (kNm)	Verification
HEA300	ULS2	0.00	−148.85	0.00	0.00	0.00	−234.52	OK
HEA300	ULS2	0.00	−159.5	0.00	0.00	0.00	−225.56	OK

Table 4.22 Verification of beams 2UPN260 in Y-direction

Cross section	Combination	N (kN)	V_y (kN)	V_z (kN)	M_T (kNm)	M_y (kNm)	M_Z (kNm)	Verification
2UPN260 s10	ULS2	0.00	−46.13	0.00	0.00	0.00	−47.52	OK
2UPN260 s10	ULS2	0.00	−45.09	0.00	0.00	0.00	−47.54	OK

Table 4.23 Verification of columns

Cross section	Combination	N (kN)	V_Y (kN)	V_Z (kN)	M_T (kNm)	M_Y (kNm)	M_Z (kNm)	Verification
HEB240	ULS1	−639.56	−42.80	−1.75	−0.01	−2.70	−63.94	OK
HEB240	ULS2	−91.53	9.09	−9.68	−0.00	19.62	101.73	OK
HEB240	ULS2	206.12	−75.21	0.36	−0.01	−4.98	−127.32	OK
HEB280	ULS2	−2220.00	−51.42	−2.45	0.00	−3.93	−86.53	OK
HEB280	ULS2	−585.45	−65.26	8.30	0.00	−18.07	−32.84	OK
HEB280	ULS2	−616.87	−90.79	−0.12	0.00	−4.78	−145.49	OK
HEB280+ p18	ULS1	−4219.92	−22.11	−4.19	−0.02	−6.98	0.00	OK
HEB280+ p18	ULS2	−1852.32	−17.06	5.52	0.01	−22.19	19.71	OK
HEB280+ p18	ULS2	−1765.70	−4.98	−1.46	0.00	−4.51	−79.62	OK

Table 4.24 Bracing connections in X direction

Storey	Profile	R_{fy} (kN)	1.1 γ_{ov} R_{fy} (kN)	Bolts and class
1	2UPN300 s12	1888	2388.32	12 M22 8.8
2	2UPN280 s12	1614	2041.71	9 M22 8.8
3	2UPN260 s12	1375	1739.38	8 M22 8.8
4	2UPN220 s12	961.7	1216.55	6 M22 8.8
5	2UPN180 s12	593.1	750.27	6 M18 8.8
6	2UPN160 s12	475.5	601.51	6 M18 8.8
7	2UPN100 s12	207.1	261.98	4 M14 8.8

Design of Bracing Connections

For bolted non dissipative connections, the resistance of the connection must satisfy the following condition according to Eurocode 8-1 (2004), Sect. 6.5.5:

$$R_d \geq 1.1\gamma_{ov}R_{fy},$$

where R_d is the resistance of the connection and R_{fy} is the plastic resistance of the connected dissipative member according to Eurocode 3-1-1 (2005). The overstrength factor γ_{ov} is defined in Sect. 6.2.3 of Eurocode 8-1 (2004). Tables 4.24 and 4.25 include the capacity design of the bolted connections. Herein s indicates the clear distance between the UPN profiles. The design taking overstrength into account guarantees the formation of the plastic hinges in the connected dissipative diagonals.

Table 4.25 Bracing connections in Y direction

Storey	Profile	R_{fy} (kN)	1.1 γ_{ov} R_{fy} (kN)	Bolts and class
1	2UPN280 s12	935	1182.78	6 M22 8.8
2	2UPN260 s12	772.7	977.47	6 M22 8.8
3	2UPN260 s12	772.7	977.47	6 M22 8.8
4	2UPN240 s12	625.3	791.01	4 M22 8.8
5	2UPN220 s12	518.1	655.40	6 M18 8.8
6	2UPN200 s12	405.5	512.96	4 M18 8.8
7	2UPN140 s12	202.3	255.91	4 M14 8.8

4.2.3 Design of Masonry Structures

Due to its local availability and its straightforward application, masonry has been a traditional way of constructing houses and residential buildings in Europe for many centuries. Sustainability and durability properties of masonry structures are second to none, and in addition, masonry meets all of today's expected requirements concerning heat and moisture transfer. From the viewpoint of statics, masonry is highly suitable to carrying vertical loads due to its high compression strength. When it comes to withstanding horizontal loads, however, the corresponding capacity of unreinforced masonry is limited due to its low tensile strength. Furthermore, masonry is a non-homogeneous, brittle and highly non-linear material. Although masonry is really simple to handle at the construction site, the modelling and the design part pose a challenge in comparison to other construction materials.

Another important fact is that masonry buildings usually do not have regular configurations both in plan and in elevation and their modelling is affected by many uncertainties. Traditionally, the seismic design of masonry structures is based on simplified linear calculations taking into account non-linear load carrying reserves by means of a conservative behaviour factor (Table 4.3). However, it is well-known that the linear, force-based design philosophy is too much conservative, and it does not correspond to the current state-of-the-art and makes almost no use of the substantial non-linear structural reserves inherently available.

Additionally, the verifications deal solely with single walls, so that single walls effectively determine the overall verifiable structural capacity. Thus, non-linear analysis of masonry structures is often required in order to exploit the strength and deformation capacities of the walls and to make use of force redistribution after local failures of single walls.

This section provides an overview of the seismic behaviour of unreinforced masonry structures and introduces linear and non-linear design concepts, specific rules, modelling approaches and calculation procedures for masonry structures subjected to seismic loading.

4.2.3.1 Requirements for Masonry Units, Mortar and Masonry Bond

The seismic design of confined, unreinforced and reinforced masonry is regulated in Sect. 9 of Eurocode 8-1 (2004). The seismic design specifications must be considered in addition to the basic design rules according to Eurocode 6-1-1 (2005). As masonry is a composite material consisting of single bricks which are connected by head and bed mortar joints, the load-carrying behaviour of masonry depends on its components (brick, mortar and brick-mortar interface) and its constructive disposition (bond and mortar bed joint thickness). Thus, Eurocode 8-1 (2004) formulates in Sect. 9.2 the following requirements for the components:

- Masonry units must provide sufficient compressive strength in order to avoid brittle failures. The recommended normalized strength values of masonry units normal to the bed joints is $f_{b,min} = 5\ \mathrm{N/mm^2}$ and parallel to the bed joints $f_{bh,min} = 2\ \mathrm{N/mm^2}$.
- The mortar must provide a minimum compressive strength. The recommended values are $f_{m,min} = 5\ \mathrm{N/mm^2}$ for unreinforced and confined masonry and $f_{m,min} = 10; \mathrm{N/mm^2}$ for reinforced masonry.
- Three different types of head joints can be executed: fully grouted head joints, ungrouted head joints and ungrouted head joints with mechanical interlocking between the masonry units.

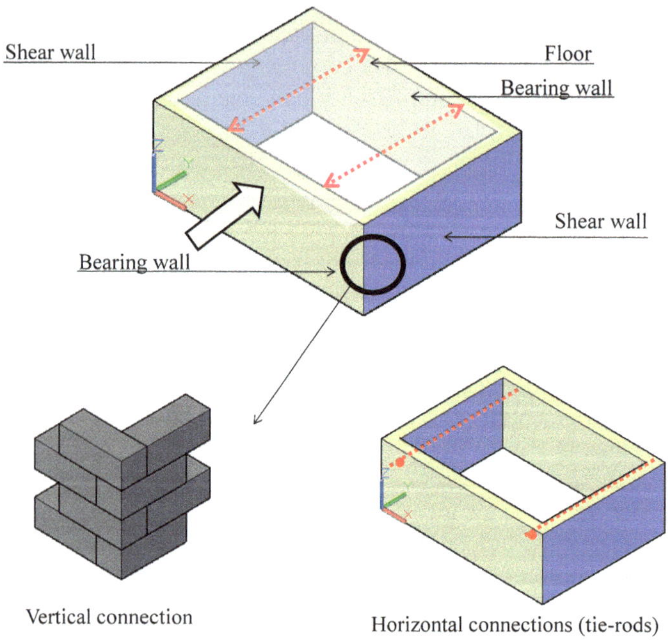

Fig. 4.57 Vertical and horizontal connections of a simple unreinforced masonry building

4.2.3.2 Construction Rules for Unreinforced Masonry Buildings

Masonry buildings are usually composed of floors and shear walls in the main building directions. The in-plane stiffness of the floors must be sufficiently large compared to the lateral stiffness of the vertical structural elements in order to be able to transfer the horizontal seismic forces to the shear wall system which in turn transfers them to the foundation system and finally to the ground. All floor types are admissible as long as their continuity and a good distribution of horizontal forces are guaranteed. The horizontal load transfer requires frictional connections between the floor diaphragms and the shear walls. Eurocode 8-1 (2004), Sect. 9.5.2 recommends the placement of ring beams with a minimum cross-sectional area of 200 mm^2 or steel tie-rods at every floor level to connect all walls together. Furthermore, a functional interconnection of intersecting walls with shear transfer in vertical direction is beneficial for the seismic resistance. Shear transfer by intersecting walls can be assumed, if the walls are bound or tied together with suitable connectors. Figure 4.57 shows a simple unreinforced masonry structure with rectangular floor plan and two walls in each direction that are connected to the floor diaphragm and are bound together at each wall intersection. Furthermore, the parallel walls are connected by additional tie-rods. The load-carrying system of the structure is able to transfer the seismic forces to the foundation as floors, shear walls and all connections are properly designed.

According to Eurocode 8-1 (2004) shear walls must satisfy geometrical requirements regarding the minimum effective wall thickness $t_{ef,min}$, the maximum ratio of the effective wall height to the effective thickness of the wall $(h_{ef}/t_{ef})_{max}$ and the minimum ratio of the wall length l to the greater clear height h of the openings adjacent to the wall. Shear walls, satisfying these geometrical requirements, must be provided in at least two orthogonal directions. The shear walls are considered as secondary elements, if these conditions are not fulfilled. Figure 4.58 shows the geometrical parameters for a simple masonry building, and Table 4.26 details the geometrical requirements for different types of masonry according to Eurocode 8-1 (2004).

The strict limitations of the minimum wall thickness and wall slenderness enhance the resistance of masonry walls against out-of-plane failure. Recent earthquakes have shown that especially the walls on the upper storeys and gable walls with low vertical stress levels are at risk to out-of-plane failures. Figure 4.59 illustrates the out-of-plane failures of a gable wall and an exterior wall at the top floor. A further stabilization of these critical walls can be achieved by providing additional horizontal reinforced concrete beams placed in the plane of the walls in order to reduce the vertical spacing and additional masonry piers or transverse walls.

4.2.3.3 Seismic Behaviour of Masonry Buildings

When unreinforced masonry buildings are subjected to seismic actions, most of their walls unavoidably experience a combination of in-plane and out-of-plane loading.

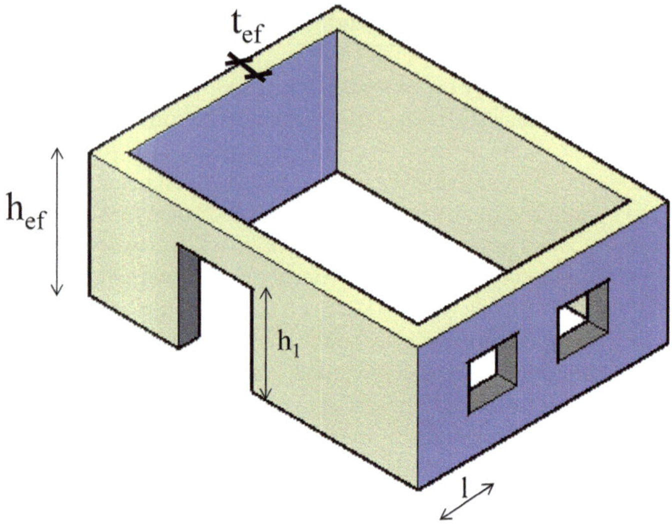

Fig. 4.58 Geometrical parameter of shear wall systems

Table 4.26 Geometric requirements for shear walls according to Eurocode 8-1 (2004), Table 9.2

Masonry type	$t_{ef,min}$ (mm)	$(h_{ef}/t_{ef})_{max}$	$(l/h)_{min}$
Unreinforced, with natural stone units	350	9	0.5
Unreinforced, with any other type of units	240	12	0.4
Unreinforced, with any other type of units, in cases of low seismicity	170	15	0.35
Confined masonry	240	15	0.3
Reinforced masonry	240	15	No restriction

with:
t_{ef} Thickness of the wall, Eurocode 6-1-1 (2005)
h_{ef} Effective height of the wall Eurocode 6-1-1 (2005)
h Greater clear height of the openings adjacent to the wall
l Length of the wall

In-plane loading results from the horizontal seismic storey forces distributed by the rigid floors, while out-of-plane loading results from out-of-plane inertial forces due to the accelerated mass of the masonry walls. Depending on the specific load situation, wall failures can occur in in-plane or out-of-plane direction. In the following, the essential in- and out-of-plane failure modes of load-bearing masonry walls are presented. The complex interaction of in- and out-of-plane loading is not presented here, as this topic is still subject of ongoing research.

Fig. 4.59 Out-of-plane failure of a gable wall and an exterior wall on the top floor, Albstadt earthquake, 1978 (Photo: Peter Doll)

4.2.3.4 In-Plane Failure Modes

In this section the failure modes of load-bearing masonry walls subjected to vertical and horizontal seismic in-plane loading are presented. Generally, the failure modes can be divided into shear, compression and tension failure modes. The formation of one of these specific failure modes depends on the wall geometry, the masonry and mortar strength and the ratio of vertical to horizontal loads. Figure 4.60 illustrates the potential shear failures concentrated in either bricks or mortar joints together with the corresponding cyclic load-displacement curves.

Shear failures of unreinforced masonry walls subjected to in-plane seismic loading are typically characterized by diagonal cracks. If the compression stresses are quite low, shear failure takes place in the mortar joints while the bricks remain undamaged. The shear resistance of the mortar joints is represented by the shear stress - shear deformation curve shown in Fig. 4.61. The curve starts with the initial shear strength f_{vk0} under zero compression stresses and increases in presence of compressive stresses σ_D up to the maximum shear stress f_{vk} according to the Mohr–Coulomb theory. The subsequent descending branch is characterised by an exponential reduction of the shear strength and the coefficient of static friction μ_F is reduced to the residual value of the coefficient of sliding friction μ_S. The corresponding cracks are either limited to single bed joints or they propagate as staircase cracks along the head and bed joints of the masonry bond. Thus, the shear strength also depends on the brick dimensions, the thickness of the bed and head joints and the specific overlapping length of the masonry bond.

For higher compression stresses, failures can also occur in the bricks as diagonal tension failures. This failure mode results from shear induced brick rotations which cause high tensile stresses in the bricks.

The cracks caused by shear failure usually appear in a crosswise pattern; they often result from a combination of both shear failure modes as seismic loads act in both directions with varying amplitudes. Figure 4.62 illustrates a typical crack pattern due to shear failure modes for two buildings hit by recent Italian earthquakes. The qualitative comparison of the hysteresis curves of both shear failure modes indicates fundamental differences with respect to their energy dissipation mechanisms. Shear failure in the mortar joints is characterized by stable and broad hysteresis loops with high energy dissipation. Furthermore, the stiffness and the maximum load of all hysteresis loops remain constant with increasing displacements (Fig. 4.60). Thus, the behaviour is similar to an elastoplastic material behaviour.

On the other hand, diagonal tension failure in the bricks is characterized by narrow hysteresis loops and moderate energy dissipation capacity. In addition, both maximum load and stiffness show a substantial reduction at higher displacement levels (Fig. 4.60).

Slender masonry walls subjected to combined vertical and horizontal loading behave more like flexural beams with the formation of cracks due to bending stresses at the bottom of the wall (Fig. 4.63). Cyclic loading activates a rocking mode of the wall with alternating tension and compression stresses at the wall ends. The induced stresses cause tension and compression failures in the corner regions at the bottom layer. In addition, vertical splitting at the head joints or, more rarely, in the bricks themselves, could take place. The corresponding hysteresis loops are severely

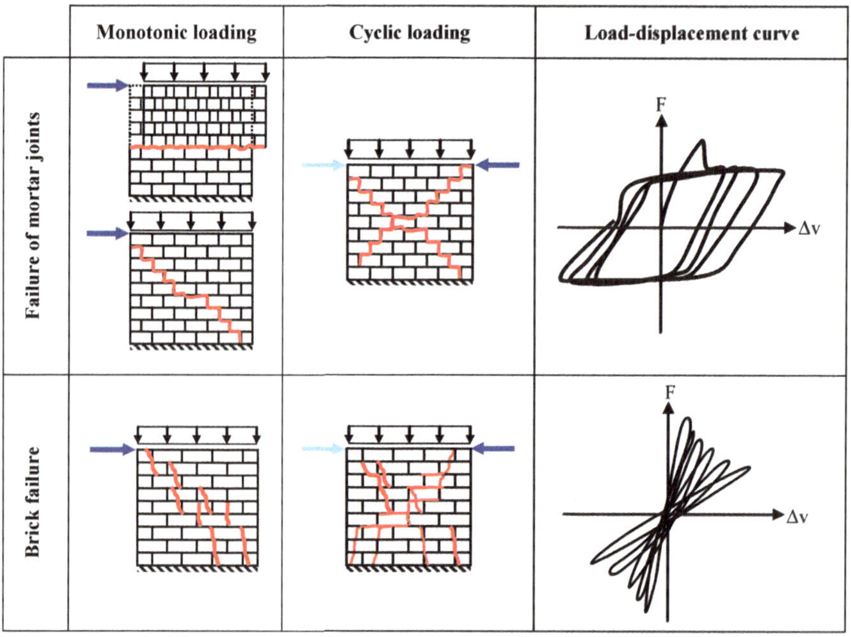

Fig. 4.60 In-plane failure modes for unreinforced masonry walls subjected to cyclic loading

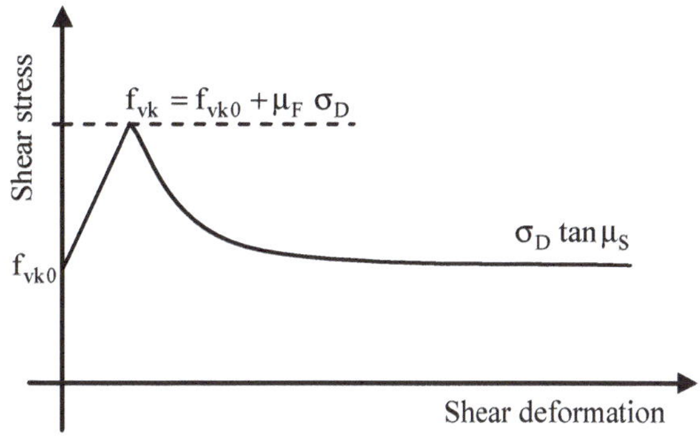

Fig. 4.61 Shear stress - shear deformation curve for mortar joints

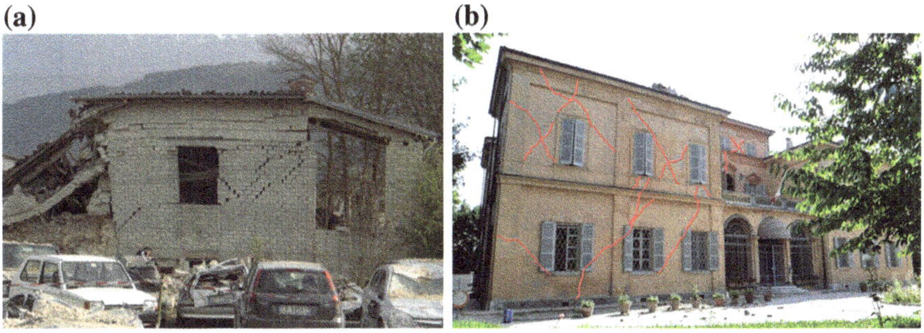

Fig. 4.62 Crack pattern caused by in-plane loading: **a** Onna (L'Aquila, Italy, 2009); **b** Rest home of Santa Maria delle Grazie in Reggiolo (Reggio Emilia, Italy, 2012)

pinched with low energy dissipation, but they exhibit a great deformation capacity. S-shaped curves result mainly from the reduction of the compressed cross-sectional area with increasing displacements.

4.2.3.5 Out-of-Plane Failure Modes

Out-of-plane failures of load-bearing and non-load-bearing masonry walls usually happen at significantly lower seismic load levels compared to walls subjected to in-plane loading. Evidence from recent earthquakes shows that out-of-plane failures often cause greater damage than in-plane failure modes. Especially walls on upper stories and gable walls with low vertical stresses are highly vulnerable to out-of-plane failure modes (Fig. 4.59). Nevertheless, the mechanisms of seismic out-of-plane response are still not fully understood and a complete verification concept has not yet been implemented in Eurocode 8-1 (2004). Out-of-plane loading generally leads to vertical and horizontal bending of the wall. Vertical and horizontal bending cause stresses perpendicular both to the head and to the bed joints. If these bending stresses exceed the masonry bending tensile strength, a crack pattern appears on the wall surface as predicted by the yield line theory. Depending on the boundary

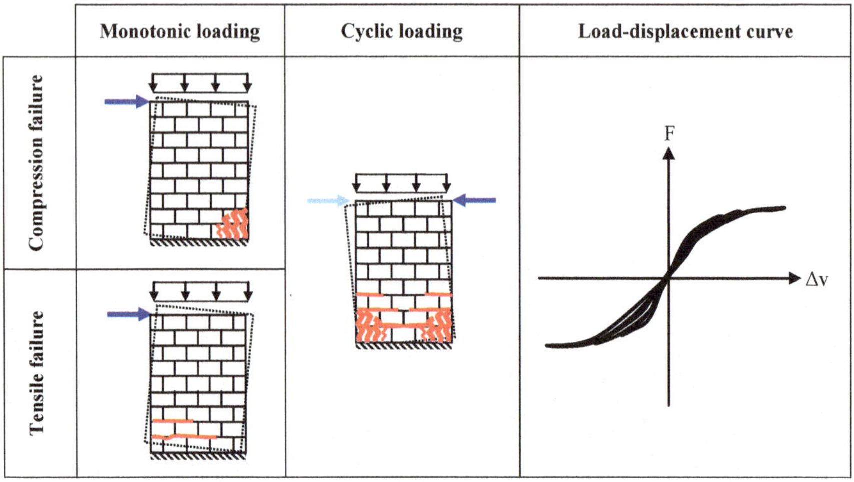

Fig. 4.63 Combined compression-tensile failure for in-plane bending

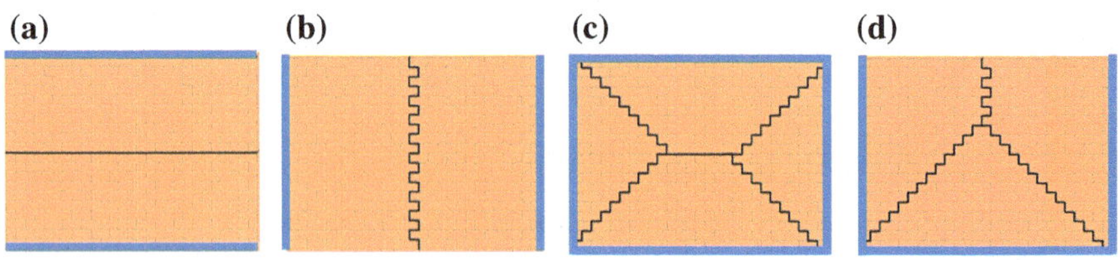

Fig. 4.64 Crack pattern under out-of-plane loading for different boundary conditions (Vaculik 2012)

conditions vertical, horizontal or combined bending occurs in the wall. Figure 4.64 shows the crack patterns for vertical and horizontal spanning walls (a, b) and four- and three-side supported walls (c, d).

A simple way to analyse out-of-plane modes is by using a kinematic approach. By defining possible kinematic sequences, the acceleration that triggers a specific mechanism can be calculated by means of a classic limit analysis. The resulting acceleration capacity is compared to the acceleration demand depending on the type of construction and the seismic actions at the building site.

Furthermore, deformation-based verification concepts for out-of-plane modes have been developed based on the results of shaking table experiments (Doherty et al. 2002; Griffith and Magenes 2003; Melis 2002). Figure 4.65 shows the result of a shaking table test with overturning of the longitudinal wall due to a complete separation from the transverse walls (Lagomarsino and Magenes 2009).

Fig. 4.65 Shaking table test with overturning of the wall (Lagomarsino and Magenes 2009)

4.2.3.6 Structural Models for the In-Plane Response of Masonry Buildings

The seismic calculation of masonry buildings can be performed with equivalent beam models, plane-frame models, pseudo three-dimensional equivalent frame models, macro-element models or detailed three-dimensional models. In each of the models either linear or non-linear material behaviour can be applied. In general, it is preferable to choose the simplest possible model with respect to the building complexity and calculation method as too detailed models are time-consuming and often useless for masonry buildings. A mechanical parameter of primary interest is the elastic modulus. As suggested by Eurocode 8-1 (2004), it is more realistic to assume the cracked bending and shear stiffness to be one half of the uncracked elastic stiffness.

Equivalent Beam Method

Regular masonry buildings can be analysed by the equivalent beam method, in which the continuous shear walls are substituted by uncoupled cantilever beams as depicted in Fig. 4.66. The resulting stiffness of the equivalent beam corresponds to the sum of the cantilever beam stiffnesses in the respective direction of the seismic action. The shear deformations of the walls are considered according to Müller and Keintzel (1984) by reducing the moment of inertia I:

$$I_E = \frac{I}{\left(1 + \frac{3.64}{h^2}\frac{EI}{GA}\right)}. \tag{4.63}$$

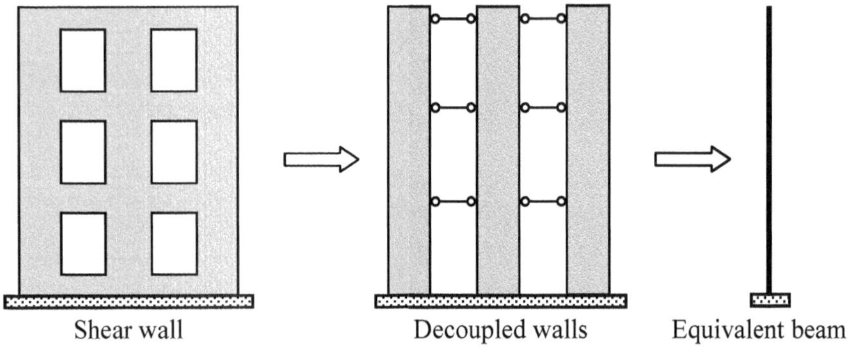

Fig. 4.66 Shear wall with openings, decoupled walls and equivalent beam

Here, E is the elastic modulus, h is the wall height, G is the shear modulus and A is the shear wall area. The interaction of walls can only be considered if a sufficiently stiff interconnection of intersecting walls with shear transfer in vertical direction is guaranteed. Shear transfer between intersecting walls can be assumed if the walls are bound or tied together with suitable connectors.

Plane-Frame Models

If plane-frame models are used, the contribution of spandrels and lintels between the continuous shear walls is considered by horizontal beams with equivalent stiffness. The resulting static system is a moment-resisting frame as shown in Fig. 4.67. The estimation of the coupling stiffness is a very complex task as it depends on the bending stiffness of the slabs, the acting vertical loads, the lengths of the shear walls and the distribution of the shear walls in plane. Müller and Keintzel (1984) developed practical approaches for the estimation of the coupling stiffness considering some of the mentioned factors. However, in reality the slab is supported over the whole length and the free length is much smaller, which leads to a substantial underestimation of the coupling stiffnesses in plane-frame models. Plane-frame models are primarily used for the application of the lateral force method and the response spectrum analysis to consider the moment distribution in the walls and the redistribution of vertical forces in a more realistic way.

Equivalent Frame Model

Another possibility of modelling is the application of pseudo three-dimensional models, which are composed of individual plane frames in each building direction. The individual frames are represented by equivalent frame models, which are discretized by piers, spandrels and lintels connected by rigid nodes. This simplified idealization

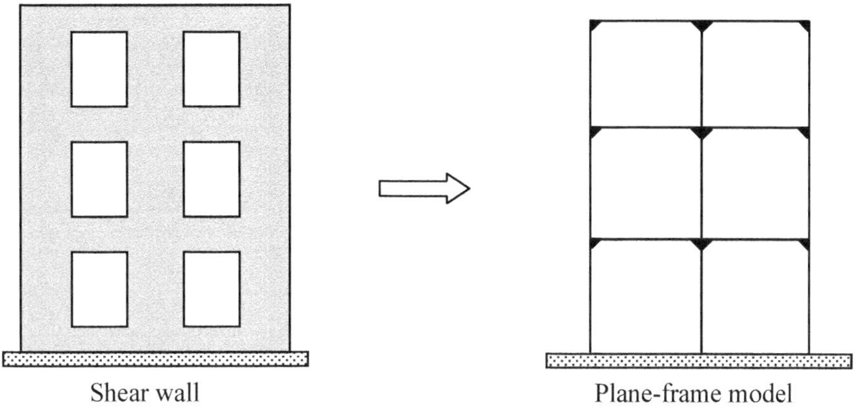

Fig. 4.67 Shear wall and plane-frame model

is based on the observation that the seismic damage of masonry structures is usually concentrated in those elements. This idea, initially developed with the so-called POR method (Tomaževič 1978), was investigated by several researchers (Magenes and Della Fontana 1998; Vanin and Foraboschi 2009) and implemented in commercial software packages (Magenes et al. 1998; Lagomarsino et al. 2013). However, these models only account for the in-plane response of masonry walls and can only be applied to structures with a box-like behaviour if rigid connections between orthogonal walls and between slabs and walls are assumed. The equivalent frames are simplified elements with concentrated plasticity at their extremities. The piers are the main structural elements that carry both vertical and horizontal loads, while the horizontal beams, representing spandrels and lintels, complete the frame effect. Usually the beams are fixed at their extremities to the piers, influencing the bending moment distribution due to the restraint of the node rotations. Rigid beams and flexible piers correspond to the "storey-mechanism" formulation introduced by Tomaževič (1978). If the beams are idealized as truss elements, the piers are uncoupled and behave as cantilever beams.

This type of modelling is used in particular for the application of non-linear static analysis, since the non-linear modelling with beam elements is simple and easy to control. The specific shear and bending failure modes are represented by non-linear joints and non-linear beam elements with elastoplastic material properties. The formulation of these nonlinearities is based on analytical formulations and experimental test data (Chen et al. 2008; Brencich et al. 1998; Magenes 2006; Morandi 2006; FEMA 306 1998; FEMA 356 2000). The non-linear static analysis using equivalent frame models can dependably mirror the failure modes and their sequence under monotonic horizontal loading (Fig. 4.68).

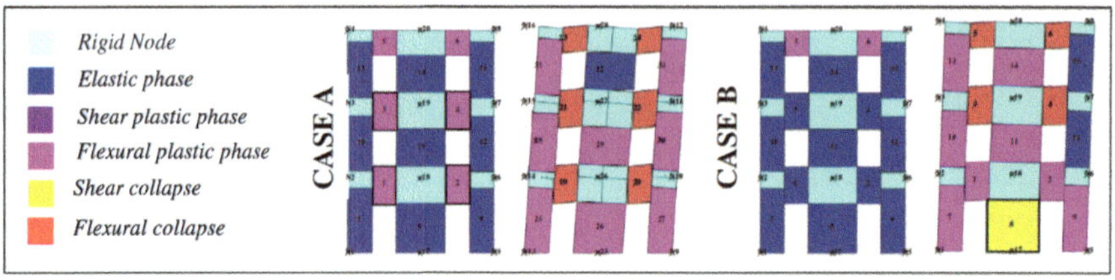

Fig. 4.68 Spandrels, piers and rigid nodes of the equivalent frame model and corresponding failure modes (Lagomarsino et al. 2013)

Macro-element Models

The second numerical strategy is based on plane macro-elements whose kinematics, governed by a discrete distribution of non-linear springs, allows a description of the in-plane behaviour of masonry walls. The macro-element simulates a portion of a masonry wall by incorporating a set of non-linear springs mounted on an articulated quadrilateral to represent the typical failure mechanisms (Fig. 4.69). This pinned quadrilateral consists of four rigid edges, in which two diagonal springs are connected to the corners to model the shear behaviour. Discrete distributions of springs normal to the sides of the macro-element simulate its interaction with adjacent macro-elements, with the purpose of evaluating the flexural response by integrating the tensile or compressive forces in the springs. Each side of the quadrangle can interact with other elements or supports by means of interfaces. A number of *n* non-linear springs are placed orthogonal on each panel edge to represent the contact of adjacent elements. Additionally, longitudinal springs are placed parallel to each panel edge to simulate a potential sliding along the macro-element sides.

Figures 4.70, 4.71and 4.72 show the flexural and shear failure modes and the representation by the macro-element model. The deformation figures and the activation of the springs illustrate that the macro-element is able to represent the typical masonry failure modes introduced in Sections “Estimation of the Total Seismic Base Shear” and “Preliminary Design of Primary Structural Members”.

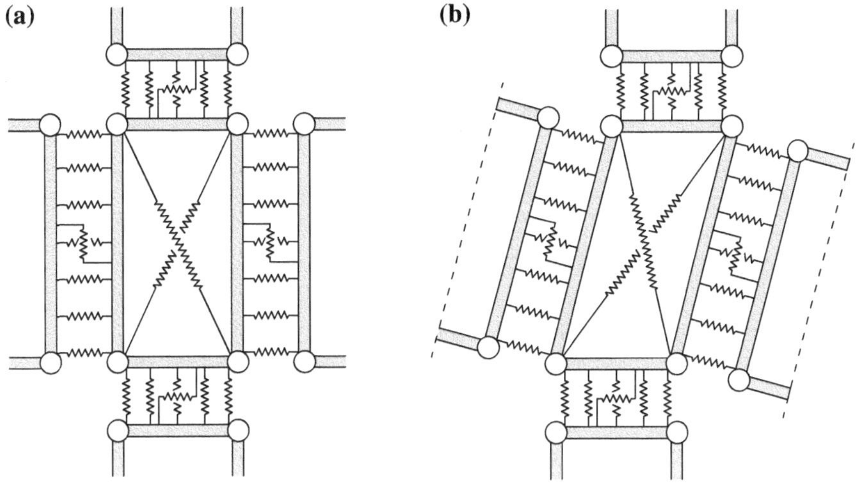

Fig. 4.69 Macro-element for masonry: **a** undeformed configuration; **b** deformed configuration (Pantò et al. 2015)

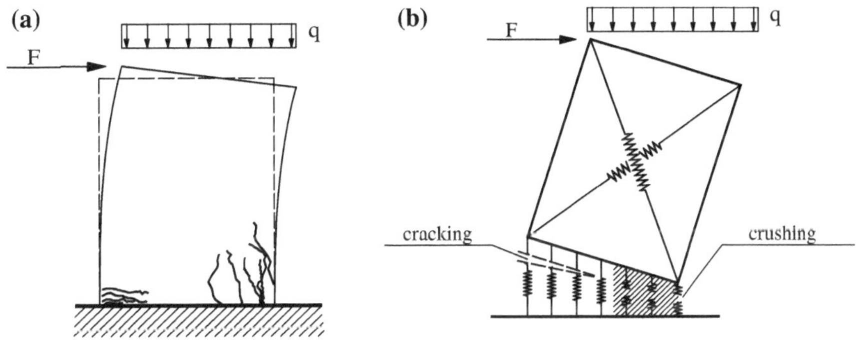

Fig. 4.70 In-plane flexural failure mode of a masonry panel: **a** qualitative response, **b** macro-element representation (Pantò et al. 2015)

The macro-element model allows the consideration of interactions along all four sides and offers the possibility to use different discretizations. This enables also the flexible modelling of masonry infills with openings and the consideration of complex interactions between reinforced concrete frames and masonry infills as shown in Fig. 4.73.

The flexural behaviour is simulated by non-linear springs placed orthogonally on each panel edge. Each spring is calibrated assuming a specific constitutive law for the masonry type by means of a simplified fibre model approach. Figure 4.74 shows the elastoplastic material law defined with the elastic modulus E, the tensile strength σ_t, the compression strength σ_c and the ultimate tensile and compression strains ε_{ut}, ε_{uc}. If the compression strength σ_c is exceeded, a complete crushing is assumed. When

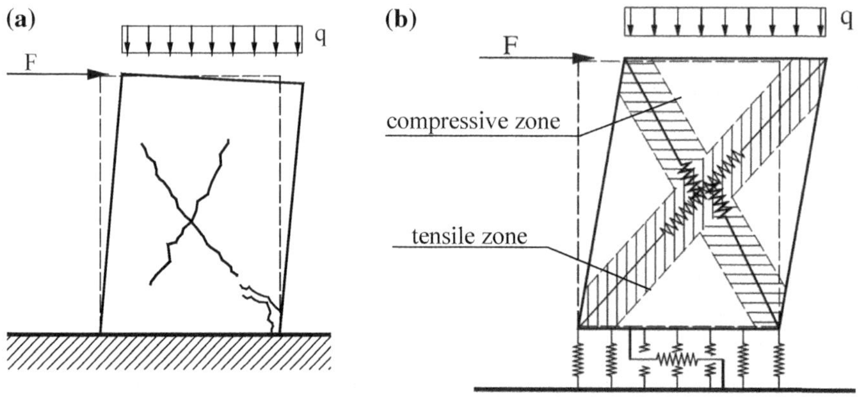

Fig. 4.71 In-plane shear failure mode (diagonal tension failure of the bricks) of a masonry panel: **a** qualitative response, **b** macro-element representation (Pantò et al. 2015)

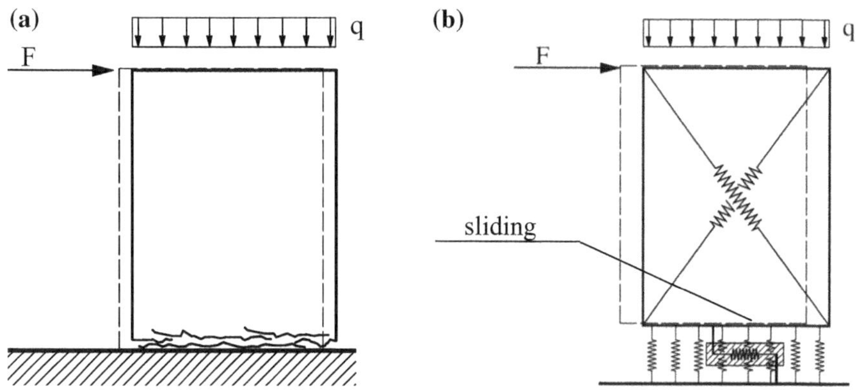

Fig. 4.72 In-plane shear-failure mode (bed joint sliding) of a masonry panel: **a** qualitative response, **b** macro-element representation (Pantò et al. 2015)

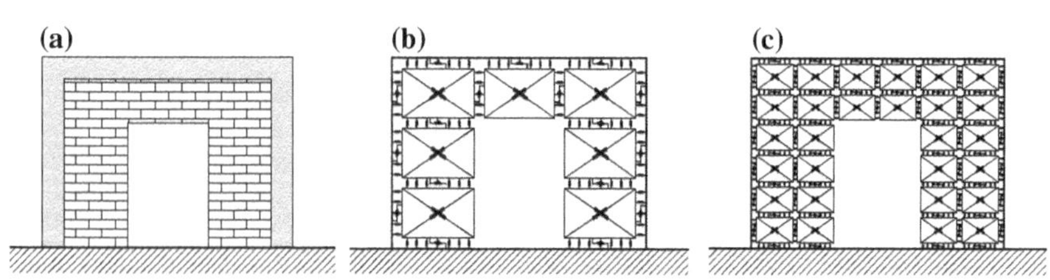

Fig. 4.73 Reinforced concrete frame with masonry infill and central door opening: **a** layout, **b** model with course discretization, **c** model with dense discretization (Caliò and Pantò 2014)

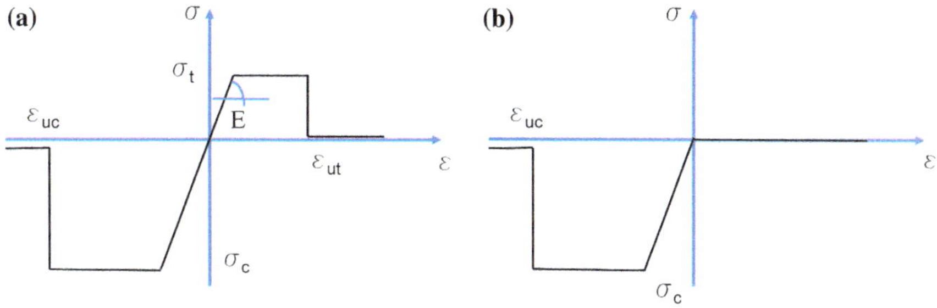

Fig. 4.74 Material law for tension and compression: **a** uncracked and **b** cracked masonry

the tensile strength σ_t is reached, the material fails in tension (cracked material), but it is still possible to transfer compression loads once the cracks are closed and the contact is reactivated.

The shear behaviour is modelled by an elastoplastic constitutive law, symmetrical in tension and compression. Again, the Mohr–Coulomb yield criterion is applied, which was already introduced in Sect. 4.2.3.4.

Three-Dimensional Finite-Elements Models

Three-dimensional finite-element models are apparently the most complete models, since they are able to reproduce the real geometry of the building. Figure 4.75 shows as an example a complex historical masonry building and the associated detailed finite-element model. Finite-element models are consistent, include vertical and horizontal load transfer, cover structural torsional effects and enable a much more realistic simulation of the vibration behaviour. Furthermore, the models allow a realistic modelling of connections between intersecting walls and between walls and slabs. Nevertheless, these models, particularly if time-history analyses are carried out, produce a huge amount of data difficult to interpret, as will be discussed in Sect. 4.2.3.10. The verifications of masonry structures modelled with finite elements are based on the comparison of the design stress resultants N_{Ed}, V_{Ed} and M_{Ed} of the seismic design situation with the resistances values of the bending and shear failure modes. Verifications directly based on the stresses σ_x, σ_y, σ_{xy} are quite difficult, because they usually have local peaks at the wall edges and these peaks can be highly variable over the wall height. Therefore, it is always recommended to evaluate the stress resultants from an integration of the stresses obtained by the calculations.

Fig. 4.75 Real historic masonry building and corresponding finite-element model

4.2.3.7 Structural Models for the Out-of-Plane Response of Masonry Buildings

Eurocode 8-1 (2004) does not contain any rule for the structural modelling and design of masonry walls subjected to out-of-plane seismic actions. Instead of a detailed verification at the ultimate limit state, the code prescribes minimum effective wall thicknesses and maximum ratios of the effective wall height to the effective thickness of the wall to prevent out-of-plane failure. The geometric requirements for the different types of masonry were already summarized in Table 4.26. However, recent earthquakes have shown the vulnerability of masonry walls to out-of-plane failures and it seems particularly important for existing structures with flexible and not properly connected diaphragms to analyse their out-of-plane behaviour considering the specific masonry properties and boundary conditions.

Kinematic Approach

The out-of-plane resistance of masonry walls can be determined through a limit analysis using the kinematic approach, which is based on the kinematic idealization of potential out-of-plane failure mechanism considered as an ensemble of rigid blocks. For each potential mechanism a load multiplier is calculated to identify the critical collapse mechanism that will occur first. The calculations are carried out according to Heyman's hypotheses (Heyman 1999) assuming infinite compressive strength, zero tensile strength and no sliding between the individual blocks. The load multipliers of all potential collapse mechanisms are calculated using the principle of virtual work or equilibrium conditions. The results of the limit analysis can be directly used for the safety verification of the out-of-plane resistance in terms of accelerations or displacements. If a linear analysis is performed the acceleration demand should not exceed the ultimate acceleration, whereas the displacement demand should not exceed the ultimate displacement in case of non-linear analysis. The kinematic approach is already included in the Italian building code NTC (2008),

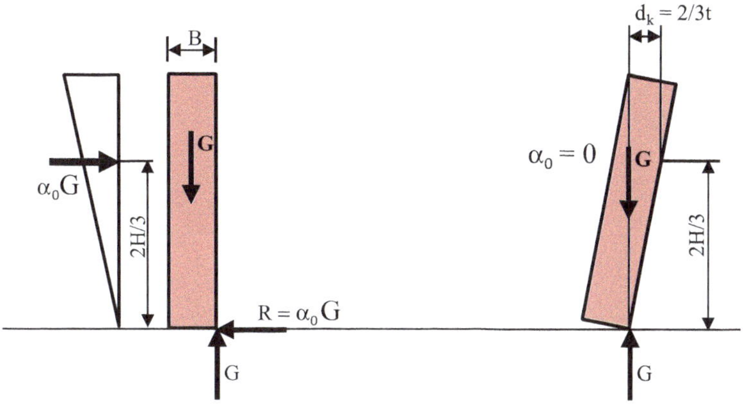

Fig. 4.76 Cantilever wall with simple overturning

because damage evaluations of recent earthquake in Italy identified the out-of-plane failure as one of the dominant failure modes of historic masonry buildings. In the following the calculation steps of the kinematic approach using linear and non-linear calculation analysis are presented.

Step 1: Determine the load multiplier

The determination of the load multiplier is illustrated for a cantilever wall with triangular load pattern (Fig. 4.76). The load multiplier α_0 can be calculated using either the principle of virtual work or moment equilibrium. If the moment equilibrium with respect to the pivot point at the base of the wall is applied, the load multiplier α_0 is calculated to:

$$\alpha_0 G \frac{2H}{3} = G\frac{B}{2} \quad \Rightarrow \quad \alpha_0 = \frac{3}{4}\frac{B}{H}. \tag{4.64}$$

Step 2: Determine the load multiplier-displacement curve

The load multiplier-displacement curve α_0 - d_k is calculated by increasing the displacement d_k of the control point (generally the centre of mass). The resulting curve is usually linear, when the forces are kept constant and the blocks are slender (Fig. 4.77).

Step 3: Transformation of the load multiplier-displacement-curve into the capacity curve in S_a-S_d coordinates

The load multiplier-displacement curve α_0 - d_k is transformed into the capacity curve in spectral acceleration - spectral displacement (S_a-S_d) coordinates by means of an equivalent single-degree-of-freedom (SDOF) oscillator. The participating mass M^* is calculated using the expression:

$$M^* = \frac{\left(\sum_{i=1}^{n+m} P_i \delta_{x,i}\right)^2}{g \sum_{i=1}^{n+m} P_i \delta_{x,i}^2}. \tag{4.65}$$

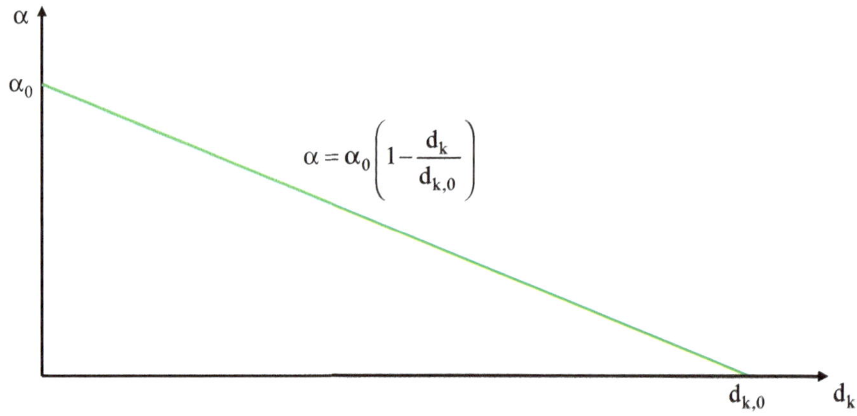

Fig. 4.77 Load multiplier-displacement curve $\alpha_0 - d_k$

Here, $n+m$ is the number of forces P_i applied to masses subjected to inertia forces, and $\delta_{x,i}$ is the virtual horizontal displacement of the control point P_i. The spectral acceleration a_0^* (relative to the activation of the mechanism) is equal to:

$$a_0^* = \frac{\alpha_0 \sum_{i=1}^{n+m} P_i}{M^* \, CF} = \frac{\alpha_0 g}{e^* CF}. \tag{4.66}$$

Here, g is the gravitational acceleration, CF is the confidence factor depending on the level of knowledge according to Section "Different Approaches for New or Existing Masonry Buildings" and e^* is the participation mass factor:

$$e^* = \frac{gM^*}{\sum_{i=1}^{n+m} P_i}. \tag{4.67}$$

Analogously, the spectral displacement is a function of the control point displacement d_k:

$$d^* = d_k \frac{\sum_{i=1}^{n+m} P_i \delta_{x,i}^2}{\delta_{x,k} \sum_{i=1}^{n+m} P_i \delta_{x,i}}. \tag{4.68}$$

The transformed capacity curve is shown in Fig. 4.78.

Step 4: Determination of the spectral yielding displacement d_y^*

If the period T^* of the SDOF oscillator is known, the corresponding spectral yielding displacement d_y^* can be calculated to:

$$d_y^* = a_0^* T^{*2} \left(4\pi^2 + \frac{a_0^* T^{*2}}{d_0^*} \right)^{-1}. \tag{4.69}$$

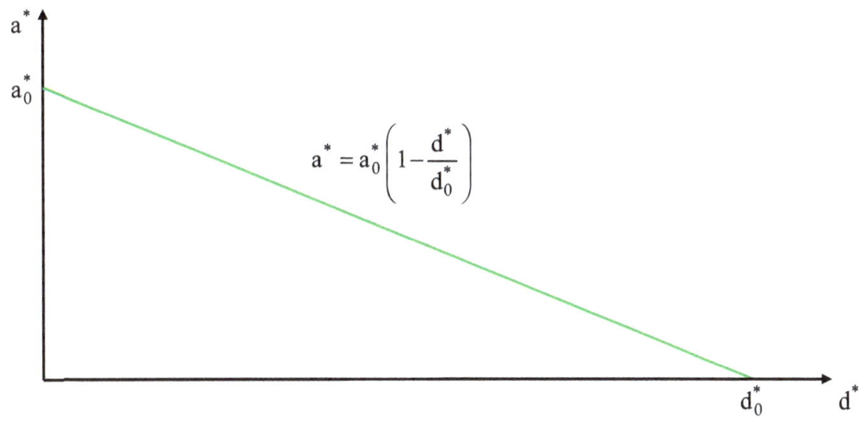

Fig. 4.78 Transformed capacity curve

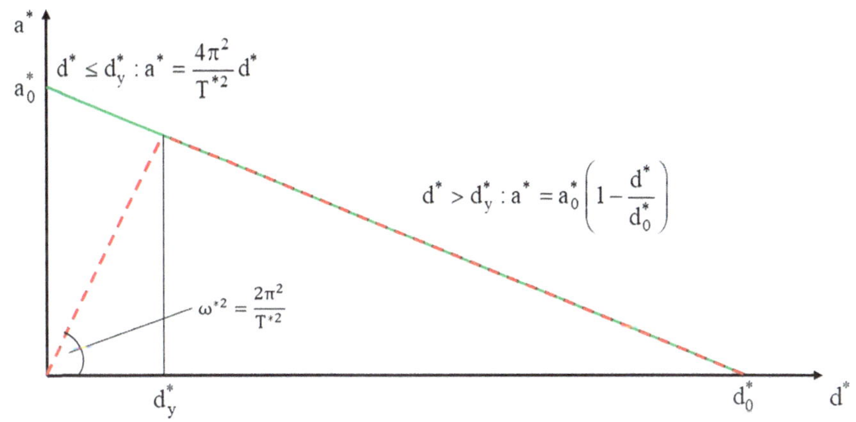

Fig. 4.79 Transformed capacity curve, spectral yield displacement d_y^* and period T^*

Figure 4.79 shows the transformed capacity curve together with the circular frequency ω^* of the equivalent SDOF oscillator and the spectral yield displacement d_y^*.

Step 5: Alternative A: Linear verification

The linear verification compares demand and the capacity in terms of spectral accelerations for the limit state of damage limitation (DLS) and the ultimate limit state or life safety (ULS). The seismic action to be taken into account for the damage limitation has a probability of exceedance P_{DLR} in 10 years and a return period T_{DLR}. Eurocode 8-1 (2004) recommends a return period of $T_{DLR} = 95$ years, but as a national parameter the return period T_{DLR} can be redefined in the national annexes of each country. If the element with the local mechanism is situated at ground level, the following condition must be verified at the damage limitation state:

$$a_0^* \geq a_g(P_{DLR}) \cdot S. \tag{4.70}$$

Here, a_g is the peak ground acceleration corresponding to the probability of exceedance P_{DLR} and S is the soil factor. If the mechanism occurs at a height Z above the level of application of the seismic action, the following verification must be fulfilled:

$$a_0^* \geq S_{ae}(T_1) \cdot \psi(Z) \cdot \gamma. \tag{4.71}$$

Herein, T_1 is the fundamental period of the building, $S_{ae}(T_1)$ is the spectral acceleration of the elastic response spectrum with return period T_{DLR} evaluated at period T_1, and $\psi(Z) = Z/H$ is an approximation of the first mode shape with the building height H and the height of the mechanism Z. Finally, γ is the corresponding modal participation coefficient, calculated as

$$\gamma = \frac{3N}{2N+1}, \tag{4.72}$$

with N being the number of floors of the building. The following conditions must be verified at the ultimate limit state:

$$a_0^* \geq \frac{a_g(P_{NCR}) \cdot S}{q} \tag{4.73}$$

and

$$a_0^* \geq \frac{S_{ae}(T_1) \cdot \psi(Z) \cdot \gamma}{q}. \tag{4.74}$$

Herein, a_g is the peak ground acceleration corresponding to the probability of exceedance P_{NCR}, $S_{ae}(T_1)$ is the spectral acceleration of the elastic acceleration spectrum with return period T_{NCR} evaluated at period T_1, and q is the behaviour factor associated with the mechanism. The behaviour factor q can be assumed equal to 2 in the absence of more precise knowledge about the dissipation properties of the specific failure mechanism. The remaining parameters correspond to the verification at the damage limitation state.

Step 5: Alternative B: Non-linear verification
A higher accuracy of the safety check at the ultimate limit state can be achieved if non-linear analysis considering the deformed configuration with large deformations is applied. In this case, the verification is performed in terms of the ultimate

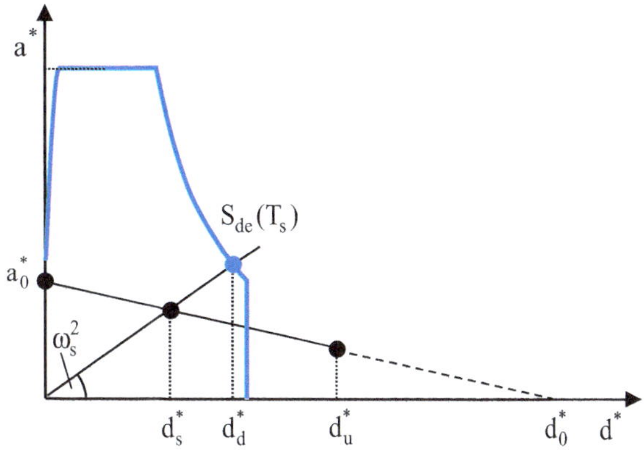

Fig. 4.80 Superposition of the demand spectra and capacity curve in S_a-S_d coordinates

displacement and the displacement demand. The ultimate displacement d_u^* is taken as 0.4 times the displacement corresponding to a null acceleration of the capacity curve (Griffith et al. 2003):

$$d_u^* = 0.4\ d_0^*. \tag{4.75}$$

The ultimate capacity d_u^* is then scaled by 40% to obtain the displacement capacity d_s^*:

$$d_s^* = 0.4\ d_u^*. \tag{4.76}$$

The displacement demand d_d^* is determined from the intersection point of the secant stiffness and the demand spectrum with return period T_{NCR} in S_a-S_d coordinates as shown in Fig. 4.80. The required secant stiffness T_s is calculated as:

$$T_s = 2\pi\sqrt{\frac{d_s^*}{a_s^*}}. \tag{4.77}$$

Here, d_s^* is the displacement capacity according to (4.76) and a_s^* is the load multiplier corresponding to d_s^*:

$$a_s^* = a_0^*\left(1 - \frac{d_s^*}{d_0^*}\right). \tag{4.78}$$

If the element with the local mechanism is situated at ground level, the following condition must be verified:

$$d_u^* \geq d_d^* = S_{de}(T_s). \tag{4.79}$$

If the mechanism occurs at height Z above the level of application of the seismic action, the following condition must be satisfied:

$$d_u^* \geq d_d^* = S_{de}(T_s)\psi(Z)\gamma \frac{\left(\frac{T_s}{T_1}\right)^2}{\sqrt{\left(1 - \frac{T_s}{T_1}\right)^2 + 0.02\frac{T_s}{T_1}}}. \tag{4.80}$$

Here, T_s is the fundamental period of the structure in the direction of the seismic action. The remaining parameters correspond to the linear verification case that was already introduced in step 5a.

The load multiplier-displacement curve α_0 - d_k is linear if the forces are constant from zero to the maximum displacement and if the block is sufficiently slender. The following parametric analysis illustrates the influence of the slenderness on the nonlinearity of the relationship between collapse multiplier and rotation. The rotation is related to the horizontal displacement of a control point, generally chosen to be the centre of mass of the block.

The collapse multiplier α of a SDOF rectangular block with height $2h$, width $2b$, and slenderness β is calculated by means of a kinematic analysis using the principle of virtual work for a generic configuration of the block inclined by ϑ (Giresini et al. 2016):

$$\alpha = \frac{\frac{b}{h} - tan|\vartheta|}{1 + \frac{b}{h}tan|\vartheta|} = tan(\beta - |\vartheta|), \text{ with } \beta = tan^{-1}\left(\frac{b}{h}\right). \tag{4.81}$$

The ultimate rotation is equal to the slenderness β since the tangent function vanishes for $\beta = |\vartheta|$. If only small rotations are considered ($tan\vartheta \cong \vartheta$) the collapse multiplier expression can be simplified to:

$$\alpha = \frac{\frac{b}{h} - |\vartheta|}{1 + \frac{b}{h}|\vartheta|}. \tag{4.82}$$

Moreover, for large rotations, the expression $tan(\beta - |\vartheta|)$ can be linearized to:

$$\alpha \cong \beta - |\vartheta|, \tag{4.83}$$

which is valid for rotation values similar to the slenderness ratio β. Figure 4.81 displays the collapse multiplier depending on the rotation angle ϑ for different *h/b* values, comparing the exact solution with the approximations for small rotations and large rotations. The difference between the solutions becomes negligible for more slender blocks. Furthermore, the comparison demonstrates that the solution for small rotations tends to be over-conservative for small slenderness values.

Another approach for evaluating the stability of rigid blocks under seismic actions is based on the kinematic model according to Housner (Housner 1963). Housner studied the 2D problem of a rigid block subjected to free vibrations, constant and

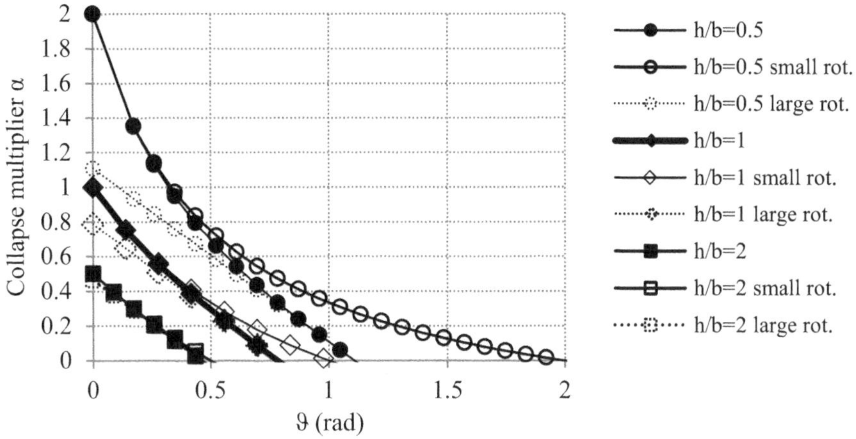

Fig. 4.81 Collapse multipliers of SDOF block with height 2 · h and width 2 · b. Solid marker: exact solution; empty marker and solid line: small rotations; empty marker and dotted line: large rotations

sinusoidal acceleration and seismic ground-motion, in order to investigate rocking behaviour. The rocking analysis, still used mainly in the academic area (Giresini et al. 2015, 2016; Giresini and Sassu 2016), is a more realistic approach for simulating the dynamic behaviour of rigid macro-elements with time-history analysis. The calculation is more precise in comparison to a simplified kinematic analysis, in which the transient response and the characteristics of the rocking phenomenon are not explicitly taken into account. Actually, the kinematic analysis is based on a SDOF oscillator with constant vibration period and elastic response, whereas a rocking block includes a rotation-amplitude-dependent vibration period and negative stiffness values. An interesting discussion on the difference between the two approaches can be found in (Makris and Kostantinidis 2001, 2003). Masonry macro-elements can represent whole façades or parts of façades. For historic buildings and churches specific macro-elements like free-standing walls, tympanums, bell-towers and apses are used. Macro-elements are generally applied to perform local analyses, but they can also be applied for global analyses in combination with energy-based approaches to evaluate seismic vulnerability (Giresini 2015).

4.2.3.8 Analysis Methods

The seismic calculation of masonry buildings can be carried out by means of the following calculation procedures:

- Lateral force method of analysis (Sect. 4.1.5.1)
- Modal response spectrum analysis (Sect. 4.1.5.2)
- Pushover analysis (Sects. 4.1.5.3–4.1.5.6)
- Non-linear time history analysis (Sect. 4.1.5.7)

Table 4.27 Types of construction and upper limits of the behaviour factor, Eurocode 8-1 (2004)

Type of masonry construction	Behaviour factor q
Unreinforced masonry in accordance with Eurocode 6-1-1 (2005)	1.5
Unreinforced masonry in accordance with Eurocode 8-1 (2004)	1.5 – 2.5
Confined masonry	2.0 – 3.0
Reinforced masonry	2.5 – 3.0

Linear Calculation Methods

The lateral force method and the modal response spectrum analysis are linear calculation methods that are applied using specific behaviour factors for the different types of masonry. Table 4.27 presents the range of behaviour factors for the different types of masonry constructions. Due to the intrinsic low ductility of masonry structures, the admissible behaviour factors are quite low in comparison to steel and reinforced concrete structures. The behaviour factor of unreinforced masonry buildings designed in accordance with Eurocode 6-1-1 (2005) is 1.5. If the masonry is designed in accordance with Eurocode 8-1 (2004) a behaviour factor between 1.5 and 2.5 may be applied. Higher values up to 3.0 are only permitted for confined and reinforced masonry. However, for all types of masonry, the recommended values of the code are the lower limits of the proposed ranges. The behaviour factors should be reduced by 20% in case of non-regularity in elevation, but need not be taken less than 1.5.

Non-linear Calculation Methods

The pushover analysis of masonry structures is an attractive alternative to linear calculation methods and allows a much better utilization of the non-linear load-bearing reserves. The safety verification is carried out by a comparison between the ultimate displacement capacity of the building and the displacement demand. The theoretical background and general application of the pushover analysis are described in detail in Sects. 4.1.5.3–4.1.5.6.

The pushover analysis of unreinforced masonry structures requires drift limits on wall level corresponding to the failure modes for shear and bending. These drift limits are not defined in Eurocode 8-1 (2004), but definitions can be found in Eurocode 8-3 (2005), the Italian code NTC (2008) and the German National Annex DIN EN 1998-1/NA (2011). The drift limits of these codes are summarized in Table 4.28.

Table 4.28 Drift limits for unreinforced masonry walls according to the Italian code NTC (2008), the German National Annex DIN EN 1998/NA (2011) and Eurocode 8-3 (2005)

Drift limits	NTC (2008)	DIN EN 1998-1/NA (2011)	Eurocode 8-3 (2005)
Bending	$0.008 \cdot H$	$0.006 \cdot H_0/L \cdot H$	$0.008 \cdot H_0/L \cdot H$
Shear	$0.004 \cdot H$	$0.004 \cdot H$ for $\sigma_d \leq 0.15\ f_k$ $0.003 \cdot H$ for $\sigma_d > 0.15\ f_k$	$0.004 \cdot H$
Parameter	H: Storey height L: Wall length H_0: Distance between the section where the flexural capacity is attained and the contraflexure point f_k: Characteristic compressive strength of masonry according to Eurocode 6-1-1 (2005) σ_d: Average compressive stress in the seismic design situation		

Drift limits for unreinforced masonry walls are defined based on comprehensive statistical evaluations of shear wall tests in the past and must be regarded as an approximation, particularly for modern masonry buildings. The definition of drift limits for modern masonry products is still a subject of ongoing research. It can be expected that the new generation of Eurocodes will include more reliable drift limits for modern masonry products. Several software packages have been recently made available for performing non-linear analyses on masonry structures: ANDILWall (2017), Lagomarsino et al. (2006), Magenes et al. (2006), 3Muri (2018), MINEA (2018). However, linear calculation methods are still the standard in practise as further developments and regulations are required to introduce pushover analysis as an alternative standard calculation procedure in the engineering practise.

The practical application of non-linear time history analyses to masonry structures is rather limited as the nonlinearities and interactions on wall and structural level are quite complex, the computational effort and calculations times are too high and the obtained time-dependent results are not easy to use for the subsequent cross-sectional design of masonry buildings. However, in some special cases detailed non-linear time-history analyses can be a reasonable choice.

Verification of "Simple Masonry Buildings"

Eurocode 8-1 (2004) considers a special class of structures, called "simple masonry buildings". The safety evaluation of buildings belonging to importance categories I or II that meet the specific requirements given in Sects. 9.2, 9.5 and 9.7.2 of Eurocode 8-1 (2004) can be performed with a simplified approach. In particular, an explicit structural verification against collapse is not mandatory. The building can be considered safe without performing specific seismic analyses.

Simple buildings must have proper connections between walls and walls/floors in order to be able to distribute the seismic forces to all elements of the horizontal system. In both orthogonal horizontal directions, the difference in mass and in the horizontal shear wall cross-sectional area between adjacent storeys should be limited to a recommended value of 20%. This limitation aims at obtaining regular mode shapes and a more predictable behaviour without local concentrations of ductility demands in just a few zones of the building.

The criteria for which a building can be considered "simple" depend on its in-plan configuration and the layout of the shear walls.

The rules about the in-plan configuration of the building are the following:

PL-1: The plan has to be approximately rectangular.
PL-2: The ratio of the shorter side to the longer one should be greater than 1:4.
PL-3: The area of projections of recesses from the rectangular shape should not exceed than a percentage p_{max} of the total floor area above the considered level. The recommended value of p_{max} is 15%.

The shear walls necessary for stiffening the building in the two main directions have to comply with the following specific criteria:

SW-1: The layout of shear walls in plan must be approximately symmetric.
SW-2: In each direction a minimum of two parallel walls should be present. The length of each shear wall must exceed 30 % of the length of the building in that direction. In cases of low seismicity the required wall length can be provided by the sum of the shear walls along a single axis, if as a minimum the length of one wall is twice the minimum shear wall length according to Table 4.26.
SW-3: The distance between parallel shear walls should not be greater than 75 % of the length of the building in the other direction.
SW-4: At least 75 % of the vertical loads should be transmitted by the shear walls.
SW-5: Shear walls should be continuous from top to bottom of the building.

Furthermore the walls in both directions should be connected with a maximum spacing of 7 m.

Eurocode 8-1 (2004) also gives a permitted number of storeys above ground for "simple masonry buildings", depending on the type of construction. In any case, unreinforced masonry buildings cannot have more than 4 storeys, whereas for reinforced and confined masonry 5 storeys are possible. A minimum percentage of the total floor area per storey $p_{A,min}$ is also defined in each direction as the sum of cross sectional area of shear walls in Table 9.3 of Eurocode 8-1 (2004). This minimum value, to be verified in order to consider the building as "simple", depends on the acceleration at site $a_g S$ and on the type of construction. According to this table, generally masonry buildings built at a site with acceleration $a_g S > 0.2g$ cannot be considered to be "simple masonry buildings". The recommended values for the number of storeys and minimum total cross-sectional areas are usually redefined in the National Annexes with country-specific values. Furthermore, the National Annexes often contain additional rules considering further parameters such as unit strengths, type of construction and geometric requirements.

Different Approaches for New or Existing Masonry Buildings

The seismic safety assessment of masonry structures varies depending on whether the building is new or an existing one. For new buildings, the design process is generally simpler because of the freedom in adopting suitable techniques so that the design process optimizes the structural conception. For instance, steel tie-rods or special connections between floors and walls can be used to avoid out-of-plane failure. The chosen structural model, described in Sect. 4.2.3.6, should be as simple as possible while retaining adequate accuracy. In order to reduce computational efforts, macro-element or equivalent frame models should be preferred, unless specific stress concentration areas have to be investigated. All analysis methods may be used, from simple linear to complex non-linear ones (static and dynamic).

The seismic assessment of existing masonry building is usually more difficult, since it involves knowledge about geometry, details and materials before the analysis itself is carried out. This collection of information is a crucial step for an effective seismic vulnerability assessment, after which, if needed, a structural intervention may follow. Specific structural interventions suggested by Eurocode 8-3 (2005) are oriented towards mitigating the out-of-plane response. Indeed, inadequate connections between floors and walls should be improved and out-of-plane horizontal thrusts against walls should be eliminated according to Eurocode 8-3 (2005), Sect. 5.1.2. In addition, non-ductile lintels should be replaced. The collection of information is necessary for defining a knowledge level, which must be classified as limited, normal or full.

At the beginning, type and degradation condition of masonry units and mortar has to be identified through non-destructive (radiography, ultrasonic tests, impact echo tests, etc.) or destructive testing (hydraulic flat jack tests, diagonal compression tests, Twin Panel test, etc.). At a local level a global assessment has to be performed by identifying the type and quality of connections between walls and between walls and floors (Sect. 4.2.3.2). This is a very important step because it directly concerns the decision about how to analyse both in-plane and out-of-plane modes. In addition, the configuration of masonry elements has to be evaluated in order to identify the load paths through the resisting lateral elements. Moreover, the presence of adjacent buildings and non-structural elements has to be considered in the seismic vulnerability assessment.

The underlying knowledge level determines both the allowable method of analysis and the confidence factor CF that are given in Table 3.1 of Eurocode 8-3 (2005). The recommended values of CF are 1.35, 1.20 or 1.0 respectively for the knowledge levels described as limited, normal or full. This confidence factor is necessary for evaluating the mechanical properties of existing materials to be used in the calculation of the capacity. As a matter of fact, the mean values of strengths obtained from in situ tests and from other sources have to be divided by the confidence factor CF.

The European reference Code for this class of buildings is Eurocode 8-1 (2004): Design of structures for earthquake resistance - Part 3: Assessment and retrofitting of buildings, Annex C. Linear analysis methods can be used only when the conditions expressed in Sect. C.3.2 of Eurocode 8-3 (2005) are met. They refer to proper distribution of resistant walls in the two main directions of the building, adequate in-plane stiffness of the floor and its connections to vertical elements, interlocked blocks in the spandrels. If these conditions are not met, non-linear analyses are recommended: in this case, capacity is defined in terms of roof displacement. In Sect. C.3.3 of Eurocode 8-3 (2005) the ultimate displacement capacity is taken as the roof displacement at which the base shear drops below 80% of the peak resistance of the structure due to progressive damage and failure of lateral load resisting elements. The structural interventions, if needed, are listed in Sect. C.5 of Eurocode 8-3 (2005), but no specific indications are given on the effectiveness of the adopted strengthening measures.

4.2.3.9 Calculation Example: Two-Storey Limestone Masonry Building

The plan of a two-storey limestone masonry building is shown in Fig. 4.82. The floor and roof of the residential building are able to develop horizontal diaphragm action. They consist of cast-in-place reinforced concrete and prefabricated elements with reinforcement cover of 40 mm and a total thickness of 28 cm. Floors and walls are connected by ring beams to ensure a proper box-behaviour. Moreover, the shear walls are continuous over the building height and the masses are equally distributed in both storeys. All walls are 25 cm thick with the exception of the internal wall W4, whose thickness is 24 cm. The interstorey height is 2.7 m and the maximum clear height of the openings is 1.9 m.

The residential building is verified using the rules for "simple masonry buildings" given in Sects. 9.2, 9.5 and 9.7.2 of Eurocode 8-1 (2004). The results of the verification are summarized in Tables 4.29, 4.30 and 4.31.

It turns out that the two-storey limestone masonry building can be regarded as a "simple masonry building" and that it meets all specific requirements according to Eurocode 8-1 (2004). Hence the residential building can be considered as earthquake resistant and additional calculative safety verifications are not required.

4.2.3.10 Calculation Example: Multifamily House Made of Calcium Silicate Units

This section presents the analysis of a multifamily house typical for Northern Europe architecture. Figures 4.83 and 4.84 show the ground plan configuration and a section of the multifamily house. The two-storey building has a plan of about 16×11 m^2, with an interstorey height of 2.635 m. The basement is designed as a rigid box made of reinforced concrete, whereas the structure consists of calcium silicate masonry units. The roof has a slope of 40°. For acoustic insolation reasons the separating walls and the walls to the stairs and neighbouring flats are 240 mm thick and are built with compression strength class 12. The remaining internal walls as well as the external walls are erected with a thickness of 175 mm and with compression strength class 20. Firstly, the multifamily house is verified using the rules for "simple masonry buildings" given in Sects. 9.2, 9.5 and 9.7.2 of Eurocode 8-1 (2004). Secondly, modal

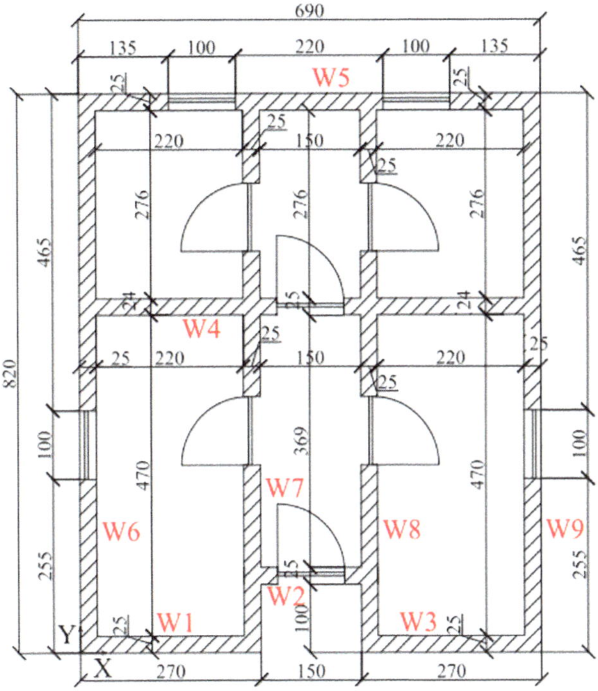

Fig. 4.82 Two-storey limestone masonry building with dimensions in (cm)

dynamic analyses are carried out on the three-dimensional structure. Additionally, a static non-linear analysis is performed. The design and the calculations are carried out with the MINEA software (2018). The input files and a demo version the software MINEA (2018) are available for free download.

Wall Characteristics

The walls are built with non-solidified head joints and thin layer mortar for the bed joints. The corresponding mean compressive strength f_k of the masonry is calculated according to Eurocode 6-1-1 (2005):

$$f_k = K \cdot f_{st}^{\alpha} \cdot f_m^{\beta}$$

where:

f_k	Characteristic compressive strength of the masonry
f_m	Characteristic compressive strength of the mortar
f_{st}	Mean compressive strength of the units
K, α, β	Parameters according to DIN EN 1996-1-1/NA (2012), Table NA.7

Table 4.29 Design criteria and construction rules according to Sect. 9.5.1 of Eurocode 8-1 (2004)

Section 9.5.1	Construction rules	Satisfied?
(1)	Masonry walls are composed of floors and walls, which are connected in two orthogonal horizontal directions and in the vertical direction	Yes
(2)	The connection between the floors and walls shall be provided by steel ties or reinforced concrete ring beams	Yes
(3)	Any type of floors may be used, provided that the general requirements of continuity and effective diaphragm action are satisfied	Yes
(4)	Shear walls shall be provided in at least two orthogonal directions	Yes
(5a)	The effective thickness of shear walls t_{ef} may not be less than a minimum value $t_{ef,min}$ = 240 mm	Yes
(5b)	The ratio h_{ef}/t_{ef} of the effective wall height to its effective thickness does not exceed the maximum value $(h_{ef}/t_{ef})_{max}$ = 12	Yes
(5c)	The ratio of the length of the wall l, to the grater clear height h of the openings adjacent to the wall is 1.35/1.90 = 0.71 > 0.40 and not less than the minimum value $(l/h)_{min}$ = 0.4	Yes

The modulus of elasticity is determined according to DIN EN 1996-1-1/NA (2012), Table NA.12:

$$E = K_E \cdot f_k = 950 \cdot f_k.$$

The initial shear strength f_{vk0} is calculated according to DIN EN 1996-1-1/NA (2012), Table NA-11 and results to f_{vk0} = 0.22 N/mm^2. The geometric properties, mechanical properties and resulting strength values are summarised in Table 4.32.

Building Location and Seismic Input Parameter

The design response spectrum is constructed according to DIN EN 1998-1/NA (2011) as the building is located close to Cologne in Germany. The building site lies in seismic zone 2 with a design ground acceleration of a_g = 0.6 m/s^2. The soil factor S and the control periods of the horizontal design spectrum are determined for the deep geology class S in combination with soil class C:

$$S = 0.75,\ T_B = 0.1\,\text{s},\ T_C = 0.50\,s\,\text{and}\,T_D = 2.0\,\text{s}.$$

The multifamily house is classified in importance category II with an importance factor of γ_I = 1.0.

Table 4.30 Rules according to Sect. 9.7.2 (1), (2) of Eurocode 8-1 (2004)

Section 9.7.2	Construction rules		Satisfied?
(1)	Minimum shear walls requirements according to Eurocode 8-1 (2004), Table 9.3		Yes
	Input values $a_{gR} = 1.0\ m/s^2$ S = 1.0 (ground type A) $\gamma_I = 1.0$ (residential building) $A_g = 56.6\ m^2$ (total floor area) n = 2 (number of storeys)		
	Required shear wall areas		
	Transverse direction (X)	**Longitudinal direction (Y)**	
	Coefficient k_x $k_x = 1 + (l_{av} - 2)/4 = 1.05 \leq 2$ (Shear walls with length > 2 m: > 70%)	*Coefficient* k_y $k_y = 1 + (l_{av} - 2)/4 = 1.4 \leq 2$ (Shear wall with lengths > 2 m: > 70%)	
	$a_g \cdot S \cdot \gamma_I = 1.0$ $0.10 \cdot k_x \cdot g = 1.03\, m/s^2$	$a_g \cdot S \cdot \gamma_I = 1.0$ $0.10 \cdot k_y \cdot g = 1.37\, m/s^2$	
	Required shear wall area A_{sx} req. $A_{Sx} = 2.5\% \cdot 56.6\ m^2 = 1.42\ m^2$	*Required shear wall area* A_{sy} req. $A_{Sy} = 2.5\% \cdot 56.6\ m^2 = 1.42\ m^2$	
	Existing shear wall areas		
	Transverse direction	**Longitudinal direction**	
	ex. $A_{sx} = 3.99\ m^2$ > req. $A_{sx} = 1.42\ m^2$	ex. $A_{sy} = 7.15\ m^2$ > req. $A_{sy} = 1.42\ m^2$	
(2a)	The plan is approximately rectangular		Yes
(2b)	The ratio between length of the small side and the length of the long side in plan 6.9/8.2 = 0.84 is not less than a minimum value $\lambda_{min} = 0.25$		Yes
(2c)	The area of projections of recesses from the rectangular shape is $1.5\ m^2$, which corresponds to 3% of the total floor area. The recommended maximum percentage $p_{max} = 15\%$ is fulfilled		Yes

Simplified Check based on Eurocode 8-1

All material and strength requirements for the calcium silicate masonry with thin layer mortar are automatically satisfied according to the European code DIN EN 771-1 (2015). The design criteria and construction rules conform to those stated for "simple masonry buildings" (Eurocode 8-1 (2004), Sect. 9.5.1), as shown in Tables 4.33, 4.34 and 4.35.

All special requirements of Eurocode 8-1 (2004) are met and the considered multi-family house can be regarded as an earthquake resistant simple masonry building. Therefore, no further calculative safety verifications are required in this case. How-

Table 4.31 Rules according to Sect. 9.7.2 (3)–(6) of Eurocode 8-1 (2004)

(3a)	The building is stiffened by shear walls arranged approximately symmetrically in plan in two orthogonal directions	Yes
(3b)	At least two parallel walls should be arranged in two orthogonal directions, the length of each wall is greater than 30% of the length of the building in the considered direction Transverse direction (X): $0.3 \cdot 6.9 = 2.07$ m < 2.70 m (W1, W4) Longitudinal direction (Y): $0.3 \cdot 8.2 = 2.46$ m < 2.55 m (W6, W9)	Yes
(3c)	For the walls in at least one direction, the distance between these walls should be greater than 75% of the length of the building in the other direction in order to provide sufficient torsional stiffness	Yes
(3d)	At least 75% of the vertical loads is supported by shear walls	Yes
(3e)	Shear walls should be continuous from the top to the bottom of the building	Yes
(5)	In both orthogonal directions the difference in mass and in the horizontal shear walls cross section area between adjacent storeys should be limited to a maximum value $\Delta_{m,max} = 20\%$ and $\Delta_{A,max} = 20\%$	Yes
(6)	For unreinforced masonry buildings, walls in one direction should be connected with walls in the orthogonal direction at a maximum spacing of 7 m	Yes

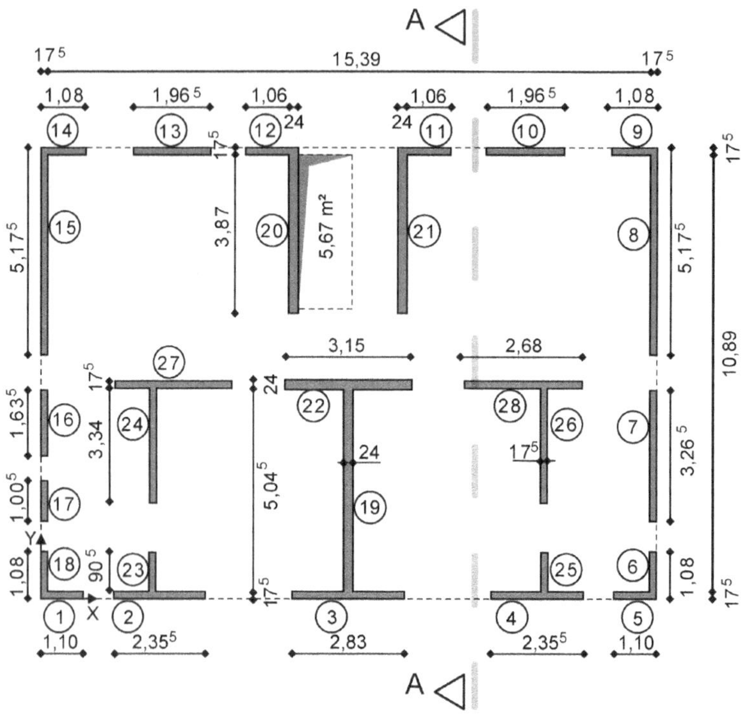

Fig. 4.83 Ground plan configuration of the considered multi-family house

Table 4.32 Geometric and mechanical properties and strength values of the calcium silicate masonry

Wall no.	l (m)	d (m)	$I_{w,red}$ (m^4)	E (N/mm^2)	EI_W (N/mm^2)	f_{st} (N/mm^2)	f_m (N/mm^2)	f_k (N/mm^2)	Strength class (−)	$f_{bt,cal}$ (N/mm^2)	f_{vk0} (N/mm^2)
1	1.01	0.175	0.01	9980.81	146378.67	25.00	10.00	10.51	20	0.80	0.22
2	2.36	0.175	0.17	9980.81	1677661.83	25.00	10.00	10.51	20	0.80	0.22
3	2.83	0.175	0.28	9980.81	2766936.63	25.00	10.00	10.51	20	0.80	0.22
4	2.36	0.175	0.17	9980.81	1677661.83	25.00	10.00	10.51	20	0.80	0.22
5	1.01	0.175	0.01	9980.81	146378.67	25.00	10.00	10.51	20	0.80	0.22
6	0.99	0.175	0.01	9980.81	137983.32	25.00	10.00	10.51	20	0.80	0.22
7	3.27	0.175	0.40	9980.81	4033665.33	25.00	10.00	10.51	20	0.80	0.22
8	5.09	0.175	1.19	9980.81	11833476.72	25.00	10.00	10.51	20	0.80	0.22
9	0.99	0.175	0.01	9980.81	137983.32	25.00	10.00	10.51	20	0.80	0.22
10	1.97	0.175	0.10	9980.81	1010662.92	25.00	10.00	10.51	20	0.80	0.22
11	1.18	0.175	0.02	9980.81	231412.53	25.00	10.00	10.51	20	0.80	0.22
12	1.18	0.175	0.02	9980.81	231412.53	25.00	10.00	10.51	20	0.80	0.22
13	1.97	0.175	0.10	9980.81	1010662.92	25.00	10.00	10.51	20	0.80	0.22
14	0.99	0.175	0.01	9980.81	137983.32	25.00	10.00	10.51	20	0.80	0.22
15	5.09	0.175	1.19	9980.81	11833476.72	25.00	10.00	10.51	20	0.80	0.22
16	1.64	0.175	0.06	9980.81	597804.49	25.00	10.00	10.51	20	0.80	0.22
17	1.01	0.175	0.01	9980.81	144249.54	25.00	10.00	10.51	20	0.80	0.22
18	0.99	0.175	0.01	9980.81	137983.32	25.00	10.00	10.51	20	0.80	0.22
19	5.25	0.240	1.74	6632.84	11560814.60	15.00	10.00	6.98	12	0.48	0.22
20	3.96	0.240	0.90	6632.84	5984424.70	15.00	10.00	6.98	12	0.48	0.22
21	3.96	0.240	0.90	6632.84	5984424.70	15.00	10.00	6.98	12	0.48	0.22
22	3.15	0.240	0.50	6632.84	3348455.97	15.00	10.00	6.98	12	0.48	0.22
23	0.99	0.175	0.01	9980.81	137983.32	25.00	10.00	10.51	20	0.80	0.22
24	3.43	0.175	0.46	9980.81	4569661.53	25.00	10.00	10.51	20	0.80	0.22
25	0.99	0.175	0.01	9980.81	137983.32	25.00	10.00	10.51	20	0.80	0.22
26	3.43	0.175	0.46	9980.81	4569661.53	25.00	10.00	10.51	20	0.80	0.22
27	2.68	0.175	0.24	9980.81	2389644.79	25.00	10.00	10.51	20	0.80	0.22
28	2.68	0.175	0.24	9980.81	2389644.79	25.00	10.00	10.51	20	0.80	0.22

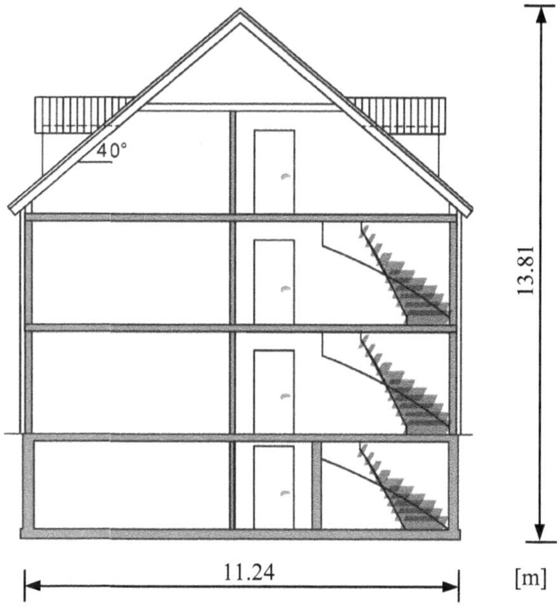

Fig. 4.84 Section A-A of the considered multi-family house

Table 4.33 Design criteria and construction rules according to Sect. 9.5.1 of Eurocode 8-1 (2004)

Section 9.5.1	Construction rules	Satisfied?
(1)	Masonry walls are composed of floors and walls, which are connected in two orthogonal horizontal directions and in the vertical direction	Yes
(2)	The connection between the floors and walls shall be provided by steel ties or reinforced concrete ring beams	Yes
(3)	Any type of floors may be used, provided that the general requirements of continuity and effective diaphragm action are satisfied	Yes
(4)	Shear walls shall be provided in at least two orthogonal directions	Yes
(5a)	The effective thickness of shear walls t_{ef} may not be less than a minimum value $t_{ef,min} = 175$ mm	Yes
(5b)	The ratio h_{ef}/t_{ef} of the effective wall height to its effective thickness does not exceed the maximum value $(h_{ef}/t_{ef})_{max} = 18$	Yes
(5c)	The ratio of the length of the wall l, to the grater clear height h of the openings adjacent to the wall may not be less than a minimum value $(l/h)_{min} = 0.27$	Yes

ever, the application of linear and non-linear calculation and design will be demonstrated in the next sections.

Table 4.34 Rules according to Sect. 9.7.2 (1), (2) of Eurocode 8-1 (2004)

<table>
<tr><th>Section 9.7.2</th><th colspan="2">Construction rules</th><th>Satisfied?</th></tr>
<tr><td rowspan="13">(1)</td><td colspan="2">Minimum shear walls areas according to DIN EN 1998-1/NA, Table NA.12 (2011)</td><td rowspan="13">Yes</td></tr>
<tr><td colspan="2">Input values*
$a_g = 0.6$ m/s^2 (DIN EN 1998-1/NA (2011), Table NA.3, Seismic zone 2)
S = 0.75 (DIN EN 1998-1/NA (2011), Table NA.4, UK C-S)
$\gamma_I = 1.0$ (DIN EN 1998-1/NA (2011), Table NA.6, residential building)
$A_g = 176.92$ m^2 (total floor area)</td></tr>
<tr><td colspan="2">Required shear wall surface</td></tr>
<tr><td>Transverse direction (X)</td><td>Longitudinal direction (Y)</td></tr>
<tr><td>Coefficient k
$k_x = 1$
(Shear walls with length > 2 m: < 70%)</td><td>Coefficient k
$k_y = 1$
(Shear walls with length > 2 m: < 70%)</td></tr>
<tr><td>Coefficient $k_{r,x}$ - Table NA.12
$k_{r,x} = 1$</td><td>Coefficient $k_{r,y}$ - Table NA.12
$k_{r,y} = 1$</td></tr>
<tr><td>Left column - Table NA.12
$a_g \cdot S \cdot \gamma_I = 0.45 \leq 0.6 \cdot k_x \cdot k_{r,x} = 0.6$</td><td>Left column - Table NA.12
$a_g \cdot S \cdot \gamma_I = 0.45 \leq 0.6 \cdot k_y \cdot k_{r,y} = 0.6$</td></tr>
<tr><td>Required shear wall area A_{sx}
req. $A_{Sx} = 2\% \cdot 176.92$ m^2 = 3.54 m^2</td><td>Required shear wall area A_{sy}
req. $A_{Sy} = 2\% \cdot 176.92$ m^2 = 3.54 m^2</td></tr>
<tr><td colspan="2">Existing shear wall area</td></tr>
<tr><td>Transverse direction</td><td>Longitudinal direction</td></tr>
<tr><td>ex. $A_{sx} = 4.81$ m^2 > req. A_{sx} = 3.54 m^2</td><td>ex. $A_{sy} = 7.87$ m^2 >req. A_{sy} = 3.54 m^2</td></tr>
<tr><td>(2a)</td><td colspan="2">The plan is approximately rectangular</td><td>Yes</td></tr>
<tr><td>(2b)</td><td colspan="2">The ratio between length of the small side and the length of the long side in plan should be not less than a minimum value λ_{min} = 0.25</td><td>Yes</td></tr>
<tr><td>(2c)</td><td colspan="2">The area of projections of recesses from the rectangular shape should be not greater than a percentage p_{max} = 15% of the total floor area above the considered level</td><td>Yes</td></tr>
</table>

Simplified Response Spectrum Analysis

The dynamic response of the building under consideration with continuous shear walls over the building height depends on the distribution of the horizontal stiffness and of the floor masses along the building height. The regular masonry building is modelled by multi-degree of freedom (MDOF) systemss with concentrated masses at each storey level in each building direction. The stiffnesses of the MDOF systems are calculated as the sum of all wall stiffnesses in the considered direction. The deter-

Table 4.35 Rules according to Sect. 9.7.2 (3)–(6) Eurocode 8-1 (2004)

(3a)	The building is stiffened by shear walls arranged approximately symmetrically in plan in two orthogonal directions	Yes
(3b)	At least two parallel walls should be arranged in two orthogonal directions, the length of each wall is greater than 30% of the length of the building in the considered direction. Longitudinal direction (X): $0.3 \cdot 11.24 = 3.37\ \text{m} < 5.175\ \text{m}$ (not satisfied) Transverse direction (Y): $0.3 \cdot 15.74 = 4.72\ \text{m} > 3.15\ \text{m}$ In longitudinal direction (X) the check is not successful. Nevertheless, for low seismicity zones, the walls may be interrupted by openings. In addition, at least one shear wall in each direction has a length twice the minimum wall length $l_{min} = 0.27 \cdot 2.635 = 0.71$ m: Central axis (Walls 22, 27, 28): $2.68 \cdot 2+3.15 = 8.51\ \text{m} > 1.42\ \text{m}$ Lower axis: (Walls 1–5): $1.1 \cdot 2+2.355 \cdot 2+2.355 = 9.74\ \text{m} > 1.42\ \text{m}$ All walls in transverse and longitudinal directions exceed the minimum shear wall length of 0.71 m	Yes
(3c)	For the walls in at least one direction, the distance between these walls should be greater than 75% of the length of the building in the other direction in order to provide a sufficient torsional stiffness	Yes
(3d)	At least 75% of the vertical loads is supported by shear walls	Yes
(3e)	Shear walls should be continuous from the top to the bottom of the building	Yes
(5)	In both orthogonal directions the difference in mass and in the horizontal shear walls cross section area between adjacent storeys should be limited to a maximum value $\Delta_{m,max} = 20\%$ and $\Delta_{A,max} = 20\%$	Yes
(6)	For unreinforced masonry buildings, walls in one direction should be connected with walls in the orthogonal direction at a maximum spacing of 7 m	Yes

mination of the equivalent seismic forces using the simplified response spectrum analysis is carried out separately in X- and Y-direction. According to Eurocode 8-1 (2004), Sect. 4.3.3.5.8 (8), for buildings satisfying regularity criteria in plan and in which walls in the two main horizontal directions are the only primary seismic elements, the seismic action may be assumed to act separately, and their effects need not be combined. According to Eurocode 8-1 (2004), the combination of the effects in two directions is unnecessary for symmetric regular wall-stiffened buildings. Although this criterion is not completely fulfilled for the ground plan, a superposition of the results in two directions is not considered in the following. This decision is based on the fact that the building is adequately stiffened in both directions and has a sufficiently high torsional stability. However, superposition effects will be considered for the three-dimensional building model in Section "Multimodal Response Spectrum Analysis with Three-Dimensional Structural Model".

Determination of seismic masses on each storey level

The seismic storey masses of the building are calculated from the floor dead and live loads and the weight of the masonry walls. Eurocode 8-1 (2004) differentiates between independently occupied storeys and storeys with correlated occupancies reflected by a factor φ varying with respect to the load categories of Eurocode 1-1-1 (2004). Corresponding to the use as residential building, each storey can be regarded as independent and the factor φ is applied with 1.0 for the roof and with 0.7 for the first floor according to DIN EN 1998-1/NA (2011).

The combination factor ψ_2 is defined in accordance with Eurocode 0 (2004) except for snow loads. In contrast to Eurocode 0 (2004), the snow loads are combined with a combination factor of $\psi_2 = 0.5$ according to DIN EN 1998-1/NA (2011). In the present case the snow load is calculated for a roof pitch of 40° in snow load zone II and at an altitude of 245 m determined according to DIN EN 1991-1-3/NA (2010).

Table 4.36 provides the floor masses of dead and live loads with the corresponding combination coefficients. The mass of the roof is equal to 24.9 t which is less than 50% of the mass of the subjacent storey. Therefore, the roof is simply considered as

Table 4.36 Determination of masses for each storey

	Floor above ground floor	Floor above first floor	Roof structure
Floor loads			
Floor area	$A_{GF} = 171.25\ m^2$	$A_{1F} = 171.25\ m^2$	$A_R = 171.25\ m^2$
Permanent loads	Reinforced concrete floor incl. floor cover	Reinforced concrete floor incl. floor cover	Span roof construction
	$g_k = 6\ kN/m^2$	$g_k = 6\ kN/m^2$	$g_k = 1.2\ kN/m^2$
Variable loads	Including loads of partitions walls	Including loads of partitions walls	Snow loads
	$q_k = 2.7\ kN/m^2$	$q_k = 2.7\ kN/m^2$	$q_k = 0.45\ kN/m^2$
φ - coefficient	0.7 (–)	1.0 (–)	1.0 (–)
ψ_2 - coefficient	0.3 (–)	0.3 (–)	0.5 ( )
Wall loads			
Total cross-sectional area of shear walls	$A_W = 12.68\ m^2$	$A_W = 12.68\ m^2$	–
Considered wall heights	h = 2.81 m (full storey height)	h = 1.405 m (half storey height)	–
Material density	$\rho_{MW} = 2\ t/m^3$	$\rho_{MW} = 2\ t/m^3$	–
Wall weight	$G_{k,MW} = 699$ kN	$G_{k,MW} = 349.5$ kN	–
Sums			
$\sum G_{ki}$	171.25·6 + 699 = 1726.50 kN	171.25·6 + 349.5 = 1377 kN	171.25·1.2 = 205.50 kN
$\sum \varphi \cdot \psi_{2i} \cdot Q_{ki}$	171.25 · (2.7 · 0.7 · 0.3) = 97.10 kN	171.25 · (2.7 · 1.0 · 0.3) = 138.71 kN	171.25 · (0.45 · 1.0 · 0.5) = 38.53 kN
$\sum G_{ki} + \sum \varphi \cdot \psi_{2i} \cdot Q_{ki}$	1.823.6 kN ~ 185.9 t	1.515.7 kN ~ 154.5 t	244 kN ~ 24.9 t < 0.5 · 155 t = 78 t

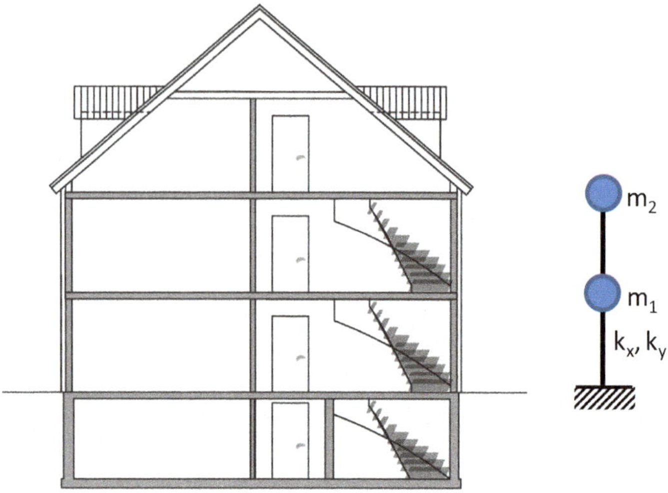

Fig. 4.85 MDOF-system of the multifamily masonry building

an additional mass on the upper storey instead of as a separate storey. Referring to Fig. 4.85, the following concentrated masses are determined:

$$\begin{aligned} m_1 &= 185.9\,\text{t} \\ m_2 &= 154.5\,\text{t} + 24.9\,\text{t} = 179.4\,\text{t} \\ m_{tot} &= 365.3\,\text{t}. \end{aligned}$$

Determination of the horizontal stiffness in each building direction

The system stiffness value for the X- and the Y-direction can be calculated from the single wall stiffnesses. To calculate the stiffness, all walls should have the requirement of minimum wall length of 0.27 multiplied by 2.635 m = 0.76 m according to DIN EN 1998-1/NA (2011). The single wall stiffness can be calculated by taking into account a reduction factor for shear deformations. For instance, for wall 1 that has a thickness of $d = 0.175$ m and a length of $l = 1.01$ m, the stiffness contribution in X-direction is calculated as

$$k_{x,1} = E \cdot I_{x,1} = E \cdot 0.175 \cdot 1.01^3/12 = E \cdot 0.015\,\text{kNm}^2.$$

The stiffness of the single wall is reduced by taking shear deformation into account using the expression given by Müller and Keintzel (1984):

$$k_{x,1} = E \cdot I_{Ex,1} = E \cdot \frac{I}{\left(1 + \frac{3.64\,EI}{h^2 \cdot GA}\right)} = E \cdot \frac{0.015}{\left(1 + \frac{3.64 \cdot 0.015}{5.62^2 \cdot 0.4 \cdot 0.175 \cdot 1.01}\right)} = E \cdot 0.0146\,\text{kNm}^2.$$

The stiffness values of the remaining wall sections can be determined analogously. Table 4.32 lists the moments of inertia and the stiffness values for each considered wall.

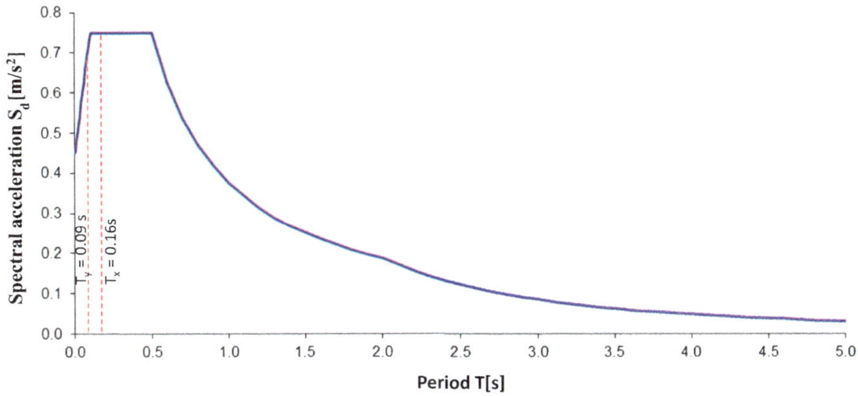

Fig. 4.86 Design response spectrum for EZ2, UK C-S, q = 1.5, DIN EN 1998-1/NA (2011)

The overall stiffnesses of the equivalent multi-degree of freedom (MDOF) systems for the X- and Y direction respectively amount to:

$$k_x = \sum_{i=1} k_{x,i} = 17302880.7 \text{ kNm}^2$$

$$k_y = \sum_{i=1} k_{y,i} = 61663593.1 \text{ kNm}^2.$$

Determination of seismic equivalent forces in X- and Y-direction

In the simplified response spectrum method, the fundamental periods of the structure are calculated in both building directions using the two-mass oscillator model and the standard modal analysis procedure. In this particular example, the modal analysis by the MINEA software (2018) determines the first vibration period equal to 0.16 s in X-direction and 0.09 s in Y-direction. The spectral accelerations associated with the determined fundamental periods are taken from the acceleration response spectrum of DIN EN 1998-1/NA (2011) of the building site. A behaviour factor of $q = 1.5$ for unreinforced masonry is assumed according to DIN EN 1998/NA (2011), Table 9.1. Figure 4.86 shows the design spectrum and the calculated periods. For both directions the maximum spectral acceleration of the plateau range of 0.75 m/s^2 is applied. Keeping in mind that the calculated periods are only approximations, the initial increasing branch of the spectrum is neglected for the calculation of the design acceleration in Y-direction. Total seismic forces in each direction are calculated as products of the total mass of the structure and the determined spectral accelerations. These products must be multiplied by the correction factor λ which equals 1.0 for structures with not more than two storeys. The total base shears in X- and Y-direction result to:

$$\text{X-direction: } F_{b,x} = S_d(T_1) \cdot M \cdot \lambda = 0.75 \cdot 365.3 \cdot 1.0 = 273.98 \text{ kN}$$

$$\text{Y-direction: } F_{b,y} = S_d(T_1) \cdot M \cdot \lambda = 0.75 \cdot 365.3 \cdot 1.0 = 273.98 \text{ kN}.$$

Table 4.37 Floor forces of distribution proportional to mass and height

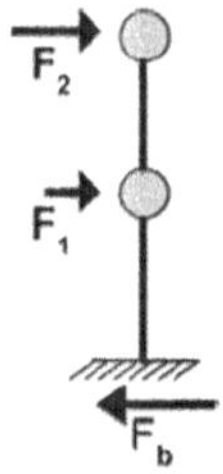

Force distribution		Seismic zone 2	
		X-direction	Y-direction
	F_2	180.5 KN	180.5 KN
	F_1	93.5 KN	93.5 KN
	F_b	274.0 KN	274.0 KN

The distribution of the total seismic forces applied to the equivalent MDOF system is assumed to be proportional to the height of each floor level and the corresponding storey mass. This corresponds to a linear approximation of the fundamental mode as described in Sect. 4.1.5.1. The lateral forces for each floor are summarized in Table 4.37.

Distribution of equivalent seismic forces on the shear walls

The equivalent seismic forces on each floor level must be distributed to the load bearing shear walls according to their individual wall stiffnesses. For this purpose, load distribution factors s_i, s_j for the stiffening elements in each storey are calculated parallel (index i) and perpendicular (index j) to the direction of the seismic action. The distribution factors correspond to the percentage of the horizontal seismic forces acting on the floor levels and can be calculated according to (4.54).

Since the plan in question features an asymmetrical distribution of the horizontal stiffness and of the mass, torsional effects must be taken into account when calculating the individual wall forces. Table 4.38 lists all conditions necessary to perform a linear elastic analysis by using two planar models, one for each main horizontal direction, even if the criteria for in-plan regularity are not satisfied. In this case, all conditions are met.

The condition d in Table 4.38 requires that the square of the torsional radius r^2 be greater than the sum of the square of the radius of gyration l_s^2 and the of square of the structural eccentricities e_0^2. Checking this condition requires a computational verification. However, if the condition is not fulfilled it is still possible to carry out the calculation with two planar models if the first three conditions of Table 4.38 are satisfied. In this case, though, all seismic action effects must be increased by 25%. The square of the radius of gyration l_s^2 for the rectangular plan of the building can be calculated as follows:

$$l_s^2 = (L^2 + B^2)/12 = (15.74^2 + 11.24^2)/12 = 31.17\,\text{m}^2.$$

Table 4.38 Check of the torsional conditions according to Eurocode 8-1 (2004)

Conditions of Eurocode 8-1 (2004), Sect. 4.3.3.1 (8)		Satisfied?
(a)	The building has well-distributed and relatively rigid cladding and partitions	Yes
(b)	The height of the building must not exceed 10 m	Yes
(c)	A rigid diaphragm behaviour can be assumed	Yes
(d)	The centres of lateral stiffness and mass shall be each approximately on a vertical line and, in the two horizontal directions of analysis, satisfy the conditions: $r_x^2 > l_s^2 + e_{ox}^2$, $r_y^2 > l_s^2 + e_{oy}^2$	Yes (Table 4.40)

The torsional stiffness depends on the square of the torsional radius in X- and Y-direction and on the translational stiffness according to:

$$k_T = \sum_{i=1}^{n} k_i \cdot r_i^2 + \sum_{j=1}^{l} k_j \cdot r_j^2 = 2234688614\,\text{kNm}^2.$$

The squares of the torsional radii of the multi-family house are calculated for both directions based on the translational and torsional stiffness of the wall system:

$$r_x^2 = \frac{k_T}{k_y} = \frac{2234688614}{61{,}663{,}593.1} = 36.24$$

$$r_y^2 = \frac{k_T}{k_x} = \frac{2234688614}{17302880.7} = 129.15.$$

The centre of mass of a storey is approximately determined as the centre of mass of the floor neglecting the influence of the walls' self-weight and the opening of the stairway:

$$x_M = 7.91\text{ m}$$

$$y_M = 5.51m.$$

The centre of stiffness is calculated using the following expressions:

$$x_S = \frac{\sum_i E_{x,i} \cdot I_{x,i} \cdot x_{s,i}}{\sum_i E_{x,i} \cdot I_{x,i}} = 8.29\,\text{m}$$

$$y_S = \frac{\sum_i E_{y,i} \cdot I_{y,i} \cdot y_{s,i}}{\sum_i E_{y,i} \cdot I_{y,i}} = 4.33\,\text{m}.$$

Table 4.39 Structural eccentricities e_0, squares of the torsional radii r and the radii of gyration l_s

Direction	e_0 (m)	e_0^2 (m^2)	r^2 (m^2)	l_s^2 (m^2)	$r^2 > l_s^2 + e_0^2$
X-direction	8.29 − 7.91 = 0.38	0.14	36.24	31.17	Yes
Y-direction	5.51 − 4.33 = 1.18	1.39	129.15	31.17	Yes

Table 4.40 Structural, accidental and additional eccentricities

Earthquake direction	Eccentricities				
	e_0	e_1	e_2	e_{min}	e_{max}
X-direction (m)	1.18	0.562	0.37	0.028	2.11
Y-direction (m)	0.38	0.78	1.32	−0.597	2.487

The structural eccentricities e_0 in each building direction are calculated as differences between the coordinates of the centre of mass and the centre of stiffness. By using the structural eccentricities e_0, the squares of the torsional radii r^2 and the squares of the radii of gyration l_s^2 the condition d in Table 4.38 is successfully verified in Table 4.39 for both building directions.

A maximum and a minimum eccentricity can be calculated from the structural eccentricity e_0, the additional eccentricity e_2 and the accidental eccentricity e_1 by using (4.53):

$$\begin{aligned} e_{\max} &= e_0 + e_1 + e_2 \\ e_{\min} &= 0.5 \cdot e_0 - e_1. \end{aligned}$$

The accidental eccentricity e_1 in X and Y-direction is calculated according to (4.51):

$$e_1 = \pm 0.05 \cdot L_i \Rightarrow \begin{aligned} e_{1,x} &= \pm 0.05 \cdot 15.74 = 0.787\,m \\ e_{1,y} &= \pm 0.05 \cdot 11.24 = 0.562\,\text{m}. \end{aligned}$$

The additional eccentricity e_2 is calculated according to (4.52):

$$e_2 = \min \begin{cases} 0,1 \cdot (L+B) \cdot \sqrt{\frac{10 \cdot e_0}{L}} \leq 0.1 \cdot (L+B) \\ \frac{1}{2 \cdot e_0}\left[l_s^2 - e_0^2 - r^2 + \sqrt{\left(l_s^2 + e_0^2 - r^2\right)^2 + 4 \cdot e_0^2 \cdot r^2}\right]. \end{cases}$$

The additional eccentricity must be calculated in the two main directions of the building. The additional eccentricity $e_{2,x}$ in X-direction is equal to:

$$\begin{cases} 0.1 \cdot (15.74 + 11.24) \cdot \sqrt{\frac{10 \cdot 0.38}{15.74}} = 1.32 \leq 0.1 \cdot (15.74 + 11.24) = 2.698\,\text{m} \\ \frac{1}{2 \cdot 0.38} \cdot \left[31.17 - 0.38^2 - 36.24 + \sqrt{(31.17 + 0.38^2 - 36.24)^2 + 4 \cdot 0.38^2 \cdot 36.24}\right] = 1.98\,\text{m}. \end{cases}$$

Analogously, an additional eccentricity in Y-direction of $e_{2,\,y} = 0.37$ m is determined. A summary of the in-plan eccentricities and their combined values e_{min} and e_{max} is presented in Table 4.40.

With the minimum and maximum eccentricities, the distribution factors s_{ix} of each single wall for the earthquake component in X-direction and s_{iy} for the earthquake component in Y-direction are calculated. As an example, the distribution factors for shear wall 1 are computed as:

$$s_{1x} = \frac{k_{1x}}{k_x} \cdot \left(1 + \frac{k_x \cdot r_{1y} \cdot e_{y,\min}}{k_T}\right) = \frac{146378.67}{17302880.7} \cdot \left(1 + \frac{17302880.7 \cdot (4.33 - 0.0875) \cdot 0.028}{2234688614}\right)$$
$$= 0.00847$$

$$s_{1y} = r_{1y} \cdot e_{x,\max} \cdot \frac{k_{1x}}{k_T} = (4.33 - 0.0875) \cdot 2.487 \cdot \frac{146378.67}{2234688614} = 0.00069.$$

Accordingly, the design forces of the shear wall 1 are determined based on the horizontal loads on the storey levels shown in Table 4.37.

The design shear force and the bending moment in X-direction are:

$$V_{ED} = 0.00847 \cdot 274 = 2.32\,\text{kN}$$
$$M_{ED} = 0.00847 \cdot (180.5 \cdot 5.62 + 93.5 \cdot 2.81) = 10.82\,\text{kNm}.$$

Analogously, the design shear force and the bending moment in Y-direction are:

$$V_{ED} = 0.0007 \cdot 274.0 = 0.12\,\text{kN}$$
$$M_{ED} = 0.0007 \cdot (180.5 \cdot 5.62 + 93.5 \cdot 2.81) = 0.89\,\text{kNm}.$$

The comparison of the design internal force resultants of wall 1 shows that the critical direction for the subsequent wall design is the X-direction.

Structural verifications according to DIN EN 1998-1/NA and DIN EN 1996-1/NA
The design value E_{dAE} of the effects of actions in the seismic design situation is determined in accordance with Eurocode 0 (2002) as presented in Sect. 4.1.6.3:

$$E_{dAE} = E\left\{\sum_{j \geq 1} G_{k,j} \oplus A_{Ed} \oplus \sum_{i \geq 1} \psi_{2,i} \cdot Q_{k,i}\right\}.$$

The combination coefficient $\psi_{2,i}$ for quasi-permanent values of a variable action i is 0.3 for variable loads and 0.5 for snow loads. The safety verifications for the walls are carried out with the software MINEA (2018) according to DIN EN 1998-1/NA (2011) and DIN EN 1996-1/NA (2012). The axial (vertical) loads on the walls are determined according to their tributary areas (Fig. 4.87).

Fig. 4.87 Tributary areas for the distributed vertical loads

Based on the calculated design internal force resultants, safety verifications are carried out in the ultimate limit state for axial, bending and shear loading at the top, centre and bottom of the walls in the ground floor.

The verification procedures include reduction factors for the wall slenderness and for in-plane- and out-of-plane eccentricities and they also take the risk of buckling into account. Table 4.41 lists the design internal force resultants N_{Ed}, V_{Ed}, M_{Ed} and the resistances N_{Rd} and V_{Rd} at the bottom of the walls, as this section is critical for the structural verification. The results in the last column of Table 4.41 show that the verification of the walls 2, 3 and 4 along the X-axis fails. Such a result is quite representative for the linear verification of unreinforced masonry buildings. The combination of the cantilever system and the linear "wall by wall" verification procedure without any force redistribution among the walls leads to rather conservative results and often cannot be successfully applied even for low to moderate seismic actions and well-stiffened buildings. In the following sections it will, however, be demonstrated that the building can be easily verified if more refined models are employed.

Multimodal Response Spectrum Analysis with Three-Dimensional Structural Model

Modal analysis

In the following, the safety verification of the multi-family house is carried out by means of a multimodal response spectrum analysis using a three-dimensional model in which walls and floors are modelled with shell elements. The spatial struc-

Table 4.41 Verifications for axial, bending and shear loading at the bottom of the walls

No.	N_{Ed} (kN)	V_{Ed} (kN)	M_{Ed} (kNm)	N_{Rd} (kN)	N_{Ed}/N_{Rd} (–)	V_{Rd} (kN)	V_{Ed}/V_{Rd} (–)
1	36.01	2.32	10.79	452.87	0.08	11.67	0.20
2	***109.30***	***26.54***	***123.70***	***100.89***	***1.08***	***13.59***	***1.95***
3	***154.49***	***43.77***	***204.01***	***208.36***	***0.74***	***26.33***	***1.66***
4	***109.30***	***26.54***	***123.70***	***100.89***	***1.08***	***13.59***	***1.95***
5	36.07	2.32	10.79	453.97	0.08	11.69	0.20
6	35.92	0.69	3.20	895.27	0.04	14.86	0.05
7	149.87	20.10	93.69	2222.48	0.07	78.75	0.26
8	239.37	58.98	274.86	3081.59	0.08	117.61	0.50
9	39.43	2.43	11.32	458.65	0.09	12.35	0.20
10	153.26	17.79	82.92	973.93	0.16	47.62	0.37
11	47.65	4.07	18.99	422.63	0.11	13.39	0.30
12	47.65	4.07	18.99	422.63	0.11	13.39	0.30
13	153.27	17.79	82.92	973.98	0.16	47.63	0.37
14	39.48	2.43	11.32	459.43	0.09	12.36	0.20
15	239.32	82.09	382.75	2086.34	0.11	100.22	0.82
16	79.10	4.15	19.34	1264.32	0.06	30.95	0.13
17	58.61	1.00	4.67	933.02	0.06	19.02	0.05
18	35.98	0.96	4.46	818.46	0.04	14.87	0.06
19	351.38	52.83	246.24	3871.12	0.09	186.14	0.28
20	244.86	29.85	139.15	2838.61	0.09	134.99	0.22
21	316.56	27.01	125.90	3181.57	0.10	154.11	0.18
22	232.52	53.91	251.28	993.95	0.23	78.75	0.68
23	48.85	0.84	3.91	915.38	0.05	17.15	0.05
24	223.84	27.80	129.59	2503.16	0.09	103.37	0.27
25	48.85	0.66	3.07	953.34	0.05	17.15	0.04
26	223.74	21.83	101.75	2777.07	0.08	103.64	0.21
27	260.16	38.48	179.33	1435.64	0.18	92.19	0.42
28	260.16	38.48	179.33	1435.64	0.18	92.19	0.42

tural model is generated and analysed with the software package MINEA (2018). Figure 4.88 shows the three-dimensional building model and Fig. 4.89 displays the automatically generated finite element model of the building. The most important aspect of the three-dimensional modelling is an appropriate idealization of the coupling between walls and floors as well as between adjacent shear walls. The connections between floors and walls are modelled with hinged connections, allowing the transmission of tensile forces. However, this does not reflect the real behaviour in the presence of gap effects between walls and floors, but must be accepted as a compromise, since the multimodal response spectra method is a linear dynamic calculation method. The walls can be considered as coupled or decoupled. In gen-

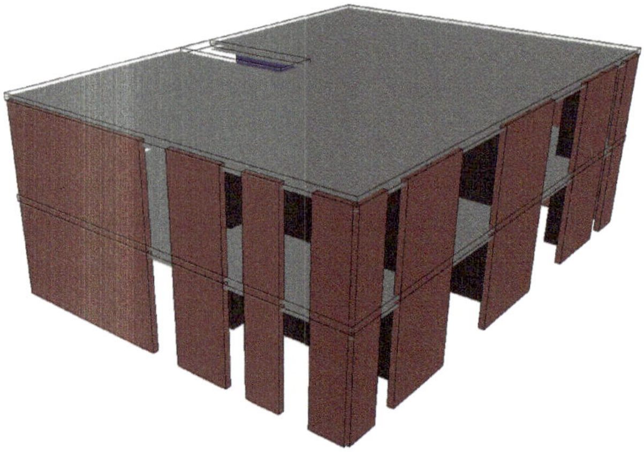

Fig. 4.88 Three-dimensional building model of the multi-family house

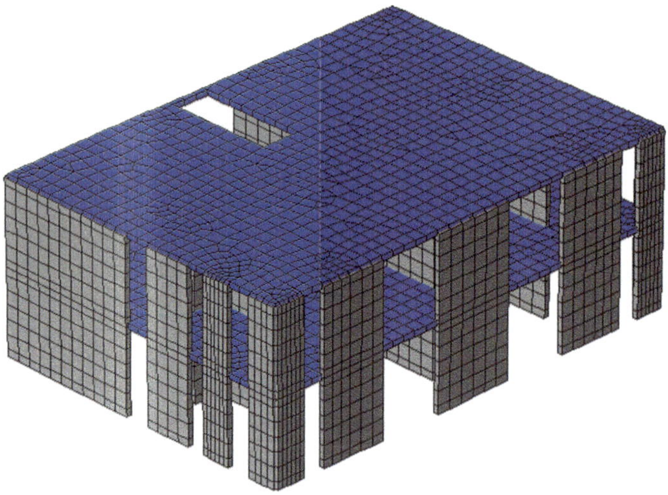

Fig. 4.89 Finite-element model of the multi-family house

eral, decoupled walls are highly recommended, since a verification of the connection joint is usually questionable due to the low quality of execution at the construction site and the preference of butt-jointed connections in practise. Moreover, complex three-dimensional models with decoupled walls are much better understandable and controllable.

To clarify the discussed aspects, selected results of the multimodal analysis with and without coupling of adjacent walls will be compared. The resulting fundamental periods and activated masses in X- and Y-direction for the two modelling approaches are compared in Table 4.42.

Table 4.42 Natural periods and activated masses with and without coupling of the walls

Mode	Uncoupled walls			Coupled walls		
	Period (s)	Activated mass (%)		Period (s)	Activated mass (%)	
		X	*Y*		*X*	*Y*
1	0.108	**80.9**	0	0.105	**80.2**	0.0
2	0.068	0	**75.6**	0.064	0	9.6
3	0.064	0	0	0.063	0	**64.6**
4	0.063	0	0.4	0.058	0	1.0
5	0.062	0	4.8	0.055	0	5.1
6	0.055	0	0	0.054	0	0
7	0.045	0	0	0.044	0	0
8	0.045	0	0	0.044	0	0
9	0.044	0	0	0.044	0.2	0.1
10	0.044	4.4	0	0.044	0.1	0.2
11	0.044	0	0.1	0.044	0.3	0
12	0.044	0	0.1	0.044	0.3	0
Sum		**85.3**	**81.0**		**81.1**	**80.6**

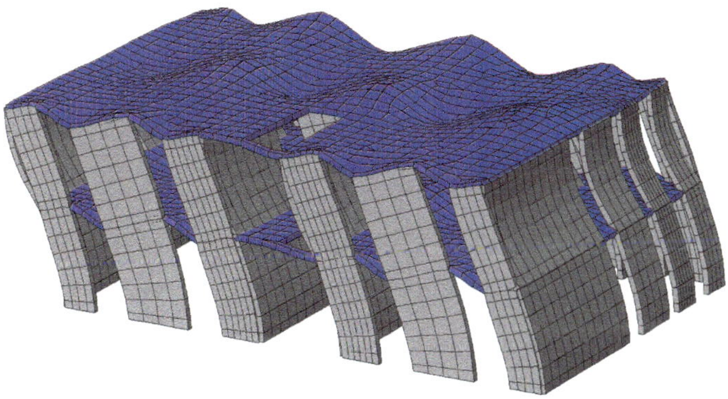

Fig. 4.90 1st mode shape in X-direction: T = 0.108 s (uncoupled walls)

The modal analysis was carried out for a total number of 12 modes of vibration. The sum of the effective modal masses for theses modes amounts to about 80% of the total mass of the structure for both modelling approaches in X- and Y-direction. Hence the recommended effective modal mass of at least 90% according to Eurocode 8-1 (2004) is not achieved. However, the alternative requirement to consider only modes of vibration significantly contributing to the global response with effective modal masses greater than 5% of the total mass is fulfilled. The higher modes of vibrations are local ceiling vibrations in the vertical direction, not important for the global dynamic response of the multi-family house.

The modes of vibration show that the fundamental periods of the model with wall couplings are, as expected, smaller than the periods of the model without wall cou-

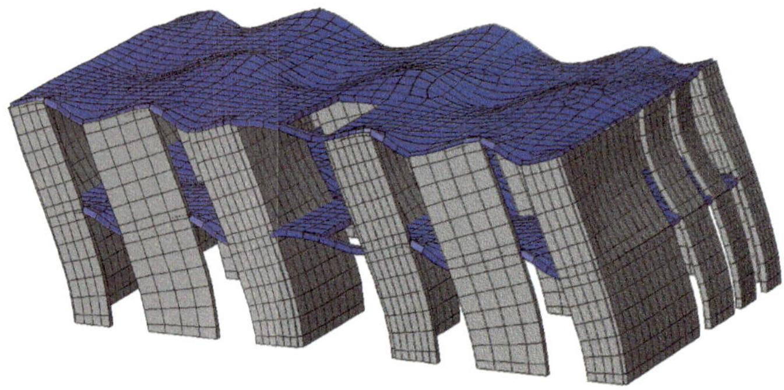

Fig. 4.91 1st mode shape in X-direction: T = 0.105 s (coupled walls)

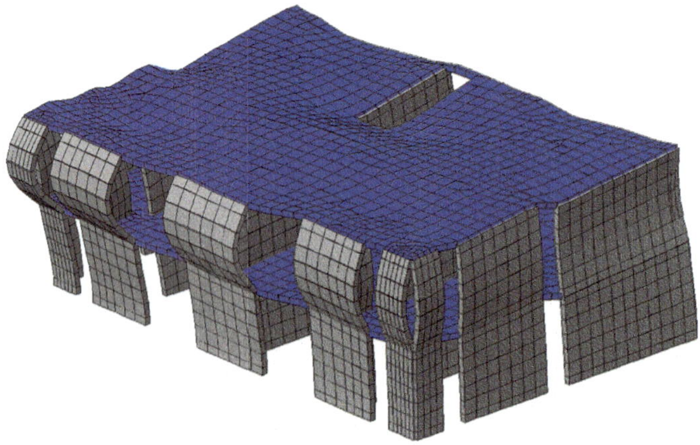

Fig. 4.92 2nd mode shape in Y-direction: T = 0.068 s (uncoupled walls)

plings. This is due to increased system stiffness because of the composite wall cross-sections. The two main translational mode shapes of the two modelling approaches in X- and Y-direction are shown in Figs. 4.90, 4.91, 4.92, 4.93 and 4.94. The mode shapes clearly show the change in the vibration behaviour in the case of composite wall cross-sections through wall couplings at wall intersections and building corners. A comparison of the natural periods with those of the simplified MDOF system shows that the stiffness of the three-dimensional model increases due to the more realistic representation of the walls and the consideration of the interactions between walls and slabs. As a result, the natural period of the MDOF system is reduced in X-direction from 0.16 to 0.108 s (0.105 s) and in Y-direction from 0.09 to 0.068 s (0.063 s). However, the change of the natural periods has no influence on the calculation of the spectral accelerations, since the plateau value of the spectrum has been extended to the period T = 0 s.

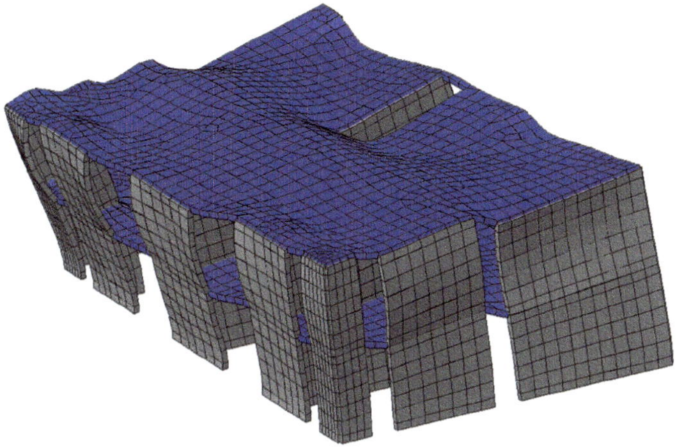

Fig. 4.93 3rd mode shape in Y-direction: T = 0.063 s (coupled walls)

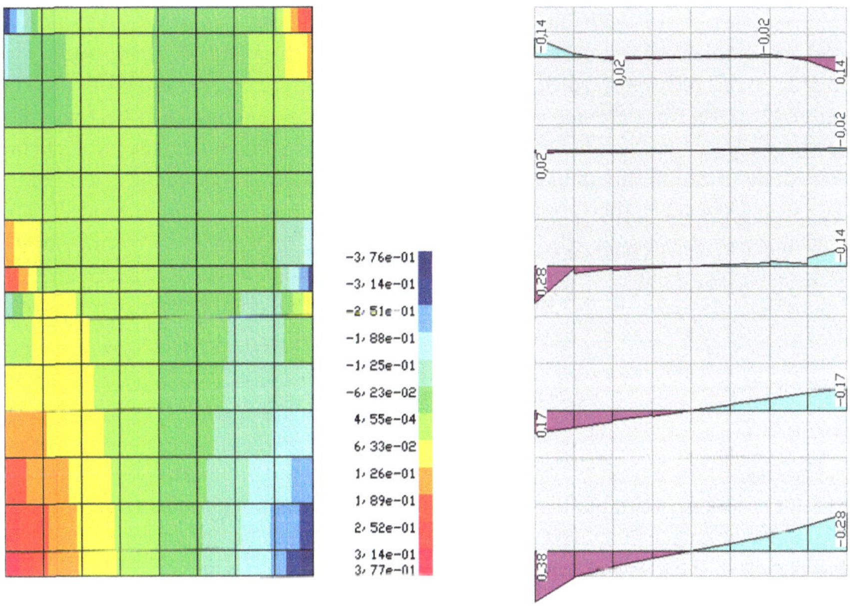

Fig. 4.94 Inner wall 22: Vertical stresses for the 1st natural mode in x-direction (N/mm^2)

Determination of the design internal forces

The biggest problem of three-dimensional models lies in the evaluation of the results and their interpretation for the design on the wall level. In masonry structures, the normative wall verifications have to be performed based on the internal force resultants N, V and M. The check based on linear stresses σ_x, σ_y, σ_{xy} is problematic, since these stresses have local peaks in the corner areas of the walls and can vary greatly over the wall height. Figure 4.94 shows a typical distribution of the vertical

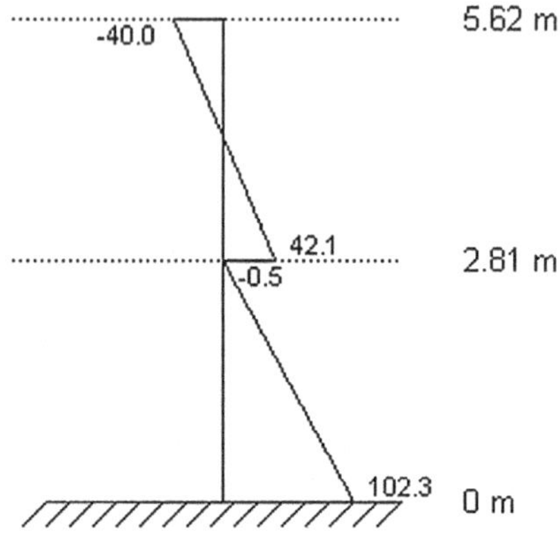

Fig. 4.95 Inner wall 22: distribution of the moment (kNm) integrated from the vertical stress distribution of the first mode shape in X-direction (uncoupled walls)

stresses over the wall length in various sections for the inner wall 22 obtained for the first natural mode in X-direction.

For the design it is sensible to integrate the in-plane stresses in wall direction in order to calculate the corresponding internal forces *N, Q* and *M* of the wall. The integration has to be performed at the base, the wall centre and the top of the wall as required by DIN EN 1996-1-1 (2010). This integration is automatically performed in MINEA (2018). For the inner wall 22, the integration of vertical stresses results in the moment distribution shown in Fig. 4.95 for the first natural mode in the X-direction. The integration was carried out for each storey at the base and top of the wall. The distribution of the internal forces between the integration results was assumed to be linear. The integrated wall internal forces of individual earthquakes directions have to be superimposed with the 30% rule for the earthquake combination:

$$E_{Edx} \oplus 0.30 \cdot E_{Edy}$$
$$0.30 \cdot E_{Edx} \oplus E_{Edy}.$$

This results in a total of six load cases that have to be superimposed with the permanent and the variable loads. The verification of each wall must be carried out separately for each of these eight combinations since the size the magnitude and the distribution of the internal forces are influenced by the following aspects:

- The frame effect leads to an increase or reduction of the axial forces depending on the direction of the seismic action.
- Even under vertical loading conditions, the walls in spatial models are subjected to not-negligible shear forces and bending moments.
- The interaction between the walls may lead to direction-dependent force redistributions among the walls.

Results for the walls 1–5 *in axis* Y=0

The influence of the spatial structural analysis on the design is demonstrated for walls 1–5 in axis $Y = 0$. The verification results determined with the MINEA software

Table 4.43 Design results for the walls 1–5 in axis (Y = 0) (uncoupled walls)

No.	N_{Ed} (kN)	V_{Ed} (kN)	M_{Ed} (kNm)	N_{Rd} (kN)	V_{Rd} (kN)	N_{Ed}/N_{Rd} (kN)	V_{Ed}/V_{Rd} (kN)	Combination
1	41.23	5.71	8.78	646.39	12.37	0.06	0.38	STAT − 1.0x − 0.3y
2	115.34	26.36	59.70	1617.59	51.21	0.07	0.51	STAT − 1.0x − 0.3y
3	146.06	30.04	88.68	1980.48	70.05	0.07	0.43	STAT + 1.0x + 0.3y
4	94.86	26.35	59.33	1353.46	42.45	0.07	0.62	STAT + 1.0x + 0.3y
5	31.35	5.63	8.69	558.48	11.42	0.06	0.49	STAT + 1.0x + 0.3y

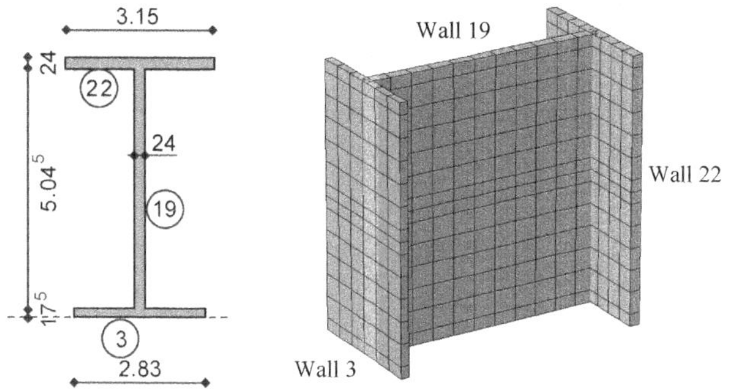

Fig. 4.96 Configuration of the walls 3, 19 and 22

(2018) on the basis of the model without wall couplings are summarized in Table 4.43. The results show that, in contrast to the simple model, the verifications of all walls in axis Y = 0 are now fulfilled. The reason for this is the consideration of the interaction between shear walls and slabs and the resulting reduction of bending moments in the shear walls. The reduction of the moments leads to larger compressed cross-sectional areas, which has in return a positive influence on the verification of the shear forces.

Effects of wall couplings

The influence of the wall coupling on the modelling is illustrated by the example of wall 3. Figure 4.96 shows the configuration of wall 3 with the transverse wall 19 and the adjacent wall 22 in the floor plan and in the finite-element model.

The critical third mode shape in Y-direction causes a rotation of wall 19, which is resisted by the walls 3 and 22 in case of coupled walls. The interaction effects are illustrated in Figs. 4.97 and 4.98 by a comparison of the deformation shapes and stress distributions of coupled and uncoupled walls. Due to the deformation compatibility, the rotation of wall 19 makes itself felt in the walls 3 and 22 and stress concentrations arise in the intersection areas of the walls. In contrast, in case of decoupled modelling, wall 19 carries the horizontal load without walls 3 and 22 being involved. This can be seen very clearly in the stress distribution with stress concentrations in the wall corners of wall 19.

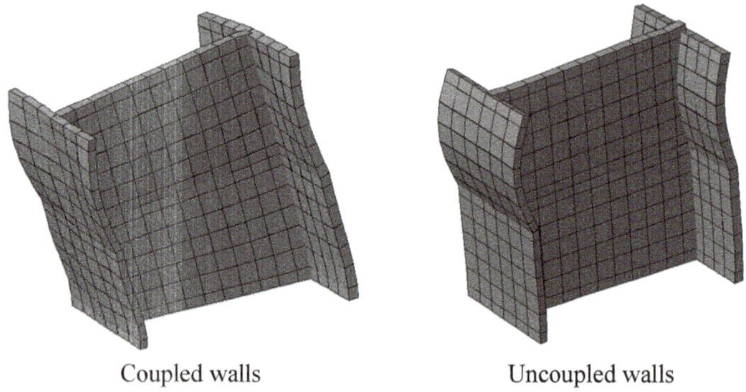

Fig. 4.97 Walls 3, 19, 22: Deformation for the 1st vibration mode in Y-direction

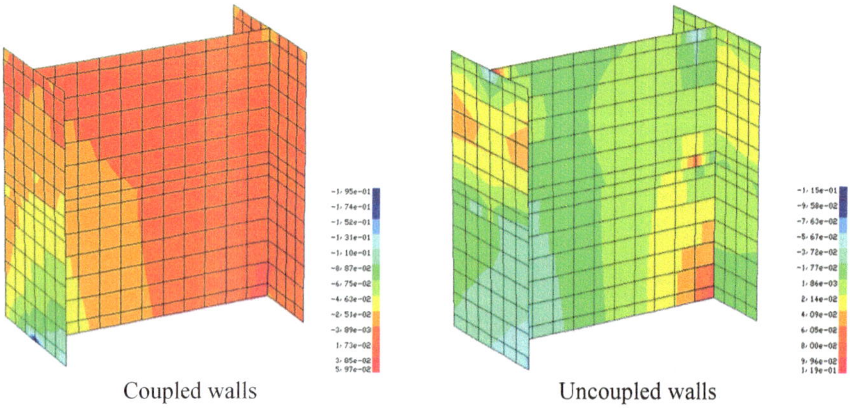

Fig. 4.98 Walls 3, 19, 22: vertical stresses (N/mm^2) for the 3rd vibration mode in Y-direction

The effects of the interaction can also be seen in the results of the integrated and superimposed internal forces for wall 3. The stress resultants are shown in Figs. 4.99 and 4.100 with and without coupling of the walls. The coupling leads to an increase of the axial forces, caused by the rotation of the perpendicular wall 19.

Effects of the interaction between walls and slabs
The three-dimensional modelling of masonry buildings considering the wall-slab interaction leads to additional loads acting on the slabs. Due to the wall rotations, forces act locally on the slabs, which must be considered and verified in the context of the bending and shear design of the slabs. If necessary, additional reinforcement must be installed to cover the local interaction effects.

Summary of results
The application of three-dimensional models as a basis for the seismic design of masonry buildings is more complex due to several overlapping structural effects. In addition, taking into account the connections between the walls and

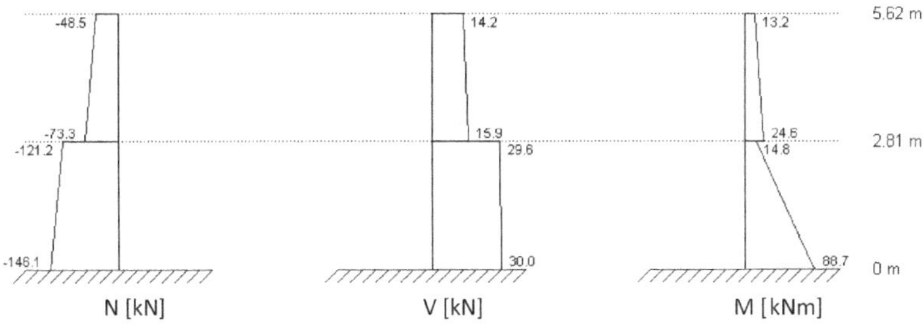

Fig. 4.99 Wall 3: stress resultants, seismic combination STAT + 1.0x + 0.3y, uncoupled walls

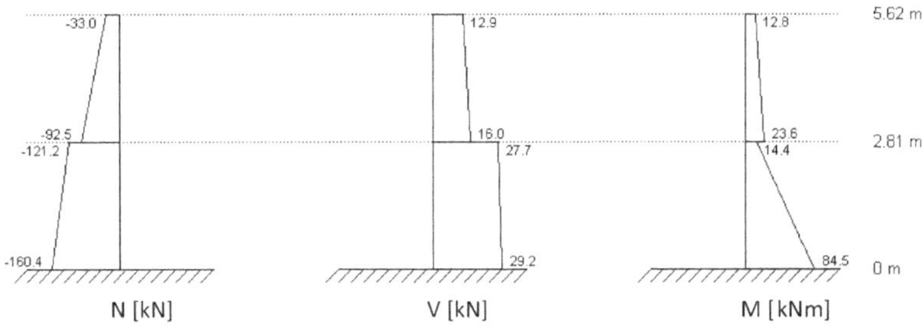

Fig. 4.100 Wall 3: stress resultants, seismic combination STAT + 1.0x + 0.3y, coupled walls

the wall-slab interaction has a non-negligible influence on the design results. In the case of wall couplings, the bond between the coupled walls must be verified. For the present case of the multifamily house, the seismic design was successfully performed for all shear walls. In contrast to the verification results based on the simple MDOF system (Section "Simplified Response Spectrum Analysis"), also the critical wall axis Y = 0 was verified without any problems.

Non-linear Static Analysis

The structural system of the multifamily house is again represented by continuous shear walls over the building height, neglecting spandrels and interlocking effects with transverse walls. This simplified modelling allows the assumption that failure will take place in the ground floor while the upper stories will mostly remain in the elastic range. Due to this, the control point for the non-linear pushover curve is chosen on the floor level of the first storey.

In contrast to the linear elastic models, the interactions of walls and slabs are considered by the level of restraint α describing the moment distribution in the wall. Since non-linear calculations are carried out, mean values of the material and strength parameters are used. The seismic input is defined by the linear-elastic design spectrum as shown in Fig. 4.86.

The vertical loads are applied as calculated in Section “Simplified Response Spectrum Analysis” for the linear elastic model. The horizontally acting seismic forces are distributed to the floor levels according to the first natural mode shape. Due to the conservative assumption of a failure in the ground floor, investigations of further load patterns are not required. All non-linear calculations are carried out with MINEA (2018).

Level of restraint

Masonry shear walls continuous over the building height or over several stories are not acting as simple cantilever arms like reinforced concrete shear walls. In case of masonry shear walls the transfer of tension is missing and the walls start to rotate and to interact with the slabs. The rotation of the walls leads to an uplift on one side and to a formation of a diagonal compression strut between two corners of the wall. The interaction activates a restoring effect that increases in the higher deformation range. The activation of the frame action with contribution of the slab influences the moment distribution in the shear wall. The moment distribution can be described by the level of restraint α as the quotient of the height h_0 at the point of zero moment to the wall height h_W. Usually the moment at the base, M_u, is greater than the moment M_o at the top and the level of restraint α is calculated to:

$$\alpha = \frac{h_0}{h_W} = \frac{M_u}{M_u - M_o}.$$

In the case of normal forces acting eccentrically, the moment at the base can be less than the moment at the top. In this case the level of restraint is defined as follows:

$$\alpha = \frac{M_u}{M_o - M_u}.$$

Comprehensive parametric studies based on time history analyses show that the realistic levels of restraint are between 0.5 and 0.75 (Gellert 2010). Therefore the calculations of the three buildings are carried out with the mean value of $\alpha = 0.625$.

Energy dissipation

The energy dissipation is taken into account by means of an effective damping ξ_{eff}, calculated as the sum of the equivalent viscous damping ξ_0 and the hysteretic damping ξ_{hyst}. The effective damping is calculated for each point of the capacity curve and used to scale the elastic response spectra by the reduction factor η proposed by Priestley and Grant (2005):

$$\eta = \sqrt{\frac{7}{2 + \xi_{eff}}} \text{ with } \xi_{eff} = \xi_0 + \xi_{hyst}.$$

As recommended in Eurocode 8-1 (2004) the equivalent viscous damping ξ_0 is considered to be 5%. The calculation of the hysteretic damping is based on a comprehensive evaluation of cyclic shear wall tests by Norda (2012). The damping is described in terms of the ductility μ depending on the dominant failure modes for bending and shear:

- Bending failure (BA)
- Shear failure due to bed joint sliding (SS)
- Shear failure due to diagonal tension (ST)

The effective damping of each individual wall is calculated according to a parametrized approach proposed by Dwairi et al. (2007) and Priestley et al. (2005):

$$\xi_{eff} = \xi_0 + c \cdot \left(\frac{\mu - 1}{\pi \cdot \mu}\right).$$

Norda (2012) proposed different constants c for each of the dominant failure types, since their amount of damping is quite different. For the calculations of the multi-family house median values of the statistical evaluation described by the following functions are applied:

$$\xi_{eff,BL,mean} = \xi_0 + 0.20 \cdot \left(\frac{\mu - 1}{\pi \cdot \mu}\right)$$

$$\xi_{eff,SS,mean} = \xi_0 + 0.90 \cdot \left(\frac{\mu - 1}{\pi \cdot \mu}\right)$$

$$\xi_{eff,SZ,mean} = \xi_0 + 0.44 \cdot \left(\frac{\mu - 1}{\pi \cdot \mu}\right).$$

As hysteretic damping occurs only in the non-linear range, hysteretic damping is not considered for ductility values less than 1. The resulting damping functions (median and 5% quantile) and the test data are shown in Figs. 4.101, 4.102 and 4.103.

According to Magenes and Calvi (1997), an additional damping of 5% is considered in case of bending failure due to the rocking of the wall. The additional damping is considered linearly for ductility values μ greater than 1.

The effective damping on the structural level is determined taking into account the damping contribution of each wall with respect to specific failure mode. The overall damping of the building is calculated as the ratio of the weighted sum of the single wall damping contributions:

$$\xi_{building} = \frac{\sum V_j \cdot x_j \cdot \xi_j}{\sum V_j \cdot x_j}$$

with:

V_j horizontal load of the wall j
x_j horizontal displacement of the wall j
ξ_j damping ratio of the wall j

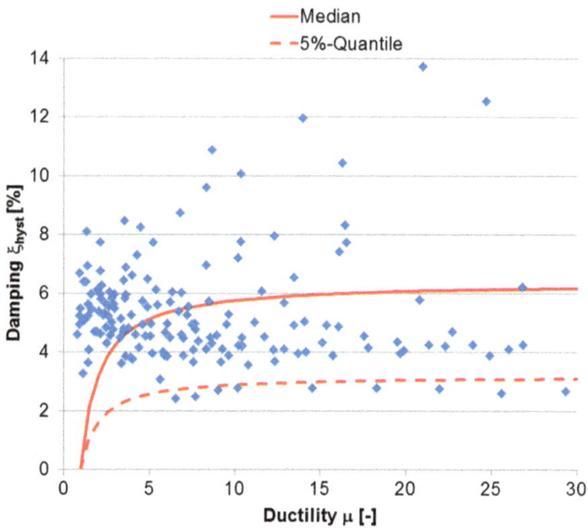

Fig. 4.101 Damping functions for failure mode due to axial load and bending (Norda 2012)

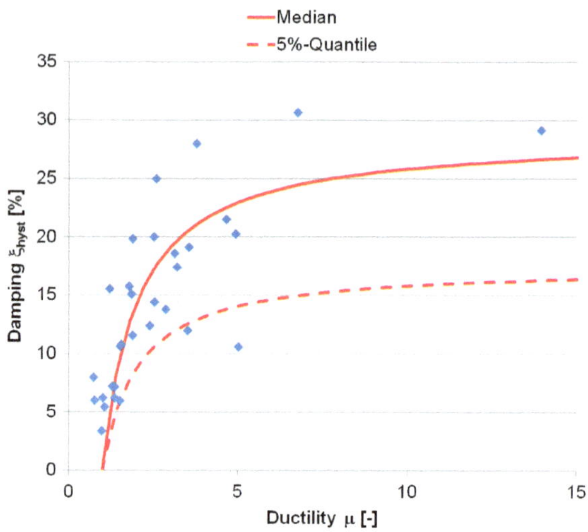

Fig. 4.102 Damping functions for shear failure due to bed-joint sliding (Norda 2012)

The overall damping of the building is used to calculate the reduction factor η along the load-displacement path. This leads to a damped spectrum with increasing damping values in the higher deformation range as qualitatively shown in Fig. 4.13.

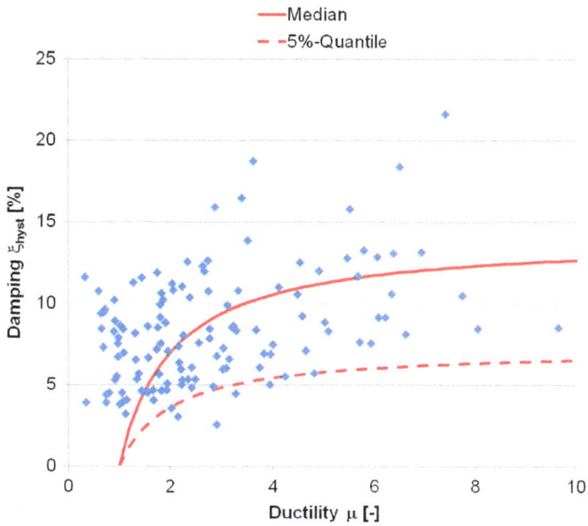

Fig. 4.103 Damping functions for shear failure due to diagonal tension (Norda 2012)

Wall capacity curve and overall building pushover curve
The non-linear wall capacity curves are calculated based on the resistances for bending and axial force (BA), shear with friction failure (SS) and shear with diagonal tension failure (ST) according to DIN EN 1996-1-1/NA (2012). For this, the drift limits (d_1, d_2) and the levels of restraints at the top of the walls are required as input parameters. The drift limits are determined with respect to the specific failure modes for shear and bending in accordance with DIN EN 1998-1/NA (2011), Sect. 9.4 (6). After reaching the drift limits, the wall fails and the capacity drops down to zero. The level of restraint is applied with $\alpha = 0.625$.

The non-linear calculation model idealizes the masonry wall as a Timoshenko beam element considering bending and shear deformations. In the elastic range, the stiffness of the wall is defined by the superposition of the shear and bending stiffness of the equivalent beam. In the nonlinear range, cracked stiffness values for shear and bending are calculated with respect to the compressed length of the wall. For each applied displacement step the corresponding shear forces are calculated as the minimum of the actual bending and shear resistances (V_{BA}, V_{SS}, V_{ST}). Figure 4.104 gives an overview of the calculation approach.

The capacities of all shear walls are superposed to obtain the overall capacity of the building, which corresponds to the pushover curve of the ground floor. The pushover curve is calculated iteratively by imposing a displacement increment in the direction of the seismic action. Afterwards the resulting forces of all shear walls are calculated using shear wall capacity curves. The typically non-symmetric wall configuration of the buildings leads to a torsional moment, which produces a rotation of the system around the centre of mass. For each imposed displacement, the system experiences rotations and translations until equilibrium is reached. A more detailed

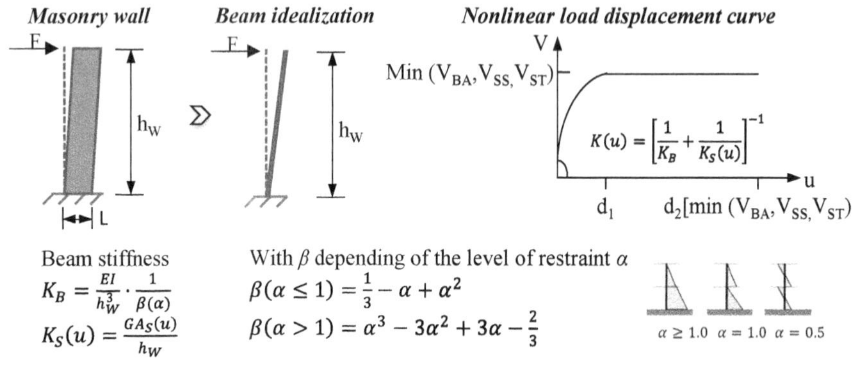

Fig. 4.104 Non-linear calculation procedure of wall capacity curves

description of the methodology is given in 4.1.5.3 and in Butenweg et al. (2010). Table 4.44 presents the failure mode, resistance and deformation capacity of each wall.

Figure 4.105 depicts the non-linear capacity curve and the bilinear idealization for the critical X-direction. Since the building capacity shows a decrease of more than 20% after the first failure of wall 22, the building capacity is reached at this point. The bilinear idealization considers this fact and the plateau range is limited to the displacement at the failure of wall 7.

Safety verification

The pushover curve and the damped spectrum are converted into acceleration-displacement response spectral ordinates. The intersection point of the two curves is called "Performance Point" and represents the maximum spectral displacement of the equivalent single degree of freedom oscillator from which the maximum displacement of the control point at ground floor level is calculated as described in Sect. 4.1.5.4. The superposition of the capacity curve and the damped spectrum is shown in Fig. 4.106. The resulting "Performance Point" is associated with a displacement at ground floor level of 0.012 cm. The safety verification is possible without any problems and shows further non-linear load bearing reserves of the multi-family house, since the "Performance Point" is obtained at a rather low displacement level.

The non-linear results obtained here are in contrast to the failed verification based on a simplified linear elastic calculation. It looks like that the combination of a cantilever system and linear verification wall by wall without force redistribution among the walls cannot be successfully applied in the case of seismic actions. The results demonstrate clearly the deficiencies of the existing linear verification concept and the need of improvements in order to arrive at more realistic design results. This is in line with remarks and recommendations of several other authors (Morandi 2006).

Table 4.44 Failure modes, resistances and deformation capacities of the shear walls

No.	VSS (kN)	VST (kN)	VBA (kN)	Vmax (kN)	d1 (mm)	d2 (mm)	α (–)	Failure (–)
w1	8.11	36.15	7.21	7.21	0.05	29.32	0.63	BA
w2	49.91	128.81	57.69	49.91	0.06	11.24	0.63	SS
w3	77.30	165.53	98.75	77.30	0.07	11.24	0.63	SS
w4	48.70	128.53	56.14	48.70	0.05	11.24	0.63	SS
w5	8.48	36.15	7.54	7.54	0.05	29.32	0.63	BA
w6	9.75	35.03	8.66	8.66	0.06	29.91	0.63	BA
w7	83.68	195.76	109.95	83.68	0.06	11.24	0.63	SS
w8	166.85	320.98	277.52	166.85	0.08	11.24	0.63	SS
w9	9.57	35.02	8.49	8.49	0.06	29.91	0.63	BA
w10	63.51	105.82	71.06	63.51	0.10	11.24	0.63	SS
w11	12.98	46.54	12.32	12.32	0.06	25.10	0.63	BA
w12	11.57	46.51	10.94	10.94	0.05	25.10	0.63	BA
w13	64.58	105.95	72.32	64.58	0.10	11.24	0.63	SS
w14	8.31	34.99	7.34	7.34	0.05	29.91	0.63	BA
w15	164.46	320.99	273.33	164.46	0.08	11.24	0.63	SS
w16	27.87	78.99	28.89	27.87	0.06	11.24	0.63	SS
w17	13.54	35.99	12.14	12.14	0.08	29.47	0.63	BA
w18	8.67	35.00	7.67	7.67	0.05	29.91	0.63	BA
w19	210.35	331.16	385.93	214.55	0.10	11.24	0.63	SS
w20	224.81	245.03	173.51	173.51	0.52	7.48	0.63	BA
w21	251.51	257.31	231.08	231.08	0.69	7.48	0.63	BA
w22	193.27	201.60	134.65	134.65	0.56	9.40	0.63	BA
w23	14.26	35.15	12.73	12.98	0.09	29.92	0.63	BA
w24	126.01	216.16	183.04	128.53	0.09	11.24	0.63	SS
w25	13.36	35.12	11.92	12.16	0.08	29.92	0.63	BA
w26	120.19	216.17	174.17	122.59	0.09	11.24	0.63	SS
w27	125.41	164.38	169.87	127.92	0.13	11.24	0.63	SS
w28	125.41	164.38	169.87	127.92	0.13	11.24	0.63	SS

4.2.3.11 Calculation Example: Existing Building Made of Rubble Masonry

This section presents the seismic vulnerability assessment of an existing two-storey masonry building made of rubble masonry (Fig. 4.107a) aiming to discuss the critical aspects for this class of structures. Figures 4.108 and 4.109 display the plan configurations of the basement, the ground floor and the first floor.

The dimensions of the building are about 20 × 16 m^2, with interstorey heights of 2.78 m (basement), 4.50 m (ground floor) and 4.34 m (first floor). The basement is not modelled, since it is partly cellared and possesses a sufficient rigidity in order

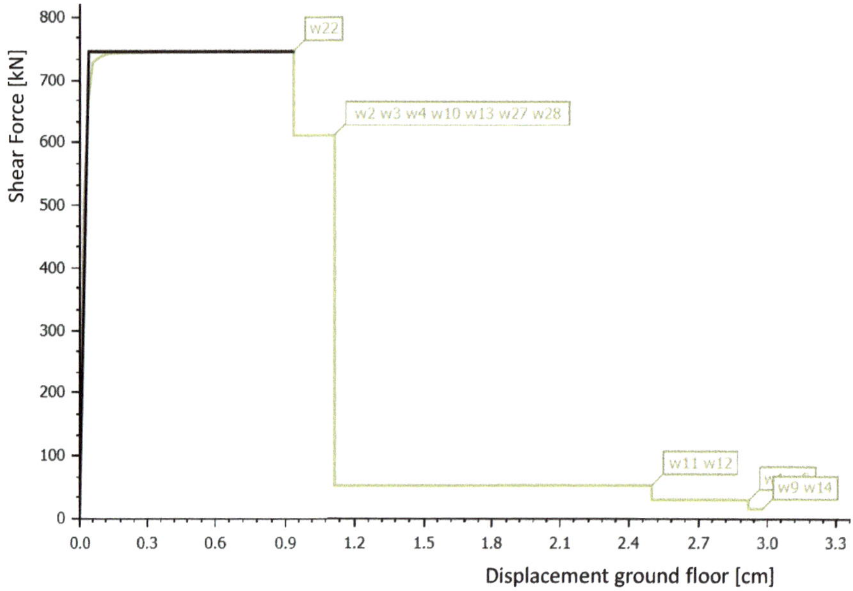

Fig. 4.105 Non-linear capacity curve with successive wall failure and bilinear idealization

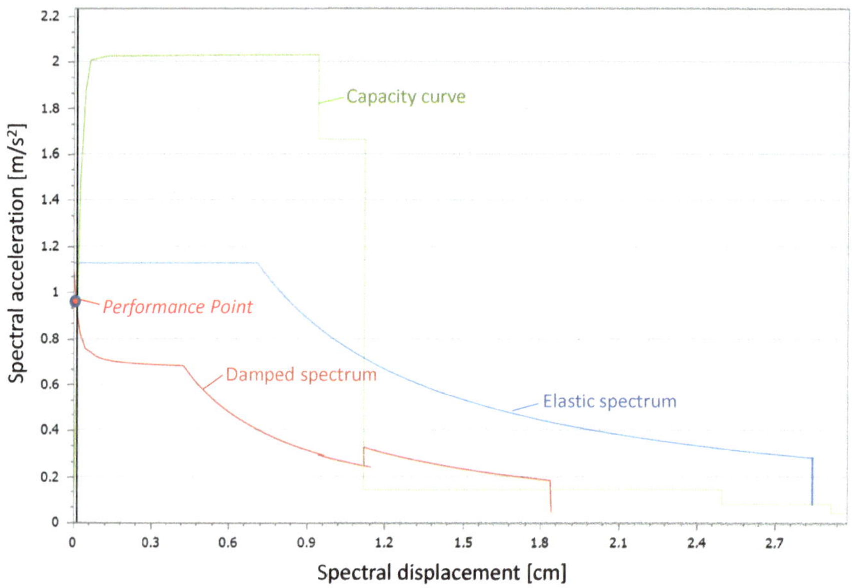

Fig. 4.106 Elastic and damped spectrum, capacity curve and "Performance Point"

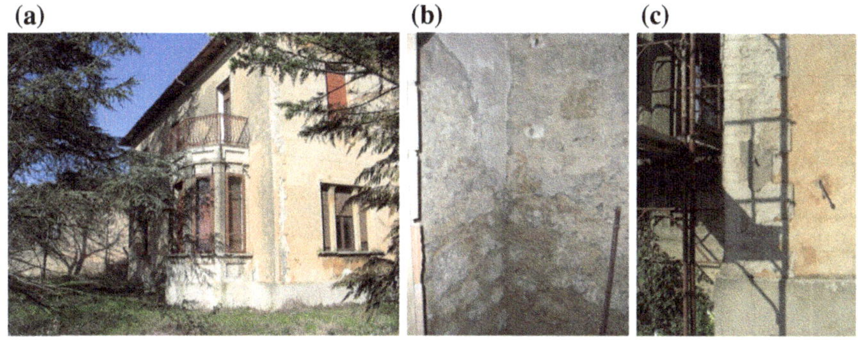

Fig. 4.107 **a** View from south-west, **b** rubble masonry, **c** high interlocking quality at the corners

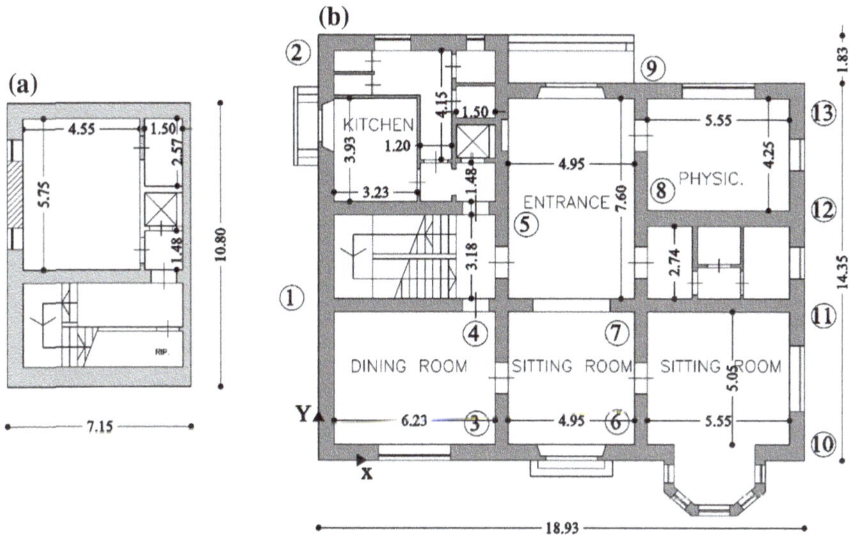

Fig. 4.108 Plan of the building: **a** basement; **b** ground floor (m)

to displace together with the surrounding soil. The floors are made of steel girders, clay hollow flat tiles and a layer of screed. The span-roof has a slope of 10°.

A structural intervention was conducted in the seventies: the original roof was replaced by a horizontal diaphragm of pre-stressed concrete beams and clay elements, covered with a 40 mm thick slab (Fig. 4.110). Above the horizontal diaphragm, clay walls support the inclined roof. The collection of original drawings, test protocols or other sources is of crucial importance in the knowledge process for a proper seismic vulnerability assessment of existing buildings (Section "Different Approaches for New or Existing Masonry Buildings"). It appears that the whole roof structure corresponds to a permanent load of 3.50 kN/m^2.

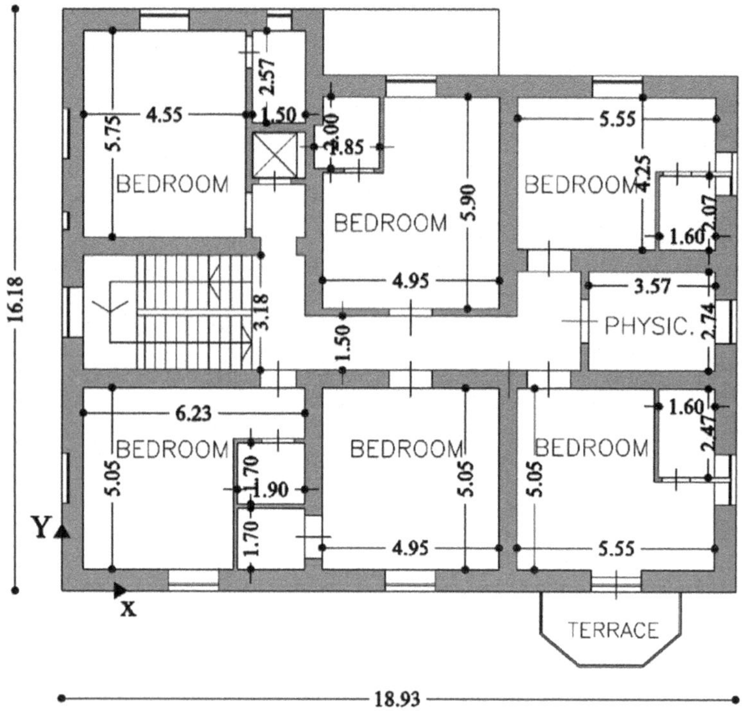

Fig. 4.109 Ground plan of the building: first floor (m)

The material mostly used is rubble masonry (Table 4.45), but some regular bricks are located at the building corners, ensuring a better interlocking between perpendicular walls (Fig. 4.107b and c). The external walls' thickness is 60 cm, while the internal walls are 50 cm thick. The partitions are assumed to have an equivalent uniformly distributed load of 1.2 kN/m^2 according to Sect. 6.3.1.2 of Eurocode 1-1-1 (2004).

As the building in question is an existing structure situated in Italy, the reference standards are Eurocode 8-3 (2005) and the Italian code NTC (2008) with its applications rules CIRC (2009). The analysis of the structure is performed with the 3DMACRO software (2018), based on the macro-element approach introduced in Sect. 4.2.3.6. Non-linear static analyses and out-of-plane analyses are carried out to define the seismic vulnerability of this existing building.

Regarding the characterization of materials, Table C8A.2.1 (CIRC 2009) indicates value ranges of mechanical parameters for different types of existing masonry buildings. This reference is particularly useful when, as in this case, experimental tests cannot be performed. Nevertheless, the designer has to be aware that the values indicated in the mentioned table were collected for specific types of Italian buildings' masonry and are, therefore, valid only for them. Only a survey of the building geometry and limited verifications concerning structural details and mechanical properties were performed for the building under examination. As a consequence, the knowl-

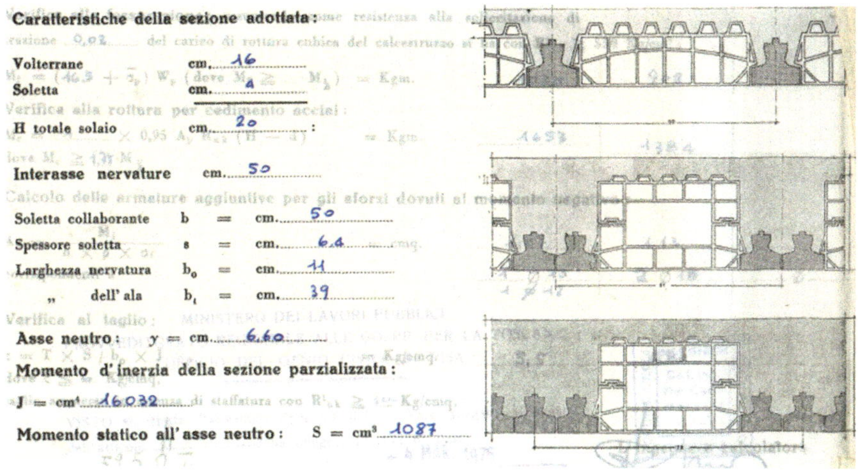

Fig. 4.110 Original drawing with the calculation of the roof floor characteristics

Table 4.45 Mechanical properties and strengths according to CIRC (2009), Table C8A.2.1

E-Modulus (kN/m^2)	G-modulus (kN/m^2)	f_m (kN/m^2)	f_{vk0} (kN/m^2)	γ (kN/m^3)
690000	230000	1000	200	19

edge level is KL1 (the lowest one) and the corresponding confidence factor CF is 1.35 (Section "Different Approaches for New or Existing Masonry Buildings"). The mean values of mechanical parameters, assumed as characteristic, must be selected differently depending on the acquired knowledge level. For KL1, the minimum mean values of the range reported in Table C8A.2.1 (CIRC 2009) are applied for the compressive strength f_m, the initial shear strength f_{vk0} under zero compression stresses and the specific weight γ provided in Table 4.45. Furthermore the elastic and shear modulus are considered for cracked masonry as specified in Table 4.45.

Determination of the Seismic Mass

The seismic mass of the building is calculated from the floor dead and live loads and the weight of the walls according to Eurocode 8-1 (2004). The combination factor ψ_2 is defined in accordance to Eurocode 0 (2002). According to Table 2.5.I of the Italian code NTC (2008) and Table A1.1 of Eurocode 0 (2002), the snow and the wind loads are combined with a combination factor $\psi_2 = 0.0$ as the site altitude is lower than 1000 m. The combination factor $\psi_2 = 0.3$ is applied for residential buildings. The coefficient φ is set to 0.8 according to Eurocode 8-1 (2004), Sect. 4.2.4.

Table 4.46 Determination of masses for each storey

Description	First floor	Roof floor
Floor loads		
Floor area	$A_{GF} = 285.11\ m^2$	$A_R = 285.11\ m^2$
Permanent loads	Steel and clay hollow flat tiles floor including concrete slab and internal partitions	Steel and clay hollow flat tiles floor incl. concrete slab
	$g_k = 3.67\ kN/m^2$	$g_k = 3.5\ kN/m^2$
Variable loads	Live loads: $q_k = 2\ kN/m^2$	Snow loads: $q_k = 1.51\ kN/m^2$
φ - coefficient	0.8 (–)	1.0 (–)
ψ_2 - coefficient	0.3 (–)	0.0 (–)
Wall loads		
Total cross-sectional area of shear walls	$A_W = 53.15\ m^2$	$A_W = 57.90\ m^2$
Considered wall heights	h = 4.50 m	h = 4.34 m
Thickness	40 – 60 cm	40 – 60 cm
Material density	$\rho_{MW} = 1.9\ t/m^3$	$\rho_{MW} = 1.9\ t/m^3$
Spandrel area		
South façade	9.54 m^2	7.24 m^2
North façade	14.16 m^2	10.78 m^2
East façade	11.66 m^2	11.66 m^2
West façade	3.50 m^2	7.63 m^2
Total	38.86 m^2	37.31 m^2
Wall weight	$G_{k,MW} = 4544$ kN	$G_{k,MW} = 4774$ kN
Spandrels weight	$G_{k,SP} = 443$ kN	$G_{k,SP} = 425$ kN
Sum		
$\sum G_{ki}$	$285.11 \cdot 3.67 + 4544 + 443 =$ 6033.35 kN	$285.11 \cdot 3.5 + 4774 + 425 =$ 6196.89 kN
$\sum \varphi \cdot \psi_{2i} \cdot Q_{ki}$	$285.11 \cdot (2 \cdot 0.8 \cdot 0.3) = 136.85$ kN	$285.11 \cdot (1.51 \cdot 1.0 \cdot 0.0) = 0.0$ kN
$\sum G_{ki} + \sum \varphi \cdot \psi_{2i} \cdot Q_{ki}$	6170 kN ~ 619 t	6197 kN ~ 622 t

Table 4.46 gives the floor masses for dead and live loads with the corresponding combination coefficients. The mass is calculated separately for the first floor, including the dead and live loads of the first storey (floor + walls) and for the second floor (roof floor + walls). The storey masses are computed for the first and the second floor:

$$m_1 = 617\,t$$
$$m_2 = 618\,t.$$

The total mass of the structure is therefore approximately 1235 t.

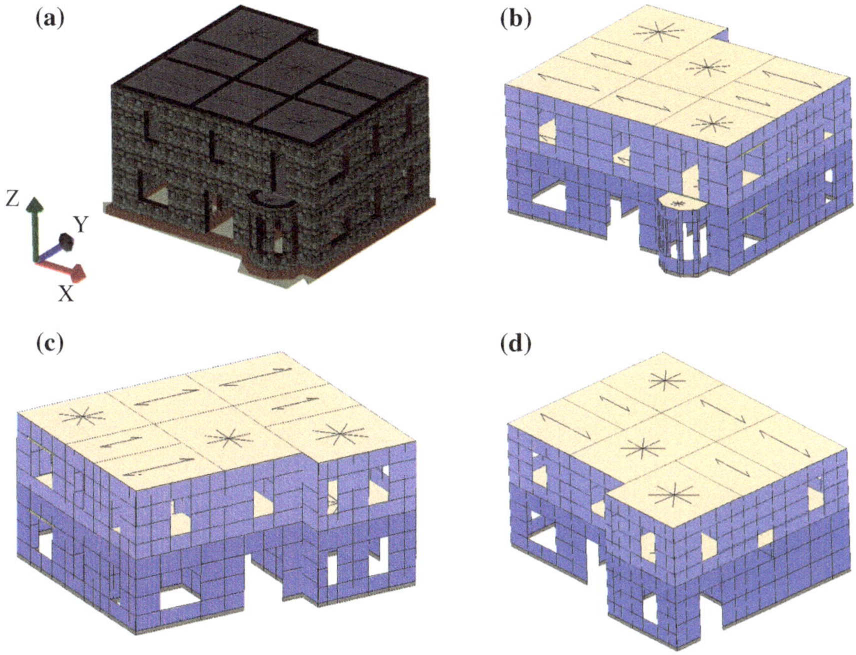

Fig. 4.111 Main axes of the building (**a**); structural model with macro-elements viewed from the south-east (**b**), north-east (**c**), north-west (**d**)

Macro-element Model

The structural model is based on the macro-element approach introduced in Section "Macro-element Models". The span directions of the floors are mainly oriented in X-direction and the vertical loads are transferred mostly to the walls in Y-direction, since the roof is additionally supported by the walls along the Y-direction. Figure 4.111 shows the building and views of the structural model.

The discretization with macro-elements is such that the maximum side of each macro-element is 2 m. Where stress concentration is expected, e.g. near the openings, the macro-elements are smaller. In order to describe the non-linear behaviour of the masonry, the constitutive laws explained in Section "Macro-element Models" are assumed.

The design values of strength and shear modulus, in case of pushover analysis, are obtained from the ratio of the mean values (Table 4.45) to the confidence factor FC according to section C8.7.1.5 in CIRC (2009). The design values are therefore indicated by subindex d. The assumed mechanical properties are those listed in

Table 4.47 Design values of strength for rubble masonry and knowledge level KL1 (pushover analysis)

Material	f_m (kN/m^2)	f_{vk0} (kN/m^2)	FC	f_d (kN/m^2)	f_{vd0} (kN/m^2)
Rubble masonry (irregular)	1000	20	1.35	740.74	14.82

Table 4.48 Modal analysis results of the two-storey building made of rubble masonry

Mode (–)	T (s)	ω (rad/s)	M_x (%)	M_y (%)	M_z (%)	$M_{x,Sum}$ (%)	$M_{y,Sum}$ (%)	$M_{z,sum}$ (%)
1	0.203	31.006	8.59	13.60	0.01	8.59	13.60	0.01
2	0.185	33.912	25.66	48.02	0.0	34.25	61.62	0.01
3	0.178	35.367	40.44	13.91	0.0	74.70	75.53	0.01
4	0.158	39.890	0.80	0.32	0.0	75.50	75.85	0.01

Table 4.47. By contrast, if a linear analysis were carried out, the design values would have been again reduced by a safety factor $\gamma = 2.0$.

Modal Analysis

The modal analysis procedure is performed with the 3D model developed in the 3DMACRO software (2018). The first four mode shapes are displayed in Fig. 4.112 whereas the periods and activated masses are summarized in Table 4.48. In particular, the table shows the vibration period *T* and the circular frequency ω of each mode, the activated modal masses *Mx, My, Mz* and the cumulative activated modal masses $M_{xsum}, M_{ysum}, M_{zsum}$.

The distribution of the modes reveals that with four modes more than 75% of the total mass is activated. The first mode is a torsional mode, while the second and third are translational modes along the X and Y direction. The fourth mode shape activates minor masses, as it is just a local vibration mode of the masonry pillars in the bow window. It is worthwhile to compare the fundamental period of the first mode to the simplified one proposed by the Italian code NTC (2008) for linear static analysis:

$$T_1 = C \cdot H^{\frac{3}{4}} = 0.05 \cdot 8.84^{\frac{3}{4}} = 0.256\,\text{s}$$

where *C* is a coefficient dependent on the building type and *H* is the total height of the building. The simplified estimation overestimates the fundamental period by 28%. However, both calculated natural periods must be regarded as an estimation for masonry buildings, since their vibration behaviour is non-linear. In the end, both values are acceptable as they lie in the plateau range of the response spectrum.

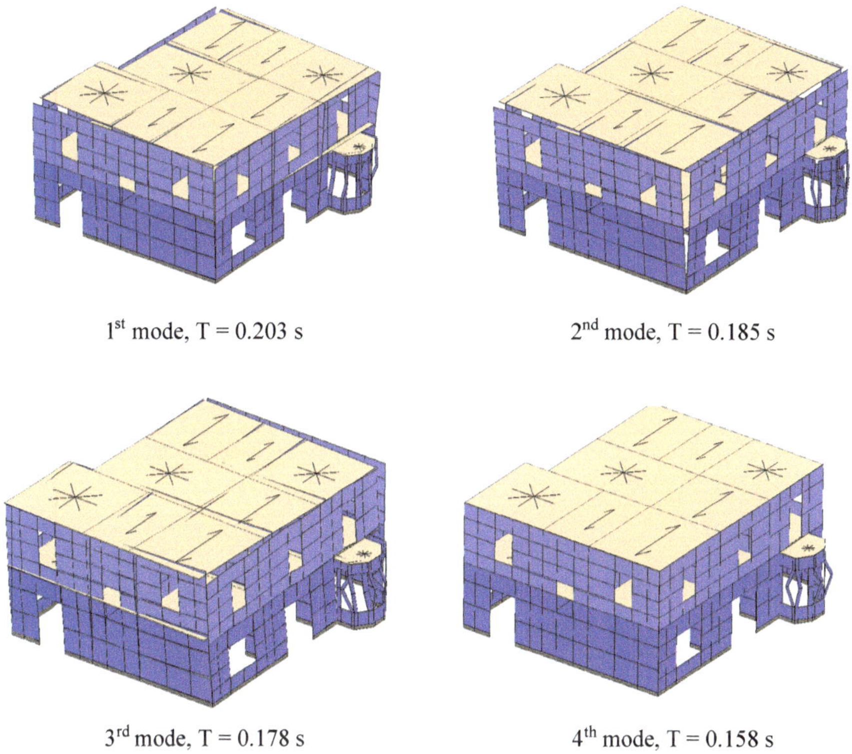

Fig. 4.112 Modal analysis results: mode shapes and corresponding periods

Pushover Analysis

A pushover analysis is performed by following the procedure described in Sect. 4.1.5.3. As first outcome of the analysis the capacity curves display the relationship between the maximum horizontal force at the base of the building and the horizontal displacement of a control point, here chosen as the centre of mass of the roof. The capacity curves are expressed in terms of the shear coefficient C_b, equal to the ratio between the total shear force at the base of the building along the considered direction to the seismic weight of the building:

$$C_b = \frac{V_b}{W}.$$

Figure 4.113 displays all the capacity curves resulting from the pushover analysis in both directions X and Y and with both force distributions (proportional to mass and to first mode shape, Sect. 4.1.5.3). In these diagrams, the number inside the circle represents a safety factor given by:

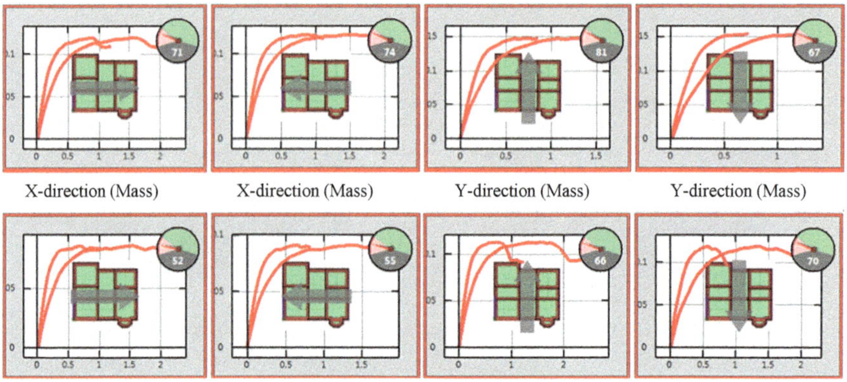

X-direction (Mass) X-direction (Mass) Y-direction (Mass) Y-direction (Mass)

X-direction (Mode shape) X-direction (Mode shape) Y-direction (Mode shape) Y-direction (Mode shape)

Fig. 4.113 Capacity curves for the pushover analysis at ULS with force distribution proportional to mass or first mode shape with knowledge level KL1

$$SF_{ULS} = \frac{d_{C,u}}{d_{D,u}} \cdot 100$$

$$SF_{DLS} = \frac{d_{C,d}}{d_{D,d}} \cdot 100$$

where d_C and d_D are, respectively, capacities and demand displacements for the two limit states (Sect. 4.2.3.8).

A parametric analysis in which the knowledge level is fictitiously increased to the maximum achievable level is also performed. The corresponding capacity curves are shown in Fig. 4.114. This operation could be useful in order to evaluate the convenience of performing more detailed investigations and experimental testing. In this specific case, an improvement of the knowledge level would not give a substantial enhancement in terms of global building response. Indeed, only the safety verifications of the pushover analysis along the Y-direction with a force distribution proportional to the mass and height are satisfied.

In the capacity curves shown in Fig. 4.113, the minimum safety factor SF refers to the analysis in direction X with a force distribution proportional to the first mode shape. For this pushover analysis, the roof floor is more ductile than the first floor (Fig. 4.115a). Moreover, the verification at the ULS is not satisfied but at the DLS it is (Fig. 4.115b).

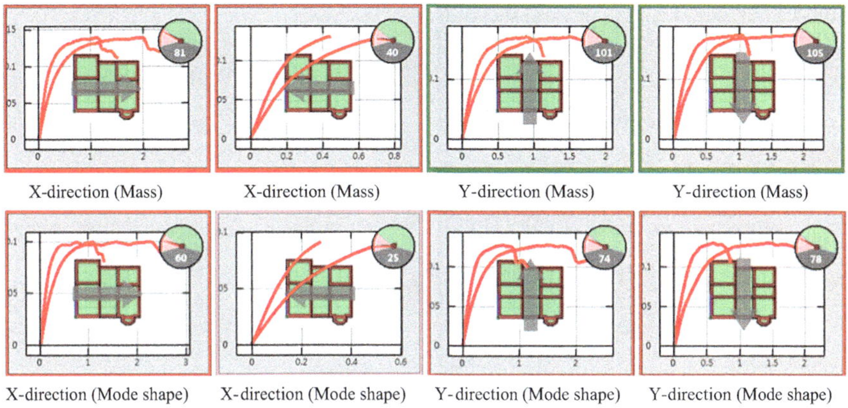

Fig. 4.114 Capacity curves for the pushover analysis at ULS with force distribution proportional to mass or first mode shape with knowledge level KL3

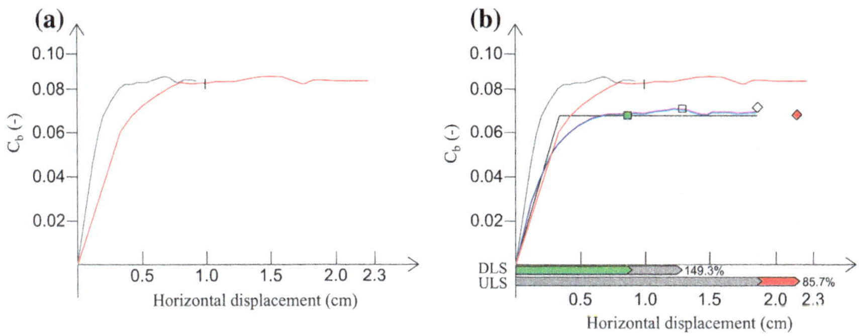

Fig. 4.115 Pushover analysis in direction X (force distribution proportional to the first mode shape): **a** capacity curves (grey: centre of mass of the first floor, red: centre of mass of the roof storey); **b** safety factors SF (grey: capacity, red: deficient demand, green: sufficient demand)

In this case, the parameters of the single degree of freedom oscillator are:

Effective mass: $M^* = 977.6$ t
Modal participation factor: $\Gamma = 1.2037$
Effective stiffness: $K^* = 252256$ kN/m
Effective period: $T^* = 0.3911$ s
Yield displacement of the control point: $d_Y^* = 0.34$ cm
Ultimate displacement of the control point: $d_u^* = 1.86$ cm
Maximum shear coefficient: $C_b^* = 0.068$ cm
Available ductility: $\mu = 5.42$
Behaviour factor: $q = 3.14$

Thus, the results of the pushover analysis are (Fig. 4.115):

Maximum shear force at the base: $V_{b,max} = 1086.3$ kN

Table 4.49 Types of failures in the pushover analysis

	Panel with shear crack		Panel with closed shear crack
	Panel with shear failure		Panel crushed for compression
	Panel cracked for tensile action		Plastic hinge open in a beam
	Plastic hinge in a beam at ¾ of the ultimate rotation		Plastic hinge in ultimate state

Seismic weight of the whole structure: $W = 12702.2$ t
Maximum shear coefficient: $C_b = 0.0855$
Maximum displacement of the control point: $d_{C,u} = 2.24$ cm

Other relevant results of the pushover curve are the failure modes that influence each structural member in the analysis steps. The failure modes are related both to masonry panels and to reinforced concrete beams present at the top of the openings. Table 4.49 shows the failure types and corresponding captions.

Figure 4.116 shows the progression of the failure modes from the DLS to the ULS in the south façade. It is possible to notice the cumulative damage in the area adjacent to the largest opening with panels crushed for compression and panels cracked for tensile actions.

The north façade exhibits a more uniform crack pattern at the ULS. The first shear cracks appear at the spandrels above the first storey openings, spreading to the second storey together with tensile failures (Fig. 4.117).

Kinematic Analysis for the Out-of-Plane Response

The kinematic analysis of out-of-plane modes is carried out for the external façades of the building (Fig. 4.118), according to the procedure illustrated in Sect. 4.2.3.7. A resume of results is presented in Table 4.50 (for ULS) and in Table 4.51 (for DLS).

The table of results reports the type of mechanism, the involved storey, the minimum acceleration that triggers the mechanism a_0^*, the demand acceleration according to the considered limit state $a_{g,D}$ and the safety factor SF ($= a_0^*/a_{g,D}$), specifying whether the verification is successful or unsuccessful. The analysis is linear as it involves only the acceleration that activates the mechanisms and not the displacements.

Let us consider, for instance, façade no. 6 shown in Fig. 4.119a. Only the self-weight W and the inertia force αW - applied to the centre of mass of the façade - are the forces involved in the kinematic chain (Fig. 4.119b).

Table 4.50 Out-of-plane analysis at ULS (°per unit of length)

FACADE	$a_{g,D}$ (g)	a_0^* (g)	SF (–)	Involved storey	Verification
Overturning					
F1	0.104	0.060	0.57	Storeys 1–2	Unsuccessful
F1	0.261	0.102	0.39	Storey 2	Unsuccessful
F2	0.104	0.101	0.97	Storeys 1–2	Unsuccessful
F2	0.261	0.102	0.39	Storey 2	Unsuccessful
F3	0.125	0.533	4.25	Storey 1	Successful
F3	0.261	0.098	0.38	Storey 2	Unsuccessful
F4 (rigid floor)	0.309	0.976	3.16	Storey 2	Successful
F4 (friction)	0.213	0.294	1.38	Storeys 1–2	Successful
F5	0.104	0.054	0.52	Storeys 1–2	Unsuccessful
F5	0.261	0.100	0.38	Storey 2	Unsuccessful
F6	0.104	0.068	0.33	Storeys 1–2	Unsuccessful
F7 (rigid floor)	0.310	0.882	2.85	Storey 2	Successful
F7 (friction)	0.213	0.291	1.36	Storeys 1–2	Successful
Flexural bending					
F7°	0.104	0.340	3.26	Storeys 1–2	Successful

Table 4.51 Out-of-plane analysis at DLS (°per unit of length)

FACADE	$a_{g,D}$ (g)	a_0^* (g)	SF (–)	Involved storey	Verification
Overturning					
F1	0.088	0.060	0.68	Storeys 1–2	Unsuccessful
F1	0.442	0.102	0.23	Storey 2	Unsuccessful
F2	0.088	0.101	1.15	Storeys 1–2	Successful
F2	0.442	0.102	0.23	Storey 2	Unsuccessful
F3	0.251	0.533	2.13	Storey 1	Successful
F3	0.442	0.098	0.22	Storey 2	Unsuccessful
F4 (rigid floor)	0.618	0.976	1.58	Storey 2	Successful
F4 (friction)	0.426	0.294	0.69	Storeys 1–2	Unsuccessful
F5	0.088	0.054	0.62	Storeys 1–2	Unsuccessful
F5	0.442	0.100	0.23	Storey 2	Unsuccessful
F6	0.088	0.068	0.81	Storeys 1–2	Unsuccessful
F7 (rigid floor)	0.215	0.882	4.10	Storey 2	Successful
F7 (friction)	0.149	0.291	1.95	Storeys 1–2	Successful
Flexural bending					
F7°	0.088	0.340	3.88	Storeys 1–2	Successful

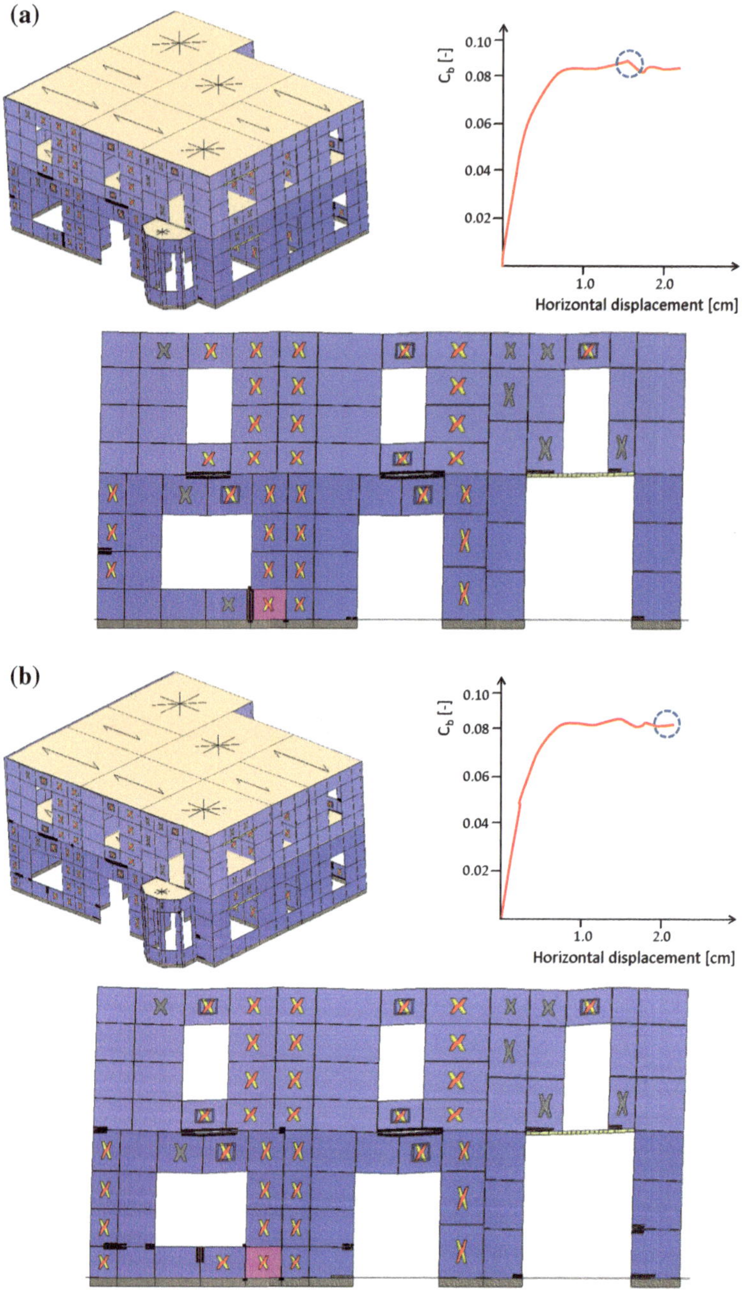

Fig. 4.116 Failures of the south façade: pushover analysis + X (First mode shape): **a** DLS; **b** ULS

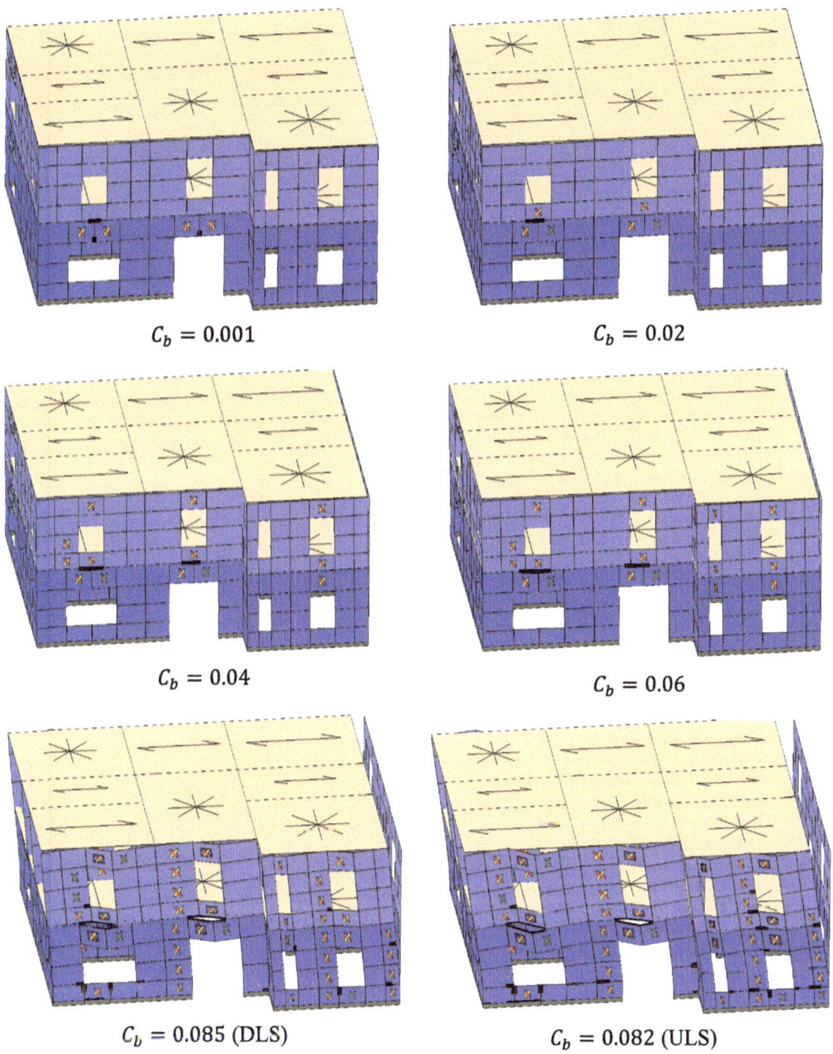

Fig. 4.117 Cumulative damage of the north façade: pushover analysis + X-direction (First mode shape) with corresponding shear coefficient at the base of the building

The acceleration that triggers the overturning mechanism of the whole façade (storeys 1–2) can be calculated with a balance of stabilizing and destabilizing momentum around a base hinge:

$$\alpha W \cdot h_G = W \cdot \frac{t}{2}$$

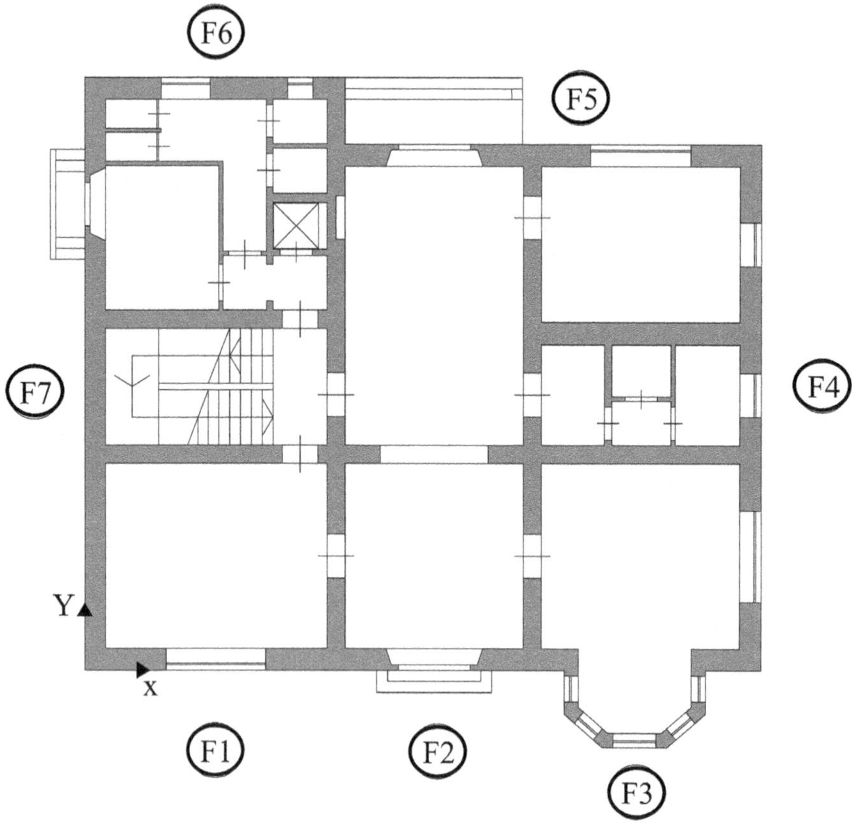

Fig. 4.118 External façades subjected to out-of-plane modes

where h_G is the depth of the centre of mass and t is the wall thickness (equal for storey 1 and 2). Therefore α can be calculated to:

$$\alpha = \frac{t}{2h_G} = \frac{0.60}{2 \cdot 4.43} = 0.068$$

and the collapse multiplier that triggers the mechanism is:

$$\alpha_0 = 0.068g.$$

The same result can be obtained from the principle of virtual works. The expressions of the displacements in horizontal (subscript x) and vertical (y) are (Fig. 4.120):

$$\delta u_x = Rcos(\beta - \vartheta)\delta\vartheta$$
$$u_x = R[sin(\beta) - sin(\beta - \vartheta)]$$

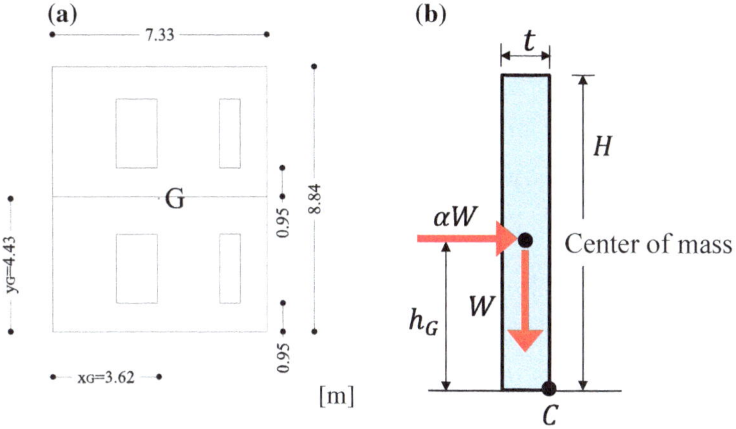

Fig. 4.119 Façade F6 (**a**) and overturning mechanism (**b**)

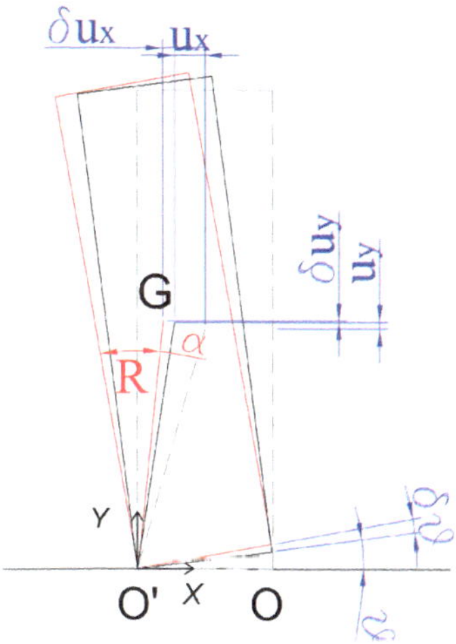

Fig. 4.120 Horizontal and vertical virtual displacements

$$\delta u_y = R\,sin(\beta - \vartheta)\delta\vartheta.$$

Consider the wall rotating from an initial configuration of rotation ϑ, to which a virtual rotation $\delta\vartheta$ applies. R is the radius vector connecting the pivot point and the centre of mass of the block and β the slenderness ratio, equal to the arctangent of the ratio thickness to height. The principle of virtual work expressed in this case is:

$$\alpha W \cdot R\,cos(\beta - \vartheta)\delta\vartheta - W \cdot R\,sin(\beta - \vartheta)\delta\vartheta = 0$$

which, for $\vartheta = 0$, gives:

$$\alpha = tan\,\beta = \frac{t}{2h_G},$$

the same expression previously obtained with momentum balance equation.

The effective mass according to Sect. 4.2.3.7.1 is:

$$M^* = \frac{\left(\sum_{i=1}^{n+m} P_i \delta_{x,i}\right)^2}{g \sum_{i=1}^{n+m} P_i \delta_{x,i}^2} = \frac{\left(W \delta_{x,max}\right)^2}{g W \delta_{x,max}^2} = \frac{(628.6 \cdot 0.30)^2}{9.81 \cdot 628.6 \cdot 0.30^2} = \frac{628.6}{9.81} = 64.08 \, \frac{\mathrm{kN\,s^2}}{\mathrm{m}}.$$

With the effective mass, it is possible to calculate the acceleration capacity:

$$a_0^* = \frac{\alpha_0 \sum_{i=1}^{n+m} P_i}{M^* \, CF} = \frac{0.068 \cdot 628.6}{64.08 \cdot 1.35} = 0.494 \, \frac{\mathrm{m}}{\mathrm{s^2}} = 0.050g.$$

If one assumes that the confidence factor is equal to 1.0, the value increases to:

$$a_0^* = \frac{\alpha_0 \sum_{i=1}^{n+m} P_i}{M^* \cdot CF} = \frac{0.068 \cdot 628.6}{64.08 \cdot 1.00} = 0.667 \, \frac{\mathrm{m}}{\mathrm{s^2}} = 0.068g.$$

The checks for the damage limit state (DLS) and for the ultimate limit state (ULS) are unsuccessful, indeed:

$$a_0^* < \frac{a_g\left(P_{V_R}\right) \cdot S}{q} \rightarrow 0.068g < \frac{0.14g \cdot 1.49}{2} = 0.104g.$$

for ULS, whereas for DLS:

$$a_0^* < \frac{a_g\left(P_{V_R}\right) \cdot S}{q} \rightarrow 0.068g < 0.059g \cdot 1.49 = 0.088g.$$

In both cases the demand exceeds the capacity of the mechanism. In order to reach a safety condition, a couple of steel-tie rods could be placed at the upper corners of the façade. In this case, the principle of virtual work has an additional stabilizing term:

$$\alpha W \cdot R\,cos(\beta - \vartheta)\delta\vartheta - W \cdot R\,sin(\beta - \vartheta)\delta\vartheta - T \cdot 2R\,cos(\beta - \vartheta)\delta\vartheta = 0$$

assuming that the radius vector of the tie-rods, whose force is T, is $2R$. Obviously, the tie rods could be placed in another position rather than at the top of the wall and therefore the radius vector changed accordingly.

For $\vartheta = 0$, the equation gives:

$$\alpha = \frac{W \sin\beta + 2T \cos\beta}{W \cos\beta}.$$

The acceleration capacity a_0^* should be set at least equal to the maximum demand (in this case that of the ULS), in order to get the verifications satisfied. The tie-force value is still unknown. Therefore, it has to be determined by equating a_0^* with the maximum demand:

$$T = \frac{W}{2}(\alpha - tan\,\beta) = \frac{W}{2}\left(\frac{a_0^* \cdot M^* \cdot CF}{\sum_{i=1}^{n+m} P_i} - tan\,\beta\right)$$
$$= \frac{628.6}{2}\left(\frac{0.104g \cdot 1.0}{g} - 0.068\right) = 11.32\,\mathrm{kN},$$

the only inertial vertical load being the self-weight. By considering $T = 15$ kN, the previous expression obtained from the principle of virtual work leads to:

$$\alpha = \frac{W\,sin\,\beta + 2T\,cos\,\beta}{W\,cos\,\beta} = tan\,\beta + \frac{2T}{W} = 0.068 + \frac{2 \cdot 15}{628.6} = 0.115,$$

which is the updated acceleration capacity. Now, the verifications are obviously satisfied:

$$a_0^* = \frac{\alpha_0 \sum_{i=1}^{n+m} P_i}{M^* \cdot CF} = \frac{0.115 \cdot 628.6}{64.08 \cdot 1.0} = 0.115g.$$

The checks for the damage limit state (DLS) and for the ultimate limit state (ULS) are finally:

$$a_0^* > \frac{a_g(P_{V_R}) \cdot S}{q} \rightarrow 0.115g > \frac{0.14g \cdot 1.49}{2} = 0.104\,g$$

for ULS, whereas for DLS:

$$a_0^* > \frac{a_g(P_{V_R}) \cdot S}{q} \rightarrow 0.115g > 0.059g \cdot 1.49 = 0.088g$$

with safety factors of:

$$SF_{DLS} = \frac{0.115}{0.088} = 1.31$$
$$SF_{ULS} = \frac{0.115}{0.104} = 1.11$$

both>1. This stabilizing force may be obtained with the action of no. 2 steel-tie rods made of steel S235 of diameter 12 mm, pre-stressed with a tension of 7.5 kN each. The yield force is:

$$F_y = f_{yd} \cdot A = \frac{2350}{1.05} \cdot \pi \cdot \frac{1.2^2}{4} = 25.3\,\mathrm{kN} > 15.0\,\mathrm{kN}$$

for each tie-rod. The pre-tension corresponds therefore to 30% of the yielding value. The change in terms of displacement capacity can be appreciated performing a kinematic non-linear analysis with the procedure illustrated in Sect. 4.2.3.7.

The capacity curve is obtained by varying the initial configuration of the block and calculating, for each configuration, the acceleration that activates the mechanism with the principle of virtual work. The dotted curve in Fig. 4.121 refers to the façade

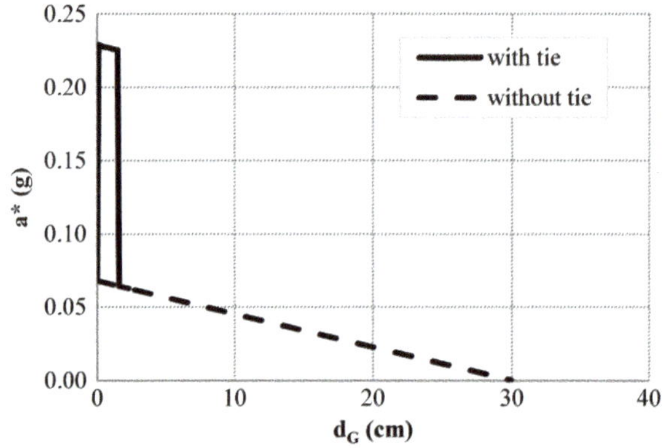

Fig. 4.121 Acceleration-displacement of centre of mass curve, with and without tie-rod

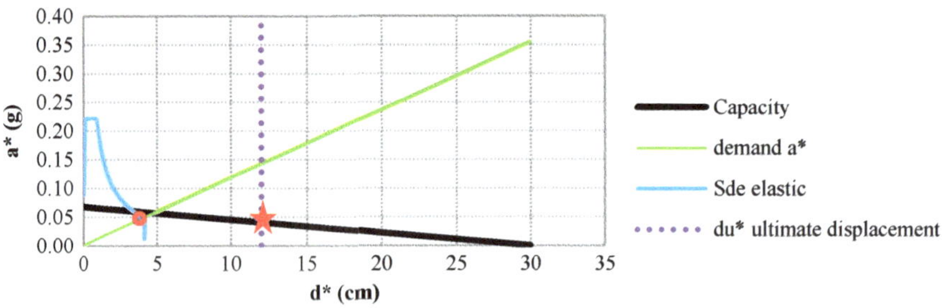

Fig. 4.122 Kinematic non-linear analysis results: displacement capacity (star) greater than displacement demand (circle) without tie-rods (see Sect. 4.2.3.7)

restrained by no. 2 steel-tie rods S235 of diameter 12 mm, placed at the top of the façade (depth of 8.60 m from the base) and pre-stressed with 7.5 kN each, whilst the continuous line represents the response of the free-standing block. When the tie-rods act to stabilize the out-of-plane mode, the capacity increases (Fig. 4.121) in the initial branch of the curve.

After a peak corresponding to the yielding of the tie, the capacity decreases tending to the curve related to the case without tie-rods. By increasing the block rotation, that is the displacement of the centre of gravity d_G, the tie strain increases as well. The yielding force, is constant (plastic phase) up to the failure of the rod at an ultimate strain (assumed 0.010). At the tie failure, the capacity suddenly drops and the curve coincides to that of the wall without restraint (Fig. 4.121). Therefore, the ultimate displacement is about 30 cm for both curves and the verification in terms of displacements does not change considering the tie-rods.

The verification of the initial state (without tie-rods) is depicted in Fig. 4.122. The ADRS, correlating the displacement with the acceleration demand, was taken in favour of safety as that corresponding to the elastic spectrum. With the verification in terms of displacements, the façade is safe even without tie-rods. The safety factor SF can be expressed as:

$$SF_{ULS} = \frac{d_u^*}{S_{De}(T_s)} = \frac{11.98}{4.14} = 2.89.$$

Risk Index

In the assessment of the seismic vulnerability of existing buildings a risk index could be useful as well. The risk index is defined as the ratio of the return period (in years) for which the structure is verified, to the return period of the demand earthquake (ultimate and damage limit states). If the risk index is lower than 1, the structure is unsafe for the considered limit state. Table 4.52 lists the risk indices calculated for the existing masonry building. It can be seen that an unsafe condition is calculated for both global and local analyses in ULS. The DLS is instead not exceeded for the global pushover analysis. It should be noticed that the out-of-plane analyses are performed as linear analyses.

Table 4.52 Risk indexes of the existing masonry building

Event	T_R (years)	$(T_{RC,ULS}/T_{RD,ULS})$	$(T_{RC,DLS}/T_{RD,DLS})$
Shear failure (masonry)	30	0.063	0.595
Yield rotation (r.c. beam)	44	0.092	0.878
3/4 ultimate rotation (r.c. beam)	152	0.319	3.038
Bending failure (beam)	152	0.319	3.038
Out-of-plane overturning (DLS)	30	–	0.595
Out-of-plane overturning (ULS)	49	0.050	–
Global vulnerability in terms of force	81.98	0.171	–
X direction	81.98	0.171	–
Y direction	147.11	0.309	–
Global vulnerability (pushover) (DLS)	113.23	–	2.260
X direction	118.64	–	2.368
Y direction	113.23	–	2.260
Global vulnerability (pushover) (ULS)	179.49	0.377	–
X direction	215.39	0.452	–
Y direction	179.49	0.377	–

Appendix

Input files for calculation examples of this chapter

The input files for the calculation of the considered multi-family house in Section "Calculation Example: Multifamily House Made of Calcium Silicate Units" are provided for the different types of analysis:

Sect. 4.2.1.1: 4-storey reinforced concrete building
Input file: RC-Building-3D.bd

Section "Simplified Response Spectrum Analysis": Simplified response spectrum analysis
Input file: MFH-2D.bd

Section "Multimodal Response Spectrum Analysis with Three-Dimensional Structural Model": Multimodal response spectrum analysis with three-dimensional models
Input file: MFH-3D-Coupled-Walls.bd (coupled shear walls)
Input file: MFH-3D-Uncoupled-Walls.bd (uncoupled shear walls)

Section "Non-linear Static Analysis": Non-linear static analysis
Input file: MFH-Pushover.bd

A free of charge version of the software MINEA (2018) can be downloaded on the website www.minea-design.com to run the example with the provided input files.

References

ATC-40: Seismic Evaluation and Retrofit of Concrete Buildings. Applied Technology Council, vol. 1 (1996)

ANDILWall 3, Program for unreinforced, reinforced or mixed masonry structures (In Italian), http://www.crsoft.it/andilwall/ (2017)

Bachmann, H.: Erdbebensicherung von Bauwerken. 2. Überarbeitete Auflage, Birkhäuser Verlag, Basel (2002)

Bertero, R.D., Bertero, V.V.: Redundancy in earthquake-resistant design. J. Struct. Eng. **125**(1), 81–88 (1999)

Brencich, A., Gambarotta, L., Lagomarsino, S.: A macroelement approach to the three-dimensional seismic analysis of masonry buildings. In: 11th European Conference on Earthquake Engineering, Niederlande, Rotterdam (1998)

Butenweg, C., Gellert, C., Meyer, U.: Erdbebenbemessung bei Mauerwerksbauten, Mauerwerk Kalender 2010, Verlag Ernst & Sohn (2010)

Caliò, I., Pantò, B.: A macro-element modelling approach of infilled frame structures. Comput. Struct. **143**, 91–107 (2014)

Chen, S.-Y., Moon, F.L., Yi, T.: A macroelement for the nonlinear analysis of in-plane unreinforced masonry piers. Eng. Struct. **30**, 2242–2252 (2008)

CIRC, Circolare esplicativa del 02.02.2009 contenente "Istruzioni per l'applicazione delle nuove norme tecniche per le costruzioni di cui al D.M. 14.01.2008" (In Italian) (2009)

Chopra, A.K.: Dynamics of Structures, Theory and Applications to Earthquake Engineering, 2nd edn. Prentice Hall, New Jersey (2001)
Department of Italian Civil Protection, http://www.protezionecivile.gov.it/jcms/en/storia.wp (2018)
DIN 4149: Bauten in deutschen Erdbebengebieten. Deutsches Institut für Normung (DIN), Berlin Beuth-Verlag, Berlin (2005)
DIN EN 1991-1-4/NA: Nationaler Anhang - National festgelegte Parameter - Eurocode 1: Einwirkungen auf Tragwerke, Teil 1-4: Allgemeine Einwirkungen, - Windlasten, Dezember 2010
DIN EN 1991-1-3/NA: Nationaler Anhang - National festgelegte Parameter - Eurocode 1: Einwirkungen auf Tragwerke - Teil 1-3: Allgemeine Einwirkungen - Schneelasten, Dezember 2010
DIN EN 1996-1-1: Eurocode 6: Bemessung und Konstruktion von Mauerwerksbauten - Teil 1-1: Allgemeine Regeln für bewehrtes und unbewehrtes Mauerwerk; Deutsche Fassung EN 1996-1-1:2005+AC:2009, Dezember 2010
DIN EN 1996-1-1/NA: Nationaler Anhang - National festgelegte Parameter - Eurocode 6: Bemessung und Konstruktion von Mauerwerksbauten - Teil 1-1: Allgemeine Regeln für bewehrtes und unbewehrtes Mauerwerk, Januar 2012
DIN EN 1998-1/NA: Nationaler Anhang - National festgelegte Parameter - Eurocode 8: Auslegung von Bauwerken gegen Erdbeben - Teil 1: Grundlagen, Erdbebeneinwirkungen und Regeln für Hochbau, Januar 2011
DIN EN 771-1: Festlegungen für Mauersteine - Teil 1: Mauerziegel, Deutsche Fassung EN 771-1:2011+A1, 2015
Doherty, K.T., Griffith, M.C., Lam, N., Wilson, J.: Displacement-based seismic analysis for out-of-plane bending of unreinforced masonry walls. Earthq. Eng. Struct. Dyn. **31**, 833–850 (2002)
Dwairi, H., Kowalsky, M., Nau, J.M.: Equivalent damping in support of direct displacement based seismic design. J. Earthq. Eng. **11**, 512–530 (2007)
Eurocode 0: Basis of structural design, European Standard, European Committee for Standardization, April 2002
Eurocode 1-1-1: Actions on structures - General actions - Densities, self-weight, imposed loads for buildings, European Standard, European Committee for Standardization, April 2004
Eurocode 1-1-4: Actions on structures - General actions - Wind actions, European Standard, European Committee for Standardization, January 2004
Eurocode 2-1-1: Design of concrete structures, General rules and rules for buildings, European Standard, European Committee for Standardization, April 2004
Eurocode 3-1-1: Design of steel structures - General rules and rules for buildings, European Standard, European Committee for Standardization, May 2005
Eurocode 6 1 1: Design of Masonry Structures: General rules for reinforced and unreinforced masonry structures, European Standard, European Committee for Standardization, November 2005
Eurocode 8-1: Design of structures for earthquake resistance, General rules, seismic actions and rules for buildings, European Standard, European Committee for Standardization, May 2004
Eurocode 8-2: Design of structures for earthquake resistance, Bridges, European Standard, European Committee for Standardization, June 2004
Eurocode 8-3: Design of structures for earthquake resistance, Assessment and retrofitting of buildings, European Standard, European Committee for Standardization, June 2005
Eurocode 8-4: Design of structures for earthquake resistance, Silos, tanks and pipelines, European Standard, European Committee for Standardization, December 2004
Eurocode 8-5: Design of structures for earthquake resistance, Foundations, retaining structures and geotechnical aspects, European Standard, European Committee for Standardization, April 2003
Eurocode 8-6: Design of structures for earthquake resistance, Towers, masts and chimneys, European Standard, European Committee for Standardization, April 2005

Fajfar, P., Fischinger, M.: N2 - A method for non-linear seismic analysis for regular buildings. In: Proceedings of the 9th World Conference on Earthquake Engineering, Tokyo-Kyoto, Japan. vol. 5, pp. 111–116 (1989)

Fajfar, P., Drobnič, D.: Nonlinear seismic analysis of the ELSA buildings. In: Proceedings of 11th European Conference on Earthquake Engineering, Paris. CD-ROM, Balkema, Rotterdam (1998)

Fajfar, P.: Capacity spectrum method based on inelastic demand spectra. Earthq. Eng. Struct. Dyn. **28** (1999)

FEMA 273: NEHRP guidelines for the seismic rehabilitation of buildings. Applied Technology Council (ATC), Redwood City, USA (1997)

FEMA 274: NEHRP commentary on the guidelines for the seismic rehabilitation of buildings. Federal Emergency Management Agency, Washington, D.C., USA (1997)

FEMA 306: Applied Technology Council (ATC), Publication No. 306, FEMA 306, Evaluation of earthquake damaged concrete and masonry wall buildings - Basic Procedures Manual. Federal Emergency Management Agency, Washington D.C., USA (1998)

FEMA 356: Applied Technology Council (ATC), Publication No. 356, Prestandard and Commentary for the seismic rehabilitation of buildings. Federal Emergency Management Agency, Washington D.C., USA (2000)

Freeman, S.A.: The capacity spectrum method as a tool for seismic design. In: Proceedings of the 11th European Conference on Earthquake Engineering (1998)

Freeman, S.A., Nicoletti, J.P. and Tyrell, J.V.: Evaluations of Existing Buildings for Seismic Risk - A Case Study of Puget Sound Naval Shipyard, Bremerton, Washington, Proceedings of the U.S. National Conference on Earthquake Engineers, EERI, pp. 113–122, Berkeley (1975)

Frilo Statik: Friedrich und Lochner GmbH (2018)

Gellert, C.: Nichtlinearer Nachweis von unbewehrten Mauerwerksbauten unter Erdbebeneinwirkung, Dissertation, RWTH Aachen University, Aachen (2010)

Giresini, L.: Energy-based method for identifying vulnerable macro-elements in historic masonry churches. Bull. Earthq. Eng. **44**(13), 2359–2376 (2015)

Giresini, L., Fragiacomo, M., Lourenço, P.B.: Comparison between rocking analysis and kinematic analysis for the dynamic out-of-plane behavior of masonry walls. Earthquake Eng. Struct. Dynam. **44**(13), 2359–2376 (2015)

Giresini, L., Sassu, M.: Horizontally restrained rocking blocks: evaluation of the role of boundary conditions with static and dynamic approaches. Bull. Earthq. Eng. **15**, 385–410 (2016)

Giresini, L., Fragiacomo, M., Sassu, M.: Rocking analysis of masonry walls interacting with roofs. Eng. Struct. **116**, 107–120 (2016)

Griffith, M., Magenes, G.: Accuracy of displacement-based seismic evaluation of unreinforced masonry wall stability. In: Pacific Conference on Earthquake Engineering, Christchurch, New Zealand, 13–15 Feb (2003)

Heyman, J.: The science of structural engineering. Imperial College Press, London (1999)

Housner, G.W.: The behavior of inverted pendulum structures during earthquakes. Bull. Seismol. Soc. Am. **53**, 403–417 (1963)

Lagomarsino, S., Penna, A., Galasco, A.: TREMURI Program: Seismic Analysis Program for 3D Masonry Buildings. University of Genoa (2006)

Lagomarsino, S., Magenes, G.: Evaluation and reduction of the vulnerability of masonry buildings, The State of Earthquake Engineering Research in Italy: The ReLUIS-DPC 2005–2008 Project, Napoli, Italy (2009)

Lagomarsino, S., Penna, A., Galasco, A., Cattari, S.: TREMURI program: an equivalent frame model for the nonlinear seismic analysis of masonry buildings. Eng. Struct. **56**, 1787–1799 (2013)

Magenes, G., Della Fontana, A.: Simplified non-linear seismic analysis of masonry buildings. Proc. British Mason. Soc. **8**, 190–195 (1998)

Magenes, G., Remino, M., Manzini, M., Morandi, P., Bolognini, D.: SAM II, Software for the Simplified Seismic Analysis of Masonry Buildings. University of Pavia and EUCENTRE (2006)

Magenes, G.: Masonry building design in seismic areas: recent experiences and prospects from a European standpoint. In: First European Conference on Earthquake Engineering and Seismology, Geneva, Switzerland (2006)

Müller, F.P., Keintzel, E.: Erdbebensicherung von Hochbauten. Ernst & Sohn, Berlin (1984)

NTC08, Nuove Norme Tecniche per le Costruzioni, D.M. 14/01/2008, approvazione delle nuove norme tecniche per le costruzioni, G. U. della Repubblica Italiana, n. 29 del 4 febbraio 2008 Supplemento Ordinario n. 30 (In Italian) (2008)

Makris, N., Kostantinidis, D.: The Rocking Spectrum and the Shortcomings of Design Guidelines: Pacific Earthquake Engineering Research Center, College of Engineering, PEER Report 2001/07 (2001)

Makris, N., Kostantinidis, D.: The rocking spectrum and the limitations of practical design methodologies. Earthq. Eng. Struct. Dyn. **32**, 265–289 (2003)

MINEA: Structural Analysis and Design of Masonry Structures. SDA-engineering GmbH, Herzogenrath, http://www.minea-design.de (2018)

Morandi, P.: Inconsistencies in codified procedures for seismic design of masonry buildings. Dissertation, Rose School, Pavia, Italy (2006)

Magenes, G., Calvi, M.: In-plane seismic response of brick masonry walls. Earthq. Eng. Struct. Dyn. **26**, 1091–1112 (1997)

Melis, G.: Displacement-based seismic analysis for out of plane bending of unreinforced masonry walls. Dissertation, Rose School, Pavia, Italy (2002)

Nensel, R.: Beitrag zur Bemessung von Stahlkonstruktionen unter Erdbebenbelastung bei Berücksichtigung der Duktilität. Dissertation, RWTH Aachen, Schriftenreihe Heft 7, Lehrstuhl für Stahlbau (1986)

Noh, S.-Y.: Beitrag zur numerischen Analyse der Schädigungsmechanismen von Naturzugkühltürmen. Dissertation, RWTH Aachen, Schriftenreihe des Lehrstuhls für Baustatik und Baudynamik, Heft 01/1 (2001)

Norda, H.: Beitrag zum statischen nichtlinearen Erdbebenbachweis von unbewehrten Mauerwerksbauten unter Berücksichtigung einer und höherer Modalformen, Dissertation, RWTH Aachen University, Aachen, Deutschland (2012)

Pantò, B., Raka, E., Cannizzaro, F., Camata, G., Caddemi, S., Spacone, E., Caliò, I.: Numerical macro-modeling of unreinforced masonry structures: a critical appraisal. In: Kruis, J., Tsompanakis, Y., Topping, B.H.V. (eds.) Proceedings of the Fifteenth International Conference on Civil, Structural and Environmental Engineering Computing, Civil-Comp Press, Stirlingshire, UK, Paper 81 (2015)

Plevris, V., Kremmyda G., Fahjan, Y.: Performance-Based Seismic Design of Concrete Structures and Infrastructures, p. 320 (2017)

Priestley, M.J.N., Grant, D.N.: Viscous damping in seismic design and analysis. J. Earthq. Eng. **9**(sup2), pp. 229, 255 (2005)

SAP2000: Integrated software for structural analysis and design, Computers and structures Inc., https://www.csiamerica.com/products/sap2000 (2018)

Tomazevic, M.: 1978: The computer program POR. Institute for Testing and Research in Materials and Structures ZRMK, Ljubljana, Slovenia (1978)

Tomazevic, M., Bosiljkov, V., Weiss, P., Klemenc, I.: Experimental research for identification of structural behaviour factor for masonry buildings. Part I - research report P 115/00-650-1. Im Auftrag der Deutschen Gesellschaft für Mauerwerksbau e.V. (DGfM). Ljubljana, Slowenien (2004)

Vaculik, J.: Unreinforced masonry walls subjected to out-of-plane seismic actions, University of Adelaide, School of Civil, Environmental & Mining Engineering, Ph.D. thesis (2012)

Vanin, A., Foraboschi, P.: Modelling of masonry panels by truss analogy - part 1. Mason. Int. **22**(1), 1–10 (2009)

Vidic, T., Fajfar, P., Fischinger, M.: Consistent inelastic design spectra: strength and displacement. Earthq. Eng. Struct. Dyn. **23**, 502–521 (1994)
Wilson, E.L., Der Kiureghian, A., Bayo, E.P.: A replacement for the SRSS method in seismic analysis. Earthq. Eng. Struct. Dyn. **9**, 187–194 (1981)
3muri: program for masonry structures: http://www.ingware.ch/3muri/index.html (2018)
3DMACRO: software for masonry buildings, http://www.murature.com/sismica/ (2018)

Chapter 5
Seismic Design of Structures and Components in Industrial Units

Christoph Butenweg and Britta Holtschoppen

Abstract Industrial units consist of the primary load-carrying structure and various process engineering components, the latter being by far the most important in financial terms. In addition, supply structures such as free-standing tanks and silos are usually required for each plant to ensure the supply of material and product storage. Thus, for the earthquake-proof design of industrial plants, design and construction rules are required for the primary structures, the secondary structures and the supply structures. Within the framework of these rules, possible interactions of primary and secondary structures must also be taken into account. Importance factors are used in seismic design in order to take into account the usually higher risk potential of an industrial unit compared to conventional building structures. Industrial facilities must be able to withstand seismic actions because of possibly wide-ranging damage consequences in addition to losses due to production standstill and the destruction of valuable equipment. The chapter presents an integrated concept for the seismic design of industrial units based on current seismic standards and the latest research results. Special attention is devoted to the seismic design of steel thin-walled silos and tank structures.

Keywords Industrial units · Seismic design · Tanks · Silos · Components
Secondary structures

Industrial facilities must be thoroughly designed to withstand seismic action as they exhibit an increased loss potential due to the possibly wide-ranging damage consequences and the valuable process engineering equipment. This boils down to design rules for the load-carrying parts, the non load-carrying machinery and equipment and also for the storage facilities. No such complete compendium of normative design rules is available in Europe, since Eurocode 8-5 (2004) is valid only for standard buildings without special risk potential. This chapter presents an integrated concept for the seismic design of industrial units based on current seismic standards and the latest research results.

K. Meskouris et al., *Structural Dynamics with Applications in Earthquake and Wind Engineering*, https://doi.org/10.1007/978-3-662-57550-5_5

Fig. 5.1 Typical industrial unit (BASF 2010)

5.1 Introduction

Industrial units consist of the primary load-carrying structure and various process engineering components (secondary structures), the latter being being by far the most important in financial terms. Figure 5.1 shows a typical industrial unit with a steel load-carrying structure.

In addition, supply structures such as free-standing tanks and silos are usually required for each plant to ensure the supply of material and product storage. Thus, in terms of earthquake-proof design of industrial plants, design and construction rules are required for the primary structures, the secondary structures and the supply structures. Within the framework of these rules, possible interactions of primary and secondary structures must also be taken into account. Importance factors are applied in seismic design to sufficiently take into account the possibly higher risk potential of the plant compared to conventional building structures.

5.2 Safety Concept Based on Importance Factors

The importance factors given in Eurocode 8 (2004) cannot be applied to industrial units because they refer solely to standard buildings with their corresponding damage potential. Choosing a suitable importance factor for an industrial unit should consider the following aspects in the case of failure:

- Impact on people, health and safety,
- Impact on the environment,
- Impact due to failure of lifeline functionalities.

Table 5.1 Importance factors γ_I

γ_I	Corresponding return period, in years	Exceedance probability in 50 years (%)	Recommended importance classes according to Eurocode 8
0.8	225	20	I (*ineligible according to VCI-Guideline*)
1.0	475	10	II
1.1	630	7.5	–
1.2	820	5.8	III
1.3	1045	4.6	–
1.4	1300	3.6	IV
1.5	1600	3.0	–
1.6	1945	2.4	–

Such a holistic approach can presently neither be found in national nor in international standards. A meaningful approach is given in the VCI Guideline "The seismic load case in plant construction" (2012) which is structured along the lines of Eurocode 8 (2004) and provides the necessary data for industrial units. According to the VCI Guideline, the importance factor may be chosen as the maximum value from Tables 5.2, 5.3 and 5.4 that consider the aspects mentioned above and should be determined in close cooperation with the owner of the facility. The values of the importance factors given in these tables were agreed upon on the basis of interdisciplinary expertise and have been checked against international standards.

Using importance factors is equivalent to a simple linear scaling of response spectra found in design standards (Table 5.1). Such a linear scaling can only constitute a rough approximation of reality, since for different return periods both spectral shapes and the boundaries of seismic zones are subject to change. It is certainly more meaningful to choose a higher return period for the seismic event to be considered and determine the corresponding spectrum from suitable seismic maps, if such are available. These seismic hazard maps mirror the variable boundaries of seismic zones with increasing return periods. To illustrate: Non-seismic zones in earthquake design codes imply there will be no seismic action, no matter how high the adopted importance factor may be, while it is clear that for longer return periods parts of non seismic zones may indeed acquire some degree of seismic hazard.

5.3 Design of Primary Structures

In general, industrial units exhibit a much higher degree of structural variability than residential buildings, since they must comply with process engineering requirements. Nevertheless, some characteristics are common and can be optimized with respect to obtaining enhanced resistance to seismic loads. Steel frames, either braced or

Table 5.2 Importance factors γ_I with reference to protection of human lives (VCI 2012)

		Consequences				
		Within the individual facility	Immediate vicinity (block within the plant)[b]	Within the plant/industrial area (with fence)	Outside the plant/industrial area	Large-scale consequences outside the plant/industrial area
Damage potential[a]	Non-volatile toxic substances Flammable and oxidizing substances	1.0	1.0	1.0	1.0	1.1
	Non-volatile highly toxic substances Highly and extremely flammable substances Oxidizing gases	1.0	1.1	1.2	1.2	1.2
	Volatile and highly volatile toxic substances Volatile highly toxic substances Explosive substances Extremely flammable liquefied gas	1.1	1.2	1.3	1.4	1.4
	Highly volatile highly toxic substances	1.2	1.3	1.4	1.5	1.6

[a] Flammable, easily flammable and highly flammable and oxidising substances include only gases and liquids
[b] A block inside a plant corresponds to an operational area according to the "Hazardous Incident Ordinance"

Table 5.3 Importance factors γ_I with reference to environmental protection (VCI 2012)

	Consequences		
	No consequences for the environment outside the plant	Minor consequences for the environment outside the plant	Large-scale consequences for the environment outside the plant
Impact on the environment	1.0	1.2	1.4

Table 5.4 Importance factors γ_I for lifeline functions (VCI 2012)

	Requirements		
	Standard requirements regarding availability	High requirements regarding availability	Very high requirements regarding availability
Retention systems, traffic routes, emergency routes	1.2	1.2	1.2
Lifeline structures (fire stations, fire-extinguishing systems, rescue-service stations, power supply, pipe bridges)	1.3	1.4	1.4
Emergency power supply[a], safety systems[a]	1.4	1.5	1.6

[a]Special systems necessary to transfer operational processes to a safe state

moment-resisting, are often chosen as basic structural elements (Fig. 5.1). Thereby, diagonally braced frames generally exhibit higher horizontal rigidity and require less material. Moment resisting frames, on the other hand, are laterally more flexible, but offer a wider scope for possible structural modifications, should such become necessary in the context of updated production techniques.

The standard structural recommendations and regularity criteria for the seismic design of buildings usually cannot be adopted for industrial units due to the requirements of the production process. As an example, massive storage vessels are sometimes installed in higher levels, often eccentrically, so that the locations of the mass and stiffness centres of the different structural levels vary strongly. Disregarding regularity criteria in plan and elevation implies higher torsional loads and a stronger contribution of higher mode shapes (Chopra and Goel 1993; Sasaki et al. 1998). In choosing the mathematical model for the analysis, care must be taken so that such effects are duly considered. Having said that, if a 2D discretization is at all possible, according e.g. to the criteria of Eurocode 8 (2004), it should be given preference in respect to 3D models, because of the usually frequent structural modifications which should not require time and again the analysis of the entire 3D system in all its complexity.

In the discretized model of the primary structure it is usually sufficient to consider secondary structures as lumped masses in the corresponding stories or levels. If strong interaction effects between them and the primary load-carrying structure are expected or suspected, models of such non load-carrying secondary components and their attachments should be included in the overall model. Figure 5.2 shows schematically the exemplary discretization of a frame storey with secondary components.

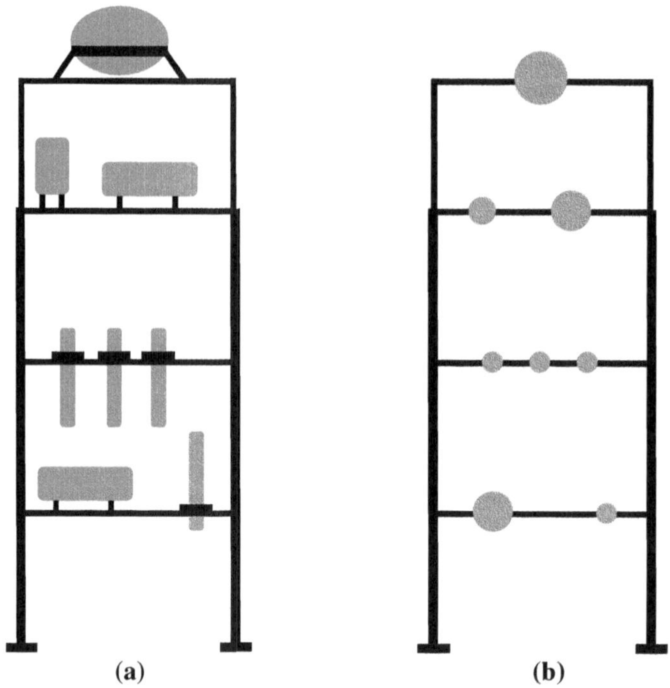

Fig. 5.2 (**a**) Frame storey with components and (**b**) System model

The correct computation of the sum of the relevant gravity loads is a central point in determining the design value A_{Ed} for the seismic actions. Here, gravity loads depend also on production parameters such as to what extent vessels and pipes are usually full or not. We obtain:

$$\sum G_{kj} \oplus \sum \psi_{Ei} \cdot Q_{ki} \quad (5.1)$$

with:

- $\oplus$ "to be combined with",
- Σ "the combined action of",
- G_{kj} characteristic value of the quasi-permanent action j,
- Q_{ki} characteristic value of the variable action i,
- Ψ_{Ei} combination coefficient for the variable action i.

Ψ_{Ei} is given by

$$\psi_{Ei} = \varphi \cdot \psi_{2,i} \quad (5.2)$$

It quantifies the probability that variable loads will not all be present at their maximum values when an earthquake occurs. For industrial units φ might be set equal to 1.0 and the combination coefficients $\psi_{2,i}$ may be obtained from Table 5.5 (from the VCI Guideline) instead of Table A.1.1 in Eurocode 0 (2002). Generally, the most unfavourable combination of gravity loads for the seismic case must be determined for the unit in question taking into account the specific production situation. This calls for a close interdisciplinary cooperation between civil and production engineers.

Standard analysis methods for industrial units are the simplified and the multi-modal response spectrum method, as in the case of residential buildings. Alternatively to these linear methods, nonlinear static methods may also be employed. Such displacement based methods (Freeman 2004) allow a much better quantitative use of the available nonlinear structural capacity than linear methods using behaviour factors. However, in using static nonlinear methods it must be ensured that the necessary inelastic deformation capacity is indeed available. This can be done by using capacity design rules, where specific regions for plastic hinge formation with adequate plastic rotational capacities are chosen. Non-dissipative regions are designed to remain elastic even after the formation of plastic hinges with their possible overstrengths.

Table 5.5 Combination coefficients $\psi_{2,i}$ after Eurocode 0, Table A.1.1 (2002) and VCI-Guideline (2012)

Action	Combination coefficient ψ_2
Live loads	
Storage areas	0.8
Operational areas	0.15
Office spaces	0.3
Vertical crane and hauled loads	0.8
Variable machine loads, vehicle loads	0.5
Brake loads, starting loads (caused by vehicles or cranes etc.)	0
Loads due to assemblage or other short time loads	0
Operational loads	
Variable operational loads	0.6[a]
Operating pressure	1.0
Operating temperature	1.0
Wind loads	0
External temperature impact (temporary)	0
Snow loads	0.5
Likely differential settlement of the foundation soil	1.0

[a]Constant operational loads are to be considered as permanent load G_k

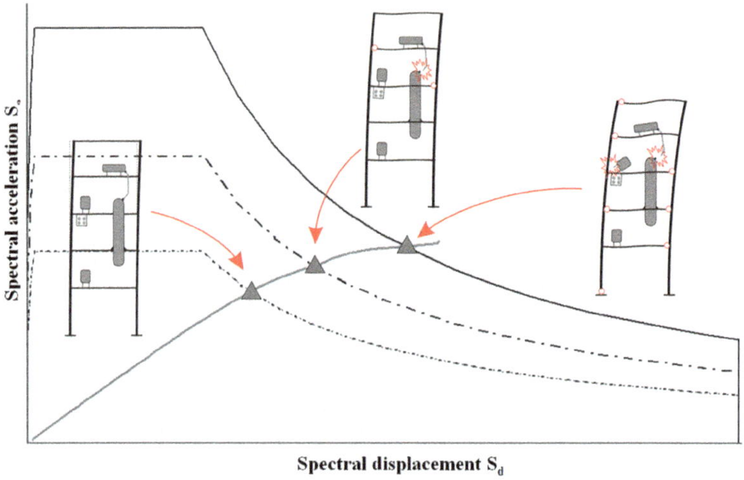

Fig. 5.3 Performance based design of industrial units (Holtschoppen 2009a)

A detailed description of the capacity design method for steel structures can be found in Chap. 4.

Once adequate rotational capacities and inelastic deformation supply have been ensured, the viability of the design can be checked by determining the point where the nonlinear pushover curve of the structure intersects the response spectrum in the spectral acceleration versus spectral displacement diagram. Annex B in DIN EN 1998-1 (2010) describes a simplified procedure for carrying out this task. Alternatively, more accurate methods can be employed that have been successfully in use for a number of years (Sasaki et al. 1998; Freeman 2004). A detailed description of these methods and corresponding application examples can be found in Chap. 4.

Nonlinear static methods offer the advantage of being able to implement a Performance Based Design. By combining the capacity curve with response spectra corresponding to different return periods it is possible to investigate different performance levels with a single analysis (Fig. 5.3). Thus, the engineer can check not only the load-carrying capacity of the structure but also, in agreement with the unit owner, its conformity with various Performance Levels regarding damage limitation and functionality stages (Bachmann 2004). This allows for an individually tailored design of the industrial unit in question, guaranteed to minimize damage and costs due to loss of functionality for seismic events of different exceedance probabilities or return periods. Nonlinear time history analyses, on the other hand, should be restricted to special cases because of the high numerical effort involved in addition to difficulties and ambiguities in checking the validity of the results. Such special cases are e.g. existing industrial units when it is imperative to mobilize all structural capacity reserves.

5.4 Secondary Structures

Past earthquakes have shown that damage to or due to secondary structures failures are often much more severe than those only involving the primary load-carrying structure (Villaverde 1997). This refers not only to irreparably damaged components but also to expensive interruptions of the production process. Industrial units with enclosed supporting structures pose a particular problem as in such structures housed components of process engineering have usually not been designed for lateral (wind) loads and, thus, their supports may not be able to safely accommodate lateral seismic loads. In the following, simplified analysis concepts and methods are presented for the seismic design of secondary structures.

5.4.1 Design Concepts

Seismic codes often present simplified expressions for determining equivalent static loads. For determining such loads for secondary components at a certain height, the ground acceleration is multiplied by a factor taking into account this height and an additional factor accounting for resonance effects between the component and the primary structure. Furthermore, the importance of the components and their energy dissipating capacity (including their attachments to the primary structure) are considered by a behaviour factor. In the following, the code provisions of DIN 4149 (2005), DIN EN 1998-1 (2010) (2004), SIA (SIA 261 2003) and FEMA 450 (2003) are presented and discussed.

DIN 4149 (2005)

$$
\begin{aligned}
F_a &= S_a \cdot m_a \cdot \gamma_a / q_a \\
S_a &= a_{gR} \cdot \gamma_I \cdot S \cdot \left[\frac{3\left(1 + \frac{z}{H}\right)}{1 + \left(1 - \frac{T_a}{T_1}\right)^2} - 0.5 \right] a_{gR} \cdot S \\
&= a_{gR} \cdot \gamma_I \cdot S \cdot [(A_a + 0.5) \cdot A_h - 0.5]\, a_{gR} \cdot S \\
A_a &= \frac{3}{1 + \left(1 - \frac{T_a}{T_1}\right)^2} - 0.5; \quad A_h = 1 + \frac{z}{H}
\end{aligned}
\tag{5.3}
$$

DIN EN 1998-1 (2010)

$$
\begin{aligned}
F_a &= S_a \cdot m_a \cdot \gamma_a / q_a \\
S_a &= a_{gR} \cdot \gamma_I \cdot S \cdot \left[\frac{3\left(1 + \frac{z}{H}\right)}{1 + \left(1 - \frac{T_a}{T_1}\right)^2} - 0.5 \right] \geq a_{gR} \cdot S \\
&= a_{gR} \cdot \gamma_I \cdot S \cdot [(A_a + 0.5) \cdot A_h - 0.5] \geq a_{gR} \cdot S \\
A_a &= \frac{3}{1 + \left(1 - \frac{T_a}{T_1}\right)^2} - 0.5; \quad A_h = 1 + \frac{z}{H}
\end{aligned}
\tag{5.4}
$$

SIA 261 (2003)*

$$F_a = S_a \cdot m_a \cdot \gamma_a / q_a$$

$$S_a = a_{gR} \cdot \gamma_I \cdot S \cdot \left[\frac{2\left(1 + \frac{z}{H}\right)}{1 + \left(1 - \frac{T_a}{T_1}\right)^2} \right] = a_{gR} \cdot \gamma_I \cdot S \cdot A_a \cdot A_h$$

$$A_a = \frac{2}{1 + \left(1 - \frac{T_a}{T_1}\right)^2}; \quad A_h = 1 + \frac{z}{H} \tag{5.5}$$

*This version of the formula to determine the equivalent static force for secondary structures has been replaced in SIA 261 (2014) by the formula of DIN EN 1998-1 (2010). It is kept in this summary, however, to compare and explain the different dynamic amplification factors (Fig. 5.5).

FEMA 450 (2003)

$$F_a = S_a \cdot m_a \cdot \frac{\gamma_a}{q_a}$$

$$S_a = 0.4 \cdot S_{a,max} \cdot A_a \cdot A_h \tag{5.6}$$

A_a as shown in Fig. 5.4; $A_h = 1 + 2\frac{z}{H}$

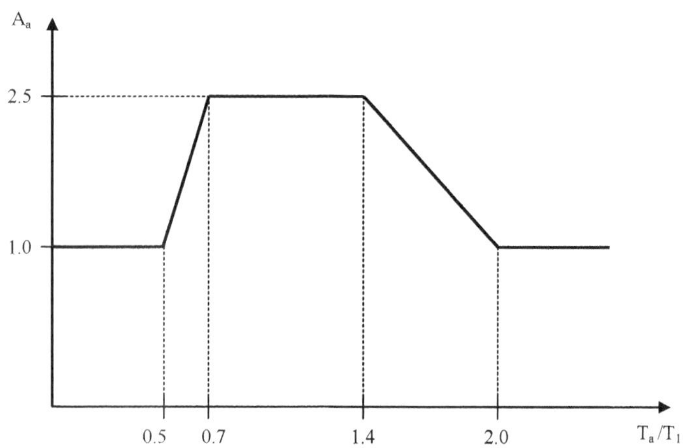

Fig. 5.4 Dynamic amplification factor A_a after FEMA 450 (2003)

FEMA 450 (2003) ***alternatively***

$$F_a = a_i \cdot m_a \cdot \gamma_a \cdot \frac{A_a \cdot A_{Ti}}{q_a}$$
$$0.3 \cdot S_{a,max} \cdot \gamma_a \cdot m_a \leq F_a \leq 1.6 \cdot S_{a,max} \cdot \gamma_a \cdot m_a \tag{5.7}$$

A_a as shown in Fig. 5.4.

The variables in (5.3) through (5.7) are:

F_a	Horizontal seismic force through the mass centre of the component (kN)
m_a	Mass of the component (t)
S_a	Spectral acceleration (m/s^2)
γ_a	Importance factor of the component (–)
q_a	Behaviour factor (–)
a_{gR}	Reference peak ground acceleration on type A ground (rock) (m/s^2)
γ_I	Importance factor for the primary structure (–)
S	Soil parameter (–)
z	Height of the component relative to the foundation or to the deck of a rigid basement (m)
H	Height of the primary structure relative to the foundation or to the deck of a rigid basement (m)
T_a	Fundamental natural period of the component (s)
T_1	Fundamental natural period of the primary structure (s)
A_a	Dynamic amplification factor (–)
A_h	Height factor (–)
A_{Ti}	Factor for considering torsional effects at structure level i (–)
a_i	Resulting acceleration at level i from the response spectrum analysis of the primary structure (m/s^2)
$S_{a,max}$	Plateau value of the elastic ground acceleration response spectrum, including the importance factor γ_I(m/s^2).

It is obvious that the expressions in DIN 4149 (2005), DIN EN 1998-1 (2010) and the first noted option of FEMA 450 (2003) are very similar. They all assume a linear increase of the floor accelerations with height, corresponding to a triangular fundamental mode shape. However, due to the typically irregular mass distributions in plan and elevation in industrial units, higher modes become important, leading to a nonlinear distribution of floor accelerations along the height. The alternative expression of FEMA 450 (2003) can be employed to take this effect into account with sufficient accuracy without having to determine floor spectra or carry out time history analyses. The concept is based on determining floor accelerations a_i by the multimodal response spectrum method. If 3D models are used for this analysis, the resulting floor acceleration a_i automatically includes the effects due to irregular mass distribution as well as irregularities in plan and elevation, and the torsional factor A_{Ti} can be set to 1.0. If 2D models are used, the torsional effects for floor i must be taken care of by the corresponding factor A_{Ti}. A conservative assumption according

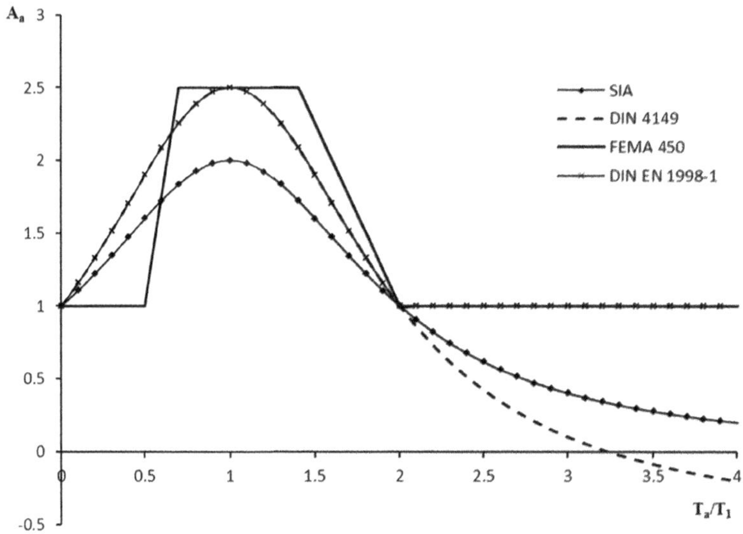

Fig. 5.5 Amplification factor A_a in different codes

to FEMA 450 (2003) is $A_{Ti} = 3$. Better and more problem-specific values can be computed based on DIN 4149 (2005), Sect. 6.2.3.3:

$$A_{Ti} = (1 \pm e \cdot r_j / r^2) \tag{5.8}$$

Herein, r^2 is the square of the torsional radius, which is defined as the ratio between torsional and lateral stiffness in the direction in question. The distance of stiffening element j to the stiffness centre is denoted by r_j; e is the assumed eccentricity which includes the actual part, the accidental part and also the additional part for taking coupled torsional and lateral natural modes into account. The single parts are explained more deeply in Chap. 4.

Resonance effects between the primary and secondary structures are considered in the codes through the ratios of the natural periods of the single components T_a to the fundamental period of the former, T_1. Figure 5.5 shows the corresponding functions for the amplification factors A_a according to different codes. It must be noted that the (clearly nonsensical) negative values in DIN 4149 (2005) for large period ratios have been rendered harmless in DIN EN 1998-1 (2010) by prescribing a minimum design force, which should also be used with DIN 4149 (2005). The A_a functions in the different codes are basically similar, with only the range around $T_a/T_1 = 1.0$ showing larger deviations.

Amplification factors were given as functions of the ratio of the fundamental natural period of the component to the fundamental period of the structure, so they can be determined only after the former is known. This can happen through free vibration tests considering the actual boundary conditions or analytically using all available

Table 5.6 Parameter A_a and q_a for components in industrial and technical units (VCI 2012)

Plantspecific components	A_a	q_a
General mechanical components		
Pressure vessels	1.5	1.5
Ovens and boilers	1.0	2.0
Slender components (e.g. chimneys)	2.5	2.0
Conveyor systems	2.5	2.0
Vibration isolated components	1.0	2.5
Piping systems		
highly deformable (e.g. pipes with expansion joints)	1.5	2.5
limited deformable	1.5	1.5
non deformable (e.g. brittle material)	1.5	1.0
Truss systems	1.5	2.5
Structural components		
Non load bearing masonry walls	1.0	1.5
Non load-bearing walls made of other materials	1.0	2.0
Atticas and parapets	2.5	2.5
Facade elements and wall claddings		
high deformability (elements and fastenings)	1.0	2.5
small deformability (elements and fastenings)	1.0	1.5
Suspended ceiling elements	1.0	2.5

Remarks
For rigid components (Frequency > 16 Hz), A_a can be assumed with a factor of 1.0. For flexible components A_a can be applied with a factor of 2.5.
For components with low plastic deformation capacity, a value of 1.0 shall be assumed for q_a. The assumption of reasonable intermediate values is allowed.

information. Since fundamental periods for components are normally not readily available, T_a/T_1 is often set equal to 1 (Resonance). As an alternative, FEMA 450 (2003) and the VCI Manual present empirical values for amplification factors A_a and behaviour factors q_a for various standard components (Table 5.6). It has been shown (Holtschoppen et al. 2008; Holtschoppen 2009a; Singh 1998) that especially for flexible moment-resisting frames and irregular mass distribution resonance effects with higher modes can occur, which must be taken into account in determining A_a. As an example, Fig. 5.6 shows the amplification effects for a 5-storey frame (Holtschoppen 2009a, b). The floor spectra show that resonance effects with the second natural frequency ($f_a \approx 3.3$ Hz) must be considered.

Resonance effects and the problem of inaccuracies of component fundamental periods can be addressed by modifying the expression for A_a in FEMA 450 (2003) in the range $T_a/T_1 < 0.7$ (Fig. 5.7). A_a is set equal to unity for components which

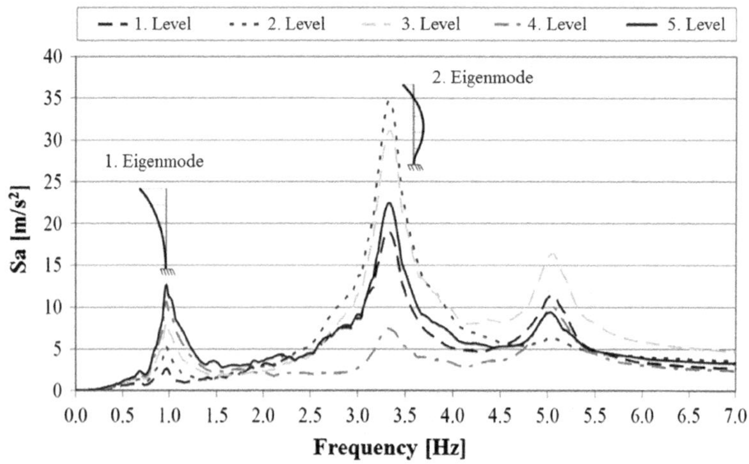

Fig. 5.6 Floor spectra: 5 storey frame with components (Holtschoppen 2009a, b)

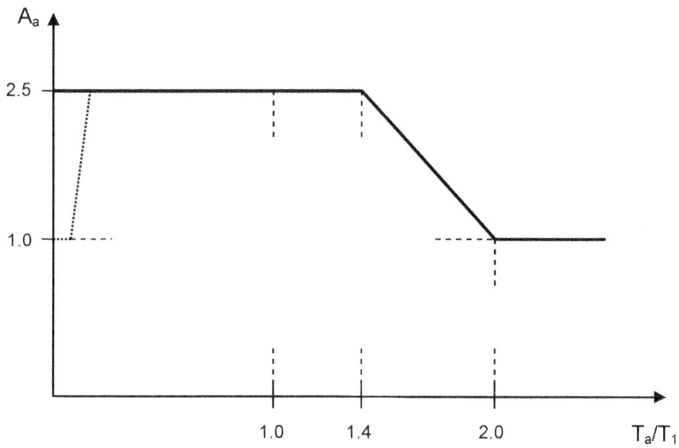

Fig. 5.7 Amplification factor A_a taking into account higher modes

are much stiffer than the primary structure, independently of the period ratio. In US codes, $T_a = 0.06$ s or $f_a = 16$ Hz is used for defining such stiff components. The A_a range for $T_a/T_1 > 1.0$ has been adopted from FEMA 450 2003), although for large ratios $T_a/T_1 > 2$ the resulting design forces are a bit too conservative. The curves in Fig. 5.7 take resonance effects due to higher modes into account.

If the period ratios are known, A_a can be determined for all relevant modes using the original curves in the codes mentioned above. For each mode i the factor A_{ai} must be determined and the lateral seismic force F_a results from the maximum amplification factor of all relevant modes.

Since in the design stage there is scant information available concerning e.g. the location of the component or the dynamic properties of the primary structure, obtaining a conservative value for the lateral seismic force F_a is of high practical importance. An upper limit is given in FEMA 450 (2003), as given here with the variables used in Eurocode 8 (2004), DIN 4149 (2005) and DIN EN 1998-1 (2010):

$$F_{a,max} = 1.6 \cdot S_{a,max} \cdot \gamma_a \cdot m_a = 1.6 \cdot \left(2.5 \cdot S \cdot a_{gR} \cdot \gamma_I\right) \cdot \gamma_a \cdot m_a$$
$$\Rightarrow F_{a,max} = 4 \cdot S \cdot a_{gR} \cdot \gamma_I \cdot \gamma_a \cdot m_a \qquad (5.9)$$

Empirical investigations have confirmed the validity of this upper limit (Holtschoppen et al. 2008) which can also be found in a series of codes (Eurocode 8 2004; FEMA 450 2003; UBC 1997; IBC 2015). It has also been adopted in the German NAD for DIN EN 1998-1 (2010) that is DIN EN 1998-1/NA (2011).

5.4.2 Example: Container in a 5-Storey Unit

The design force for the container shown in is to be determined. Its fundamental natural frequency is 5.0 Hz and its centre of mass lies 1.0 m above the floor. A 5 m wide strip of the unit is considered, and all components are modelled as lumped masses at girder height, while permanent and variable loads are assumed to be concentrated at the roof levels. Both girders and columns are made of S235 steel. The configuration of the steel profiles and the layout of the components are shown in Fig. 5.8.

Loads	
Floor self-weight	8.6 (kN/m^2)
Floor variable load	15 (kN/m^2)
Self-weight frame	197.5 (kN)
Response spectrum	DIN EN 1998-1/NA, Zone 3, 5% damping Ground type C-R
Importance factors	Structure: $\gamma_I = 1.4$; Components $\gamma_a = 1.6$
Behaviour factors	Structure: q = 1.0; Components: $q_a = 1.0$
Mass and stiffness distribution	Irregular

The elastic response spectrum (q = 1) of DIN EN 1998-1/NA (2010) is shown in Fig. 5.9. Floor loads consisting of self-weight and 30% of the variable loads are assumed as lumped floor masses:

$$M_{EG+V} = \left(5\text{m} \cdot 9\text{m} \cdot \left(8.6\ \text{kN/m}^2 + 0.3 \cdot 15\ \text{kN/m}^2\right)/9.81\ \text{m/s}^2\right)/2 = 30\,\text{t}$$

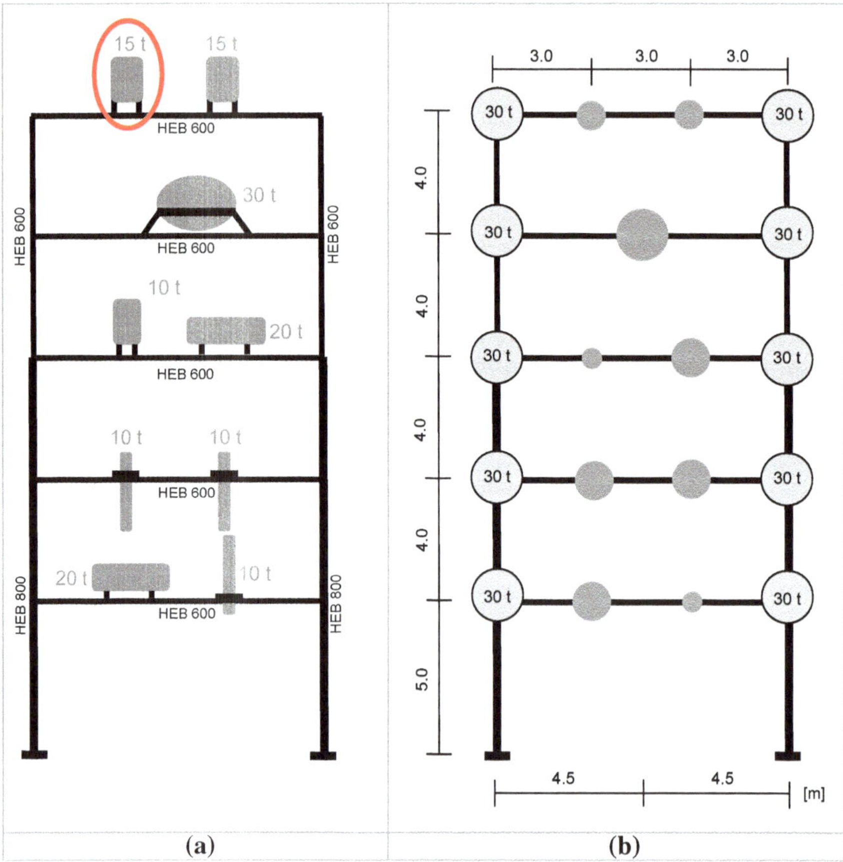

Fig. 5.8 5-storey unit: (**a**) Frame with components. (**b**) Model used

Modal analysis of the frame, floor acceleration

Figure 5.10 shows the computed first three vibration modes of the frame.

For applying the multimodal approach of FEMA 450 (2003) the floor acceleration at roof level must be determined. This is carried out here by two different methods:

- From the equivalent static loads using multimodal response spectrum analysis
- From the first three modal shapes and the pertinent participation factors.

The resulting equivalent static load is determined using the SRSS rule for the first three modal contributions, with Fig. 5.11 showing the single values. Using the values at roof level and dividing through the pertinent mass yields for the roof acceleration a_5:

$$a_5 = \frac{a_{SRSS}}{(M_{EG+V} + M_{EG,Rahmen})} = \frac{\sqrt{42.36^2 + 48.12^2 + 25.10^2}}{30 + 0.31} = 2.27\ \text{m/s}^2$$

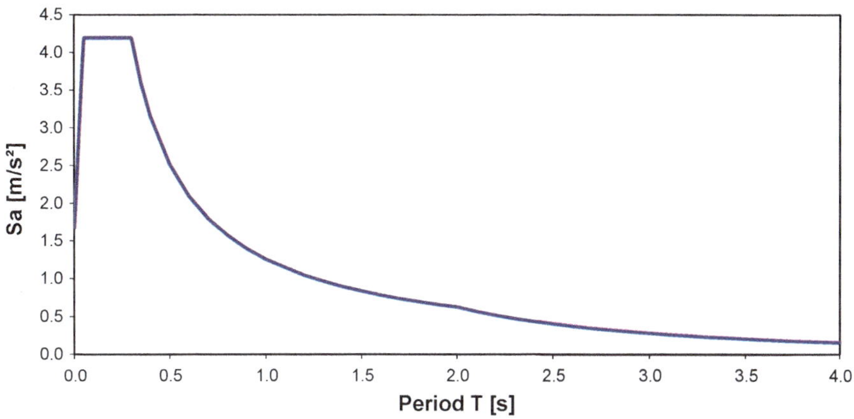

Fig. 5.9 Design response spectrum, DIN EN 1998-1/NA, Zone 3, ground type C-R, $\gamma_I = 1.4$

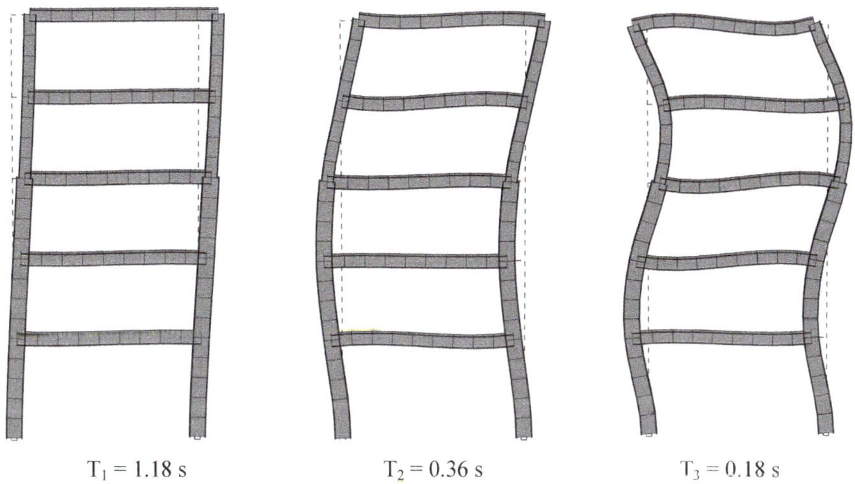

Fig. 5.10 First, second and third natural mode of the 5-storey frame

The floor accelerations may also be computed using the i-th mode shapes $\underline{\Phi}_i$, participation factors β_i and spectral accelerations $S_{a,i}$ as:

$$\underline{a}_i = S_{a,i} \cdot \beta_i \cdot \underline{\Phi}_i$$

Using the SRSS rule yields:

$$\underline{a}_{res} = \sqrt{(\underline{a}_1)^2 + (\underline{a}_2)^2 + (\underline{a}_3)^2}$$

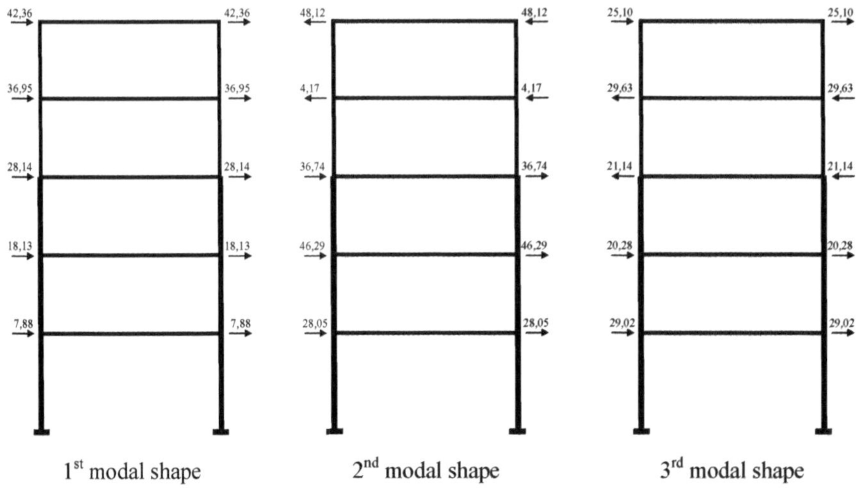

Fig. 5.11 Static equivalent loads at floor level

For the calculation of the acceleration values in each modal contribution the participation factors β_i must be determined:

$$\beta_i = \underline{\Phi}_i^T \cdot \underline{M} \cdot \underline{r},$$

$\underline{M}$ is the diagonal mass matrix and $\underline{r}$ the vector containing the displacements in the single master degrees of freedom due to a unit base displacement in the direction of the seismic excitation. If not output by the software used, they may be determined from the effective modal masses, which, in this case are given by:

$$M_1 = 391.88\,\mathrm{t}$$
$$M_2 = 56.78\,\mathrm{t}$$
$$M_3 = 18.91\,\mathrm{t}$$

The sum of these three effective modal masses equals 467.57 t, corresponding to 97% of the total effective mass. If the modal shapes are normalized to unit modal mass ($\underline{\Phi}^T \cdot \underline{M} \cdot \underline{\Phi} = 1$), the participation factors can be extracted directly from the effective modal masses M_i:

$$\beta_1 = \sqrt{391.88} = 19.795$$
$$\beta_2 = \sqrt{56.78} = 7.535$$
$$\beta_3 = \sqrt{18.91} = 4.349$$

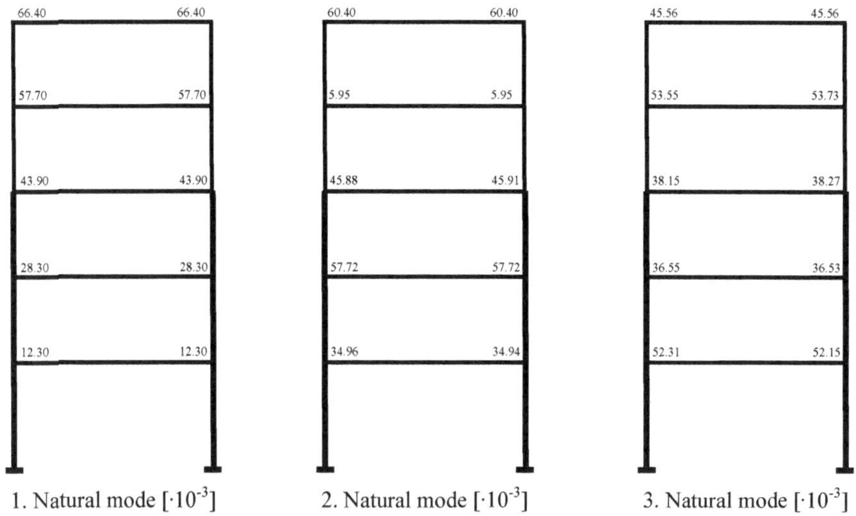

Fig. 5.12 Modal shape ordinates (normalized to unit mass), absolute values

Spectral accelerations are obtained from the response spectrum:

$$S_{a,1} = 1.066$$
$$S_{a,2} = 3.505$$
$$S_{a,3} = 4.200$$

The ordinates of the modal shapes (here normalized to unit modal mass) at roof level are shown in Fig. 5.12:

$$\Phi_{5,1} = 0.0664$$
$$\Phi_{5,2} = 0.0604$$
$$\Phi_{5,3} = 0.0456$$

The resulting floor acceleration at roof level is finally given by:

$$\begin{aligned} \boldsymbol{a}_5 &= \sqrt{(\boldsymbol{S}_{a,1} \cdot \beta_1 \cdot \Phi_{5,1})^2 + (\boldsymbol{S}_{a,2} \cdot \beta_2 \cdot \Phi_{5,2})^2 + (\boldsymbol{S}_{a,3} \cdot \beta_3 \cdot \Phi_{5,3})^2} \\ &= \sqrt{(1.066 \cdot 19.795 \cdot 0.0664)^2 + (3.505 \cdot 7.535 \cdot 0.0604)^2 + (4.200 \cdot 4.349 \cdot 0.0456)^2} \\ &= 2.28\ \text{m/s}^2 \end{aligned}$$

Design forces for the vessel at roof-level (storey 5)

In the following, the design forces for the vessel considered are computed according to different codes. The following values hold:

Peak ground acceleration

$$a_{gR} = 0.8\,\text{m/s}^2$$

Frame with loads	Vessel
$z = H = 21.0$ m	$q_a = 1.0$
$T_1 = 1.18$ s	$\gamma_a = 1.6$
$\gamma_I = 1.4$	$m_a = 15$ t
$S = 1.5$	$T_a = 0.2$ s
$a_5 = 2.28$ m/s^2	

DIN 4149 (2005)/***DIN EN 1998-1*** (2010)

$$F_a = S_a \cdot m_a \cdot \gamma_a / q_a = 5.46 \cdot 15 \cdot 1.6/1.0 = 131.0\ \text{kN}$$

$$S_a = a_{gR} \cdot \gamma_I \cdot S \cdot (A_a + 0.5) \cdot A_h - 0.5 = 0.8 \cdot 1.4 \cdot 1.5 \cdot (1.275 + 0.5) \cdot 2.0 - 0.5 = 5.46\ \text{m/s}^2$$

$$A_a = \frac{3}{1 + \left(1 - \frac{T_a}{T_1}\right)^2} - 0.5 = 1.275; \quad A_h = 1 + \frac{z}{H} = 2.0$$

SIA 261 (2003)

$$F_a = S_a \cdot m_a \cdot \gamma_a / q_a = 3.98 \cdot 15 \cdot 1.6/1.0 = 95.52\ \text{kN}$$

$$S_a = a_{gR} \cdot \gamma_I \cdot S \cdot A_a \cdot A_h = 0.8 \cdot 1.4 \cdot 1.5 \cdot 1.184 \cdot 2.0 = 3.98\ \text{m/s}^2$$

$$A_a = \frac{2}{1 + \left(1 - \frac{T_a}{T_1}\right)^2} = 1.184; \quad A_h = 1 + \frac{z}{H} = 2.0$$

FEMA 450, alternative approach (2003)

$$\boldsymbol{F_a} = \boldsymbol{a_5} \cdot \boldsymbol{m_a} \cdot \frac{\boldsymbol{A_{T5}} \cdot \boldsymbol{A_a} \cdot \gamma_a}{\boldsymbol{q_a}} = 2.28 \cdot 15 \cdot \frac{1.0 \cdot 1.0 \cdot 1.6}{1.0} = 54.72\ \text{kN}$$
$$\geq 0,3 \cdot \boldsymbol{S_{a,max}} \cdot \gamma_a \cdot \boldsymbol{m_a} = 0.3 \cdot 4.2 \cdot 1.6 \cdot 15 = 30.24\ \text{kN}$$
$$\leq 1,6 \cdot \boldsymbol{S_{a,max}} \cdot \gamma_a \cdot \boldsymbol{m_a} = 1.6 \cdot 4.2 \cdot 1.6 \cdot 15 = 161.28\ \text{kN}$$

$A_a = 1.0$ as shown in Fig. 5.4

$$S_{a,max} = 2.5 \cdot a_{gR} \cdot \gamma_I \cdot S = 2.5 \cdot 0.8 \cdot 1.4 \cdot 1.5 = 4.2\ \text{m/s}^2$$

FEMA 450, alternative approach, A_a-factor modified (2009)

$$\boldsymbol{F_a} = \boldsymbol{a_5} \cdot \boldsymbol{m_a} \cdot \frac{\boldsymbol{A_{T5}} \cdot \boldsymbol{A_a} \cdot \gamma_a}{\boldsymbol{q_a}} = 2.28 \cdot 15 \cdot \frac{1.0 \cdot 2.5 \cdot 1.6}{1.0} = 136.80\ \text{kN}$$
$$\geq 0,3 \cdot \boldsymbol{S_{a,max}} \cdot \gamma_a \cdot \boldsymbol{m_a} = 0.3 \cdot 4.2 \cdot 1.6 \cdot 15 = 30.24\ \text{kN}$$
$$\leq 1,6 \cdot \boldsymbol{S_{a,max}} \cdot \gamma_a \cdot \boldsymbol{m_a} = 1.6 \cdot 4.2 \cdot 1.6 \cdot 15 = 161.28\ \text{kN}$$

$A_a = 2.5$ according to Fig. 5.7

$$S_{a,max} = 2.5 \cdot a_{gR} \cdot \gamma_I \cdot S = 2.5 \cdot 0.8 \cdot 1.4 \cdot 1.5 = 4.2\ \text{m/s}^2$$

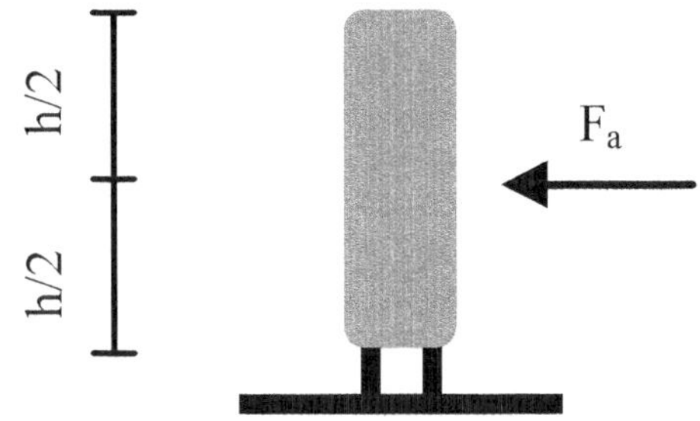

Fig. 5.13 Model of the vessel and the equivalent static load

The results vary over a wide range. In order to gain a deeper insight, a modal analysis is carried out, in which the vessel is modelled as an eccentric mass (with its centre of mass 1 m above the girder), attached to it through a beam element HEB 160. Assuming a fixed connection, the resulting natural frequency of this substructure is 5.15 Hz, a good approximation of the vessel fundamental natural frequency of 5 Hz. The modal analysis result for the lateral equivalent static force is 89.6 kN, using the first 5 modal contributions. Here, the influence of higher modes must be taken into account since their natural frequencies are near the fundamental frequency of the vessel. This explains why the force given by FEMA 450 (2003) is non-conservative, while results after DIN 4149 (2005), DIN EN 1998-1 (2010) and the modified FEMA 450 (2009) approach are on the safe side. The SIA 261 (2003) value (95.52 kN) is also on the safe side.

The computed equivalent static load is applied to the vessel as shown in Fig. 5.13 for dimensioning its attachment to the frame. For the design of the vessel itself, this load can be distributed along its height according to the mass and stiffness distribution. Hydrodynamic effects due to liquid vessel contents may be neglected in small vessels.

5.5 Silos

Silos generally function as buffers between supply and demand for various goods and their structural safety has long been of interest to the civil engineering profession (Martens 1998). This is especially true for dynamically loaded silos, e.g. under seismic loads. Typical damage types are local buckling phenomena of the silo shell induced by concentrated loads (Fig. 5.14a) and at the silo bottom (Fig. 5.14b). Such failure modes are due to critical combinations of vertical compressive stresses, hoop tensile stresses and high shear stresses due to seismic loading and are commonly referred to as "elephant foot buckling". Silos on columns are prone to tilting, usually due to faulty anchoring or foundation failure (Fig. 5.15). Another reason is not considering seismic loads in dimensioning the columns.

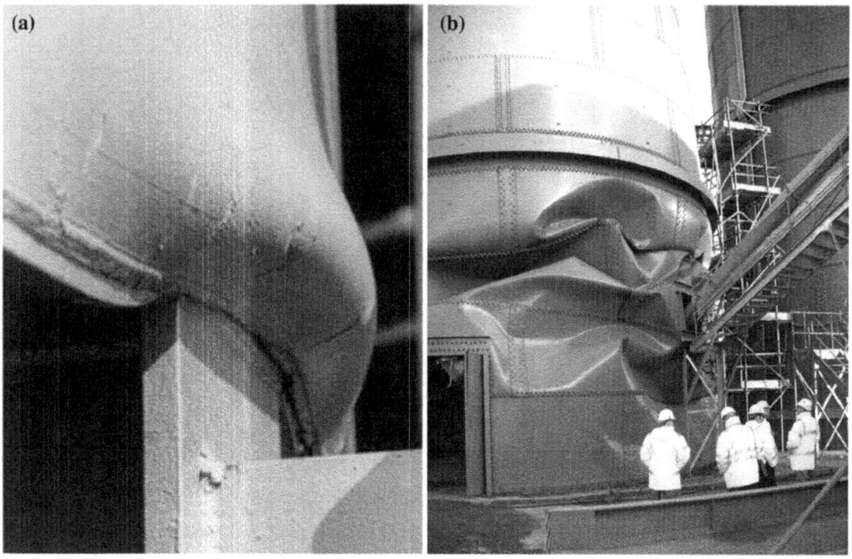

Fig. 5.14 (**a**) Buckling adjacent to the support (Guggenberger 1998) and (**b**) Buckling of a skirt support (Schmidt 2004; photos by J. M. Rotter, Edinburgh)

Fig. 5.15 Silo collapse due to column failure (Bruneau 1999)

Seismically excited silo structures have long been the subject of intensive research, which aimed at understanding the interaction between filling and silo wall and to deriving applicable calculation and design concepts. Rotter and Hall (1989) investigated the problem of compact cylindrical silos, identified the main failure modes and derived design criteria for steel silos based on a numerical model. Yokota et al.

(1983), Shimamato et al. (1984) and Sakai et al. (1984) carried out vibration tests on cylindrical model silos filled with coal and obtained basic insights about the vibration behaviour of silos. Younan and Veletsos (1998) developed an analytical formulation for describing the seismic response of material-filled silos with rigid walls for constant accelerations, harmonic excitations and stochastic seismic effects, which were also extended to flexible tank shells. Bauer (1992) and Braun (1997) dealt with the material behaviour of bulk materials and their behaviour under dynamic loads. The current version of the Eurocode 8-4 (2006) is essentially based on the formulation of Younan and Veletsos (1998) and the work of Rotter and Hall (1989). The complex interaction between filling material and silo shell, which is dependent on the pressure conditions as well as ground and wall friction coefficients, is not explicitly taken into account in the standard. Holler and Meskouris (2006) showed that the loading approaches are too conservative in the case of squat silos, whereas loads on slender silos are well-represented. Recent shaking table tests (Silvestri et al. 2016) have shown that the approaches in Eurocode 8-4 (2006) are too conservative and the dynamic response is strongly dependent on the wall friction coefficient. Better agreement with the experimental results is obtained with the analytical approach according to Silvestri et al. (2012). The same results are obtained by Pieraccini et al. (2015) with an improved approach based on the theory of Silvestri et al. (2012).

According to Eurocode 8-4 (2006) the design of dry material silos may be carried out in two different ways. In the first, the seismic loading is modelled through statically equivalent loads acting on the shell. Alternatively, a time history analysis might be carried out, in which nonlinear phenomena due to the contents as well as to the interaction between the shell and the silo contents might be taken into account.

5.5.1 Equivalent Static Force Approach After Eurocode 8-4 (2006)

The determination of internal forces due to seismic loads is carried out using the code design response spectrum and the equivalent static force approach. In the following, the procedure after Eurocode 8-4 (2006) is described in more detail.

Horizontal seismic loads for cylindrical silos

The silo walls are subject to a horizontal radial compressive pressure equal to the product of fill mass and seismic acceleration. Additionally, an annular pressure (Fig. 5.16) is developed, which for cylindrical silos is given by

$$\Delta_{ph,s} = \Delta_{ph,so} \cdot \cos\theta \tag{5.10}$$

Fig. 5.16 Equivalent static load for cylindrical silos after Eurocode 8-4 (2006)

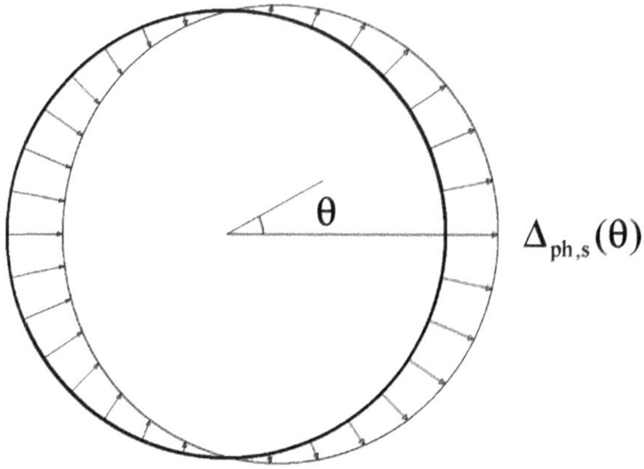

Herein, $\Delta_{ph,so}$ is the reference pressure acting normal to the silo wall at a distance x from the flat bottom or from the tip of a conical or pyramidal funnel:

$$Silo\ wall: \Delta_{ph,so} = \alpha(z) \cdot \gamma \cdot \min(r_s^*, 3x) \quad (5.11)$$

$$Silo\ funnel: \Delta_{ph,so} = \alpha(z) \cdot \gamma \cdot \min(r_s^*, 3x)/\cos\beta \quad (5.12)$$

Further variables:

$\alpha(z)$ Acceleration of the silo at a depth z from the equivalent surface of the fill mass (g)
γ Characteristic value for the specific weight of the fill mass
r_s^* $r_s^* = \min(h_b, d_c/2)$
h_b Total height of the silo from a flat bottom or a funnel tip to the equivalent surface of the fill mass
d_c Interior dimension of the silo parallel to the horizontal component of the seismic excitation (interior diameter d_c in cylindrical silos or silo compartments, horizontal interior dimension b in rectangular silos or silo compartments)
θ Angle relative to the direction of the seismic excitation ($0 \leq \theta \leq 360°$)
β Slope of the funnel wall relative to a vertical axis or maximum wall slope (relative to the vertical) for pyramidal funnels.

Total equivalent static forces are usually distributed along the wall according to a cosine function. Generally, combinations of these forces with those resulting from filling or emptying the container must be considered in order to ensure that no negative (pointing to the interior) forces arise and the contact between the bulk material and the silo wall is maintained at all times.

Some marked differences arise in the various codes concerning the distribution of equivalent static forces along the height. In Eurocode 1-4 (2006), a constant distribution was chosen, to be combined with the forces resulting from filling. The same approach was adopted in DIN 1055-6 (2005). In the latest version of Eurocode 8-4 (2006) a variable distribution described by a function α(z) was prescribed, for which the variation of the acceleration along the height must be determined beforehand. If the latter is not known, $\alpha(z)$ may be substituted by the value of the acceleration acting at the height of the mass centre.

New assumptions have also been introduced for the equivalent seismic loading in the upper and lower parts of the silos. In prEN 1998-4 (2004), zero loading is assumed near the bottom of the silo, up to a height determined by the silo geometry. In the upper part of the silo, a linear distribution with zero value at the silo top is assumed, in order to take the formation of the angle of repose into account (Holler 2006). In the latest version of Eurocode 8-4 (2006), this linear decrease is revoked; in the lower part of the silo, a linear function is assumed up to a certain height determined by the silo geometry. Table 5.7 summarizes the different loading distributions for equivalent seismic loads and filling loads, giving also as an example their combinations for a 10 m high silo with a diameter of 5 m.

The reason for this drastic revision of loading assumptions lies in the results obtained by recent research efforts (Braun 1997; Wagner 2009) which have been seamlessly integrated in the new code versions, e.g. in Eurocode 8-4 (2006). These updated loading assumptions take into account that for squat silos a large part of the lateral seismic force is taken up by internal friction in the bulk material and does not affect the silo wall. Loading assumptions for funnels are also a part of the latest code versions, which should be used exclusively.

Lateral seismic loads for rectangular silos

The pressure acting on the averted wall face perpendicular to the direction of the seismic excitation is given by:

$$\Delta_{ph,s} = \Delta_{ph,so} \tag{5.13}$$

The wall facing the lateral seismic component experiences a negative (directed inwards) pressure and is given by:

$$\Delta_{ph,s} = -\Delta_{ph,so} \tag{5.14}$$

No additional pressure is generated for walls parallel to the lateral seismic component:

$$\Delta_{ph,s} = 0 \tag{5.15}$$

The reference pressure value $\Delta_{ph,so}$ is evaluated as for cylindrical silos and is given in (5.11) and (5.12). Combinations of the seismic forces with those resulting from filling or emptying the silo must also be considered in order to ensure that no negative (pointing inwards) forces arise and the contact between the bulk material and the silo wall is maintained at all times.

Vertical seismic loads

Eurocode 8-4 (2006) stipulates that vertical seismic loads must be considered in addition to the lateral seismic loads but does not prescribe how to apply them to the

Table 5.7 Different assumptions for seismic loads in cylindrical silos

Static equivalent load	Filling loads	Superposition
Eurocode 1-4 (1995)		
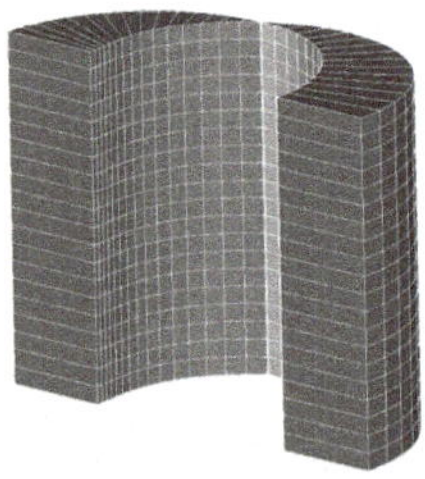	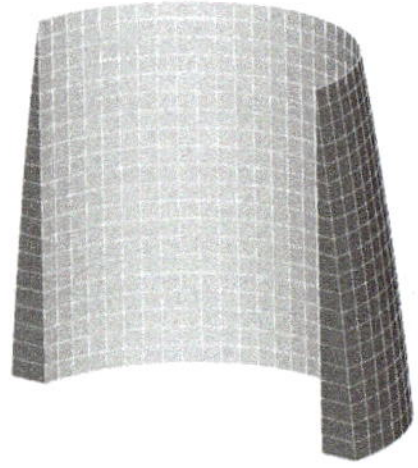	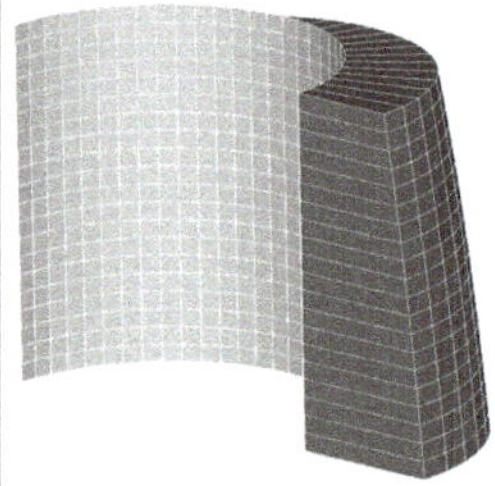
prEN 1998-4 (2004)		
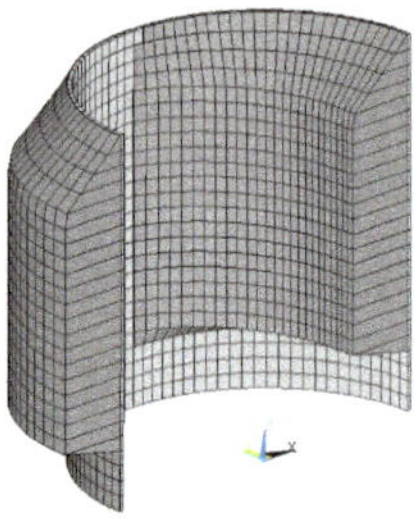	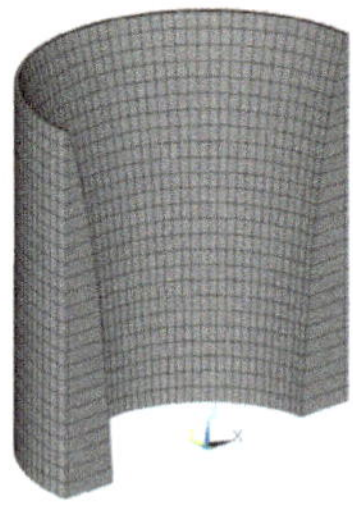	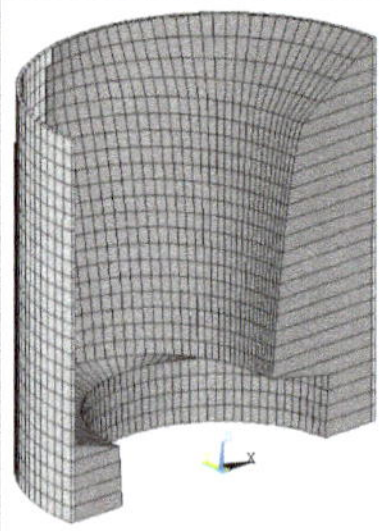
Eurocode 8-4 (2006) with constant acceleration		
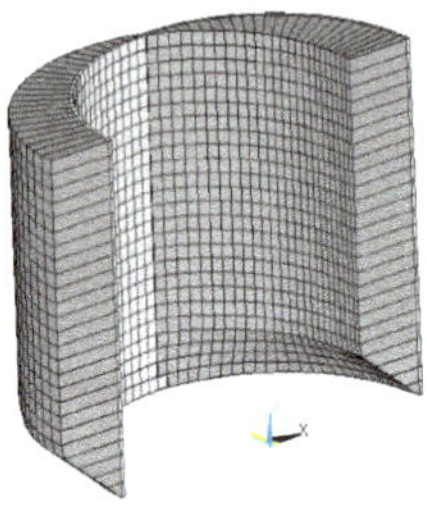		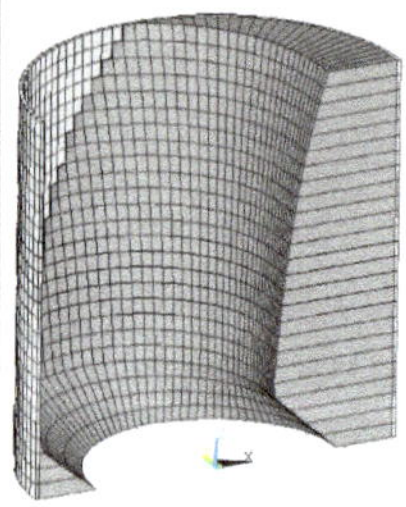
Eurocode 8-4 (2006) with variable acceleration distribution		
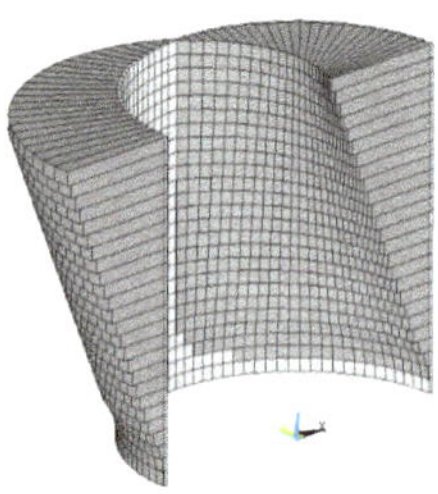		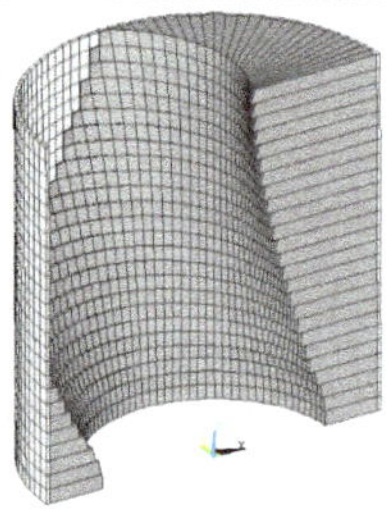

silo walls. A meaningful approach consists in deriving the additional dynamic forces directly from the static forces due to filling and emptying. To that effect a scaling factor C_d is determined as the spectral acceleration for the fundamental vertical natural mode, given in units of g. Measurements and numerical results have shown that the fundamental vertical natural mode lies in the low-period range of the response spectrum, so that the plateau spectral acceleration value $S_{av,\max}$ can be assumed to be on the safe side. We obtain:

$$C_d = \frac{S_{av,\max}}{g} \tag{5.16}$$

C_d is then used for computing additional friction forces p_w for walls and vertical forces p_v for silo floors and funnels after Sects. 5.5 and 5.6 respectively of Eurocode 1-4 (2006). As an example, the following seismically induced forces must be considered for a slender silo (height/diameter ≥ 2) with horizontal floor:

$$\begin{aligned} p_{wf,av}(z) &= C_d \cdot \mu \cdot p_{ho} \cdot Y_J(z) \\ p_{vf,av}(z) &= C_d \cdot p_{vf} = C_d \cdot p_{ho}/\, K \cdot Y_J(z) \end{aligned} \tag{5.17}$$

where:

$$p_{ho} = \gamma \cdot K \cdot z_o, \qquad z_o = \frac{A}{K \cdot \mu \cdot U}, \qquad Y_J(z) = 1 - e^{-z/z_o}$$

- γ Characteristic value for the specific weight of the fill mass
- μ Characteristic value of the wall friction coefficient
- K Characteristic value of the lateral force ratio
- z Depth of the fill mass below the equivalent fill mass surface
- A Inner section area of the silo
- U Circumference of the inner section area of the silo
- $Y_J(z)$ Depth variation function of the Janssen theory.

Due to the seismically induced vertical loads we obtain additional lateral forces for the silo wall:

$$p_{hf,av}(z) = C_d \cdot p_{ho} \cdot Y_J(z) \tag{5.18}$$

The additional silo wall forces can be computed in a similar way for squat silos and silos with more or less steep funnels. The corresponding expressions can be found in Eurocode 1-4 (2006) and are not reproduced here. It should also be mentioned that additional seismically induced forces may be computed for the loads arising during silo emptying - it must be decided on a case by case basis, which load combination is the most unfavourable. The importance of vertical seismic forces increases with higher seismic actions, since the additional forces are no longer covered by the safety factors for the serviceability limit states.

Combination of lateral and vertical seismic loads
Generally, lateral (in two mutually orthogonal directions) and vertical seismic loads applied to silos must be considered to act jointly. This effect may be taken into account approximately through the standard 30% rule in Sect. 4.3.3.5 of Eurocode 8-1 (2004):

$$\begin{aligned} &1.0 \cdot \boldsymbol{E}_{\boldsymbol{Edx}} \oplus 0.3 \cdot \boldsymbol{E}_{\boldsymbol{Edy}} \oplus 0.3 \cdot \boldsymbol{E}_{\boldsymbol{Edz}} \\ &0.3 \cdot \boldsymbol{E}_{\boldsymbol{Edx}} \oplus 1.0 \cdot \boldsymbol{E}_{\boldsymbol{Edy}} \oplus 0.3 \cdot \boldsymbol{E}_{\boldsymbol{Edz}} \\ &0.3 \cdot \boldsymbol{E}_{\boldsymbol{Edx}} \oplus 0.3 \cdot \boldsymbol{E}_{\boldsymbol{Edy}} \oplus 1.0 \cdot \boldsymbol{E}_{\boldsymbol{Edz}} \end{aligned} \tag{5.19}$$

with:

$\oplus$	"combined with"
$E_{Edx}, E_{Edy}, E_{Edz}$	Design values of the action effects due to the horizontal (x, y) and vertical (z) components of the seismic action.

On the other hand, Eurocode 8-4 (2006) stipulates that for axisymmetric silos it is sufficient to consider a single lateral component together with the vertical component.

5.5.2 Nonlinear Simulation Model

The hypoplastic material law is used to describe the behaviour of the granular material. Different formulations for hypoplasticity have been investigated: the hypoplasticity based on the formulation of Gudehus (1996), two modified versions using time history functions according to Bauer (1992) and Braun (1997) and the intergranular strain approach developed by Niemunis and Herle (1997). Comparisons of the different approaches with soil mechanic cyclic tests clarifies that the intergranular strain approach according to Niemunis and Herle (1997) leads to the most realistic results (Holler and Meskouris 2006; Wagner 2009). Therefore, this approach was applied within the overall model.

The set-up of the nonlinear simulation model is presented in Fig. 5.17. The foundation slab is placed on the soil, which is regarded as an elastic half space represented by the well-known cone model according to Wolf (1994). The granular material is modeled by 20-node solid elements incorporating the intergranular strain approach according to Niemunis and Herle (1997a, b). The silo wall is represented by eight-node shell elements and connected with contact elements to the granular material. The contact elements transfer compression and friction forces and allow for a separation between the filling and the silo wall. The proposed model was validated by shaking table tests of scaled steel silo models (Holler and Meskouris 2006).

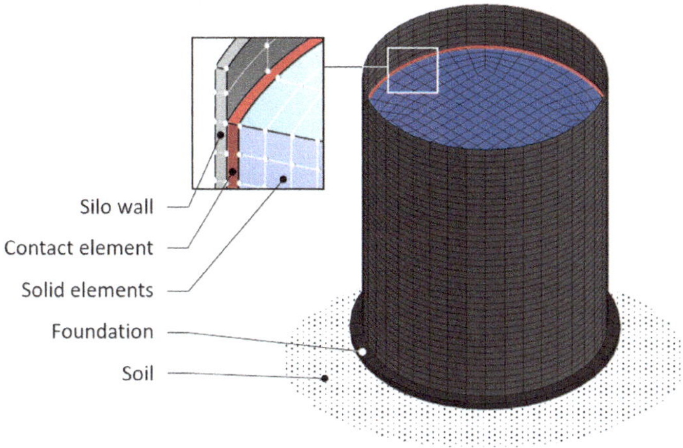

Fig. 5.17 Nonlinear calculation model

5.5.3 Determination of the Natural Frequencies of Silos

5.5.3.1 Silos that Rest Directly on Foundations Embedded in the Ground

Such silos behave approximately as built-in cantilever beams with constantly distributed mass. The expressions given by Nottrott (1963) or Rayleigh (see Petersen 2000) can be used for determining the lateral natural frequencies.

According to Rayleigh, the fundamental horizontal natural frequency of a cantilever beam with constant mass and stiffness, taking into account 2nd order effects and a compressive axial force D_A which increases linearly towards the built-in end section (denoted by A), is given by:

$$f = \frac{1}{2\pi} \cdot 3.530 \cdot \frac{1}{h^2} \cdot \sqrt{\frac{EI}{m_L}} \sqrt{1 - \frac{4.451}{12.461} \frac{D_A \cdot h^2}{EI}} \tag{5.20}$$

with:

- h Silo height (m)
- E Young's modulus of the silo wall (kN/m^2)
- I Moment of inertia of the silo wall section (m^4)
- m_L Mass per unit length (t/m)
- D_A Compressive force at the built-in silo bottom: $m_L \cdot h \cdot 9.81$ (kN).

According to Nottrott (1963), the fundamental lateral natural frequency for silos with piecewise constant wall thickness is given by:

$$f = \frac{1}{2\pi} \cdot \sqrt{\frac{k_{eq}}{m_{eq}}} \tag{5.21}$$

with:

k_{eq} Stiffness of the equivalent SDOF system considering the silo wall alone
m_{eq} Mass of the equivalent SDOF system considering both the silo and its contents.

The stiffness k_{eq} and the mass m_{eq} of the equivalent SDOF system are determined as follows:

$$k_{eq} = \frac{3 \cdot EI_u}{h^3} + \frac{2 \cdot (1+F)}{3+F} \tag{5.22}$$

$$m_{eq} = \frac{1}{4} \cdot \left(A_{u,Silo} \cdot \rho_{Silo} \cdot \frac{1+F}{3-F} + A_{content} \cdot \rho_{content} \right) \cdot h \tag{5.23}$$

with:

E Young's modulus of the silo wall (kN/m^2)
I_u Moment of inertia of the silo wall, bottom section (m^4)
h Silo height (m)
t_o minimum wall thickness (top) (m)
t_u maximum wall thickness (bottom) (m)
F t_o/t_u (−)
$A_{u,Silo}$ Section area of the silo wall at the bottom: $2\,\pi\, d_{cm}/2\, t_u$ (m^2)
d_{cm} Mean diameter of the silo wall (m)
$A_{content}$ Section area of the silo content: $\pi\,(d_c/2)^2$ (m^2)
ρ_{Silo} Specific mass, silo wall (t/m^3)
$\rho_{content}$ Specific mass, silo content (t/m^3).

These expressions yield good results for slender silos with fundamental modes essentially due to bending vibration. However, they should not be used for squat silos with $d_c/h > 0.5$, since they do not consider shear effects.

An accurate determination of the natural frequencies can be carried out using Finite Elements. For a full silo, the wall can be modelled with shell elements while the granular filling is depicted with volume elements, yielding a linear model in which the contents are considered to be fixed to the wall. For the empty silo, employing shell elements generally yields a plethora of local modes, so that it is advisable to use a beam element idealization instead. In this case, shear deformations should be taken into account when dealing with squat silos.

Table 5.8 shows results for four silos according to the formulas of Rayleigh and Nottrott and also results of FE-simulations. Silos 1, 2 and 3 have a flat bottom; silo 4 features a funnel with a fill height of 20 m. For this silo a cantilever length of 20 m, equal to the fill height, was assumed. Results show good agreement for the slender silos; however, for the squat silo the simplified expressions yield much too high values for the natural frequencies, since they do not consider shear effects.

Table 5.8 Mode shapes and natural frequencies (Hz) of four exemplary silos

	Silo 1 h = 10 m, d = 5 m	Silo 2 h = 20 m, d = 5 m	Silo 3 h = 30 m, d = 3 m	Silo 4 h = 25 m, d = 3.60 m
Natural mode				
FE-Solution	6.18 (37.14)	2.22 (11.46)	1.10 (6.20)	2.06 (12.38)
Nottrott	10.59 (50.41)	2.65 (12.60)	1.29 (6.72)	2.23 (9.07)
Rayleigh	10.79 (51.37)	2.69 (12.84)	1.32 (6.85)	2.27 (9.25)

Bulk material: $\rho = 1.70$ t/m^3, E = 60,000 kN/m^2
Silo wall: t = 0.01 m, $\rho = 7.85$ t/m^3, E = 210,000,000 kN/m^2
Values in brackets: Natural frequencies of empty silos

Generally, it can be stated that the approximate formulas given above are sufficiently accurate for the usual silos with fixed connection to the ground. In the case of variable wall thickness, the pertinent expressions given by Notrott should be used.

The approximate formulas yield overly high natural frequencies when applied to squat silos. If the fundamental natural period thus determined lies in the period range corresponding to the increasing branch of the design spectrum, the plateau spectral value should be used in order to stay on the safe side.

5.5.3.2 Silos on Supports

So far, only silos resting directly on the ground have been discussed. However, silos containing dry bulk material usually rest on a suitable support structure, thus facilitating the emptying procedure. If this support structure is very stiff compared to the silo itself, a built-in boundary condition at the lower silo section may be assumed and the formulas given in Sect. 5.5.3.1 can be employed. If, however, the support structure strongly influences the overall dynamic behaviour, the system must be considered as a whole, since it cannot be confidently stated beforehand how the impact of the natural frequencies of the silo or of its support will be. Also, if the natural frequencies of the silo and its support structure are close to each other, a higher load for the shell may ensue. If the regularity criteria of Eurocode 8-1 (2004) are satisfied, lateral seismic loads may be considered independently of each other, employing planar models. It is usually sufficient to use a beam model both for the support structure and the silo.

The importance of the interaction between a silo and its support will be illustrated by a simple example: Assume Silo 4 of Table 5.8 rests on a support structure. The entire system is modelled by beam elements. Thereby, the one representing the silo

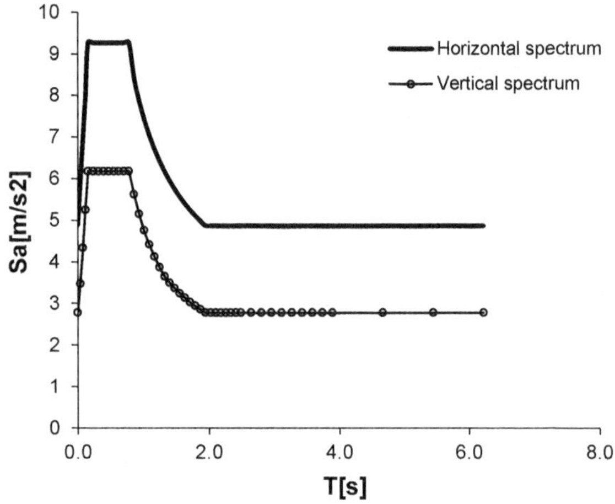

Fig. 5.18 Design spectra for Chang Bin (Taiwan)

features additional mass corresponding to the silo contents. The critical natural frequency in the direction of the strong axis (x direction) corresponds to a vibration of the silo itself (f = 0.63 Hz), which involves 70% of the effective modal mass. Additionally, a translational mode shape of the system appears at f = 2.38 Hz, in which 28% of the effective total mass are activated. In the direction of the weak axis (y direction) we obtain the critical mode shape at f = 0.35 Hz, in which 98% of the effective modal mass are activated. Finally, in the vertical direction (z direction) the critical natural frequency is 2.65 Hz, with 97% of the effective modal mass being activated. The system will be investigated by the response spectrum - modal analysis method for the Chang Bin (Taiwan) spectrum according to the 2005 Taiwanese Code shown in Fig. 5.18.

The resulting accelerations along the silo height were computed employing the SRSS rule. They are shown in Table 5.9 together with the minima and maxima in the silo wall along the x, y and z axis. The maximum accelerations occur in x direction corresponding to a local natural mode with f = 0.63 Hz and lie between 7.98 m/s^2 (bottom) and 11.96 m/s^2 (top of the silo shell). In the y direction the corresponding values are 4.48 m/s^2 (bottom) and 6.71 m/s^2 (top). The maximum acceleration values in both lateral directions must be combined along the entire height using the SRSS or the 30% rule. The resulting horizontal accelerations are then employed for the calculation of equivalent static loads which are variable along the silo height. Alternatively, the equivalent static loads can be computed approximately by using the acceleration value at the gravity centre. In the vertical direction, an acceleration of 6.32 m/s^2 is determined, leading to a dynamic amplification factor of $C_d = 1.64$ for the filling and emptying loads.

If the supporting structure is not considered, natural frequencies of 2.23 Hz (Nottrott) or 2.27 Hz (Rayleigh) are determined, yielding a maximum spectral acceleration of 9.27 m/s^2 (Fig. 5.18). In the vertical direction a maximum spectral acceleration of 6.18 m/s^2 may be employed in this case.

Table 5.9 Silo with support construction: Decisive accelerations in the silo wall from the response spectrum analyses in x-, y- and z-direction

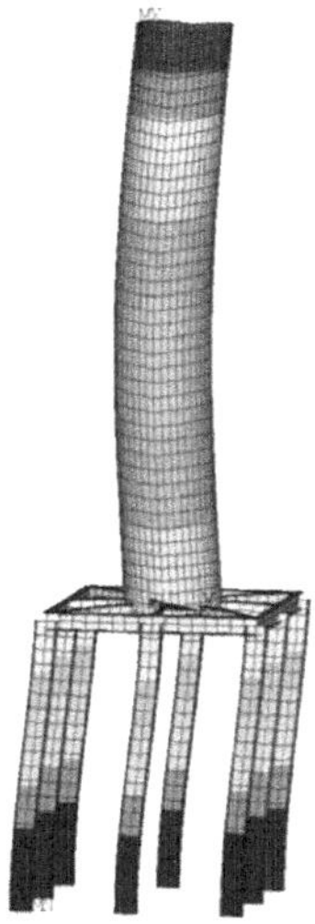	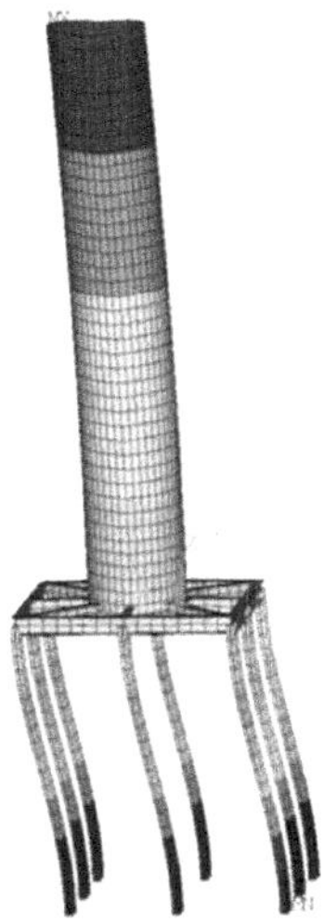	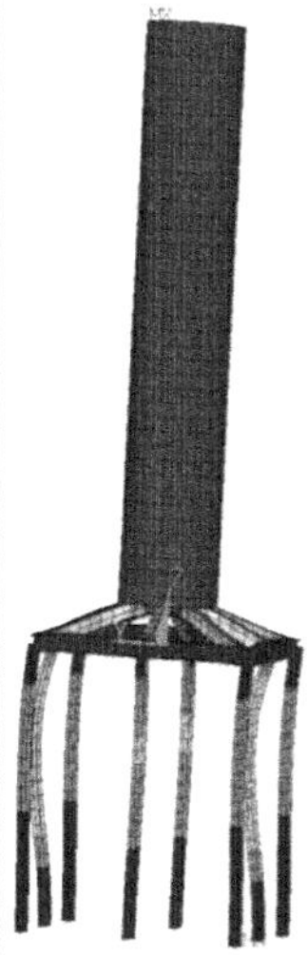
x-direction $a_{xmax} = 11.96\ m/s^2$ $a_{xmin} = 7.98\ m/s^2$	y-direction $a_{ymax} = 6.71\ m/s^2$ $a_{ymin} = 4.48\ m/s^2$	z-direction $a_{zmax} = 6.32\ m/s^2$(constant over the height)

It can generally be stated that for designing silos that rest on support structures the use of a simple beam model and the standard response spectrum modal analysis approach yields dependable and sufficiently accurate results. Interaction effects are automatically considered and the internal forces for designing the support structures are also determined.

Modelling the silo as a cantilever beam usually leads to conservative results since it yields higher natural frequencies which are nearer to the plateau region of the response spectra. Such an assumption is often unavoidable because details about the support structure are not known at the time the silo is being designed. It is however recommended that in the case of closely spaced natural frequencies the system (silo + support structure) be analysed as a whole in order to make sure that no critical interaction effects are overlooked.

5.5.3.3 Silos in Silo Batteries

A much higher degree of complexity is encountered when the natural frequencies of silos which form part of silo batteries are to be determined. This is because of the interaction between the single silos, the peculiarities of the support structure, variable fill levels and coupled natural modes due to the silos being joined to each other at the top. A first classification of the natural modes to be expected is as follows:

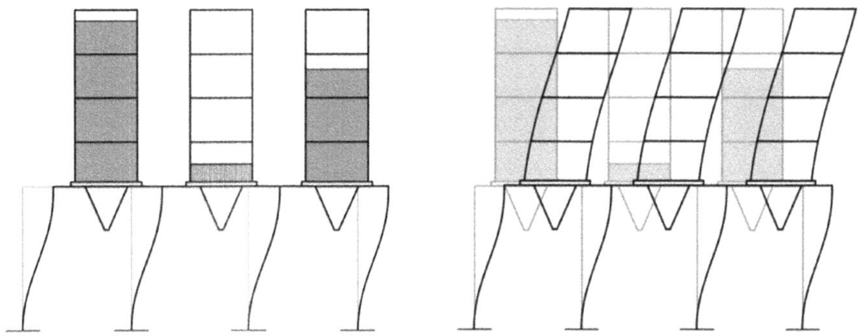

Fundamental frequency of support
$f_{silo,i} = f_{silo,j} = f_{uk}$

(a)

Fundamental frequency of entire system
$f_{silo,i} \neq f_{silo,j} \neq f_{uk}$ (not joined at top)
$f_{silo,i} = f_{silo,j} = f_{uk}$ (joined at top)

(b)

Fig. 5.19 Vibration of silos is batteries (Rinkens 2007)

- Local natural modes for single silos
- Global mode shapes for the whole battery, with the silos acting almost like rigid bodies on the flexible support structure
- Global mode shapes for the whole battery including various natural modes of the single silos.

If the entire structure is almost rigid and the single silos are not joined at the top, local natural modes of the single silos are dominant and their natural frequencies may be determined using the formulas of Sect. 5.5.3.1.

If the support structure plays an important role, two different situations may arise. In the first situation, in which the lateral stiffness of the support structure is much less than that of the structure as a whole, the fundamental natural frequency f_{uk} of the support controls the vibration performance and the silos act as rigid bodies. In this case, the natural frequencies of the single silos and the support structure are almost the same (Fig. 5.19a). In the second situation, a coupled vibration of support structure and silos occurs. If the single silos are joined at the top, a global vibration involving all silos and the support structure occurs (Fig. 5.19b); if this is not the case, the natural frequencies f_{silo} of the single silos are different.

It can be stated that due to the plethora of contributing factors the determination of the natural frequencies relevant for design purposes is not an easy task and yields results with large variability. For safety reasons it is therefore recommended to choose the natural frequency controlling the design from the three cases given above.

5.5.4 *Damping Values for Silos*

Eurocode 8-4 (2006), Sect. 2.3.2 contains some recommendations for damping values for silos, bulk material and foundations which can be used to compute an weighted average for the whole system as discussed in the following.

5.5.4.1 Damping in the Structure

For the ultimate limit state damping should be set to 5% according to Eurocode 8-1 (2004). This value includes material damping in the structure and energy dissipation through friction in the joints.

5.5.4.2 Damping of Foundation and Subsoil

Material damping in the soil can be assumed according to Eurocode 8-5 (2004), Table 4.1 as a function of the pertinent strains while radiation damping should be determined experimentally or numerically (e.g. using 1D wave propagation theory) for the site in question. However, since response spectra already include (viscous) damping, it is recommended to abstain from introducing additional soil damping.

5.5.4.3 Damping of Bulk Material

In the absence of more accurate data, damping for granular materials can be roughly assumed to be about 10%. A closer determination of the true damping value is possible by the following tests:

- "Resonant Column" test,
- Test with vibrating shear cells.

These tests are quite complex and call for specific expertise of highly specialized experimental facilities. Therefore, this part of damping will normally also be neglected. Regarding further details on damping of bulk material, the reader is referred to the literature, such as Haack and Tomas (2003) or Yanagida et al. (2003).

5.5.4.4 Weighted Damping Approach

A weighted damping value can be determined according to Eurocode 8-4 (2006), Sect. 2.3.2.4 considering the damping values of the structural damping and the damping of the bulk material. For each mode shape a weighted equivalent modal damping D_j is computed as a function of the work W_i stored in the structure and in the bulk material. The work is determined from the mode shape $\underline{\Phi}$ and the stiffness matrix $\underline{k}_i$ (i denoting the bulk material or the structure):

$$D_j = \frac{\sum_i D_i W_{i,j}}{\sum_i W_{i,j}}, \; with \; W_{i,j} = \frac{1}{2} \cdot \underline{\Phi}_j^T \cdot \underline{k}_i \cdot \underline{\Phi}_j \quad (5.24)$$

A prerequisite for computing a weighted damping value is taking into account the bulk material within the mathematical model. Since this is normally associated with a large numerical effort, it is only in special cases that the weighted damping approach is used.

5.5.5 Soil-Structure Interaction

Eurocode 8-5 (2004), Chap. 6, prescribes the consideration of soil-structure interaction effects in the following cases:

- Slender, high structures
- Structures with massive or deep foundations
- Structure with marked 2nd order theory effects
- Structures on very soft ground.

This means that for slender silos soil-structure interaction should be taken into account. Since such structures are routinely analysed by the response spectrum-modal analysis method, soil-structure interaction effects must be considered by using linear spring-dashpot models, such as the one introduced by Wolf (1994), where the ground is idealized as a homogeneous, linear elastic, semi-infinite medium. This model can be used for foundations on homogeneous and also on stratified media. In its simplest form, dynamic spring stiffness values are attributed to the translational and rotational degrees of freedom in combination with a simplified radiation damping approach (Smoltczyk 1991).

5.5.6 Calculation Examples: Squat and Slender Silo

The design approach of Eurocode 8-4 (2006) is based on equivalent static loads representing additional inertial forces due to the acceleration of the material. This assumption leads, in the case of slender silos, to a satisfactory representation of the actual load-bearing behaviour. In the case of squat silos, the internal friction of the granular material near the ground and its load bearing behaviour are taken into account only by a short linear increase in the lower region of the silo. For this reason, this approach should be validated by a comparison with the non-linear

Table 5.10 Geometry and material parameter of the slender and squat silo

Silos			
Height (squat/slender)	h	10.0/30.0	m
Inner diameter (squat/slender)	d_c	10.0/6.0	m
Thickness silo wall (squat/slender)	t	10.0/8.0	mm
Young's modulus	E	210.000	N/mm^2
Poisson's ratio	υ	0.3	(−)
Bulk Material			
Bulk unit weight	γ	15.0	kN/m^3
Horizontal load ratio	K	0.45	(−)
Wall friction angle	μ	0.40	(−)
Amplification factor	c_{pf}	1.00	(−)
Amplification factor pressure on the bottom	c_b	1.00	(−)

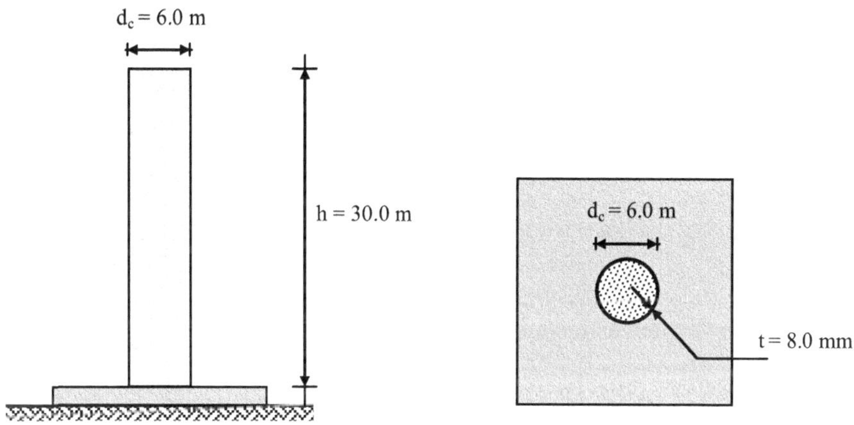

Fig. 5.20 Slender silo with cylindrical cross section

simulation model. For the slender silo, the influence of the use of constant and variable acceleration profiles over the silo height is investigated in the following. The geometry and material parameters for the slender and the squat silo are summarized in Table 5.10.

5.5.6.1 Slender Silo

For the slender silo shown in Fig. 5.20 maximum response values for seismic loading are to be determined. It is 30 m high, with an inner diameter of 6.0 m and a constant wall thickness of 8 mm.

Table 5.11 Static equivalent forces according to Eurocode 1-4 (2006)

z(m)	Y(z)	$p_{hf}(kN/m^2)$	$p_{wf}(kN/m^2)$	$p_{vf}(kN/m^2)$	$P_{wf}(kN/m)$
0	0.00	0.00	0.00	0.00	0.00
2	0.21	12.00	4.80	26.67	4.99
4	0.38	21.44	8.58	47.65	18.52
6	0.51	28.87	11.55	64.16	38.77
8	0.62	34.71	13.88	77.14	64.29
10	0.70	39.31	15.72	87.35	93.97
12	0.76	42.92	17.17	95.38	126.92
14	0.81	45.77	18.31	101.70	162.45
16	0.85	48.00	19.20	106.67	199.99
18	0.88	49.76	19.91	110.58	239.12
20	0.91	51.15	20.46	113.66	279.51
22	0.93	52.24	20.89	116.08	320.88
24	0.94	53.09	21.24	117.98	363.03
26	0.96	53.77	21.51	119.48	405.78
28	0.97	54.30	21.72	120.66	449.01
30	0.97	54.71	21.89	121.58	492.62

Filling loads and corresponding stresses

The filling loads are calculated according to Eurocode 1-4 (2006).

Table 5.11 summarizes the horizontal pressure p_{hf} after filling, the wall frictional traction p_{wf} after filling and the summarized wall frictional P_{wf} traction after filling. The load distributions are given as functions over the silo height.

The circumferential and meridian stresses due to the filling loads are calculated analytically and additionally with a finite element model consisting of shell elements. The finite element model takes advantage of system symmetry for the calculation of the resulting stresses and is illustrated in Fig. 5.21. The distributions and excellent agreement of the stresses in circumferential and meridian direction for both calculation approaches are depicted in Fig. 5.22.

Seismic loads and corresponding stresses

To determine the acceleration profiles, the silo is idealized by a beam model with 15 lumped masses. These masses consider the self-weight of the silo shell and the granular material (Fig. 5.23). The fundamental frequency is determined for the multi-degree of freedom (MDOF) oscillator at 1.0 Hz. A spectral acceleration of 5 m/s^2 is assigned for a location in New Zealand in accordance with the spectrum (NZS 2004) set in Fig. 5.24, if the ascending branch in the response spectrum is neglected.

With the spectral acceleration of 5 m/s^2 and the total mass of 1756.26 t, the base shear is $F_b = 8781.3$ kN. From this, the linear acceleration profile is determined

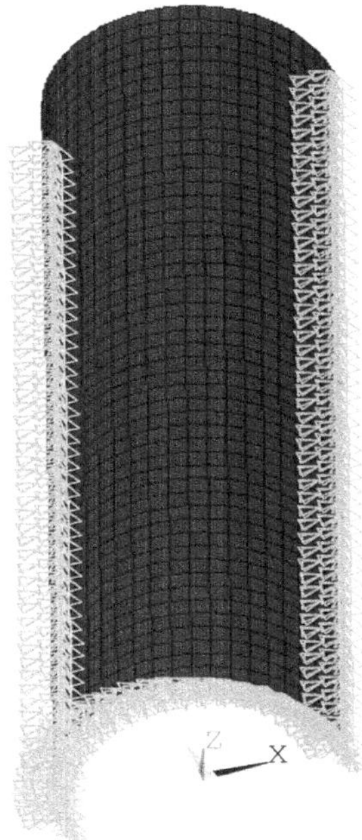

Fig. 5.21 Finite-element model with symmetrical boundary conditions

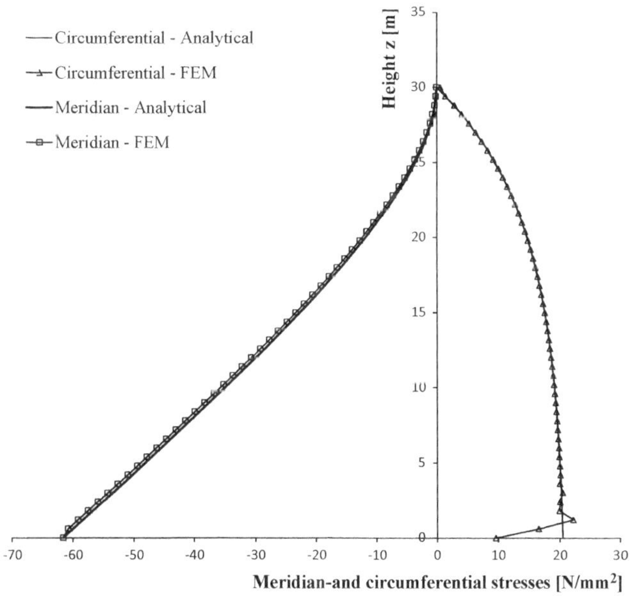

Fig. 5.22 Stresses in circumferential and meridian direction due to filling

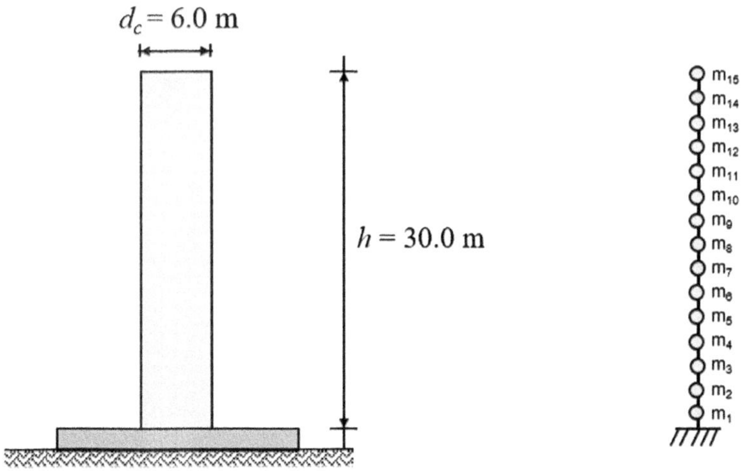

Fig. 5.23 Slender silo and idealization as multiple mass oscillator

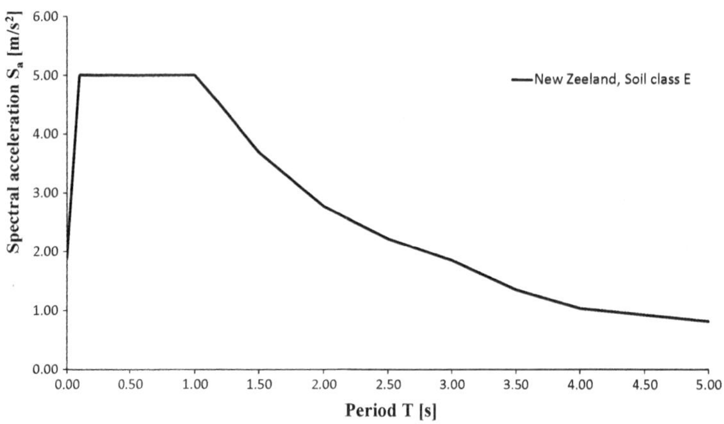

Fig. 5.24 Design response spectrum, Seismic code of New Zeeland, Soil Class E (NZS 2004)

by the height- and mass-proportional distribution of the forces on the heights of the masses. In addition, the MDOF system is used to perform a calculation using the multimodal response spectrum method, taking into account 10 mode shapes. The computed accelerations from the multimodal calculation compared to the linear approach are shown in Fig. 5.25.

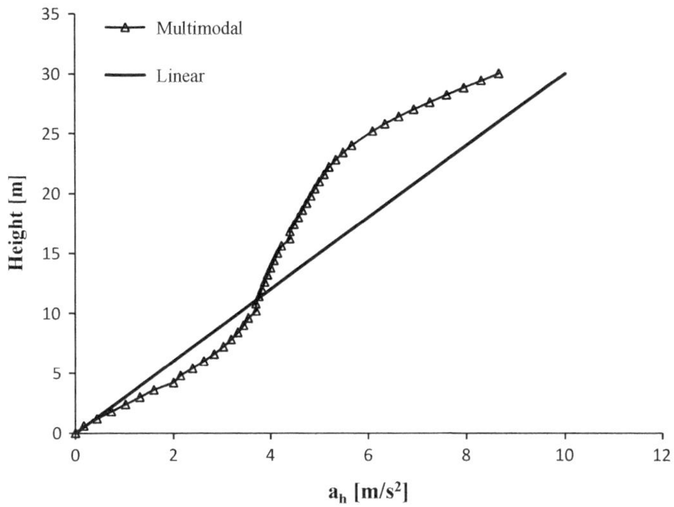

Fig. 5.25 Linear and multimodal acceleration profiles

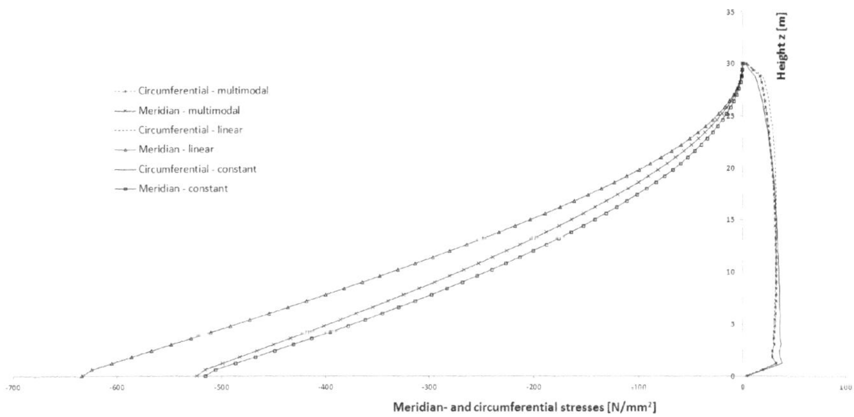

Fig. 5.26 Circumferential and meridian stresses as results of the different approaches of the acceleration

The horizontal seismic pressure values acting on the silo wall are calculated according to Sect. 5.5.1 using the variable acceleration profiles (Fig. 5.25). They are applied as static equivalent loads together with the static pressures on a finite element model of shell elements. In addition, the seismic loads due to vertical seismic excitation with $a_v = 0.7\ a_h = 3.5\ \text{m/s}^2$ according to Sect. 5.5.1 are taken into account in the model. Figure 5.26 shows the resulting circumferential and meridian stresses for a constant acceleration as well as for the acceleration profiles given in Fig. 5.25.

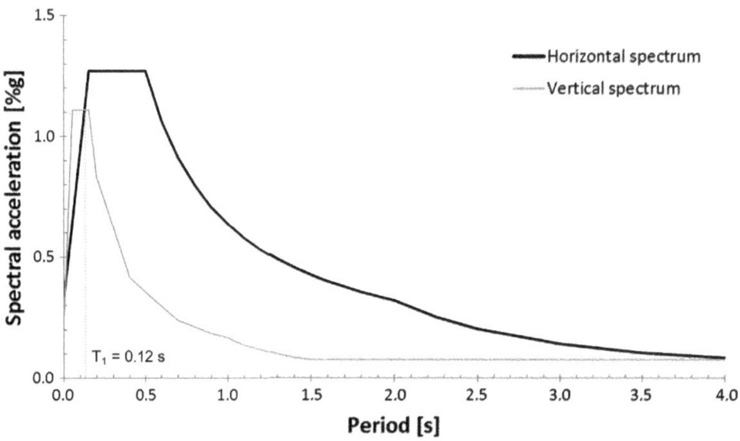

Fig. 5.27 Elastic response spectra for the location in Istanbul according to Eurocode 8-1 (2004)

5.5.6.2 Squat Silo

According to Table 5.11, the squat silo considered ($h/d_c < 1.0$) has a height of 10 m, an inner diameter of 10 m and a constant wall thickness of 8 mm. The connection to the foundation is assumed to be rigid. A simplified model of the foundation as a rigid reinforced concrete block with a density of 25 kN/m^3 is used. The connection to the lower edge of the foundation is represented by a stiffness matrix that models the elastic half space under the foundation for a shear wave velocity of 500 m/s based on the truncated cone model of Wolf (1994). The associated mass fractions and damping ratios are also taken into account in the model. The silo is located in Istanbul, for which a reference peak ground acceleration a_g of 4.16 m/s^2 must be assumed. The corresponding design response spectra in the horizontal and vertical direction according to Eurocode 8-1 (2004) for Spectrum Type I and Soil Class B are shown in Table 5.26.

Due to the thin walls of the silo and the associated risk of shell buckling, a behaviour factor will not be used (q = 1.0). A linear-elastic continuum model will be used to determine the first natural period. The bulk material is idealized using volume elements, and the silo shell is modelled with shell elements. The contact area between the bulk material and the wall of the silo is assumed to be rigid. This model yields a first natural period of $T_1 = 0.12$ s taking the interaction between the soil and the structure into account. The corresponding spectral accelerations are $S_{ah} = 10.9$ m/s^2 in the horizontal direction and $S_{av} = 11.0$ m/s^2 in the vertical direction. The vertical acceleration yields a scaling factor of $C_d = 1.12$ according to Eq. (5.16). Since we are dealing with a squat silo, the horizontal spectral acceleration is assumed to act on the centre of mass and is considered to be constant along the height of the silo as an approximation. Using the formulas for the static equivalent loads of Sect. 5.5.1, the circumferential and meridian stresses presented in Figs. 5.28, 5.29 and

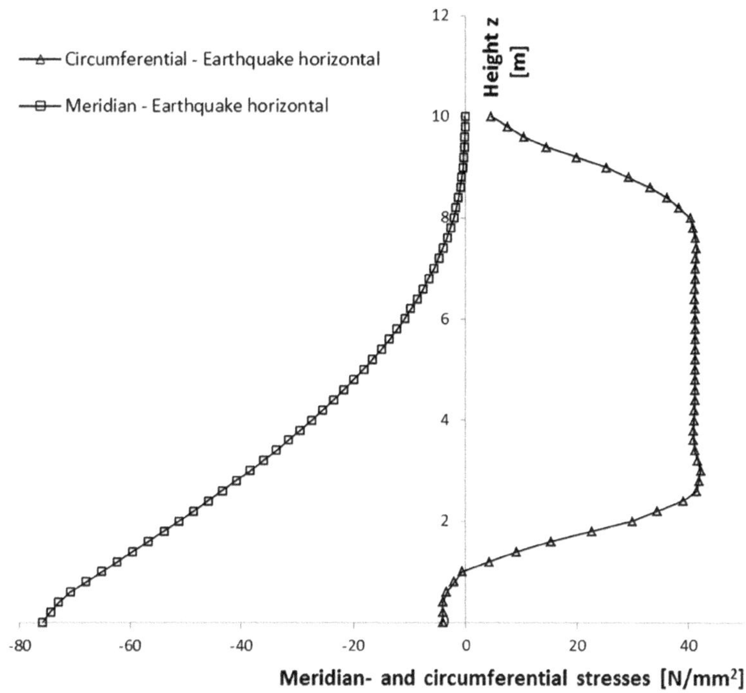

Fig. 5.28 Circumferential and meridian stresses due to horizontal seismic excitation

5.31 are obtained as a result of the horizontal and vertical effects of the earthquake. Figure 5.29 additionally shows the comparison of the resulting meridian and ring stresses using the approach according to Silvestri et al. (2012) for a constant spectral acceleration $S_{av} = 11.0\ m/s^2$. Evidently, the approach of Silvestri et al. (2012) yields more conservative results for the same input values, because they utilize a linear function instead of an exponentially distributed pressure according to the theory of Jansen (1895).

In order to enable the comparison with transient analyses, synthetic acceleration time histories are generated in horizontal and vertical direction (Fig. 5.30) based on the elastic response spectra shown in Fig. 5.27. The transient simulations are carried out using the nonlinear simulation model introduced in Sect. 5.5.2. A comparison of the dynamic stresses computed by the equivalent load method and the non-linear simulation model is shown in Fig. 5.26. Since in the non-linear simulation model the seismic loading is considered acting in both a horizontal and a vertical direction, it is necessary to carry out the comparison with the earthquake combinations in both directions. The combination of the results in both directions is carried out using the 30% rule. Additionally, a full superposition of the results of both directions is carried out.

The results show considerable differences. The amplification factors that were calculated on the basis of the equivalent load method result in a 2–3 times higher

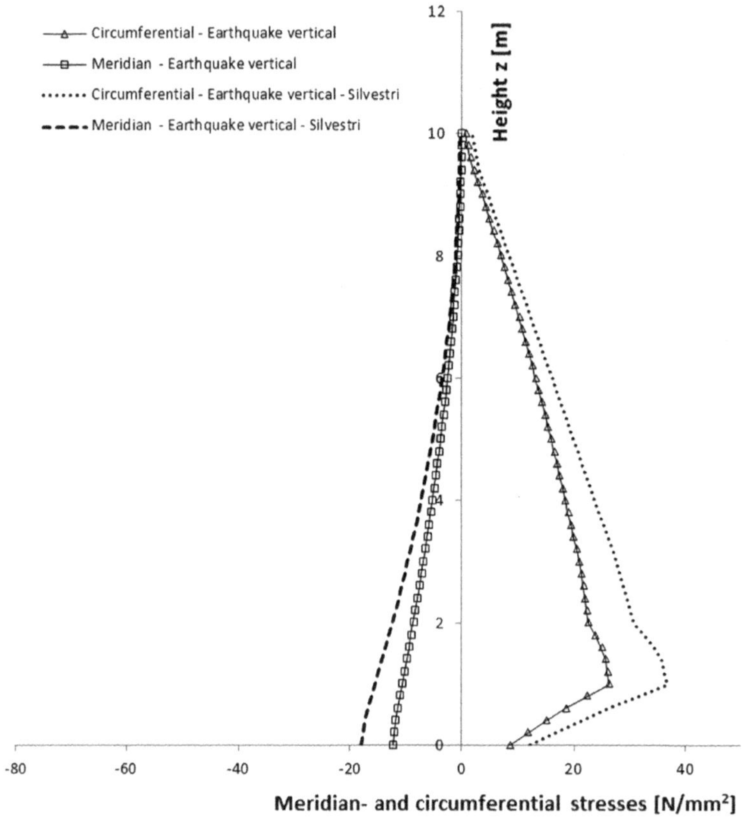

Fig. 5.29 Circumferential and meridian stresses due to vertical seismic excitation

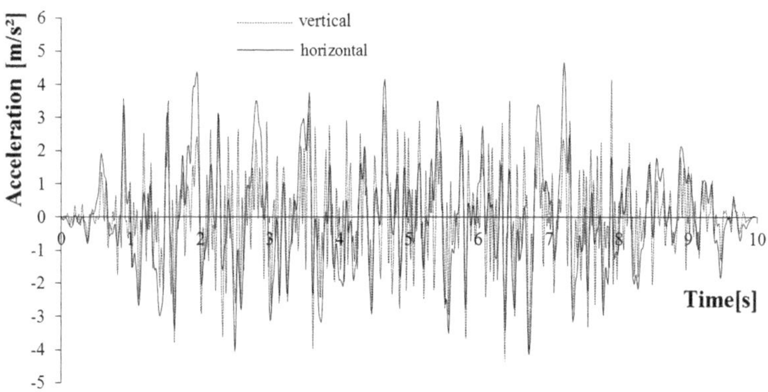

Fig. 5.30 Synthetically generated acceleration time histories

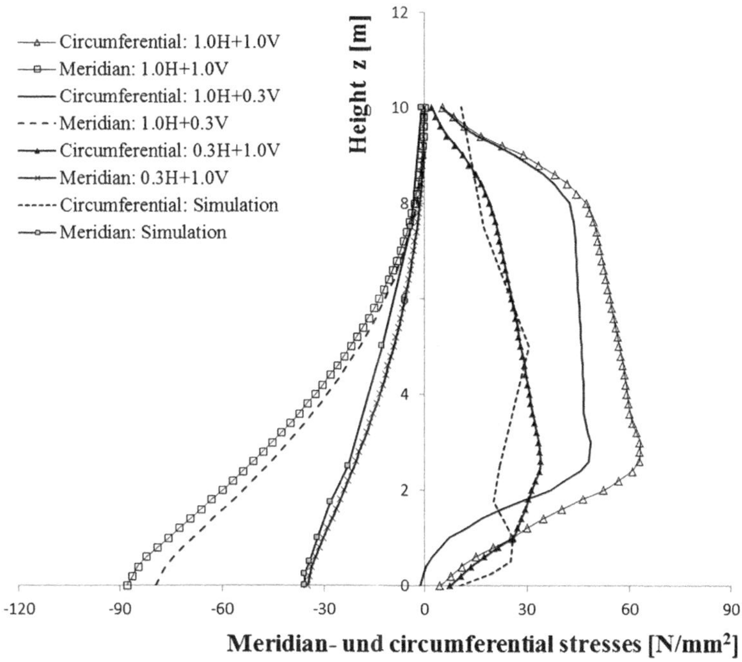

Fig. 5.31 Circumferential and meridian stresses due to seismic excitation

factor when compared to the transient simulation results on the basis of the 30% rule, in which the combination of 1.0 times the horizontal loads with 0.3 times the vertical loads is decisive. The full superposition provides even more conservative results when compared to the transient simulation result. This is mainly due to the load assumption of the Eurocode 8-4 (2006), which simply assumes that the horizontal inertia forces of the bulk materials is basically transferred through the silo shell to the foundation. Contrary to this, it is clear from the simulation results that a substantial part of the horizontal loads due to the acceleration of the material is transferred directly by friction to the foundation.

This effect is especially pronounced in the squat silo considered here. Its importance decreases with increasing H/D ratio. In the equivalent load method, an attempt is made to consider this effect by a linear scaling of the cosine-shaped earthquake load applied from the base of the silo up to a defined height. With this reduction carried out for the horizontal loads at the base of the silo considered here, a comparison with the non-linear simulation results obtained by the equivalent load method still leads to conservative results. When applied to slender silos, the differences between the calculation methods decrease significantly, with the equivalent load method reflecting the dynamic stress distribution well. This has been proven by Holler and Meskouris (2006) using advanced experimental and numerical studies.

5.6 Tank Structures

5.6.1 *Introduction*

Tank structures under seismic loading are subject to the inertia forces of the tank structure itself, to those of the fluid content and also to the interaction forces between the fluid and the shell of the tank. A rigorous investigation of these effects is quite demanding in mathematical terms, rendering simplified calculation methods necessary for the solving of practical problems.

One of these simplified methods was developed by Housner in 1963. Housner's method assumes a rigid tank and completely neglects the influence of interaction effects between the tank and the fluid. The load is calculated simply on the basis of the rigid body movement of the tank and the fluid (impulsive load component) and the sloshing oscillation of the fluid (convective load component). Calculations yield only the seismically induced foundation shear and the overturning moment applied to the base of the tank. An exact calculation of the shell stresses is not possible with Housner's method.

Numerous analyses of damage caused by earthquakes have shown that tank dimensioning using the Housner method does not result in earthquake-proof tank designs, this being especially the case for slender tanks. It follows that Housner's approach does not mirror the current state-of-the-art. An alternative tank calculation concept has been described in the informative Annex A to Eurocode 8-4 (2006); however, due to its complexity it has not been widely used in the dimensioning practise so far. Furthermore, understanding the underlying concept requires background information that it not fully presented in the normative annex.

With this in mind, the theoretical background of the equivalent load method is explained in the following together with a systematic summary of the individual pressure components that result from seismic loads. This is followed by a description of the pressure components using tabulated coefficients, with the help of which the engineer is able to carry out a tank calculation without the assistance of complex mathematics software (Holtschoppen 2011). The case most frequently encountered in practise is an above-ground, upright, cylindrical tank anchored to the ground and subject to atmospheric pressure. Two examples of tanks subject to seismic loading show the applicability of the presented calculation methods that are initially derived theoretically.

5.6.2 *Basics: Cylindrical Tank Structures Under Earthquake Loading*

The seismic actions to typical industrial buildings basically result from the inertia forces of the primary load-carrying structure and various process engineering components. The dynamic behaviour of liquid filled tanks subjected to seismic loading is much more complex and requires an adequate consideration of the tank structure and the content. This is important due to the different mechanisms involved in the generation of inertia forces of the fluid content and of the tank structure itself and to their interaction. It is especially important to consider this interaction in the design of flexible tanks (e.g. slender steel tanks), because the joint flexural vibration of the fluid and the tank strongly influences the seismic forces acting on the tank shell. All seismically induced components for tanks can be derived from the streaming potential Φ for fluids, for which the following conditions hold (Sigloch 2009):

$$\vec{v} = grad\ \Phi = \nabla\Phi = \begin{pmatrix} \partial\Phi/\partial x \\ \partial\Phi/\partial y \\ \partial\Phi/\partial z \end{pmatrix} \tag{5.25}$$

with:

$\vec{v}$ velocity vector of the liquid
∇ Nabla-Operator; $\nabla = \vec{e_x} \cdot \frac{\partial}{\partial x} + \vec{e_y} \cdot \frac{\partial}{\partial y} + \vec{e_z} \cdot \frac{\partial}{\partial z}$

The assumption of a frictionless and irrotational flow of the liquid leads to:

$$\text{rot}\ \vec{v} = \nabla \times \vec{v} = 0 \tag{5.26}$$

From the assumption of the incompressibility of the liquid (and thus constant liquid density), the continuity equation is derived:

$$\text{div}\ \vec{v} = \nabla\vec{v} = 0 \tag{5.27}$$

Using the continuity equation, the Laplace equation for source-free potential flows is obtained by substituting (5.25)

$$\Delta\Phi = \frac{\partial^2\Phi}{\partial x^2} + \frac{\partial^2\Phi}{\partial y^2} + \frac{\partial^2\Phi}{\partial z^2} = 0 \tag{5.28}$$

Δ Laplace-Operator; $\Delta = \nabla(\nabla) = \frac{\partial}{\partial x}\left(\frac{\partial}{\partial x}\right) + \frac{\partial}{\partial y}\left(\frac{\partial}{\partial y}\right) + \frac{\partial}{\partial z}\left(\frac{\partial}{\partial z}\right)$

For cylindrical tanks it is expedient to introduce cylindrical coordinates according to Fig. 5.32 in (5.28). Using these coordinates, the Laplace equation can be expressed as follows (Habenberger 2001):

$$\Delta\Phi = \frac{\partial^2\Phi}{\partial\xi^2} + \frac{1}{\xi} \cdot \frac{\partial\Phi}{\partial\xi} + \frac{1}{\xi^2} \cdot \frac{\partial^2\Phi}{\partial\theta^2} + \frac{1}{\gamma^2} \cdot \frac{\partial^2\Phi}{\partial\zeta^2} = 0 \tag{5.29}$$

Here "ξ = r/R" and "ζ = z/H" represent dimensionless coordinates, θ the angle in circumferential direction and "γ = H/R" the tank slenderness (Fig. 5.32). Equation (5.29) is not to be confused with the representation of the Laplace equation,

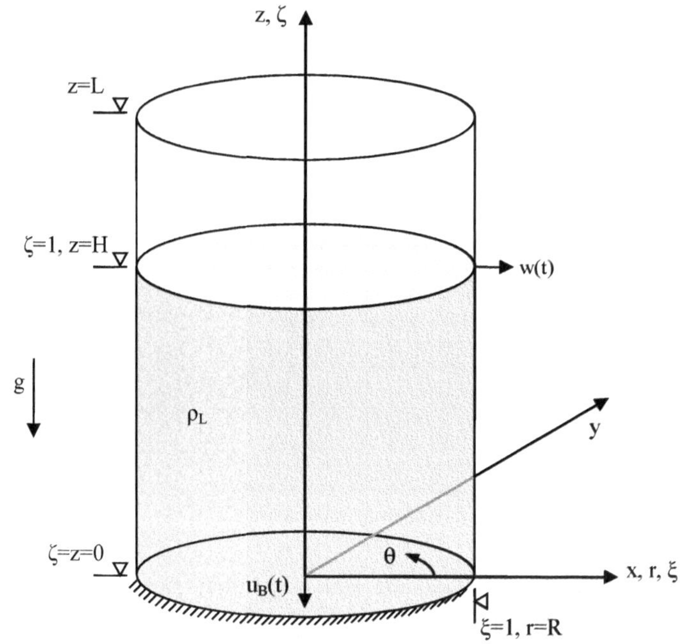

Fig. 5.32 Coordinate system of cylindrical, anchored tanks

which is occasionally used in the literature without the slenderness γ in the last term. This use is only valid when dimensioned cylindrical coordinates r, z, and θ are employed.

The hydrodynamic pressure of the liquid can be represented by the time derivative of the velocity potential (Habenberger 2001):

$$p(\xi, \zeta, \theta, t) = -\rho_L \frac{\partial \Phi}{\delta t} \tag{5.30}$$

It has to be pointed out, that especially the requirements for (5.26) (frictionless tank wall) are not generally given in reality. However, the assumption of a frictionless liquid is most often acceptable in establishing the velocity potential because the contact area between tank wall and fluid is quite small compared to the entire tank cross-section for typical tank geometries. As far as tanks with an extremely small radius are concerned, the derivation of the differential equation needs to be reconsidered in this regard - instead of the Euler approach, an approach according to Navier-Stokes must then be selected (Sigloch 2009). Contrary to this, the assumption of the incompressibility is generally sufficiently precise for fluids (this being evidently not the case for gases).

Solutions need to be found for the differential Eq. (5.29) that fulfil the stipulated boundary conditions. The boundary conditions for a horizontal tank excitation can be formulated as follows:

BC 1:
Radial velocity along the silo wall

$$\frac{\partial \Phi}{\partial r} = \frac{1}{R}\frac{\partial \Phi}{\partial \xi} = \frac{\partial w}{\partial t} \quad \text{for } \xi = 1 \tag{5.31}$$

BC 2:
Axial velocity at the silo bottom

$$\frac{\partial \Phi}{\partial z} = \frac{1}{H}\frac{\partial \Phi}{\partial \zeta} = -\frac{\partial u_B}{\partial t} = 0 \quad \text{for } \zeta = 0 \tag{5.32}$$

BC 3a:
"Sloshing constraint" at the liquid surface (determined from the axial velocity at the liquid surface $\frac{1}{H}\cdot\frac{\partial \Phi}{\partial \zeta} = \frac{\partial u_{\zeta=1}}{\partial t}$ (see BC 2) and linearized Bernoulli energy Eq. (5.43)

$$\frac{\partial^2 \Phi}{\partial t^2} + \frac{g}{H}\frac{\partial \Phi}{\partial \zeta} = 0 \quad \text{for } \zeta = 1 \tag{5.33}$$

BC 3b:
Pressure at the free surface

$$p = -\rho_L \frac{\partial \Phi}{\partial t} = 0 \quad \text{for } \zeta = 1$$

In order to fully include the above boundary conditions in solving of the differential equation, the velocity potential is subdivided into three partial potentials, one having an inhomogeneous boundary condition (right-hand side ≠ 0) and two having homogeneous boundary conditions (right-hand side = 0):

$$\Phi = \Phi_1 + \Phi_2 + \Phi_3 \tag{5.34}$$

Additionally, the Fourier product approach is employed (Bronstein and Semendjaev 1996), so that each factor depends only on one variable:

$$\Phi(\xi, \zeta, \theta, t) = P(\xi) \cdot S(\zeta) \cdot Q(\theta) \cdot F(t) \tag{5.35}$$

At this point, the assumption is made that the velocity potential is symmetrical. This enables a Fourier series with exclusively cosine terms to be used for the partial function Q(θ) in circumferential direction:

$$Q(\theta) = \sum_{m=0}^{\infty} Q_m \cdot \cos(m \cdot \theta) \tag{5.36}$$

Hereby, only the first circumferential wave is activated for the perfect cylindrical shell under horizontal seismic excitation, so that the summation over m circumferential waves in (5.36) can be omitted (Fischer and Rammerstorfer 1982). In practice, building-related deviations from a perfect cylindrical shell normally do exist. This results in high order natural modes in circumferential direction (Clough 1977). However, their contribution to the overall vibration behaviour is normally small so that higher order natural modes can be neglected. While the time-dependent function F(t) is determined by the corresponding inhomogeneous boundary condition (Habenberger 2001), two ordinary decoupled differential equations remain in S(ζ) and P(ξ):

$$\frac{1}{S(\zeta)} \cdot \frac{d^2 S(\zeta)}{d\zeta^2} = \lambda^2 \tag{5.37}$$

$$\xi^2 \cdot \frac{d^2 P(\xi)}{d\xi^2} + \xi \cdot \frac{dP(\xi)}{d\xi} + \left(\lambda^2 \xi^2 - \mathrm{m}^2\right) \cdot P(\xi) = 0 \tag{5.38}$$

λ Root of the characteristic equation of the cylindrical shell (see explanatory notes to (5.39)).

Equation (5.38) describing the radial component is a Bessel differential equation. For solving Eq. (5.38) the Bessel function ("cylindrical function") must be used, as included in numerous mathematical software packages or spreadsheet applications. Both of the decoupled differential equations can only be solved for defined boundary conditions, i.e. for specifically defined functions of the wall and ground deformations w or u_B in Eqs. (5.31) and (5.32). Some important boundary conditions for seismically induced tank and liquid vibrations are considered in the following. By applying these boundary conditions, the pressure components on the tank wall and the tank bottom are derived. By way of simplification, the Bessel components for diverse forms of tank slenderness (filling height to radius ratio) are analysed in Sect. 5.6.3. The results of these investigations are summarised in tabular form in Sect. 5.6.12.

5.6.3 *One-Dimensional Horizontal Seismic Action*

5.6.3.1 Convective Pressure Component (Sloshing)

A sloshing vibration is generated if the tank is subjected to a horizontal excitation and the surface of the fluid is able to move freely. The existence of a rigid tank can be assumed if the (long) natural periods of the sloshing modes and the (relatively short) tank natural periods are far enough apart so that the oscillations are decoupled. A qualitative presentation of the sloshing effects with the corresponding pressure distribution is shown in Fig. 5.33. The pressure is mainly acting on the upper part of the wall, but there is also a dynamic pressure at the bottom level of the tank, from which additional moments result for the foundation.

Using (5.36), only the first harmonic circumferential mode is taken into account for all pressure components - the circumferential contribution is thus limited to cos $(1 \cdot \theta)$. Corresponding to (5.31)–(5.33) the following boundary conditions apply:

BC 1 with $w(\zeta) = const.$ (Assumption of a moving rigid wall)
BC 2 (Anchored tank; no soil-structure interaction)
BC 3a (Sloshing of the fluid surface occurs).

This results in the convective pressure component:

$$p_k(\xi, \zeta, \theta, t) = \sum_{n=1}^{\infty} \frac{2 \cdot R \cdot \rho_L}{(\lambda_n^2 - 1)} \left[\frac{J_1(\lambda_n \cdot \xi)}{J_1(\lambda_n)} \right] \left[\frac{\cosh(\lambda_n \cdot \gamma \cdot \zeta)}{\cosh(\lambda_n \cdot \gamma)} \right] [\cos(\theta)][a_{kn}(t) \cdot \Gamma_{kn}] \tag{5.39}$$

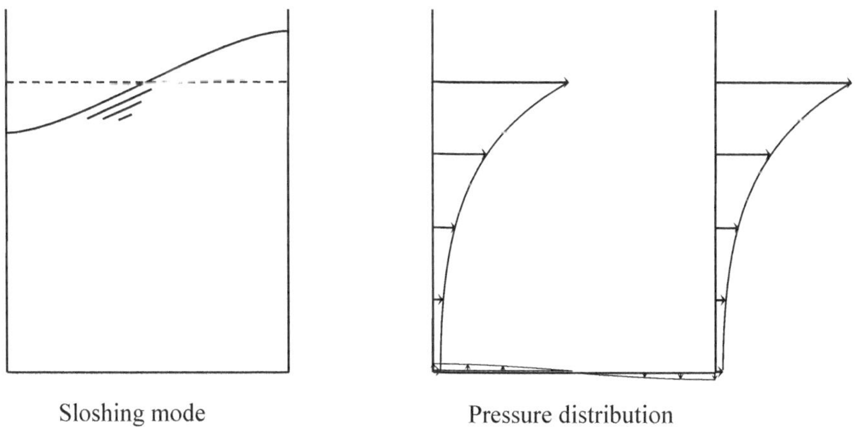

Fig. 5.33 Sloshing mode and qualitative pressure distribution on wall and bottom

with:

p_k Convective pressure component due to horizontal excitation
n Summation index; Number of considered modes for sloshing
R Inner radius of the tank wall
ρ_L Density of the liquid
J_1 First order Bessel function; Bronstein and Semendjajew (1996): $J_1(\lambda_n \cdot \xi)= \sum_{k=0}^{\infty} \frac{(-1)^k}{k! \cdot \Gamma(1+k+1)} \cdot \left(\frac{\lambda_n \cdot \xi}{2}\right)^{2k+1}$
λ_n Roots of the derivative of the Bessel function J_1: $\lambda_1 = 1.841$, $\lambda_2 = 5.331$, $\lambda_3 = 8.536$
ξ Dimensionless radius: $\xi = r/R$
ζ Dimensionless height: $\zeta = z/H$
θ Circumferential angle
γ Slenderness of the tank, respectively of the, "tank content": $\gamma = H/R$
$a_{kn}(t)$ Horizontal response acceleration time history of the equivalent single-degree-of-freedom oscillator with the period T_{kn} of the n-th sloshing natural mode. When applying the response spectra method (Sect. 5.6.6), the spectral accelerations are to be determined with the natural periods T_{kn} using the elastic acceleration response spectrum. Usually it is sufficient to take into account only the fundamental mode, which can be calculated according to Eq. (5.42). The damping of the elastic response spectrum ranges between 0 and 0.5% (damping of sloshing fluid).
Γ_{kn} Participation factor of the convective pressure component corresponding to natural mode n.

If only the fundamental natural mode of the oscillating liquid is taken into account ($n = 1$) and the pressure distribution is restricted to the tank shell ($\xi = 1$), the summation is omitted and (5.39) simplifies to:

$$p_k(\xi = 1, \zeta, \theta, t) = R \cdot \rho_L \left[0.837 \cdot \frac{\cosh(1.841 \cdot \gamma \cdot \zeta)}{\cosh(1.841 \cdot \gamma)}\right][\cos(\theta)][a_{k1}(t) \cdot \Gamma_{k1}]. \tag{5.40}$$

Due to the hydrostatic stress state in the liquid, the pressure p_k acts normally on the tank wall. The participation factor Γ_{k1} for the first sloshing mode results as follows according to Fischer et al. (1991):

$$\Gamma_{k1}=\frac{2 \cdot \sinh(\lambda_1 \cdot \gamma) \cdot [\cosh(\lambda_1 \cdot \gamma) - 1]}{\sinh(\lambda_1 \cdot \gamma) \cdot \cosh(\lambda_1 \cdot \gamma) - \lambda_1 \cdot \gamma} \tag{5.41}$$

with:

Γ_{k1} Participation factor of the convective pressure component corresponding to the fundamental natural mode
λ_1 Root of the derivative of the first order Bessel function J_1: $\lambda_1 = 1.841$
γ Slenderness of the tank, respectively of the tank content: $\gamma = H/R$.

The natural period of the n-th sloshing mode, which is necessary for the determination of the response acceleration a_{kn}, can be calculated according to Fischer and Rammerstorfer (1982) or Stempniewski (1990) with the acceleration of gravity g:

$$T_{kn} = \frac{2\pi}{\sqrt{\frac{g \cdot \lambda_n \cdot \tanh(\lambda_n \cdot \gamma)}{R}}} \tag{5.42}$$

The other variables are defined in the explanations to Eq. (5.39). With regard to the calculation formula for the vibration periods of the sloshing modes, it should be noted that the hyperbolic tangent converges towards 1 asymptomatically, so that the fundamental natural period of the basic sloshing mode T_{k1} can be assumed to $T_{k1} = 1.478 \cdot \sqrt{R}$ for slenderness values $\gamma \geq 1,5$. The maximum height $d = u_{\max}$ of the first sloshing mode can be derived from (5.40) in combination with the linearised Bernoulli equation:

$$\frac{\partial \Phi}{\partial t} + \frac{p_0}{\rho_L} + g \cdot u = 0 \tag{5.43}$$

with:

$\frac{\partial \Phi}{\partial t}$ Derivative with respect to time; $\frac{\partial \Phi}{\partial t} = 0$ for maximum deflection
p_0 Atmospheric pressure on the free surface; the pressure corresponds to the pressure ordinate p_k for $\zeta = 1$
g Acceleration of gravity: 9.81 m/s^2
u Axial movement of the liquid at the surface.

Thus, the maximum height d is obtained by considering the fundamental natural mode:

$$d = 0,837 \cdot \frac{S_a(T_{k1})}{g} \cdot R \tag{5.44}$$

with:

S_a Spectral acceleration of the elastic response spectrum determined with damping values in the range of 0–0.5%
T_{k1} Fundamental natural period of the first sloshing mode T_{k1} according to (5.42).

5.6.3.2 Impulsive Rigid Pressure Component (Rigid Body Movement)

The impulsive rigid pressure component is derived from the horizontal movement of the tank (assumed to be rigid) together with the liquid. A qualitative presentation of the tank movement and the resulting pressure distribution is shown in Fig. 5.34. This results in a pressure distribution on the tank wall and a moment action on the tank bottom.

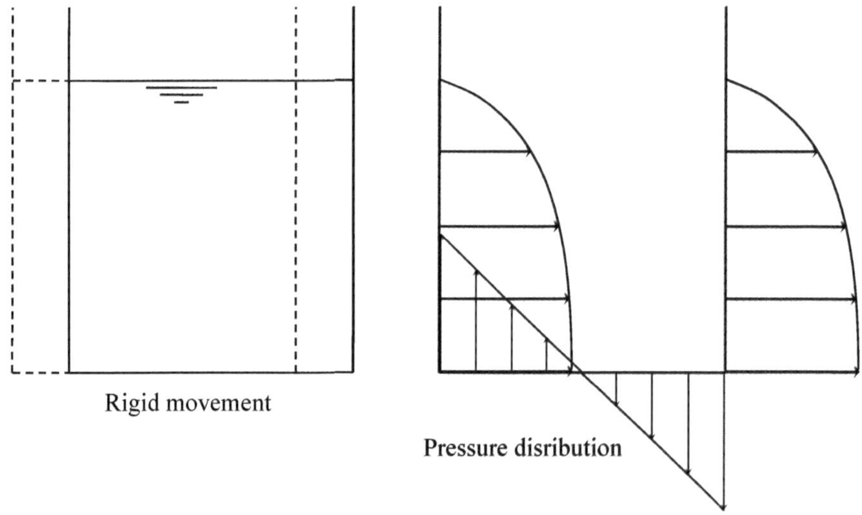

Fig. 5.34 Horizontal rigid movement and qualitative pressure distribution on wall and bottom

The sloshing mode introduced in Sect. 5.6.3.1 is suppressed from a mathematical point of view by the applied boundary condition 3b (5.33), as the sloshing component has already been taken into account separately in the convective pressure component (5.39). The following boundary conditions after (5.31)–(5.33) hold for the impulsive rigid component:

BC 1 with $w(\zeta) = const.$ (Assumption of a moving rigid wall)
BC 2 = 0 (Anchored tank; no soil-structure interaction)
BC 3b = 0 (Sloshing of the fluid surface does not occur).

This results in the impulsive rigid pressure component:

$$p_{is,h}(\xi,\zeta,\theta,t) = \sum_{n=0}^{\infty} \frac{2 \cdot R \cdot \gamma \cdot \rho_L \cdot (-1)^n}{\nu_n^2} \left[\frac{I_1\left(\frac{\nu_n}{\gamma} \cdot \xi\right)}{I_1'\left(\frac{\nu_n}{\gamma}\right)} \right] [\cos(\nu_n \cdot \zeta)][\cos(\theta)]\left[a_{is,h}(t) \cdot \Gamma_{is,h}\right] \tag{5.45}$$

with:

$p_{is,h}$ Impulsive rigid pressure component due to horizontal excitation
n Summation index
R Inner radius of the tank wall
ρ_L Density of the liquid
ν_n Coefficient: $\nu_n = \frac{2n+1}{2} \cdot \pi$
I_1 Modified first order Bessel function; Bessel function with purely imaginary argument; Bronstein and Semendjajew (1996):

$$I_1\left(\frac{\nu_n}{\gamma}\cdot\xi\right)=\frac{J_1\left(i\cdot\frac{\nu_n}{\gamma}\cdot\xi\right)}{i^n}=\sum_{k=0}^{\infty}\frac{1}{k!\cdot\Gamma(1+k+1)}\cdot\left(\frac{\frac{\nu_n}{\gamma}\cdot\xi}{2}\right)^{2k+1}$$

I_1' Derivative of the first order Bessel function (Eurocode 8-4 2006)

$$I_1'\left(\frac{\nu_n}{\gamma}\cdot\xi\right)=I_0\left(\frac{\nu_n}{\gamma}\cdot\xi\right)-\frac{I_1\left(\frac{\nu_n}{\gamma}\cdot\xi\right)}{\left(\frac{\nu_n}{\gamma}\cdot\xi\right)}$$

$$I_1'\left(\frac{\nu_n}{\gamma}\cdot\xi\right)=\sum_{k=0}^{\infty}\frac{1}{k!\cdot\Gamma(0+k+1)}\cdot\left(\frac{\frac{\nu_n}{\gamma}\cdot\xi}{2}\right)^{2k+0}-\frac{\sum_{k=0}^{\infty}\frac{1}{k!\cdot\Gamma(1+k+1)}\cdot\left(\frac{\frac{\nu_n}{\gamma}\cdot\xi}{2}\right)^{2k+1}}{\frac{\nu_n}{\gamma}\cdot\xi}$$

γ Slenderness of the tank, respectively of the "tank content": $\gamma = H/R$

ξ Dimensionless radius: $\xi = r/R$

ζ Dimensionless height: $\zeta = z/H$

θ Circumferential angle

$a_{is,h}(t)$ Horizontal acceleration time history for rigid movement (free field acceleration). When applying the response spectrum method, $a_{is,h}(t)$ is replaced by the spectral acceleration S_a at the period T = 0 s. According to Eurocode 8-1 (2004) it follows: $S_a(T = 0) = a_{gR}\cdot S\cdot\gamma_I$

a_{gR} Reference peak ground acceleration on type A ground (rock)

S Soil parameter

γ_I Importance factor of the tank according to Eurocode 8-1 (2004) or Eurocode 8-4 (2006), respectively

$\Gamma_{is,h}$ Participation factor of the impulsive rigid pressure component: $\Gamma_{is,h} = 1.0$. The rigid tank moves together with the subsoil.

By limiting the consideration of the pressure distribution on the tank wall only ($\xi = 1$), Eq. (5.45) simplifies to:

$$p_{is,h}(\xi=1,\zeta,\theta,t)=R\cdot\rho_L\cdot\sum_{n=0}^{\infty}\left[\frac{2\cdot\gamma\cdot(-1)^n}{\nu_n^2}\cdot\frac{I_1\left(\frac{\nu_n}{\gamma}\right)}{I_1'\left(\frac{\nu_n}{\gamma}\right)}\cdot\cos(\nu_n\cdot\zeta)\right][\cos(\theta)]\left[a_{is,h}(t)\cdot\Gamma_{is,h}\right] \quad (5.46)$$

5.6.3.3 Impulsive Flexible Pressure Component (Bending Vibration)

In contrast to the impulsive rigid pressure component, the deformability of the tank wall, which may be considerable (e.g. in the case of steel tanks), is taken into account by means of the impulsive flexible pressure component. This component corresponds to a joint vibration of the tank wall and the fluid moving simultaneously. Figure 5.35 shows a qualitative presentation of the joint bending vibration with the corresponding pressure distribution resulting in a pressure distribution on the wall and a moment action on the tank bottom.

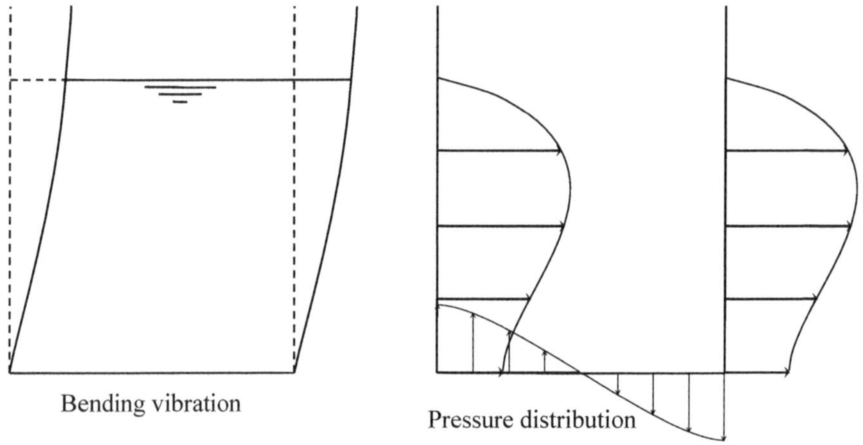

Fig. 5.35 Bending vibration of tank and liquid with qualitative pressure distribution on wall and bottom

In order to solve the potential equation, the deformation shape has to be known as function $w = f(\zeta, t)$, or it has to be determined iteratively. As the deformation shape w depends on the specific tank geometry and the filling condition, the function is usually not known a priory. Thus, the fundamental natural period and the corresponding mode shape of the tank-liquid system have to be calculated by solving the eigenvalue problem using a suitable method. A possible method, that is also suggested in Annex A of Eurocode 8-4 (2006), modifies the mass of the tank shell by means of added mass components that are to be determined from the liquid that moves simultaneously with the shell. Then the deformation shape of the "dry" shell with increased density is calculated. This so-called "added-mass-model" is also recommended by Fischer et al. (1991). The iteration procedure for an unknown interaction natural mode of the tank-liquid system is introduced after the summary of the calculation formulas that are required for the determination of the pressure distribution. The following boundary conditions need to be satisfied for the velocity potential in dependency on (5.31)–(5.33):

BC 1 with $w(\zeta) \neq const.$ (Deformation of the silo wall over the height)
BC 2 (Anchored tank; no soil-structure interaction)
BC 3b (Sloshing of the fluid surface does not occur).

This results in the impulsive flexible pressure component:

$$p_{if,h}(\xi, \zeta, \theta, t) = \sum_{n=0}^{\infty} 2 \cdot R \cdot \rho_L \cdot \left[\frac{I_1\left(\frac{\nu_n}{\gamma} \cdot \xi\right)}{\frac{\nu_n}{\gamma} \cdot I_1'\left(\frac{\nu_n}{\gamma}\right)} \right] \left[\cos(\nu_n \cdot \zeta) \int_0^1 f(\zeta) \cdot \cos(\nu_n \cdot \zeta) d\zeta \right] [\cos(\theta)] \left[a_{if,h}(t) \cdot \Gamma_{if,h} \right] \tag{5.47}$$

with:

$p_{if,h}$ Impulsive flexible horizontal pressure component corresponding to a joint bending vibration of the tank and the liquid due to horizontal excitation
n Summation index
R Inner radius of the tank wall
ρ_L Density of the liquid
I_1 Modified first order Bessel function; Bessel function with purely imaginary argument; Bronstein and Semendjajew (1996):

$$I_1\left(\frac{\nu_n}{\gamma}\cdot\xi\right)=\frac{J_1\left(i\cdot\frac{\nu_n}{\gamma}\cdot\xi\right)}{i^n}=\sum_{k=0}^{\infty}\frac{1}{k!\cdot\Gamma(1+k+1)}\left(\frac{\frac{\nu_n}{\gamma}\cdot\xi}{2}\right)^{2k+1}$$

I_1' Derivatative of the first order Bessel function I_1 (Eurocode 8-4 2006)

$$I_1'\left(\frac{\nu_n}{\gamma}\cdot\xi\right)=\sum_{k=0}^{\infty}\frac{1}{k!\cdot\Gamma(0+k+1)}\cdot\left(\frac{\frac{\nu_n}{\gamma}\cdot\xi}{2}\right)^{2k+0}-\frac{\sum_{k=0}^{\infty}\frac{1}{k!\cdot\Gamma(1+k+1)}\left(\frac{\frac{\nu_n}{\gamma}\cdot\xi}{2}\right)^{2k+1}}{\frac{\nu_n}{\gamma}\cdot\xi}$$

ν_n Coefficient: $\nu_n=\frac{2n+1}{2}\cdot\pi$
γ Slenderness of the tank, respectively of the "tank content": $\gamma=H/R$
ξ Dimensionless radius: $\xi=r/R$
$f(\zeta)$ Bending curve of the first joint (anti-symmetric) vibration of tank and liquid. Contributions of higher natural modes are neglected.
ζ Dimensionless height: $\zeta=z/H$
θ Circumferential angle
$a_{if,h}(t)$ Horizontal acceleration time history (relative acceleration time history) of the equivalent single-degree-of-freedom oscillator of the joint bending vibration of tank and liquid. When applying the response spectrum method (Sect. 5.6.6), the spectral acceleration is to be determined for the fundamental period $T_{if,h,1}$ according to (5.50). Alternatively, $T_{if,h,1}$ can be calculated with an iterative finite-element calculation of the joint bending vibration of tank and liquid. The damping for tanks made of steel can be assumed with 2% according to Kettler (2004).
$\Gamma_{if,h}$ Participation factor of the impulsive flexible pressure component.

The horizontal response acceleration $a_{if,h}(t)$ of the tank-liquid system relative to the foundation level is relevant for the impulsive flexible pressure component, as the rigid body acceleration is already included in the impulsive pressure component. In this context it has to be considered that the normative acceleration spectra depict absolute spectral accelerations. As is well-known, however, relative and absolute acceleration spectra differ from each other in the higher period range. Since the fundamental natural period $T_{if,h,1}$ for usual tank geometries does not lie in the high period range, the impulsive flexible pressure component can be calculated with an adequate precision using normative absolute acceleration spectra. This recommendation is also made by Scharf et al. (1991) and Scharf (1990). If relative acceleration spectra are used, the part of the free surface acceleration must be subtracted. The participation factors $\Gamma_{if,h}$ for the impulsive flexible pressure component can be calculated according to Fischer and Rammerstorfer (1982):

$$\Gamma_{if,h}=\frac{\int_0^1 \frac{p_{if,h}(\zeta)}{s(\zeta)}d\zeta}{\int_0^1 \frac{f(\zeta)}{s(\zeta)}\cdot p_{if,h}(\zeta)d\zeta},\ \text{for variable wall thicknesses s}(\zeta)$$

$$\Gamma_{if,h}=\frac{\int_0^1 p_{if,h}(\zeta)d\zeta}{\int_0^1 f(\zeta)\cdot p_{if,h}(\zeta)d\zeta},\ \text{for constant wall thicknesses s}(\zeta)=\text{const.} \tag{5.48}$$

with:

- $\Gamma_{if,h}$ Participation factor of the impulsive flexible pressure component (bending vibration) due to horizontal excitation
- $f(\zeta)$ Fundamental joint natural bending mode shape of tank and liquid. The contributions of higher natural modes are neglected
- $p_{if,h}(\zeta)$ Pressure function of the impulsive flexible pressure component over the filling height
- $s(\zeta)$ Wall thickness of the tank
- ζ Dimensionless height: ζ = z/H.

The pressure distribution on the tank walls ($\xi = 1$) results from (5.47) to:

$$p_{if,h}(\xi=1,\zeta,\theta,t)$$
$$=R\cdot\rho_L\sum_{n=0}^{\infty}\left[2\cdot\frac{I_1\left(\frac{\nu_n}{\gamma}\cdot 1\right)}{\frac{\nu_n}{\gamma}\cdot I_1'\left(\frac{\nu_n}{\gamma}\right)}\cdot\cos(\nu_n\cdot\zeta)\int_0^1 f(\zeta)\cdot\cos(\nu_n\cdot\zeta)d\zeta\right][\cos(\theta)][a_{if,h}(t)\cdot\Gamma_{if,h}] \tag{5.49}$$

The fundamental natural period $T_{if,h,1}$, which is necessary for determining the response acceleration $a_{if,h}(t)$ of the common vibration of tank and liquid, can be calculated according to Rammerstorfer et al. (1988), Rammerstorfer and Fischer (2004), Eurocode 8-4 (2006) and Sakai et al. (1984) as follows:

$$T_{if,h,1}=2\cdot F(\gamma)\sqrt{\frac{W_L}{\pi\cdot g\cdot E\cdot s(\zeta=1/3)}}=2\cdot R\cdot F(\gamma)\sqrt{\frac{H\cdot\rho_L}{E\cdot s(\zeta=1/3)}} \tag{5.50}$$

with:

- W_L Weight of the total liquid mass: $W_L=\pi\cdot R^2\cdot H\cdot\rho_L\cdot g$
- $F(\gamma)$ Statistically determined correction factor: $F(\gamma)=0.1567\cdot\gamma^2+\gamma+1.49$ according to Rammerstorfer und Fischer (2004)
- $s\left(\zeta=\frac{1}{3}\right)$ Tank wall thickness at 1/3 of the filling height.

It has to be considered that the statistically determined correction factor $F(\gamma)$ in Rammerstorfer et al. (1988) is usually only valid for tanks with a slenderness of $\gamma \leq 4$. However, own studies have shown that the approximate formula also yields satisfactory results for larger slenderness values.

It has already been stated that the shape of the joint natural bending mode $w = f(\zeta)$ is not generally known so that the mode shapes must be determined iteratively. At first a bending mode shape satisfying the following conditions is assumed:

$$f\text{max} = 1 \; and \; f(\zeta = 0) = 0 \tag{5.51}$$

With this assumed bending mode shape, the resulting pressure distribution on the tank wall is determined in accordance with Eq. (5.49). This in turn serves as a basis for the calculation of an added mass or density component $\Delta\rho$ representing the additional mass of the liquid on the tank shell. The inertia forces of this added mass applied to an infinitesimal tank ring are equivalent to the dynamically activated fluid pressure due to horizontal excitation. The added density component $\Delta\rho^j(\zeta)$ results from the "added-mass approach" (Holl 1987):

$$\Delta\rho^j(\zeta) = \frac{p_f^j(\zeta)}{2 \cdot s(\zeta) \cdot f^j(\zeta)} \frac{1}{\cos(\theta) \cdot a_{f,hi}(t) \cdot \Gamma_{if,h}} \tag{5.52}$$

Herein, $p_f^j(\zeta)$ is the pressure and $f^j(\zeta)$ is the bending mode shape at dimensionless height ζ in the current iteration step j. As it is assumed that the mass distribution on the cantilever arm remains constant across the circumference, the circumferential cosine for the iteration can be removed from the pressure. The division by the acceleration value $a_{if,h}(t)$ is necessary because the pressure distribution $p_f^j(\zeta)$ is divided by the acceleration distribution of the considered eigenmode. In other references Eq. (5.52) can be found in slightly different forms due to different formulations or normalisations, respectively, regarding the pressure distribution (Rammerstorfer et al. 1988). With $\rho_S(\zeta)$ as a real construction-related shell density, the effective density $\rho^j(\zeta)$ of the tank shell is calculated in each iteration step j:

$$\rho^j(\zeta) = \rho_S(\zeta) + \Delta\rho^j(\zeta). \tag{5.53}$$

A repeated eigenvalue analysis of the tank is to be carried out using this modified shell density to correct the bending function $f^j(\zeta)$. The iteration is continued until the new bending mode shape no longer exhibits any relevant changes when compared to the previous iteration step. This is normally already the case after the completion of four or five iteration steps.

The use of the iteration method is very time consuming and hardly suitable for practical use as the solution requires a coupling of specialised mathematics software (for the determination of the current pressure distribution) and a sophisticated finite-element software package for the calculation of the mode shapes of a specific tank structure. Interface programming makes sense for the transfer of the calculation between the mentioned software components.

The iterative calculation and coupling of mathematics software with a FE software package can be avoided if it is possible to assume that the natural bending mode shape is known as a result of the selection of a suitable approximate function.

In the literature (e.g. Kettler 2004; Habenberger 2001), a possible approximation is given for the bending shape by means of three functions depending on the slenderness of the tank. The suggested functions are a sine-shaped [$f(\zeta) = \sin((\pi/2) \cdot \zeta)$ for squat tanks], a linear [$f(\zeta) = \zeta$ for slender tanks] and a cosine-shaped bending curve [$f(\zeta) = 1 - \cos((\pi/2) \cdot \zeta)$ for very slender tanks]. The disadvantage however is that - contrary to a precise iterative method - in the transition zones of the tank slenderness the actual natural mode shape (and, thus, the resulting impulsive flexible pressure component) is reproduced only with limited accuracy.

Cornelissen (2010) analysed and compared diverse parametric functions with the aim of creating a realistic and reliable reproduction of the joint bending vibration shape of the tank and the liquid by mans of a single approximate function. As a result of his analyses, he suggests using a sine function comprising four free parameters a, b, c and d, so that the function can be scaled and moved along both coordinates directions. The function is given by:

$$f(\zeta) = a \cdot \sin\left(\frac{\pi}{2} \cdot (\zeta - b) \cdot c\right) + d \tag{5.54}$$

With this shape function, a good conformity with numerical calculations was achieved in numerous parametric analyses when taking the interaction vibration into account. It was seen from the analyses that the natural bending mode shape is not only influenced by the tank slenderness, but also by the Poisson's ratio ν of the tank wall, the ratio of the mass of the fluid to the mass of the tank shell and a non-uniform wall thickness over the height of the tank. However, the latter influences are comparatively small and it is sufficient to determine the parameters with the assumptions of a Poisson's ratio of $\nu = 0.3$ (steel), a constant wall thickness and neglecting the mass of the tank shell as long as the following condition holds:

$$\frac{R}{t} \geq 60 \cdot \frac{\rho_S}{\rho_L} \tag{5.55}$$

Herein, R is the inner radius of the tank, t the average wall thickness, ρ_S the density of the tank wall and ρ_L the liquid density. When adhering to condition (5.55), the free parameters can be determined depending on the tank slenderness, by conducting parameter analyses with numerical tank models and applying the Least Squares method. The resulting values for the free parameters are summarized in Table 5.12. When using parameters a, b, c and d, it is possible to use the sine function as known bending mode shape so that a time-consuming iteration is no longer necessary.

Table 5.12 Free parameters a, b, c, d of the parametrized sine function of Eq. (5.54) as approximate shape of the joint natural bending mode of tank and liquid

γ (–)	a (–)	b (–)	c (–)	d (–)
1.0	−116.7041	−34.2801	0.0851	−115.7037
1.5	2.2033	−0.6111	0.6003	−1.2004
2.0	1.1024	−0.0588	0.8382	−0.0852
3.0	0.6986	0.3384	1.0660	0.3750
4.0	0.6360	0.5073	1.1519	0.5052
5.0	0.6333	0.5978	1.1866	0.5684
6.0	0.6440	0.6521	1.2010	0.6070
8.0	0.6681	0.7113	1.2086	0.6519
10.0	0.6859	0.7414	1.2084	0.6767
12.0	0.6981	0.7589	1.2066	0.6920

5.6.3.4 Simplification of the Pressure Components for Practical Application

The pressure functions derived in the previous sections cannot be readily used in everyday practise because a coupling of mathematics software with sophisticated FE-software is needed to handle the complex mathematical functions and to solve the nonlinear interaction problem between the tank and the liquid. For practical purposes the three pressure functions are simplified as follows:

Convective pressure component (Sloshing):

$$p_k(\xi = 1, \zeta, \theta, t) = R \cdot \rho_L \sum_{n=1}^{\infty} \frac{2}{(\lambda_n^2 - 1)} \left[\frac{\cosh(\lambda_n \cdot \gamma \cdot \zeta)}{\cosh(\lambda_n \cdot \gamma)} \right] [\cos(\theta)][a_{kn}(t) \cdot \Gamma_{kn}] \tag{5.56}$$

Impulsive rigid pressure component (Rigid body movement):

$$p_{is,h}(\xi = 1, \zeta, \theta, t) = R \cdot \rho_L \sum_{n=0}^{\infty} \left[\frac{2 \cdot \gamma \cdot (-1)^n}{\nu_n^2} \cdot \frac{I_1\left(\frac{\nu_n}{\gamma}\right)}{I_1'\left(\frac{\nu_n}{\gamma}\right)} \cos(\nu_n \cdot \zeta) \right] \cdot [\cos(\theta)]\left[a_{is,h}(t) \cdot \Gamma_{is,h}\right] \tag{5.57}$$

Impulsive flexible pressure component (Joint bending vibration of tank and liquid):

$$p_{if,h}(\xi = 1, \zeta, \theta, t) = R \cdot \rho_L \sum_{n=0}^{\infty} \left[2 \frac{I_1\left(\frac{\nu_n}{\gamma}\right)}{\frac{\nu_n}{\gamma} \cdot I_1'\left(\frac{\nu_n}{\gamma}\right)} \cos(\nu_n \cdot \zeta) \int_0^1 f(\zeta) \cdot \cos(\nu_n \cdot \zeta) d\zeta \right] \cdot [\cos(\theta)]\left[a_{if,h}(t) \cdot \Gamma_{if,h}\right] \tag{5.58}$$

The three pressure functions include the inner radius R of the tank, the liquid density ρ_L, series expansions of hyperbolic cosine and Bessel functions, the peripheral cosine cos (θ), an acceleration value a(t) and a participation factor Γ. A simplification and summary of the three pressure components can be achieved by tabulating the series expansions in the form of coefficients $C_j(\zeta, \gamma)$. The convective pressure component can be simplified if only the fundamental natural mode is taken into account. This results with $a_k(t) = a_{k1}(t)$ and $\Gamma_k = \Gamma_{k1}$ to:

$$p_j(\xi = 1, \zeta, \theta, t) = R \cdot \rho_L \cdot C_j(\zeta, \gamma) \cdot \cos(\theta) \cdot a_j(t) \cdot \Gamma_j \tag{5.59}$$

with:

p_j	Pressure component j; $j = \{k; is, h; if, h\}$
$C_j(\zeta, \gamma)$	Tabulated coefficients of pressure component j; the coefficient corresponds to the normalized pressure component. The tables are compiled in Sect. 5.6.12.
$a_j(t)$	Acceleration of pressure component j
Γ_j	Participation factor of pressure component j.

The pressure ordinates p_j can be applied to a finite element model as static loads. The calculation results of the linear calculation model can be used as a basis for the tank design. The basis of the tabulated coefficients for the impulsive flexible pressure component is the assumption that a closed solution can be found for the integral in Eq. (5.49). This is possible if the non-parametrised approximate functions $f(\zeta) = \sin((\pi/2) \cdot \zeta)$, $f(\zeta) = \zeta$ and $f(\zeta) = 1 - \cos((\pi/2) \cdot \zeta)$ for the fundamental natural bending mode shape in Sect. 5.6.3.3 are used. As an alternative, the parametrised sine approach (5.54) can also be used to find a closed solution and the coefficient can also be tabulated accordingly.

The graphical representation of the coefficients in form of diagrams is shown Figs. 5.36, 5.37, 5.38, 5.39 result. Therein, the abscissa values do not represent the absolute pressure values but must be multiplied by the coefficients given in (5.59).

The coefficients $C_j(\zeta, \gamma)$ are tabulated in Sect. 5.6.12 for diverse tank slenderness values in Tables 5.20, 5.21 and 5.22, wherein the bending curves for the impulsive flexible pressure component are associated to specific slenderness ranges. Following Kettler (2004) and Habenberger (2001) a sine bending curve was applied to compact tanks ($\gamma < 3$), a linear one to medium slender tanks ($3 \leq \gamma \leq 8$) and a cosine one to very slender tanks ($\gamma > 8$) and the corresponding tabulated coefficients were determined on this basis. As an alternative, the coefficients $C_j(\zeta, \gamma)$ can also be determined for the parametric sine approach according to (5.59) by using Table 5.23.

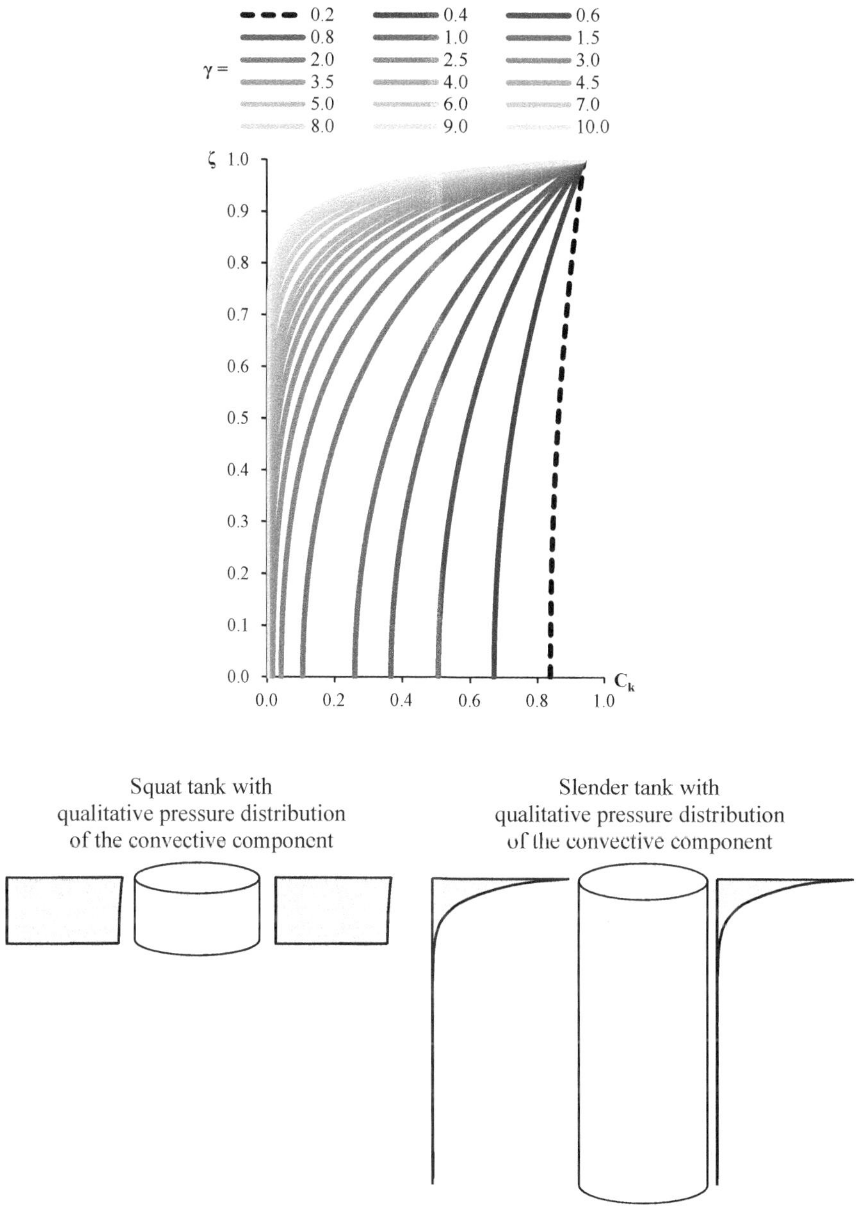

Fig. 5.36 Normalized convective pressure component $\mathbf{C_k}$ according to (5.59), to be multiplied with $\mathbf{R} \cdot \rho_\mathbf{L} \cdot \cos(\boldsymbol{\theta}) \cdot \mathbf{a_k(t)} \cdot \boldsymbol{\Gamma}_\mathbf{k}$

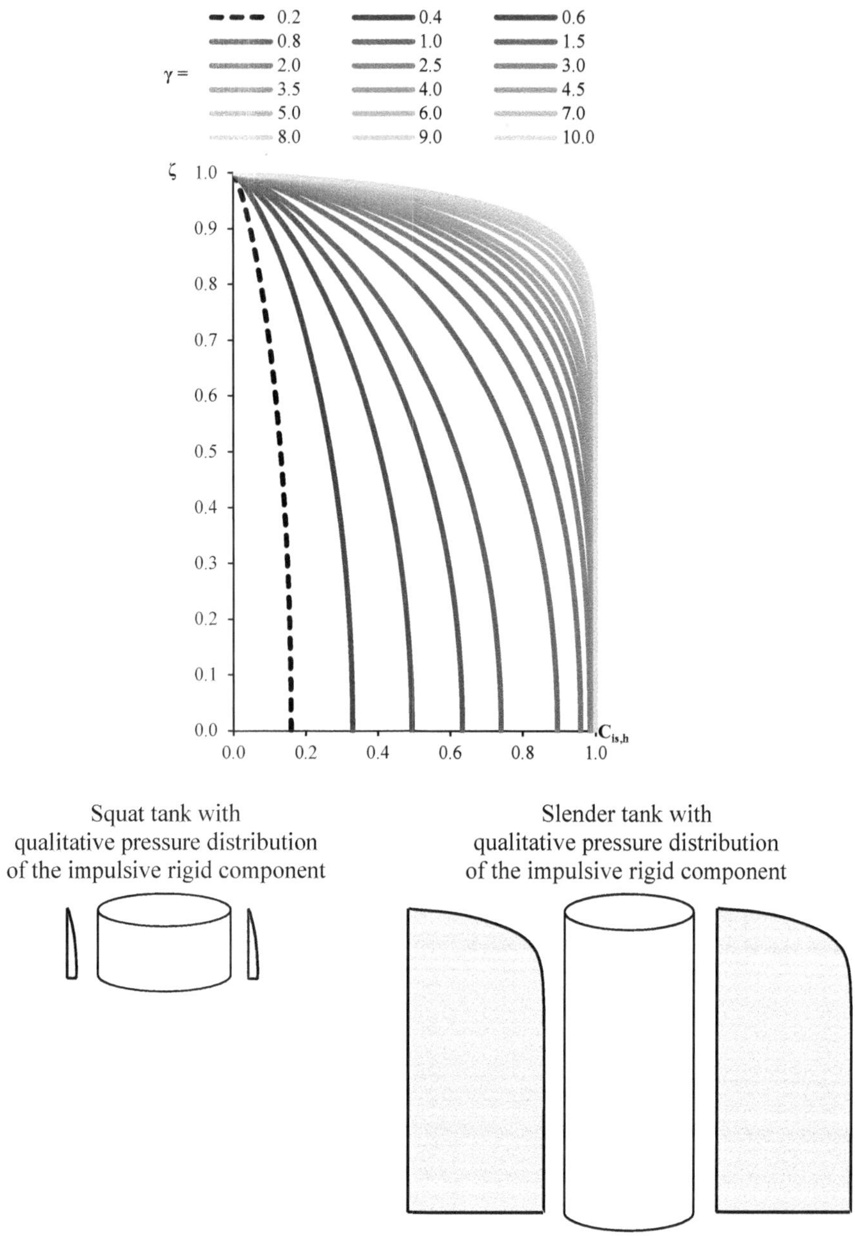

Fig. 5.37 Normalized impulsive rigid pressure component $\mathbf{C}_{\mathbf{is,h}}$ according to (5.59), to be multiplied with $\mathbf{R} \cdot \rho_L \cdot \cos(\boldsymbol{\theta}) \cdot \mathbf{a}_{\mathbf{is,h}}(\mathbf{t}) \cdot \boldsymbol{\Gamma}_{\mathbf{is,h}}$

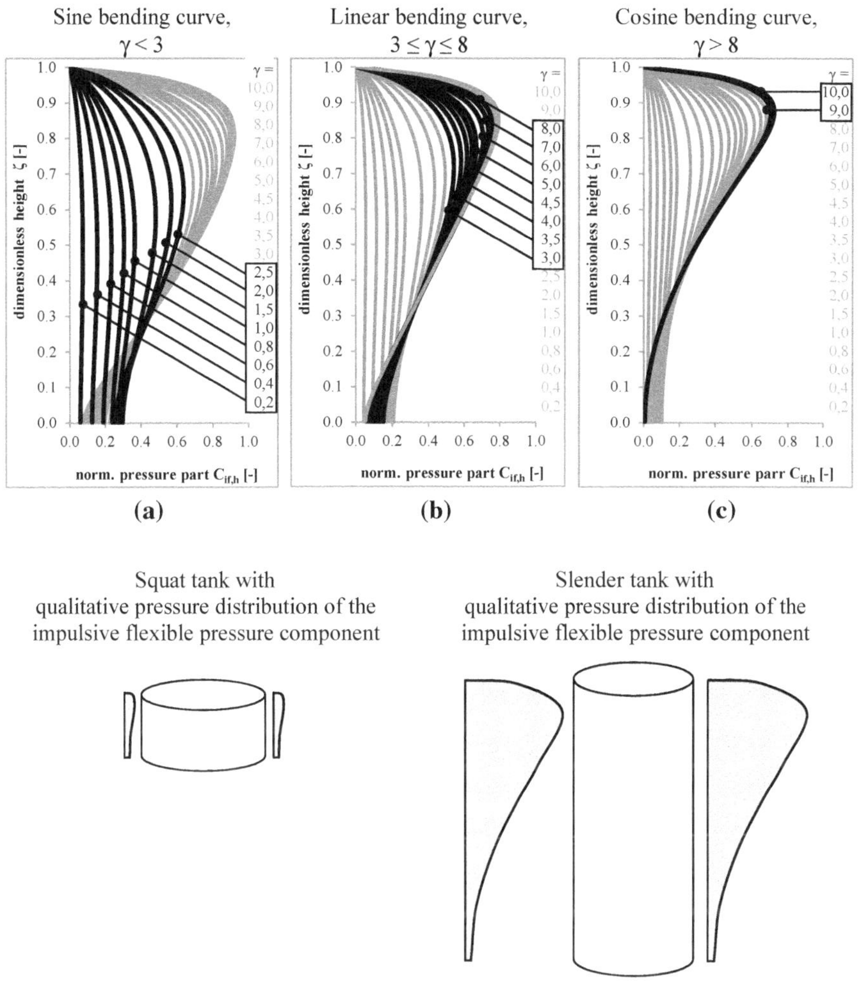

Fig. 5.38 Normalized impulsive flexible pressure component $\mathbf{C}_{\mathbf{if,h}}$ according to (5.59), to be multiplied with $\mathbf{R} \cdot \rho_{\mathbf{L}} \cdot \cos(\boldsymbol{\theta}) \cdot \mathbf{a}_{\mathbf{Is,h}}(\mathbf{t}) \cdot \boldsymbol{\Gamma}_{\mathbf{is,h}}$
(**a**) Sine bending curve for: $\gamma < 3$, $\mathbf{f}(\zeta) = \mathbf{sin}\left(\frac{\pi}{2} \cdot \zeta\right)$
(**b**) Linear deflection curve for: $3 \leq \gamma \leq 8$, $\mathbf{f}(\zeta) = \zeta$
(**c**) Cosine bending curve: $\gamma > 8$, $\mathbf{f}(\zeta) = 1 - \cos\left(\frac{\pi}{2} \cdot \zeta\right)$

5.6.3.5 Superposition for One-Dimensional Horizontal Seismic Action

As far as time integration is concerned, each of the pressure components follows the acceleration time histories applied in the corresponding spatial directions, so that the relevant frequencies and natural modes are excited independently from each other. This enables the pressure components to be applied simultaneously in each of the load steps.

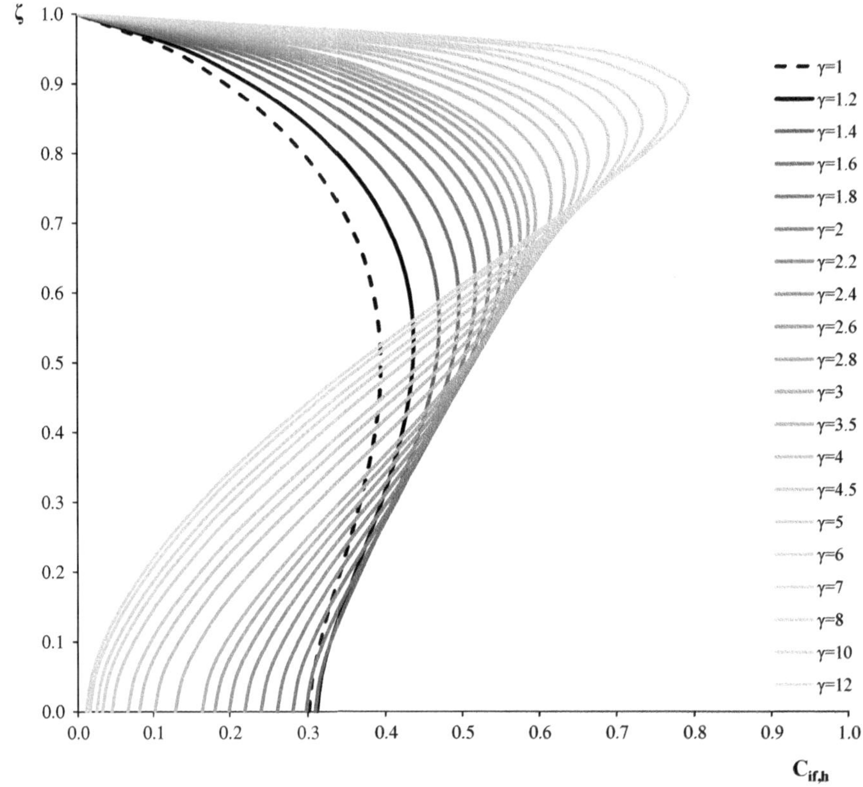

Fig. 5.39 Normalized impulsive flexible pressure component $\mathbf{C_{if,h}}$ according to (5.59) for the parametric sine approach based on Eq. (5.54), to be multiplied with $\mathbf{R} \cdot \rho_L \cdot \cos(\boldsymbol{\theta}) \cdot \mathbf{a_{is,h}(t)} \cdot \boldsymbol{\Gamma}_{\mathbf{is,h}}$

Contrary to this, when applying the response spectrum method, the maximal values of the accelerations corresponding to the natural periods are applied as load components. It has to be checked that the superposition of the single components results in a realistic overall loading of the tank. In conservative terms, the resulting horizontal pressure applied to the tank shell can be calculated as a square root of the sum of the squared pressure components according to Rammerstorfer und Fischer (2004):

$$p_{h,max}(\zeta)=\sqrt{(p_k(\zeta))^2+\left(p_{is,h}(\zeta)\right)^2+\left(p_{if,h}(\zeta)\right)^2} \tag{5.60}$$

with:

$p_{h,max}$ Resulting pressure of the response spectrum analysis as a result of the superposition of the single pressure components due to one-dimensional horizontal seismic action.

It should be noted here that the large frequency gap between the fundamental natural frequency of the ground movement and the fundamental natural frequency of the convective component can lead to an underestimation of the seismic response when applying the SRSS rule. A conservative pressure value can be obtained be summing up the maxima of the impulsive component and the convective component. As the convective component is much smaller in comparison to the impulsive component, the usual superimposition applying the SRSS rule seems sufficient. In cases where there are large stresses from the convective component, the influence of an additive superimposition on the design should be investigated.

5.6.4 *Vertical Seismic Actions*

The pressure components from vertical seismic excitations can also be derived from the streaming potential Φ for fluids with corresponding boundary conditions and from the equation of motion of the tank shell. Due to the analogy to Sect. 5.6.2, reference is made to Luft (1984), Fischer and Seeber (1988) and Tang (1986) for further detailed mathematical descriptions. As the pressure components from the vertical excitation are rotationally symmetric, they do not have an influence on the overturning moment, but they influence the stress distribution and the buckling behaviour of the tank shell to a considerable extent.

5.6.4.1 Impulsive Rigid Pressure Component

The impulsive pressure component due to vertical seismic actions corresponds to a vertical rigid body movement of the tank, similar to the rigid-pressure component in horizontal direction (Sect. 5.6.3.2). The pressure component is rotationally symmetric and the pressure distribution corresponds to the hydrostatic pressure component. Figure 5.40 shows the vertical tank movement together with the corresponding qualitative pressure distribution resulting in a rotationally symmetrical pressure distribution on the tank wall and a moment action on the tank bottom. According to Fischer et al. (1991), the pressure components with the liquid density and the vertical ground acceleration $a_v(t)$ can be described as being a function over the height of the tank:

$$p_{is,v}(\zeta, t) = \rho_L[H \cdot (1 - \zeta)][a_v(t) \cdot \Gamma_{is,v}] \tag{5.61}$$

with:

$p_{is,v}$ Impulsive rigid pressure component due to vertical seismic action
ρ_L Density of the liquid
H Filling height

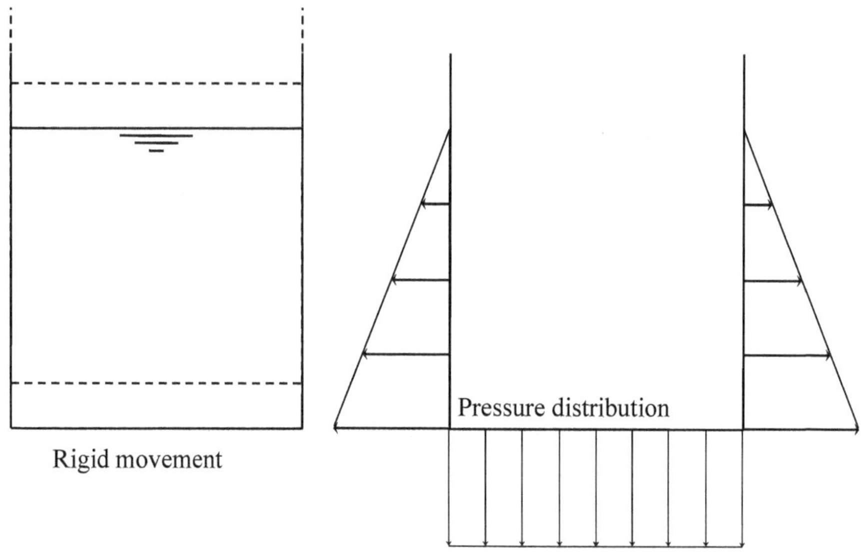

Fig. 5.40 Rigid body movement of tank and liquid with qualitative pressure distribution on wall and bottom

ζ Dimensionless height: $\zeta = z/H$

$a_v(t)$ Vertical acceleration time history for rigid movement (free field acceleration); when applying the response spectrum method the vertical acceleration $a_v(t)$ is replaced by the spectral acceleration S_{av} at period T = 0 s. According to Eurocode 8-1 (2004) it follows: $S_{av}(T = 0) = a_{gRv} \cdot S \cdot \gamma_I$

a_{gRv} Reference peak ground acceleration on type A ground (rock) in vertical direction. For spectrum type I: $a_{gRv} = 0.9 \cdot a_{gR}$; for spectrum type II: $a_{gRv} = 0.45 \cdot a_{gR}$

S Soil parameter for the vertical design spectrum: S = 1.0

γ_I Importance factor of the tank according to Eurocode 8-1 (2004) or Eurocode 8-4 (2006), respectively

$\Gamma_{is,v}$ Participation factor of the impulsive rigid pressure component in vertical direction: $\Gamma_{is,v} = 1.0$. The entire tank with liquid follows the ground movement.

5.6.4.2 Impulsive Flexible Pressure Component

The pressure component resulting from the vertical seismic excitation is also rotationally symmetrically distributed when taking the elasticity of a flexible tank shell into account (Habenberger 2001; Tang 1986). Figure 5.41 shows the bending vibration of the vertical flexible pressure component with the corresponding qualitative pressure distribution resulting in a rotationally symmetrical pressure distribution on the tank wall and a moment action on the tank bottom.

As in the case of the horizontal excitation in (5.47), the load on the tank shell is again represented by a function depending on the dimensionless coordinate ζ over the tank height:

$$p_{if,v}(\xi,\zeta,t)=\sum_{n=0}^{\infty}2R\cdot\rho_L\left[\frac{I_0\left(\frac{\nu_n}{\gamma}\cdot\xi\right)}{\frac{\nu_n}{\gamma}\cdot I_1\left(\frac{\nu_n}{\gamma}\right)}\right]\left[\cos(\nu_n\cdot\zeta)\int_0^1 f(\zeta)\cdot\cos(\nu_n\cdot\zeta)d\zeta\right]\left[a_{if,v}(t)\cdot\Gamma_{if,v}\right] \tag{5.62}$$

with:

$p_{if,v}$ Impulsive flexible pressure component due to vertical seismic action
n Summation index
R Inner radius of the tank wall
ρ_L Density of the liquid
I_0 Modified Bessel function of zero order; Bronstein and Semendjajew (1996):

$$I_0\left(\frac{\nu_n}{\gamma}\cdot\xi\right)=\frac{J_0\left(i\cdot\frac{\nu_n}{\gamma}\cdot\xi\right)}{i^n}=\sum_{k=0}^{\infty}\frac{1}{k!\cdot\Gamma(0+k+1)}\left(\frac{\frac{\nu_n}{\gamma}\cdot\xi}{2}\right)^{2k+0}$$

with:

I_1 Modified Bessel function of first order, as defined in (5.47)
ν_n Coefficient: $\nu_n=\frac{2n+1}{2}\cdot\pi$
γ Slenderness of the tank, respectively of the "tank content": $\gamma=H/R$
ξ Dimensionless radius: $\xi=r/R$

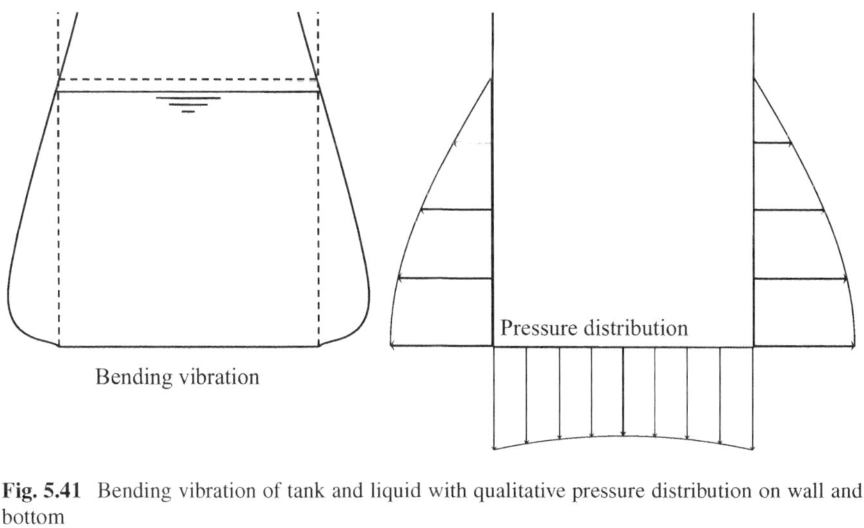

Fig. 5.41 Bending vibration of tank and liquid with qualitative pressure distribution on wall and bottom

ζ Dimensionless height: $\zeta = z/H$

$f(\zeta)$ Bending curve of the fundamental joint natural mode (rotationally symmetric) of the tank and the liquid; higher mode shapes are neglected

$a_{if,v}(t)$ Vertical acceleration time history (relative acceleration) of the joint vibration of the tank and the liquid. When applying the response spectrum method (Sect. 5.6.6) the vertical acceleration $a_v(t)$ is replaced by the spectral acceleration S_{av} at period $T = T_{if,v}$. The spectral acceleration S_{av} is to be determined for the fundamental period $T_{if,v,1}$ according to (5.63). Alternatively, $T_{if,v,1}$ can be calculated with an iterative finite-element calculation of the joint vertical vibration of tank and liquid. The damping for the calculation can be assumed with 3% for hard soil conditions and 6% for soft soil conditions as recommended by Rammerstorfer und Fischer (2004)

$\Gamma_{if,v}$ Participation factor of the impulsive flexible pressure component in the vertical direction (5.65).

The fundamental natural period $T_{if,v,1}$ of the joint vertical vibration of the tank and the liquid required for the determination of the response acceleration $a_{if,v}$ can be roughly calculated according to Rammerstorfer and Fischer (2004) and Habenberger (2001):

$$T_{if,v,1} = 4 \cdot R \cdot \sqrt{\frac{\pi \cdot \rho_L}{2} \cdot \frac{1-\nu^2}{E} \cdot \frac{H}{s\left(\zeta = \frac{1}{3}\right)} \cdot \frac{I_0\left(\frac{\pi}{2\cdot\gamma}\right)}{I_1\left(\frac{\pi}{2\cdot\gamma}\right)}} \quad (5.63)$$

with:

ν Poisson‘s ratio of the tank shell

E Young’s modulus of the tank shell

$s\left(\zeta = \frac{1}{3}\right)$ Thickness of the tank shell at 1/3 of the filling height.

Considering only the first joint bending vibration and assuming a bending curve of $f(\zeta) = \cos\left(\frac{\pi}{2} \cdot \zeta\right)$, the integral for the first term of the series is equal to 0.5 and all other terms vanish. Since again only the pressure distribution on the tank wall (“ξ = 1”) is of interest, Eq. (5.62) can be simplified to:

$$p_{if,v}(\zeta, t) = \frac{2}{\pi} \cdot H \cdot \rho_L \cdot \left[\frac{I_0\left(\frac{\pi}{2\cdot\gamma}\right)}{I_1\left(\frac{\pi}{2\cdot\gamma}\right)}\right] \cdot \left[\cos\left(\frac{\pi}{2} \cdot \zeta\right)\right] \cdot \left[a_{if,v}(t) \cdot \Gamma_{if,v}\right] \quad (5.64)$$

The participation factor $\Gamma_{if,v}$ for the impulsive flexible pressure component can be calculated using a cosine approach according to Habenberger (2001):

$$\Gamma_{if,v} = \frac{4}{\pi} \cdot \frac{I_1\left(\frac{\pi}{2\cdot\gamma}\right)}{I_0\left(\frac{\pi}{2\cdot\gamma}\right)} \quad (5.65)$$

with:

I_0, I_1 Modified Bessel functions of zero or first order; Bronstein and Semendjajew (1996), as defined in (5.47), (5.62)

γ Slenderness of the tank, respectively of the "tank content": $\gamma = H/R$.

Investigations have shown that the pressure distribution is strongly influenced by the degree of clamping at the tank bottom (Scharf 1990). Habenberger (2001) recommends a modification of the pressure function using a correction factor $\beta(\gamma)$ in order to take the clamping effect into account. Applying the correction factor $\beta(\gamma)$ to (5.64) leads to the corrected impulsive flexible pressure component $p_{if,v,corr}$ due to vertical seismic actions:

$$p_{if,v,corr}(\zeta,t) = \beta(\gamma)\cdot\frac{2}{\pi}\cdot R\cdot\gamma\cdot\rho_L\cdot\left[\frac{I_0\left(\frac{\pi}{2\cdot\gamma}\right)}{I_1\left(\frac{\pi}{2\cdot\gamma}\right)}\right]\cdot\left[\cos\left(\frac{\pi}{2}\cdot\zeta\right)\right]\cdot\left[a_{if,v}(t)\cdot\Gamma_{if,v}\right] \tag{5.66}$$

The correction factor for taking account of the clamping effect is defined by Habenberger as:

$$\beta(\gamma) = \begin{cases} 1.0 & \text{for } \gamma < 0.8 \\ 1.078 + 0.274\,ln(\gamma) & \text{for } 0.8 \le \gamma \le 4.0 \end{cases}$$

The correction factor $\beta(\gamma)$ has been empirically determined and is, so far, only available for tanks with slenderness values of $\gamma \le 4$. It is listed in Table 5.25 for typical slenderness values. All other variables are given in (5.62). The corrected impulsive flexible pressure component due to vertical seismic action can also be simplified analogously to (5.59):

$$p_{if,v}(\xi = 1,\zeta,t) = R\cdot\rho_L\cdot 1.0\cdot a_{if,v}(t)\cdot C_{if,v}(\zeta,\gamma)\cdot\Gamma_{if,v}. \tag{5.67}$$

The normalized coefficient $C_{if,v}(\zeta,\gamma)$ including the correction factor $\beta(\gamma)$ is shown in Fig. 5.42 and summarized in Table 5.24 for typical slenderness values of liquid filled tanks. If the pressure function $p_{if,v}$ according to (5.66) is substituted by $\tilde{p}_{if,v} = p_{if,v}\cdot\Gamma_{if,v}$ the final pressure function is obtained:

$$\tilde{p}_{if,v}(\zeta) = \frac{8}{\pi^2}\cdot\beta(\gamma)\cdot H\cdot\rho_L\cdot\left[\cos\left(\frac{\pi}{2}\cdot\zeta\right)\right]\cdot S_{a,if,v} \tag{5.68}$$

Herein, $S_{a,if,v}$ is the spectral acceleration for the impulsive flexible pressure component due to vertical seismic action. Similar expressions for the impulsive flexible pressure functions are given in Eurocode 8-4 (2006), Scharf (1990) and Fischer et al. (1991).

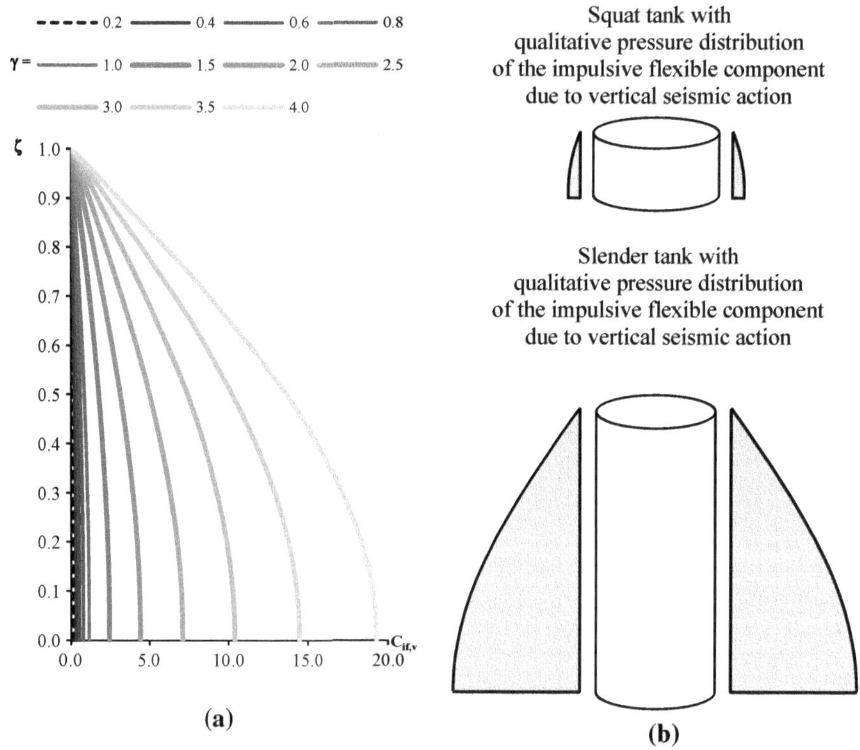

Fig. 5.42 (**a**) Normalized impulsive flexible pressure $\mathbf{C_{if,v}}$ including correction factor β(γ) according to (5.66), to be multiplied with $\mathbf{R} \cdot \boldsymbol{\rho}_\mathbf{L} \cdot \mathbf{1.0} \cdot \mathbf{a_{if,v}(t)} \cdot \boldsymbol{\Gamma}_{\mathbf{if,v}}$.
(**b**) Qualitative pressure distributions of vertical bending vibrations for squat and slender tanks

5.6.4.3 Superposition of the Vertical Pressure Components

The superposition of the two pressure components due to vertical seismic actions is carried out as described in Sect. 5.6.3.5 by applying the SRSS rule (Eurocode 8-4 2006; Scharf 1990):

$$p_{v,max}(\zeta)=\sqrt{\left(p_{is,v}(\zeta)\right)^2+\left(p_{if,v}(\zeta)\right)^2} \tag{5.69}$$

5.6.5 Superposition for Three-Dimensional Seismic Excitation

The resulting pressure distributions for the one-dimensional horizontal and vertical seismic excitation were derived separately in Sects. 5.6.3 and 5.6.4. Due to the

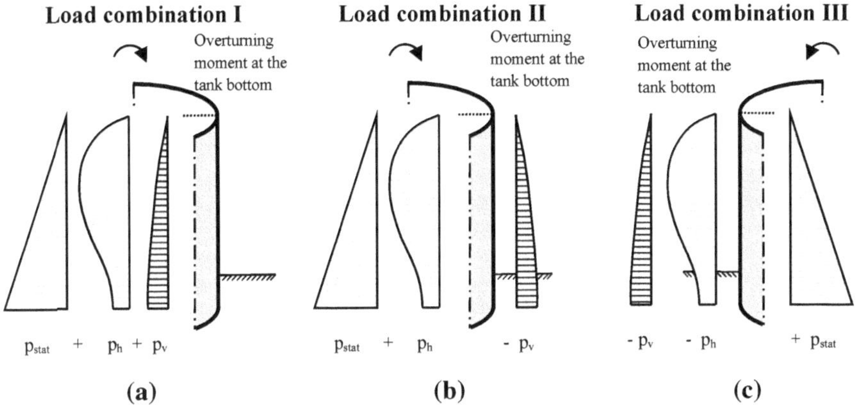

Fig. 5.43 Load combinations consisting of liquid pressure, horizontal and vertical dynamic pressures

spatial characteristic of ground movements, the horizontal pressure distribution must be assumed to act simultaneously in two orthogonal directions. Thus, considering horizontal and vertical seismic actions means that a total of three time-dependent pressure distributions has to be applied in all spatial directions simultaneously.

If a time history calculation is carried out, three stochastically independent time histories can be applied in the three spatial directions. In this case, the pressure superposition is automatically included in each time step of the calculation.

Scharf (1990) conducted time history analyses with two- and three-dimensional seismic actions. In the case of two-dimensional excitation he showed that the simultaneous occurrence of pressure maxima in the horizontal excitation directions is not covered by the superposition with the SRSS rule. That's why he proposed an additive superposition rule for the earthquake directions. However, since the conducted parameter studies included only a limited number of analyses, this conclusion cannot be generalized.

Concerning the superposition of the resulting maximum horizontal pressure distribution with the vertical pressure distribution, it is possible that the maximum or minimum pressure due to the horizontal earthquake excitation and the maximum or minimum pressure due to the vertical earthquake excitation occur simultaneously. The various combinations of the maximum and minimum pressures p_h and p_v can be related to the typical damage patterns (Rammerstorfer and Scharf 1990):

- Coincidence of the maximum pressure due to horizontal seismic excitation and the maximum pressure due to vertical seismic excitation $[p(\zeta) = p_{stat}(\zeta) + p_h(\zeta) + p_v(\zeta)]$, results in maximum circumferential tensile stresses at the base of the tank, where also large axial pressure stresses are induced from the overturning moment. This can cause plastic buckling at the base of the wall also being known as "elephant foot buckling" (Figs. 5.43a and 5.44a).

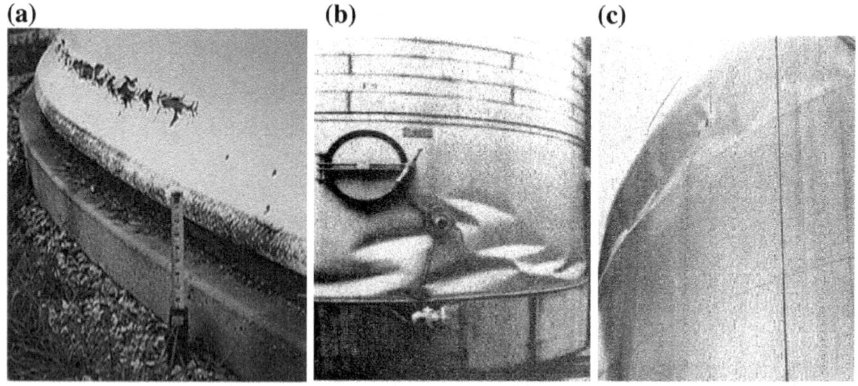

Fig. 5.44 Damage: (**a**) "Elephant foot buckling" (El-Zeiny 2000). (**b**) "Diamond shaped buckling" (Shih 1981), (**c**) Buckling due to pressure in circumferential direction and missing internal pressure (Shih 1981)

- Coincidence of the maximum pressure due to horizontal seismic excitation with the minimum pressure due to vertical seismic excitation $[p(\zeta) = p_{stat}(\zeta) + p_h(\zeta) - p_v(\zeta)]$, reduces the stabilising inner pressure. As a consequence, axial load buckling in the form of "diamond shaped buckling" (Figs. 5.43b and 5.44b) can occur.
- The same load configuration as in the previous case provokes a pressure on the opposite side of the tank that can be calculated from the hydrostatic pressure reduced by the seismically induced pressure components in both directions: $[p(\zeta) = p_{stat}(\zeta) - p_h(\zeta) - p_v(\zeta)]$. This particularly results in the formation of negative pressure areas in the upper part of the tank, which in turn can generate buckling in circumferential direction for tanks with small wall thicknesses (Fig. 5.43c and 5.44c).

In contrast to the special superposition rules for tanks proposed by Scharf (1990), another approach is given in Eurocode 8-4 (2006) concerning superposition rules for standard building constructions. According to Eurocode 8-1 (2004), Sect. 4.3.3.5, the superposition shall be executed either by means of a quadratic superposition or by applying the 30% rule, taking all three earthquake components into account.

All superposition rules mentioned are approximations which aim at considering the effects of complex spatial ground waves in a simplified but conservative manner in the course of the design process. With the help of comprehensive parameter studies conducted on numerous single-degree-of-freedom oscillators, Smoczynski und Schmitt (2011) demonstrated that the application of the superposition rules given in Eurocode 8-1 (2004) fulfils this criterion. Therefore, it can be recommended to use these superposition rules for tank structures as well.

5.6.6 Development of the Spectra for the Response Spectrum Method

When applying the response spectrum method, design spectra are required for each of the pressure components. These design spectra must be set up separately due to the different damping properties of the pressure components. Damping is taken into account by means of the damping correction coefficient η:

$$\eta = \sqrt{10/(5 + \xi)} \geq 0.55, \tag{5.70}$$

Herein, ξ is the viscous damping ratio expressed as a percentage of critical damping. Table 5.13 provides reference damping values for each pressure component. By applying these damping values the energy dissipation behaviour resulting from friction in the fluid and the dissipative potential of the tank structure are adequately described.

As an alternative to the application of viscous damping, the spectrum can also be reduced by a behaviour factor q according to Eurocode 8-1 (2004). Therein, the impulsive pressure components may be reduced with $q \geq 1.5$ while for the convective pressure component the behaviour factor is limited to $q = 1$. When applying a behaviour factor for the impulsive pressure components, the tank must actually possess a corresponding energy dissipation capacity. A high energy dissipation capacity can usually not be assumed in thin walled steel tanks as there is a risk of the occurrence of buckling effects as described in in Sect. 5.6.5.

For this reason, it is recommended to always design thin-walled steel tanks with a behaviour factor of $q = 1.0$. In contrast, reinforced concrete tanks can acquire an enhanced energy dissipation capacity by being designed accordingly. However, it does not seem too sensible to design tank structures with a high degree of dissipation capacity since they normally serve as important supply structures the structural integrity of which must be guaranteed at all times. It is thus recommended to choose a behaviour factor of $q \leq 1.5$ in all cases.

Table 5.13 Damping values of the pressure components (Scharf 1990; DIN EN 1998-4 2007)

Horizontal pressure components			Vertical pressure components	
Convective	Impulsive rigid	Impulsive flexible	Impulsive rigid	Impulsive flexible
$\xi = 0$	$\xi = 0 - 0.5\%$	Steel: $\xi = 1 - 2\%$ Concrete: $\xi = 5\%$	$\xi = 0$	Soft soil[a]: $\xi = 6\%$ Hard soil[a]: $\xi = 3\%$

[a]Soft soil: shear wave velocity ≈ 250 m/s; Hard soil: shear wave velocity ≈ 1000 m/s

5.6.7 Base Shear and Overturning Moment

5.6.7.1 Computation by Integration of the Pressure Functions

Knowledge of the base shear and overturning moment resulting from earthquake loads is necessary in order to dimension the tank foundation. The components of these effects can be determined by integrating the corresponding horizontal pressure components for the base shear and the additional multiplication with the relevant cantilever arm for the overturning moment.

Herein, it must be taken into account that the pressure components act according to Sects. 5.6.3 and 5.6.4 perpendicularly to the tank wall so that they must be transformed for the calculation of the foundation shear and the overturning moment through multiplication with $\cos(\theta)$ in the direction of the earthquake.

It must also to be taken into account that the pressure components on the tank bottom deliver a contribution to the overturning moment in the case of a horizontal earthquake excitation. The earthquake load from a vertical excitation is rotationally symmetric, rendering it irrelevant with regard to the foundation shear and overturning moment. The following term is given for the base shear components of the three pressure components:

$$F_{b,j} = \int_0^1 2 \int_{-\frac{\pi}{2}}^{+\frac{\pi}{2}} \left[p_j(\xi = 1, \zeta, \theta) \cdot \cos(\theta)\right] R\, d\theta\, H\, d\zeta$$

$$= \pi \cdot R \cdot H \int_0^1 p_j(\xi = 1, \zeta, \theta = 0)\, d\zeta \tag{5.71}$$

with:

j Index of the pressure component; $j = \{k; is, h; if, h\}$;
p_k after (5.40) with summation index n = 1
$p_{is,h}$ after (5.46)
$p_{if,h}$ after (5.49).

The overturning moment results from the pressures components acting on the tank wall $p(\xi = 1)$ and tank bottom $p(\zeta = 0)$:

$$M_j = \int_0^1 2 \int_{-\frac{\pi}{2}}^{+\frac{\pi}{2}} (H \cdot \zeta)\left[p_j(\xi = 1, \zeta, \theta) \cdot \cos(\theta)\right] R\, d\theta\, H\, d\zeta$$

$$+ \int_0^1 2 \int_{-\frac{\pi}{2}}^{+\frac{\pi}{2}} (R \cdot \xi)^2 \cdot \left[p_j(\xi, \zeta = 0, \theta) \cdot \cos(\theta)\right] R\, d\theta\, d\xi$$

$$= \pi \cdot R \cdot H^2 \int_0^1 \zeta \cdot p_j(\xi = 1, \zeta, \theta = 0)\, d\zeta + \pi \cdot R^3 \int_0^1 \xi^2 \cdot p_j(\xi, \zeta = 0, \theta = 0)\, d\xi \tag{5.72}$$

with:

j Index of the pressure component: $j = \{k; is, h; if, h\}$
$p_{k,h}$ after (5.40) with summation index n = 1
$p_{is,h}$ after (5.46)
$p_{if,h}$ after (5.49).

The calculation of the integrals in (5.71) and (5.72) for each of the pressure components can be facilitated much easier for practical applications by using tabulated coefficients. This requires four coefficients C and a participation factor Γ for each of the pressure components j = {k; is, h; if, h}:

Base shear	$C_{F,j}$
Overturning moment due to pressure on the tank wall	$C_{MW,j}$
Overturning moment due to pressures on the tank bottom	$C_{MB,j}$
Overturning moment due to pressures on the tank wall and the tank bottom	$C_{M,j}$
Participation factor	Γ_j

Using the coefficients for each of the pressure components leads to the following base shears and overturning moments:

Base shear $F_{b,j}$:

$$F_{b,j} = \pi \cdot R^2 \cdot H \cdot \rho_L \cdot a_j(t) \cdot \Gamma_j \cdot C_{F,j} = m_L \cdot a_j(t) \cdot \Gamma_j \cdot C_{F,j} \tag{5.73}$$

Overturning moment $M_{W,j}$ due to pressure on the tank wall

$$M_{W,j} = \pi \cdot R^2 \cdot H^2 \cdot \rho_L \cdot a_j(t) \cdot \Gamma_j \cdot C_{MW,j} \tag{5.74}$$

Overturning moment $M_{B,j}$ due to pressure on the tank bottom:

$$M_{B,j} = \pi \cdot R^4 \cdot \rho_L \cdot a_j(t) \cdot \Gamma_j \cdot C_{MB,j} \tag{5.75}$$

Overturning moment $M_{G,j}$ due to pressures on the tank wall and the tank bottom:

$$\begin{aligned} M_{G,j} &= \pi \cdot R^4 \cdot \rho_L \cdot a_j(t) \cdot \Gamma_j \cdot (\gamma^2 \cdot C_{MW,j} + C_{MB,j}) \\ &= \pi \cdot R^4 \cdot \rho_L \cdot a_j(t) \cdot \Gamma_j \cdot C_{M,j} \end{aligned} \tag{5.76}$$

The coefficients of each of the pressure components are tabulated in Sect. 5.7 together with the participation factors. Table 5.27 contains the coefficients for the convective and the impulsive rigid pressure components. In addition to this, the coefficients are given for the impulsive flexible pressure component using sine-shaped,

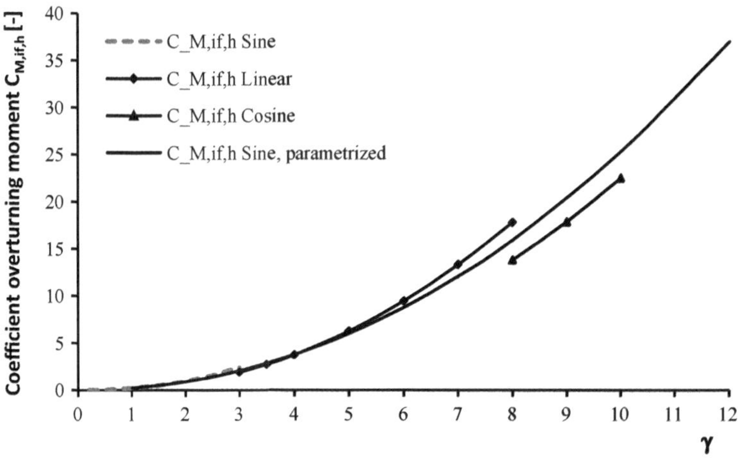

Fig. 5.45 Coefficient $C_{M,if,h}$ of the impulsive flexible pressure component

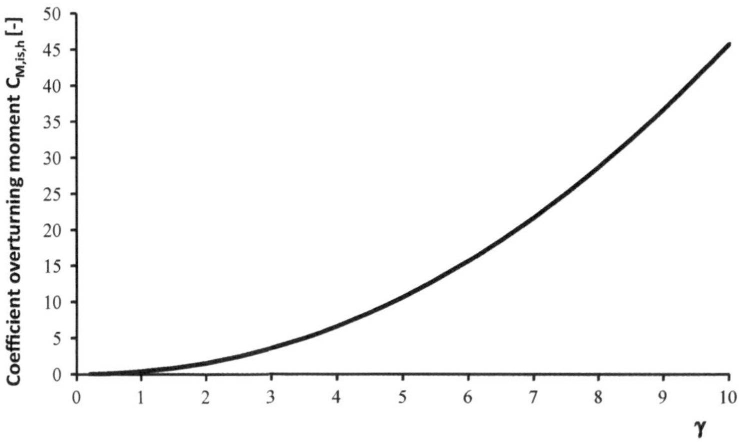

Fig. 5.46 Coefficient $C_{M,is,h}$ of the impulsive rigid pressure component

linear and cosine-shaped bending curves describing the joint flexural vibration of the fluid and the tank. Table 5.28 provides the coefficients for the parametrized sine-shaped bending curve. A comparison of the coefficient $C_{M,j}$ for the overturning moment $M_{G,j}$ due to pressures on the tank wall and the tank bottom is shown in Figs. 5.45, 5.46 and 5.47 for each of the pressure components. The comparison of the bending curves defined for the three ranges shows a good agreement with the parametrized sine function. The total base shear and the overall overturning moment are calculated by the superposition of the pressure components using the SRSS rule:

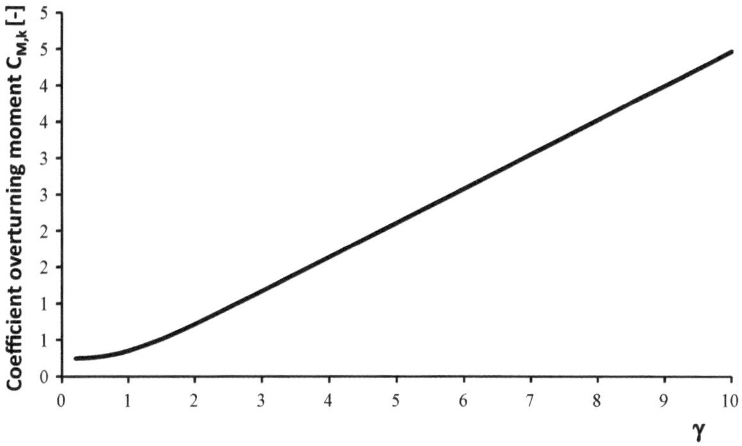

Fig. 5.47 Coefficient $C_{M,k}$ of the convective pressure component

$$F_b = \sqrt{\left(F_{b,k}\right)^2 + \left(F_{b,is,h}\right)^2 + \left(F_{b,if,h}\right)^2} \tag{5.77}$$

$$M_G = \sqrt{\left(M_{G,k}\right)^2 + \left(M_{G,is,h}\right)^2 + \left(M_{G,if,h}\right)^2} \tag{5.78}$$

The presented formulas can be used to calculate the base shear, the overturning moments resulting from the pressure on the tank wall and the overturning moments resulting from pressures acting on the tank wall and tank bottom. The overturning moments resulting from the pressure on the tank wall and those resulting from the pressures on the tank wall and tank bottom differ due to additional moments $M_{B,j}$ resulting from the dynamic pressure on the tank bottom. The moments $M_{B,j}$ are not taken into account for the calculation. The overturning moments result only from the pressure on the tank wall.

5.6.7.2 Simplified Method According to Eurocode 8-4 (2006)

A simplified method for the calculation of the base shear and the overturning moments is included in Annex A.3.2.2 of Eurocode 8-4 (2006). This method allows a simple calculation of the resulting seismic loads for the design of the substructure and the foundations without a time-consuming determination of all pressure distributions on wall and bottom. In this method, the fundamental natural periods are initially determined for the impulsive and the convective pressure components:

$$T_{ih} = C_i \cdot \frac{\sqrt{\rho_L} \cdot H}{\sqrt{s/R} \cdot \sqrt{E}} \tag{5.79}$$

$$T_k = C_k \sqrt{R} \tag{5.80}$$

with:

H Filling height for the relevant design situation
R Inner tank radius
s Equivalent constant wall thickness (weighted average over the filling height of the tank; the weighting can be carried out proportional to the strains in the tank shell, the maximum values of which occur at the tank bottom)
ρ_L Density of the liquid
E Young's modulus of the tank wall.

The coefficients C_i and C_k are given in Table 5.14. The coefficient C_i is dimensionless, the coefficient C_k has the unit (s/m$^{1/2}$) and the impulsive and convective masses m_i and m_k are normalized by the total mass m. The corresponding equivalent heights h_i, h_k or h_{iu}, h_{ku} of the impulsive and convective resulting forces are measured from the tank bottom or from the rear side of the slab, respectively.

The total base shear can be calculated as follows:

$$F_b = (m_i + m_W + m_D) \cdot S_e(T_i) + m_k \cdot S_e(T_k) \tag{5.81}$$

with:

F_b Total base shear
m_W Mass of the tank wall
m_D Mass of the tank roof
$Se\ (T_i)$ Impulsive spectral acceleration at period T_i using the elastic response spectrum with 5% damping
$Se\ (T_k)$ Convective spectral acceleration at period T_k using the elastic response spectrum with 0.5% damping.

The overturning moment due to pressure on the silo wall is calculated using the heights of the mass centres of the tank wall h_W and the tank roof h_D:

$$M_W = (m_i \cdot h_i + m_W \cdot h_W + m_D \cdot h_D) \cdot S_e(T_i) + m_k \cdot h_k \cdot S_e(T_k) \tag{5.82}$$

Table 5.14 Coefficients C_i and C_k for fundamental periods, masses m_i and m_k, heights h_i, h_k, h_{iu}, h_{ku}

H/R	$C_i(-)$	$C_k(s/m^{1/2})$	m_i/m	m_k/m	h_i/H	h_k/H	h_{iu}/H	h_{ku}/H
0.3	9.28	2.09	0.176	0.824	0.400	0.521	2.640	3.414
0.5	7.74	1.74	0.300	0.700	0.400	0.543	1.460	1.517
0.7	6.97	1.60	0.414	0.586	0.401	0.571	1.009	1.011
1.0	6.36	1.52	0.548	0.452	0.419	0.616	0.721	0.785
1.5	6.06	1.48	0.686	0.314	0.439	0.690	0.555	0.734
2.0	6.21	1.48	0.763	0.237	0.448	0.751	0.500	0.764
2.5	6.56	1.48	0.810	0.190	0.452	0.794	0.480	0.796
3.0	7.03	1.48	0.842	0.158	0.453	0.825	0.472	0.825

The overturning moment due to pressures on the tank wall and the tank bottom is:

$$M_G = (m_i \cdot h_{iu} + m_W \cdot h_W + m_D \cdot h_D) \cdot S_e(T_i) + m_k \cdot h_{ku} \cdot S_e(T_k) \quad (5.83)$$

5.6.7.3 Simplified Housner Approach

The simplified method for the calculation of liquid filled tanks proposed by Housner (1963) still forms the basis for most of the standards and guidelines around the world (e.g. API 650 2003). The method is simple and can be applied quickly as it is based on the assumption of a rigid tank anchored to the ground and subjected to horizontal seismic excitation. From the restrictive assumption of tank rigidity it follows that only the convective and the impulsive rigid pressure components must be taken into account. However, past earthquakes have clearly shown that tanks dimensioned using the Housner method exhibit a safety deficit especially for slender tanks as the impulsive flexible pressure component is not negligible. Irrespective of this, the method is still in use as it can be easily applied in simple calculations "by hand". In addition to its simplicity it is also applicable to both cylindrical and rectangular tanks. The Housner method is presented here for the sake of completeness, in spite of knowing quite well that it may provide inadequate results when used for tank structures with significant joint bending vibrations of the tank and the liquid. The method should only be applied for rigid tanks.

The Housner method differentiates between squat and slender tanks. Tanks are classified as squat tanks if the ratio H/R is less than 1.5 for cylindrical tanks and H/L less than 1.5 for rectangular tanks (Fig. 5.48). Herein, H is the filling height of the tank, R is the inner tank radius of a cylindrical tanks and L is half the width of a rectangular tank in the earthquake direction. The rigid and anchored tank is subjected to a maximum peak ground acceleration $\ddot{u}_B$, which equals to the spectral acceleration of the response spectrum at period T = 1.0 s.

The impulsive mass m_0, the convective mass m_1, the corresponding lever arms h_1^k, h_1^g and the impulsive lever arms h_0^k, h_0^g with and without bottom pressure can be calculated for squat tanks according to Table 5.15. For slender tanks it is additionally assumed that the upper part of the liquid content moves at a height of $\overline{H} = 1.5 \cdot R$ or $1.5 \cdot L$ respectively. The remaining liquid volume with a height of $\overline{H} - H$ is simply added to the impulsive mass. Taking the subdivision of the liquid into account, both the impulsive and the convective masses and the corresponding lever arms can be calculated for slender tanks in accordance with Tables 5.16 and 5.17. Furthermore, the dynamic effect of the tank mass m_T must be taken into account for both the squat tank and the slender tank. It can be assumed that the centre of the tank mass corresponds to the height of the mass centre of the impulsive liquid mass.

In the model, the convective mass is connected to the tank wall via an equivalent spring, whereas the impulsive masses are rigidly connected to the tank wall

Table 5.15 Pressure components for squat tanks with H/R < 1.5 (cylindrical tanks) and H/L < 1.5 (rectangular tanks)

	Cylindrical tank	Rectangular tank
Liquid mass m_w(t)	$m_w = \rho_L \cdot H \cdot R^2 \cdot \pi$	$m_w = \rho_L \cdot H \cdot 2L \cdot B$ (L = half the tank width in earthquake direction; B = other width of the tank)
Impulsive liquid mass m_0(t)	$m_0 = m_w \cdot \left[\frac{\tanh(\sqrt{3}\cdot\frac{R}{H})}{\sqrt{3}\cdot\frac{R}{H}}\right]$	$m_0 = m_w \cdot \left[\frac{\tanh(\sqrt{3}\cdot\frac{L}{H})}{\sqrt{3}\cdot\frac{L}{H}}\right]$
Impulsive lever arm h_0^k(m) without bottom pressure	$h_0^k = \frac{3}{8} \cdot H$	$h_0^k = \frac{3}{8} \cdot H$
Impulsive lever arm h_0^g (m) with bottom pressure	$h_0^g = \frac{H}{2\cdot\left[\frac{\tanh(\sqrt{3}\cdot\frac{R}{H})}{\sqrt{3}\cdot\frac{R}{H}}\right]} - \frac{H}{8}$	$h_0^g = \frac{H}{2\cdot\left[\frac{\tanh(\sqrt{3}\cdot\frac{L}{H})}{\sqrt{3}\cdot\frac{L}{H}}\right]} - \frac{H}{8}$
Impulsive equivalent force P_0(kN)	$P_0 = \ddot{u}_0 \cdot (m_0 + m_T)$	$P_0 = \ddot{u}_0 \cdot (m_0 + m_T)$
Convective liquid mass m_1 (t)	$m_1 = m_w \cdot 0.318 \cdot \frac{R}{H} \cdot \tanh(1.84 \cdot \frac{H}{R})$	$m_1 = m_w \cdot 0.527 \cdot \frac{L}{H} \cdot \tanh(1.58 \cdot \frac{H}{L})$
Convective lever arm h_1^k(m) without bottom pressure	$h_1^k = H \cdot \left[1 - \frac{\cosh\left(1.84\frac{H}{R}\right)-1}{1.84\frac{H}{R}\cdot\sinh\left(1.84\frac{H}{R}\right)}\right]$	$h_1^k = H \cdot \left[1 - \frac{\cosh\left(1.58\frac{H}{L}\right)-1}{1.58\frac{H}{L}\cdot\sinh\left(1.58\frac{H}{L}\right)}\right]$
Convective lever arm h_1^g (m) with bottom pressure	$h_1^g = H \cdot \left[1 - \frac{\cosh\left(1.84\frac{H}{R}\right)-2{,}01}{1.84\frac{H}{R}\cdot\sinh\left(1.84\frac{H}{R}\right)}\right]$	$h_1^g = H \cdot \left[1 - \frac{\cosh\left(1.58\frac{H}{L}\right)-2}{1.58\frac{H}{L}\cdot\sinh\left(1.58\frac{H}{L}\right)}\right]$
Natural circular eigenfrequency ω(rad/s)	$\omega^2 = \frac{1.84\cdot g}{R} \cdot \tanh(1.84 \cdot \frac{H}{R})$	$\omega^2 = \frac{1.58\cdot g}{L} \cdot \tanh(1.58 \cdot \frac{H}{L})$
Maximum horizontal displacement y_{max}(m)	$y_{max} = \frac{S_v}{\omega}$	$y_{max} = \frac{S_v}{\omega}$
Angle θ_h(rad)	$\theta_h = 1.534 \cdot \frac{y_{max}}{R} \cdot \tanh(1.84 \cdot \frac{H}{R})$	$\theta_h = 1.58 \cdot \frac{y_{max}}{L} \cdot \tanh(1.58 \cdot \frac{H}{L})$
convective equivalent force P_1(kN)	$P_1 = 1.2 \cdot m_1 \cdot g \cdot \theta_h \cdot \sin(\omega \cdot t)$	$P_1 = m_1 \cdot g \cdot \theta_h \cdot \sin(\omega \cdot t)$
maximum vertical displacement d_{max}(m)	$d_{max} = \frac{0.408\cdot R\cdot\coth(1.84\cdot\frac{H}{R})}{\frac{g}{\omega^2\cdot\theta_h\cdot R}-1}$	$d_{max} = \frac{0.527\cdot L\cdot\coth(1.58\cdot\frac{H}{L})}{\frac{g}{\omega^2\cdot\theta_h\cdot L}-1}$

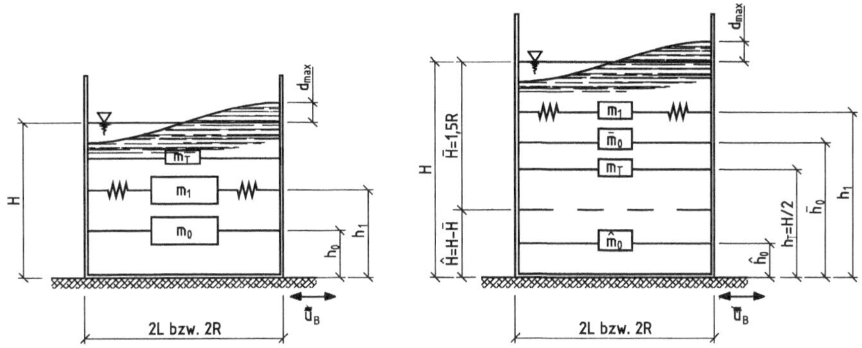

Fig. 5.48 Geometry, rigid and impulsive masses and corresponding heights for squat and slender tanks (Meskouris et al. 1999)

(Fig. 5.48). The sloshing oscillation of the fluid is described by a sloshing height, which can be calculated using Tables 5.15, 5.16 and 5.17. The calculation is based on the relationship between the pseudo relative velocity (PSV) ordinates S_v, the pseudo absolute acceleration (PSV) ordinates S_a and the spectral displacement S_d according to

$$S_v = S_d \cdot \omega = \frac{S_a}{\omega} \tag{5.84}$$

The overturning moments resulting from the wall pressure can be calculated with the impulsive equivalent force P_0, the convective equivalent force P_1 and the corresponding lever arms. The overturning moment M_W resulting from the pressure on the tank wall and the overturning moment resulting from the pressures on the tank wall and tank bottom M_G are calculated using the superposition according to the SRSS rule:

$$M_W = \sqrt{\left(P_0 \cdot h_0^k\right)^2 + \left(P_1 \cdot h_1^k\right)^2} \tag{5.85}$$

$$M_G = \sqrt{\left(P_0 \cdot h_0^g\right)^2 + \left(P_1 \cdot h_1^g\right)^2} \tag{5.86}$$

5.6.8 Seismic Design Situation and Actions for the Tank Design

The seismic design of a tank structure requires the consideration of all relevant combinations of actions for seismic design situations according to Eurocode 0 (2002). In the following sections permanent and variable actions frequently considered in

Table 5.16 Impulsive rigid pressure component for slender tanks with H/R ≥ 1.5 (cylindrical tanks) and H/L ≥ 1.5 (rectangular tanks)

	Cylindrical tank	Rectangular tank
Liquid mass m_w (t)	$m_w = \rho_L \cdot H \cdot R^2 \cdot \pi$	$m_w = \rho_L \cdot H \cdot 2L \cdot B$ (L = half the tank width in earthquake direction; B = other width of the tank)
"Free" depth $\overline{H}$(m)	$\overline{H} = 1.5 \cdot R$	$\overline{H} = 1.5 \cdot L$
"Fixed" level $\hat{H}$(m)	$\hat{H} = H - \overline{H}$	$\hat{H} = H - \overline{H}$
"Fixed" liquid mass $\hat{m}_0$(t)	$\hat{m}_0 = \rho_L \cdot \widehat{H} \cdot R^2 \cdot \pi$	$\hat{m}_0 = \rho_L \cdot \hat{H} \cdot 2L \cdot B$
"Fixed" lever arm $\hat{h}$(m)	$\hat{h}_0 = \frac{\hat{H}}{2}$	$\hat{h}_0 = \frac{\hat{H}}{2}$
"Free" liquid mass $\overline{m}$(t)	$\overline{m} = \rho_L \cdot \overline{H} \cdot R^2 \cdot \pi$	$\overline{m} = \rho_L \cdot \overline{H} \cdot 2L \cdot B$
Corresponding impulsive liquid mass $\overline{m}_0$(t)	$\overline{m}_0 = \overline{m} \cdot \left[\frac{\tanh(\sqrt{3} \cdot 0.667)}{\sqrt{3} \cdot 0.667}\right]$ $= \overline{m} \cdot 0.7095$	$\overline{m}_0 = \overline{m} \cdot \left[\frac{\tanh(\sqrt{3} \cdot 0.667)}{\sqrt{3} \cdot 0.667}\right]$ $= \overline{m} \cdot 0.7095$
Impulsive level arm $\overline{h}_0^k$(m) without bottom pressure	$\overline{h}_0^k = \frac{3}{8} \cdot 1.5 \cdot R + \hat{H}$	$\overline{h}_0^k = \frac{3}{8} \cdot 1.5 \cdot L + \hat{H}$
Impulsive lever arm $\overline{h}_0^g$(m) with bottom pressure	$\overline{h}_0^g = \frac{1.5 \cdot R}{8} \cdot \left[\frac{4}{0.7095} - 1\right] + \hat{H}$	$\overline{h}_0^g = \frac{1.5 \cdot L}{8} \cdot \left[\frac{4}{0.7095} - 1\right] + \hat{H}$
Impulsive liquid mass m_0(t)	$m_0 = \hat{m}_0 + \overline{m}_0$	$m_0 = \hat{m}_0 + \overline{m}_0$
Impulsive level arm h_0^k(m) without bottom pressure	$h_0^k = \frac{\hat{m}_0 \cdot \hat{h}_0 + \overline{m}_0 \cdot \overline{h}_0^k}{m_0}$	$h_0^k = \frac{\hat{m}_0 \cdot \hat{h}_0 + \overline{m}_0 \cdot \overline{h}_0^k}{m_0}$
impulsive lever arm h_0^g(m) with bottom pressure	$h_0^g = \frac{\hat{m}_0 \cdot \hat{h}_0 + \overline{m}_0 \cdot \overline{h}_0^g}{m_0}$	$h_0^g = \frac{\hat{m}_0 \cdot \hat{h}_0 + \overline{m}_0 \cdot \overline{h}_0^g}{m_0}$
Impulsive equivalent force P_0(kN)	$P_0 = \ddot{u}_0 \cdot m_0$	$P_0 = \ddot{u}_0 \cdot m_0$

engineering practise are introduced. Further actions must be taken into account with respect to the specific use of the tank, the discussion of which is not possible here due to their variability.

5.6.8.1 Permanent Actions Due to Self-weight

The self-weight of the tank includes the weight of the wall, the bottom, the roof, further super structural components (e.g. platforms) and also any installed equipment necessary for the operation.

Table 5.17 Convective pressure component for slender tanks with H/R ≥ 1.5 (cylindrical tanks) and H/L ≥ 1.5 (rectangular tanks)

	Cylindrical tank	Rectangular tank
Calculation of the convective pressure component		
Convective liquid mass $m_1(t)$	$m_1 = m_w \cdot 0.318 \cdot \frac{R}{H} \cdot \tanh(1.84 \cdot \frac{H}{R})$	$m_1 = m_w \cdot 0.527 \cdot \frac{L}{H} \cdot \tanh(1.58 \cdot \frac{H}{L})$
Convective lever arm h_1^k (m) without bottom pressure	$h_1^k = H \cdot \left[1 - \frac{\cosh(1.84 \cdot \frac{H}{R}) - 1}{1.84 \cdot \frac{H}{R} \cdot \sinh(1.84 \cdot \frac{H}{R})}\right]$	$h_1^k = H \cdot \left[1 - \frac{\cosh(1.58 \cdot \frac{H}{L}) - 1}{1.58 \cdot \frac{H}{L} \cdot \sinh(1.58 \cdot \frac{H}{L})}\right]$
Convective lever arm h_1^g (m) with bottom pressure	$h_1^g = H \cdot \left[1 - \frac{\cosh(1.84 \cdot \frac{H}{R}) - 2.01}{1.84 \cdot \frac{H}{R} \cdot sinh(1.84 \cdot \frac{H}{R})}\right]$	$h_1^g = H \cdot \left[1 - \frac{\cosh(1.58 \cdot \frac{H}{L}) - 2}{1.58 \cdot \frac{H}{L} \cdot \sinh(1.58 \cdot \frac{H}{L})}\right]$
Natural circular eigenfrequency ω^2 ($1/s^2$)	$\omega^2 = \frac{1.84 \cdot g}{R} \cdot \tanh(1.84 \cdot \frac{H}{R})$	$\omega^2 = \frac{1.58 \cdot g}{L} \cdot \tanh(1.58 \cdot \frac{H}{L})$
Maximum horizontal displacement y_{max} (m)	$y_{max} = \frac{S_v}{\omega}$	$y_{max} = \frac{S_v}{\omega}$
Angle θ_h(rad)	$\theta_h = 1.534 \cdot \frac{y_{max}}{R} \cdot \tanh(1.84 \cdot \frac{H}{R})$	$\theta_h = 1.58 \cdot \frac{y_{max}}{L} \cdot \tanh(1.58 \cdot \frac{H}{L})$
Convective equivalent force P_1(kN)	$P_1 = 1.2 \cdot m_1 \cdot g \cdot \theta_h \cdot \sin(\omega \cdot t)$	$P_1 = m_1 \cdot g \cdot \theta_h \cdot \sin(\omega \cdot t)$
Maximum vertical displacement d_{max}(m)	$d_{max} = \frac{0.408 \cdot R \cdot \coth(1.84 \cdot \frac{H}{R})}{\frac{g}{\omega^2 \cdot \theta_h \cdot R} - 1}$	$d_{max} = \frac{0.527 \cdot L \cdot \coth(1.58 \cdot \frac{H}{L})}{\frac{g}{\omega^2 \cdot \theta_h \cdot L} - 1}$

5.6.8.2 Hydrostatic Pressure

The hydrostatic pressure results from the filling height of the content and is distributed linearly over the filling height with its maximum being at the tank bottom. The maximum filling height is typically relevant for the seismic design situation.

5.6.8.3 Wind Action

The wind pressure profile of cylindrical tanks is calculated using an external pressure coefficient which is multiplied with a height-dependent velocity pressure. The pressure coefficient takes into account the wind direction, tank slenderness and the Reynolds number of the cross-section. Figure 5.49a shows height-dependent velocity pressure and Fig. 5.49b represents the qualitative wind pressure distribution $q_w(\varphi)$ in circumferential direction (Eurocode 1-1-4 2004). By way of simplification, the wind load distribution can be replaced by a constant rotationally symmetric pressure, which is acting in the inward directeion perpendicular to the tank wall (Eurocode 3-

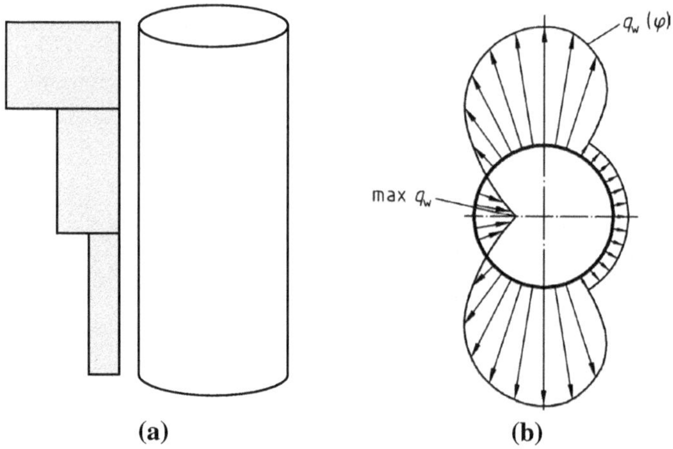

Fig. 5.49 Qualitative wind pressure distribution: (**a**) Distribution of the wind pressure over the height. (**b**) Distribution of the wind pressure in circumferential direction

1-6 2007). Wind actions need to be taken into account in the permanent and variable design situation but can be neglected in the seismic design situation.

5.6.8.4 Snow Action

Snow loads are applied to the tank roof and can be calculated depending on the roof shape and the geopraphical location of the tank (Eurocode 1-1-3 2003). Snow load for the relevant location can be determined using the snow maps generally provided in the national design codes.

5.6.8.5 Settlements

Non-uniform settlements can cause critical states of stress in the tank wall and the tank bottom. Therefore, it is necessary to consider expected non-uniform settlements during the life-span of the tank for the relevant design combinations. The order of the settlements must be estimated on the basis of the geotechnical report, which takes the local soil conditions into account.

5.6.8.6 Temperature

Operating temperature loads can be relevant for the tank design, if the tank content is significantly warmer or colder than the environment. They must be taken into account in the seismic design situation as additional load cases and the strength properties of the materials must be adjusted as functions of the operating temperature.

5.6.8.7 Prestress

The prestress load case must be considered for tank structures made of prestressed reinforced concrete. Prestress effects must be taken into account in the seismic deign situation.

5.6.8.8 Internal Pressure

Operational internal pressures can be considered as internal pressures perpendicular to the wall in outward direction. Internal pressures must be taken into account in the seismic design situation.

5.6.8.9 Seismic Design Situation

The superposition rules for the seismic design situation are given in Eurocode 0 (2002). A detailed description of the seismic design situation and its application can be found in Chap. 4.

5.6.9 Sample Calculation 1: Slender Tank

The presented calculation methods are illustrated by the example of a slender steel tank under seismic excitation using the response spectrum method. The basis of the investigation is a tank structure that has already been analysed by Gehrig (2004), which enables a comparison of the calculation results. Initially, the pressure curves are calculated using the iterative "added mass" model and the tabulated coefficients in Sect. 5.6.12. In a subsequent step, the base shear and the overturning moments resulting from pressures on the tank wall and bottom are calculated using different calculation approaches. Stresses in the tank wall are computed by integrating the pressure curves according to Sect. 5.6.7.1, based on the Housner method (1963) according to Sect. 5.6.7.3 and the calculation method proposed by Gehrig (2004). The tabulated coefficients are determined using the approximation of a linear bending curve for the joint vibration modes of the tank and the liquid. A calculation using the simplified method according to Eurocode 8-4 (2006) as described in Sect. 5.6.7.2 must be carried out since the slenderness ratio of the tank does not lie within the valid range of application.

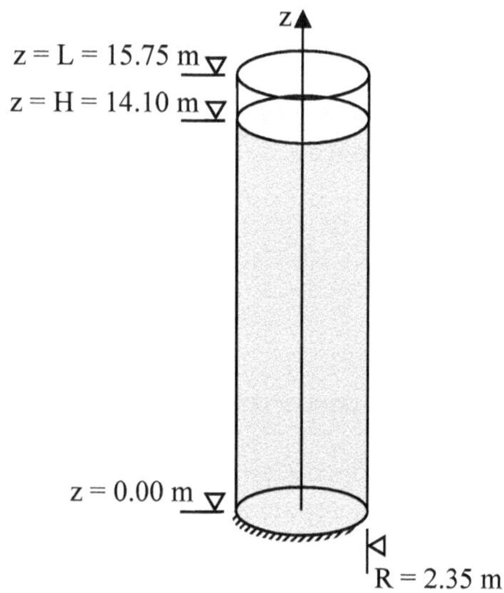

Fig. 5.50 Geometry and filling height of the analysed slender tank

5.6.9.1 Tank Description

The investigated tank has an inner radius of $R = 2.35\ m$, a height of $L = 15.75\,m$ and a filling height of $H = 14.80\,m$. This results in a tank slenderness of $\gamma = H/R = 6.3$. As the lower part of the tank has a short skirt support with dished head, a slenderness value of $\gamma = H/R = 6$ in accordance with the specifications given by Gehrig (2004) is assumed. This corresponds to an equivalent filling height of 14.1 m (Fig. 5.50).

The tank wall is subdivided into segments with different wall thicknesses, but the calculations are carried out with an averaged thickness of $t = 3.7\,mm$. The material is stainless steel with a Young's modulus of $E = 17.000\,\text{kN/cm}^2$ and the tank is used to store fruit juice concentrate with a density of $\rho_L = 1.35\,\text{t/m}^3$. The importance factor is chosen as $\gamma_I = 1.2$ and the self-weight of the tank is 14 t. The design response spectra for the pressure components are determined according to the German National Annex DIN EN 1998-1/NA (2011) using the specific damping values proposed in Sect. 5.6.6. The following input values are used to compile the design response spectra:

Earthquake zone 2	$a_g = 0.6\,\text{m/s}^2$ (DIN EN 1998-1/NA 2011)
Ground type C-S	Soil factor $S = 0.75$
Control periods	$T_A = 0\,s$ | $T_B = 0.1\ s$ | $T_C = 0.5\,s$ | $T_D = 2.0\ s$
Amplification factor	$\beta_0 = 2.5$
Damping for convective pressure	0.5 % => $\eta = 1.348$ after (5.70)

Damping for impulsive flexible pressure 2.5 % => $\eta = 1.155$ after (5.70).

5.6.9.2 Finite Element Modelling of the Tank

The modelling of the tank structure is carried out with shell elements using the FE program ANSYS. The tank support is assumed to be fully clamped at the bottom.

5.6.9.3 Calculation of the Pressure Components

The pressure components are calculated using two alternative approaches

- Application of the bending curve of the joint vibration of the tank and the liquid according to Sect. 5.6.3. The analysis is carried out with ANSYS coupled to the mathematics software Maple (Cornelissen 2010). The calculation of the horizontal impulsive flexible pressure component is carried out iteratively using the "added-mass-model" explained in Sect. 5.6.3.3. The calculation of the vertical impulsive flexible pressure component is based on a linear bending curve due to the tank slenderness of 6.0.
- The pressure components are calculated on the basis of the tabulated coefficients given in Sect. 5.7 assuming a linear bending curve for the joint vibration of the tank and the liquid.

The pressure components are determined for the horizontal and the vertical seismic excitations as follows:

Convective pressure component due to horizontal seismic excitation
The fundamental natural period for sloshing is calculated according to Eq. (5.42):

$$T_{k1} - \frac{2\pi}{\sqrt{\frac{g \cdot \lambda_n \cdot \tanh(\lambda_n \cdot \gamma)}{R}}} = \frac{2\pi}{\sqrt{\frac{9.81 \cdot 1.841 \cdot \tanh(1.841 \cdot 6)}{2.35}}} = 2.27\, s$$

The spectral acceleration is determined according to DIN EN 1998-1/NA (2011) for a spectrum with 0.5% damping:

$$a_{k1} = a_g \cdot \gamma_I \cdot S \cdot \eta \cdot \beta_0 \cdot \frac{T_C \cdot T_D}{T^2}$$

$$= 0.6 \cdot 1.2 \cdot 0.75 \cdot 1.348 \cdot 2.5 \cdot \frac{0.5 \cdot 2.0}{2.27^2} = 0.353\,\text{m/s}^2$$

The participation factor Γ_{k1} of the fundamental natural vibration mode is calculated with $\lambda_1 = 1.841$ after Eq. (5.41):

$$\Gamma_{k1} = \frac{2 \cdot \sinh(\lambda_1 \cdot \gamma) \cdot [\cosh(\lambda_1 \cdot \gamma) - 1]}{\sinh(\lambda_1 \cdot \gamma) \cdot \cosh(\lambda_1 \cdot \gamma) - \lambda_1 \cdot \gamma} = 2.0$$

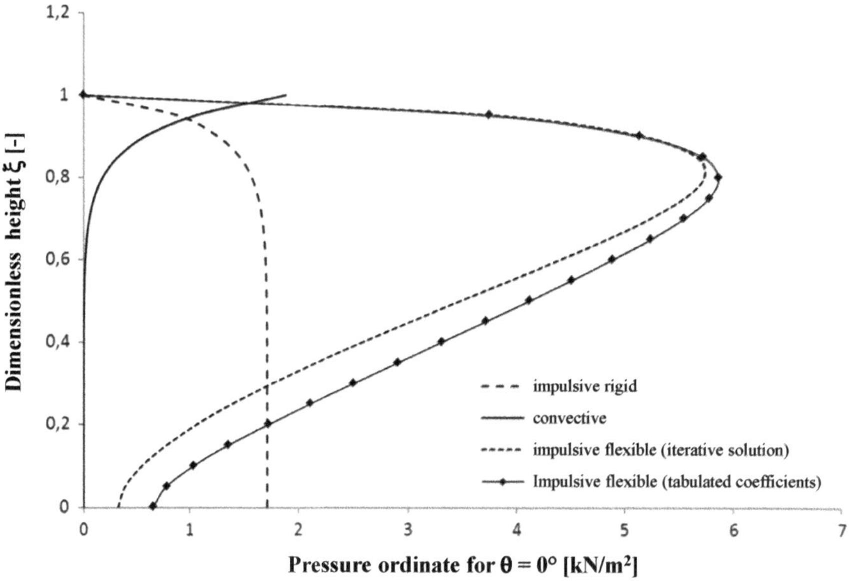

Fig. 5.51 Dynamic pressure components due to horizontal seismic excitation

The coefficients C_k for the calculation of the height-dependent pressure ordinates are determined using Table 5.20 for the slenderness value $\gamma = 6$. By using these values, the maximum pressure ordinate at the maximum filling height is:

$$p_k(\xi = 1, \zeta = 1, \theta = 0) = R \cdot \rho_L \cdot C_k(\zeta, \gamma) \cdot \cos(\theta) \cdot a_{k1} \cdot \Gamma_{k1}$$
$$= 2.35 \cdot 1.35 \cdot 0.8371 \cdot 1.0 \cdot 0.353 \cdot 2.0 = 1.87\,\text{kN/m}^2$$

The resulting convective pressure distribution is shown in Fig. 5.51 for $\theta = 0$.

Impulsive rigid pressure component due to horizontal seismic excitation

The impulsive rigid pressure component due to horizontal seismic excitation is calculated using the spectral acceleration of the design response spectrum at period T = 0 s according to DIN EN 1998-1/NA (2011)

$$a_{is,h} = a_g \cdot \gamma_I \cdot S = 0.6 \cdot 1.2 \cdot 0.75 = 0.54\,\text{m/s}^2$$

The height-dependent pressure ordinates are calculated with the coefficients $C_{is,h}$ according to Table 5.21 for a slenderness of $\gamma = 6$:

$$p_{is,h}(\xi = 1, \zeta = 0, \theta = 0) = R \cdot \rho_L \cdot C_{is,h}(\zeta, \gamma) \cdot \cos(\theta) \cdot a_{is,h}$$
$$= 2.35 \cdot 1.35 \cdot 1.0 \cdot 1.0 \cdot 0.54 = 1.71\,\text{kN/m}^2$$

The resulting impulsive rigid pressure distribution is shown in Fig. 5.51 for $\theta = 0$.

Impulsive flexible pressure component due to horizontal seismic excitation
The fundamental natural period of the joint bending vibration of the tank and the liquid due to horizontal seismic excitation is calculated with the FE model to $T_{if,h,1}$= 0.373 s. Alternatively, the fundamental period can be approximated using Eq. (5.50):

$$T_{if,h,1} = 2 \cdot R \cdot F(\gamma)\sqrt{\frac{H \cdot \rho_L}{E \cdot s(\zeta = 1/3)}} = 2 \cdot 2.35 \cdot 13.142 \cdot \sqrt{\frac{14.1 \cdot 1.35}{170e6 \cdot 0.0037}} = 0.34\,s$$

By using the fundamental period, the spectral acceleration according to DIN EN 1998-1/NA (2011) can be obtained:

$$a_{if,h} = a_g \cdot \gamma_I \cdot S \cdot \eta \cdot \beta_0 = 0.6 \cdot 1.2 \cdot 0.75 \cdot 1.155 \cdot 2.5 = 1.56\,\text{m/s}^2$$

The coefficients $C_{if,h}$ for the calculation of the height-dependent pressure ordinates are determined using Table 5.22 for $\gamma = 6$ and the corresponding participation factor $\Gamma_{if,h}$ is given in Table 5.27. Using these values, the maximum pressure ordinate at the dimensionless height $\zeta = 0.8$ is computed to:

$$p_{if,h}(\xi = 1, \zeta = 0,8, \theta = 0) = R \cdot \rho_L \cdot C_{if,h}(\zeta, \gamma) \cdot \cos(\theta) \cdot a_{if,h} \cdot \Gamma_{if,h}$$
$$= 2.35 \cdot 1.35 \cdot 0.7079 \cdot 1.0 \cdot 1.56 \cdot 1.6348 = 5.73\,\text{kN/m}^2$$

The resulting impulsive flexible pressure curve is shown in Table 5.22 for $\theta = 0$. The impulsive flexible pressure curve determined on the basis of a linear bending curve shows a good agreement with the FE results that were calculated iteratively. As expected, the convective and the impulsive rigid pressure curves calculated according to Section Table 5.27 and using the tabulated coefficients are similar to each other as they are basically derived from the same set of equations.

Impulsive rigid pressure component due to vertical seismic excitation
The impulsive rigid pressure component due to vertical seismic excitation is calculated using 70% of the spectral acceleration of the design response spectrum at period T = 0 s. This assumption results in the following impulsive rigid pressure:

$$\begin{aligned} p_{is,v}(\zeta, t) &= \rho_L \cdot [H \cdot (1 - \zeta)][a_v(t) \cdot \Gamma_{is,v}] \\ &= 1.35 \cdot [14.1 \cdot (1 - \zeta)] \cdot 0.378 \cdot 1.0 \\ &= 7.195 \cdot (1 - \zeta) \end{aligned}$$

The corresponding pressure curve is shown in Fig. 5.52.

Impulsive flexible pressure component due to vertical seismic excitation
The fundamental period of the joint bending vibration of the tank and the liquid due to vertical seismic excitation is calculated according to (5.63):

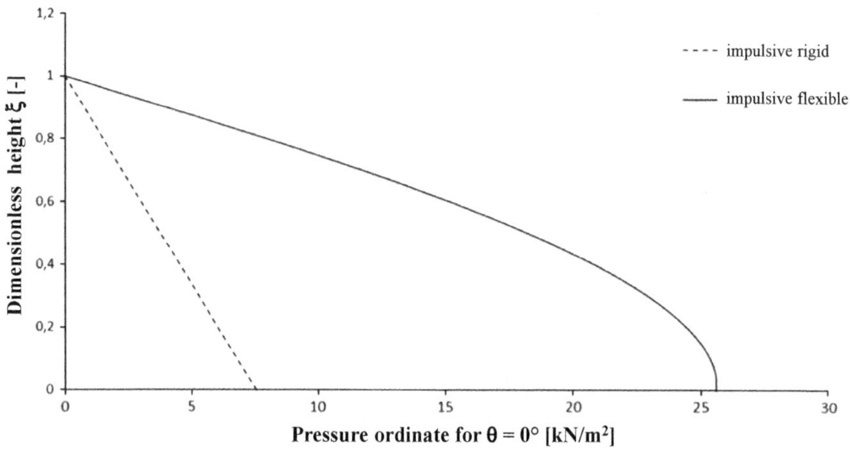

Fig. 5.52 Dynamic pressure components due to vertical seismic excitation

$$T_{if,v,1}=4\cdot R\sqrt{\frac{\pi\cdot\rho_L}{2}\cdot\frac{1-\nu^2}{E}\cdot\frac{H}{s\left(\zeta=\frac{1}{3}\right)}\cdot\frac{I_0\left(\frac{\pi}{2\cdot\gamma}\right)}{I_1\left(\frac{\pi}{2\cdot\gamma}\right)}}$$

$$=4\cdot 2.35\sqrt{\frac{\pi\cdot 1.35}{2}\cdot\frac{1-0.3^2}{170e6}\cdot\frac{14.10}{0.0037}\cdot\frac{I_0\left(\frac{\pi}{2\cdot\gamma}\right)}{I_1\left(\frac{\pi}{2\cdot\gamma}\right)}}=0.18\,s$$

The fundamental natural period lies in the plateau range bounded by the control periods $T_{B,v}=0.1\,s$ and $T_{C,v}=0.2\,s$ of the vertical design response spectrum. The evaluation of the 2.5% damped design response spectrum according to DIN EN 1998-1/NA (2011) at period $T_{if,v,1}$ yields the following spectral acceleration:

$$a_{if,v}=(0.6\cdot 0,7)\cdot 1.2\cdot 0.75\cdot 1.155\cdot 2.5=1.09\ \text{m/s}^2$$

The participation factor $\Gamma_{if,v}$ is calculated with (5.65):

$$\Gamma_{if,v}=\frac{4}{\pi}\cdot\frac{I_1\left(\frac{\pi}{2\cdot\gamma}\right)}{I_0\left(\frac{\pi}{2\cdot\gamma}\right)}=0.1653$$

The values I_0 and I_1 of the Bessel function can be determined by means of tables from literature, suitable mathematical software systems or spreadsheets (e.g. Microsoft Excel). Alternatively, the participation factor can be taken from Table 5.26. The coefficients $C_{if,v}$ are taken from Table 5.24 for a slenderness of $\gamma=6$. This yields the maximum ordinate of the height-dependent pressure distribution for the dimensionless height coordinate $\zeta=0$ according to (5.67):

$$p_{if,v}(\xi = 1, \zeta = 0) = R \cdot \rho_L \cdot 1.0 \cdot a_{if,v}(t) \cdot C_{if,v}(\zeta, \gamma) \cdot \Gamma_{if,v}$$
$$= 2.35 \cdot 1.35 \cdot 1.0 \cdot 1.09 \cdot 46.1736 \cdot 0.1653 = 26.39\,\text{kN/m}^2$$

The resulting impulsive flexible pressure distribution due to vertical excitation is shown in Fig. 5.52. The pressure distribution using tabulated coefficients corresponds exactly to the pressure distributions which can be obtained with the set of equations given in Sect. 5.6.4.

5.6.9.4 Base Shear and Overturning Moments with Accurate Pressure Curves

The foundation shear, the overturning moments resulting from the wall pressure and those resulting from the pressures on the wall and the bottom are numerically calculated on the basis of a program routine developed by Cornelissen (2010). This program routine is based on a coupling of the FE program ANSYS with the mathematics software MAPLE, which allows a simple implementation of the iterative "added mass" model. The numerical calculation of the pressure curves on the basis of Sects. 5.6.3.1–5.6.3.3 ensures a higher accuracy in comparison to the simplified approaches.

5.6.9.5 Base Shear and Overturning Moments with Tabulated Pressure Curves

When using the simplified method on the basis of the tabulated pressure curves, the base shear, the overturning moments resulting from the wall pressure and those resulting from the pressures on the wall and the bottom are calculated using the formulas given in Sect. 5.6.7 in combination with the tabulated coefficients in Table 5.27. Alternatively, the pressure curves that are calculated on the basis of the tables can also be applied to a FE model as equivalent static loads. The base shear and the overturning moment can be determined as reaction forces using the calculation model.
The tabulated coefficients are determined from Table 5.27 for $\gamma = 6$:

$c_{F,k}$	$c_{MW,k}$	$c_{MB,k}$	$c_{M,k}$	Γ_k
0.0786	0.0717	0.0	2.5801	1.9999

$c_{F,is}$	$c_{MW,is}$	$c_{MB,is}$	$c_{M,is}$	Γ_{is}
0.9209	0.4278	0.25	15.6508	1.0

$c_{F,if}$	$c_{MW,if}$	$c_{MB,if}$	$c_{M,if}$	Γ_{if}
0.4276	0.2616	0.0224	9.4396	1.6348

Convective pressure component

$$F_{b,k} = m_L \cdot a_k(t) \cdot \Gamma_k \cdot C_{F,k} = 244.63 \cdot 1.35 \cdot 0.353 \cdot 2.0 \cdot 0.0786 = 18.33\,\text{kN}$$

$$M_{W,k} = \pi \cdot R^2 \cdot H^2 \cdot \rho_L \cdot a_k(t) \cdot \Gamma_k \cdot C_{MW,k}$$
$$= 3449.24 \cdot 1.35 \cdot 0.353 \cdot 2.0 \cdot 0.0717 = 235.71\,\text{kNm}$$

$$M_{B,k} = \pi \cdot R^4 \cdot \rho_L \cdot a_k(t) \cdot \Gamma_k \cdot C_{MB,k} = 129.35 \cdot 0.353 \cdot 2.0 \cdot 0.0 = 0$$

$$M_{G,k} = \pi \cdot R^4 \cdot \rho_L \cdot a_k(t) \cdot \Gamma_k \cdot C_{M,k} = 129.347 \cdot 0.353 \cdot 2.0 \cdot 2.5801 = 235.61\,\text{kNm}$$

Impulsive rigid pressure component

$$F_{b,is,h} = m_L \cdot a_{is,h}(t) \cdot \Gamma_{is,h} \cdot C_{F,is,h} = 244.63 \cdot 1.35 \cdot 0.54 \cdot 1.0 \cdot 0.9209 = 164.23\,\text{kN}$$

$$M_{W,is,h} = \pi \cdot R^2 \cdot H^2 \cdot \rho_L \cdot a_{is,h}(t) \cdot \Gamma_{is,h} \cdot C_{MW,is,h} = 3449.24 \cdot 1.35 \cdot 0.54 \cdot 1.0 \cdot 0.4278$$
$$= 1075.70\,\text{kNm}$$

$$M_{B,is,h} = \pi \cdot R^4 \cdot \rho_L \cdot a_{is,h}(t) \cdot \Gamma_{is,h} \cdot C_{MB,is,h} = 129,35 \cdot 0.54 \cdot 1.0 \cdot 0.25 = 17.46\,\text{kNm}$$

$$M_{G,is,h} = \pi \cdot R^4 \cdot \rho_L \cdot a_{is,h}(t) \cdot \Gamma_{is,h} \cdot C_{M,is,h} = 129.347 \cdot 0.54 \cdot 1.0 \cdot 15.6508$$
$$= 1093.17\,\text{kNm}$$

Impulsive flexible pressure component

$$F_{b,if,h} = m_L \cdot a_{if,h}(t) \cdot \Gamma_{if,h} \cdot C_{F,if,h} = 244.63 \cdot 1.35 \cdot 1.56 \cdot 1.6348 \cdot 0.4276 = 360.14\,\text{kN}$$

$$M_{W,if,h} = \pi \cdot R^2 \cdot H^2 \cdot \rho_L \cdot a_{if,h}(t) \cdot \Gamma_{if,h} \cdot C_{MW,if,h}$$
$$= 3449.24 \cdot 1.35 \cdot 1.56 \cdot 1.6348 \cdot 0.2616 = 3106.59\,\text{kNm}$$

$$M_{B,if,h} = \pi \cdot R^4 \cdot \rho_L \cdot a_{if,h}(t) \cdot \Gamma_{if,h} \cdot C_{MB,if,h} = 129.35 \cdot 1.56 \cdot 1.6348 \cdot 0.0224 = 7.39\,\text{kNm}$$

$$M_{G,if,h} = \pi \cdot R^4 \cdot \rho_L \cdot a_{if,h}(t) \cdot \Gamma_{if,h} \cdot C_{M,if,h}$$
$$= 129.347 \cdot 1.56 \cdot 1.6348 \cdot 9.4396 = 3113.86\,\text{kNm}$$

5.6.9.6 Base Shear and Overturning Moments According to Housner (1963)

In the following the base shear and overturning moments are calculated according to Housner (1963) as summarized in Sect. 5.6.7.3.

Basic quantities for the calculation:

Tank slenderness:

$$\frac{H}{R} = 6 > 1.5 \;\Rightarrow\; slender\ tank$$

Liquid mass:

$$m_w = \rho_L \cdot V = 1.35 \cdot \pi \cdot 2.35^2 \cdot 14.10 = 330.25\ t$$

Coefficient:

$$1.84 \cdot H/R = 11.04$$

Subdivision of the filling height with "free" depth $\overline{H}$ and "fixed" depth $\overline{H}$:

$$\overline{H} = 1.5 \cdot R = 3.53\ m$$
$$\widehat{H} = 14.10 - 3.53 = 10.57\ m$$

Calculation of the impulsive pressure part

"Fixed" liquid part:

$$\hat{m}_0 = \rho_L \cdot \pi \cdot R^2 \cdot \hat{H} = 1.35 \cdot \pi \cdot 2.35^2 \cdot 10.57 = 247.57\ \ t$$
$$\hat{h}_0 = \frac{10.57}{2} = 5.29\ \ m$$

"Free" liquid part:

$$\overline{m} = 1.35 \cdot \pi \cdot 2.35^2 \cdot 3.53 = 82.68\ t$$
$$\overline{m}_0 = \overline{m} \cdot \left[\frac{\tanh(\sqrt{3} \cdot 0.667)}{\sqrt{3} \cdot 0.667}\right] = 82.68 \cdot 0.7095 = 58.66\ t$$
$$\overline{h}_0^k = \frac{3}{8} \cdot 1.5 \cdot R + \hat{H} = 1.32 + 10.57 = 11.89\ \ m$$
$$\overline{h}_0^g = \frac{1.5 \cdot R}{8} \cdot \left[\frac{4}{0.7095} - 1\right] + \hat{H} = 0.1875 \cdot 2.35 \cdot 4.634 + 10.57 = 12.61\ m$$
$$m_0 = \hat{m}_0 + \overline{m}_0 + m_{Tank} = 247.57 + 58.66 + 14 = 320.23\ t$$
$$h_0^k = \frac{\hat{m}_0 \cdot \hat{h}_0 + \overline{m}_0 \cdot \overline{h}_0^k + m_{Tank} \cdot \frac{H}{2}}{m_0}$$
$$= \frac{247.57 \cdot 5.29 + 58.66 \cdot 11.89 + 14.0 \cdot 7.05}{320.23} = 6.58\ \ m$$
$$h_0^g = \frac{\hat{m}_0 \cdot \hat{h}_0 + \overline{m}_0 \cdot \overline{h}_0^g + m_{Tank} \cdot \frac{H}{2}}{m_0}$$
$$= \frac{247.57 \cdot 5.29 + 58.66 \cdot 12.61 + 14.0 \cdot 7.05}{320.23} = 6.71\ \ m$$

Equivalent static force calculated with the spectral acceleration at period T = 0 s:

$$S_a(T = 0s) = 0.54\ \text{m/s}^2 :$$
$$P_0 = 0.54 \cdot m_0 = 0.54 \cdot 320.23 = 172.92\ \ \text{kN}$$

Overturning moment due to impulsive pressure on the silo wall:

$$M_0^k = 172.92 \cdot 6.58 = 1137.81 \text{ kNm}$$

Overturning moment due to impulsive pressure on the wall and the bottom:

$$M_0^g = 172.92 \cdot 6.71 = 1160.29 \text{ kNm}$$

Calculation of the convective part

$$m_1 = m_w \cdot 0.318 \cdot \frac{R}{H} \cdot \tanh(1.84 \cdot \tfrac{H}{R}) = 330.25 \cdot 0.318 \cdot 0.1667 \cdot \tanh(11.04) = 17.51 \ t$$

$$h_1^k = H \cdot \left[1 - \frac{\cosh(1.84 \cdot \frac{H}{R}) - 1}{1.84 \cdot \frac{H}{R} \cdot \sinh(1.84 \cdot \frac{H}{R})}\right] = 14.10 \cdot \left[1 - \frac{\cosh(11.04) - 1}{11.04 \cdot \sinh(11.04)}\right] = 12.82 \ m$$

$$h_1^g = H \cdot \left[1 - \frac{\cosh(1.84 \cdot \frac{H}{R}) - 2.01}{1.84 \cdot \frac{H}{R} \cdot \sinh(1.84 \cdot \frac{H}{R})}\right] = 14.10 \cdot \left[1 - \frac{\cosh(11.04) - 2.01}{11.04 \cdot \sinh(11.04)}\right] = 12.82 \ m$$

Fundamental natural frequency:

$$\omega^2 = \frac{1.84 \cdot g}{R} \cdot \tanh(1.84 \cdot \tfrac{H}{R}) = \frac{1.84 \cdot 9.81}{2.35} \cdot \tanh(11.04) = 7.68/s^2 \qquad \Rightarrow \omega = 2.77 \ Hz$$

$$T = \frac{2 \cdot \pi}{\omega} = 2.27 \ s$$

Spectral acceleration:

The spectral acceleration of the design response spectrum with T = 2.27 s and 0.5% damping is determined according to DIN EN 1998-1/NA (2011):

$$S_a = a_g \cdot \gamma_I \cdot S \cdot \eta \cdot \beta_0 \cdot \frac{T_C \cdot T_D}{T^2} = 0.6 \cdot 1.2 \cdot 0.75 \cdot 1.348 \cdot 2.5 \cdot \frac{0.5 \cdot 2.0}{2.27^2} = 0.35 \text{ m/s}^2$$

$$S_v = \frac{S_a}{\omega} = \frac{0.35}{2.77} = 0.13 \text{ m/s}$$

Maximum of the liquid surface:

$$y_{\max} = \frac{S_v}{\omega} = \frac{0.13}{2.77} = 0.047 \ m$$

$$\theta_h = 1.534 \cdot \frac{y_{\max}}{R} \cdot \tanh(1.84 \cdot \tfrac{H}{R}) = 1.534 \cdot \frac{0.047}{2.35} \cdot \tanh(11.04) = 0.031 \ rad$$

Vertical motion of the liquid surface:

$$d_{\max} = \frac{0.408 \cdot R \cdot \coth(1.84 \cdot \frac{H}{R})}{\frac{g}{\omega^2 \cdot \theta_h \cdot R} - 1} = \frac{0.408 \cdot 2.35 \cdot \coth(11.04)}{\frac{9,81}{7.68 \cdot 0.031 \cdot 2.35} - 1} = 0.055 \ m$$

Equivalent static force:

$$P_1 = 1.2 \cdot m_1 \cdot g \cdot \theta_h \cdot \sin(\omega \cdot t) = 1.2 \cdot 17.51 \cdot 9.81 \cdot 0.031 \cdot \sin(2.77 \cdot t)$$
$$= 6.39 \text{ kN} \cdot \sin(2.77 \cdot t)$$

$$\max P_1 = 6.39 \text{ kN}$$

Overturning moment due to convective pressure on the silo wall:

$$M_1^k = 6.39 \cdot 12.82 = 81.92 \text{ kNm}$$

Overturniung moment due to convective pressure on the wall and the bottom:

$$M_1^g = 6.39 \cdot 12.82 = 81.92 \text{ kNm}$$

Superposition of the impulsive and convective pressures using the SRSS rule:

$$P = \sqrt{(P_0)^2 + (P_1)^2} = \sqrt{172.79^2 + 6.39^2} = 172.91 \text{ kN}$$

$$M_W = \sqrt{\left(M_0^k\right)^2 + \left(M_1^k\right)^2} = \sqrt{1137.81^2 + 81.92^2} = 1140.76 \text{ kNm}$$

$$M_G = \sqrt{\left(M_0^g\right)^2 + \left(M_1^g\right)^2} = \sqrt{1160.29^2 + 81.92^2} = 1163.18 \text{ kNm}$$

5.6.9.7 Base Shear and Overturning Moments According to Gehrig (2004)

The following calculation of the base shear and overturning moment is based on tables and calculation formulas proposed by Gehrig, evaluated for a slenderness of $\gamma = 6$. The nomenclature used in the following is identical to the one in the publication by Gehrig (2004).

Total mass of the content: m = 3308/9.81 = 337.21 t

Convective pressure component

$$m_k = 0.076 \cdot m = 25.63\, t$$
$$h_k = 0.909 \cdot H = 12.82\, m$$
$$a_k = 0.353\, m/s^2$$
$$\Gamma_k = 2.0$$
$$Q_k = m_k \cdot a_k \cdot \Gamma_k = 18.09\, \text{kN}$$
$$M_k = (m_k \cdot a_k \cdot \Gamma_k) \cdot h_k = 231.98\, \text{kNm}$$

Impulsive rigid pressure component

$$m_{is} = 0.921 \cdot m = 310.57\, t$$
$$h_{is} = 0.464 \cdot H = 6.54\, m$$
$$a_{is} = 0.54\, m/s^2$$
$$\Gamma_{is} = 1.0$$
$$Q_{is} = m_{is} \cdot a_{is} \cdot \Gamma_{is} = 167.71\, \text{kN}$$
$$M_{is} = (m_{is} \cdot a_{is} \cdot \Gamma_{is}) \cdot h_{is} = 1096.81\, \text{kNm}$$

Impulsive flexible pressure component

$$
\begin{aligned}
m_{if} &= 0.428 \cdot m = 144.33\,t \\
h_{if} &= 0.612 \cdot H = 8.63\,m \\
a_{if} &= 1.56\,m/s^2 \\
\Gamma_{if} &= 1.635 \\
Q_{if} &= m_{if} \cdot a_{if} \cdot \Gamma_{if} = 368.13\,\text{kN} \\
M_{if} &= \left(m_{if} \cdot a_{if} \cdot \Gamma_{if}\right) \cdot h_{if} = 3176.95\,\text{kNm}
\end{aligned}
$$

Resulting base shear F_b and overturning moment M_W by superposition using the SRSS rule:

$$F_b = \sqrt{(Q_k)^2 + (Q_{is})^2 + \left(Q_{if}\right)^2} = \sqrt{18.09^2 + 167.71^2 + 368.13^2} = 404.94\ \text{kN}$$

$$M_W = \sqrt{(M_k)^2 + (M_{is})^2 + \left(M_{if}\right)^2} = \sqrt{231.98^2 + 1096.81^2 + 3176.95^2} = 3368.95\ \text{kNm}$$

A calculation of the overturning moment due to pressures on the wall and the bottom is not provided by Gehrig (2004) since the pressures on the bottom are neglected.

5.6.9.8 Comparison of the Results: Base Shear and Overturning Moment

A summary of the results obtained is given in Table 5.18. It presents values for the base shears and the overturning moments resulting from the wall pressure and those resulting from the wall and bottom pressure for the different calculation methods. Only the pressure components resulting from horizontal seismic excitation are taken into account as the pressure components from the vertical seismic excitation do not have an influence on the base shear and the overturning moment (Sect. 5.6.4) because of their rotationally symmetric pressure distribution.

The results show a good agreement between the iterative calculation with ANSYS and the simplified calculation using the tabulated coefficients. The results also agree well with the method proposed by Gehrig (2004), if only the wall pressures are considered. The results according to Housner (1963) underestimate both the base shear and the overturning moment as the impulsive flexible component is not taken into account.

It can also be derived from the results that the convective pressure component plays a minor role for the base shear and overturning moment when compared to the impulsive rigid and the impulsive flexible pressure components. Furthermore, it can be concluded that the bottom pressure has only a relatively small influence on the overturning moment in case of slender tanks. Table 5.18 clarifies this fact, since the overturning moments from the wall pressure and the overturning moment from the wall and bottom pressure do not differ from each other.

Table 5.18 Summary of the calculation results for the analysed slender tank

Pressure component	Result	Unit	ANSYS	Tables	Housner	Gehrig
Convective	Base shear	(kN)	17.90	18.33	6.39	18.09
	Overturning moment: wall pressure	(kNm)	229.41	235.61	81.92	231.98
	Overturning moment: wall and bottom pressure	(kNm)	229.41	235.61	81.92	–
	Acceleration	(m/s^2)	0.353	0.353	0.353	0.353
	Participation factor	(–)	2.0	2.0	2.0	2.0
Impulsive rigid	Base shear	(kN)	160.90	164.23	172.92	167.71
	Overturning moment: wall pressure	(kNm)	1049.60	1075.70	1137.81	1096.81
	Overturning moment: wall and bottom pressure	(kNm)	1066.50	1093.17	1160.29	–
	Acceleration	(m/s^2)	0.54	0.54	0.54	0.54
	Participation factor	(–)	1.0	1.0	1.0	1.0
Impulsive flexible	Base shear	(kN)	316.90	360.14	–	368.13
	Overturning moment: wall pressure	(kNm)	2850.1	3106.59	–	3176.95
	Overturning moment: wall and bottom pressure	(kNm)	2854.6	3113.86	–	–
	Acceleration	(m/s^2)	1.56	1.56	–	1.56
	Participation factor	(–)	1.6733	1.6348	–	1.635
Superposed with SRSS rule	Base shear	(kN)	355.86	396.24	172.91	404.94
	Overturning moment: wall pressure	(kNm)	3045.88	3295.99	1140.76	3368.95
	Overturning moment: wall and bottom pressure	(kNm)	3055.94	3308.57	1163.18	–

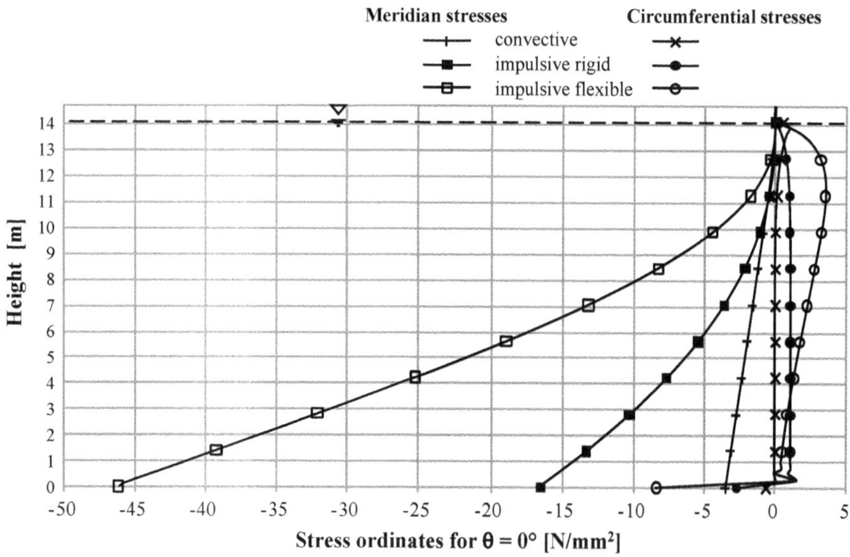

Fig. 5.53 Circumferential and meridian stresses of the pressure components due to horizontal seismic excitation

5.6.9.9 Evaluation of the Stresses in the Tank Shell

The pressure curves that were calculated in Sect. 5.6.9.3 are applied to the FE tank model as equivalent static loads. The stresses in the tank shell are calculated with a linear elastic calculation. Figures 5.53, 5.54 and 5.55 show exemplarily the decisive stress curves over the tank height for each of the pressure components as a result of a combined horizontal and vertical seismic excitation. These stresses serve as the basis for a normative tank design (not presented here).

5.6.10 Sample Calculation 2: Tank with Medium Slenderness

A second tank structure of medium slenderness is analysed under horizontal seismic action to demonstrate the practical application of the calculation methods presented in the previous sections. The base shears and the overturning moments for the different pressure components are calculated with the numerical model in ANSYS, with the tabulated coefficients in Sect. 5.6.12 and with the calculation method provided in Annex A.3.2.2 of DIN EN 1998-4 (2007). Finally, the results are compared and discussed.

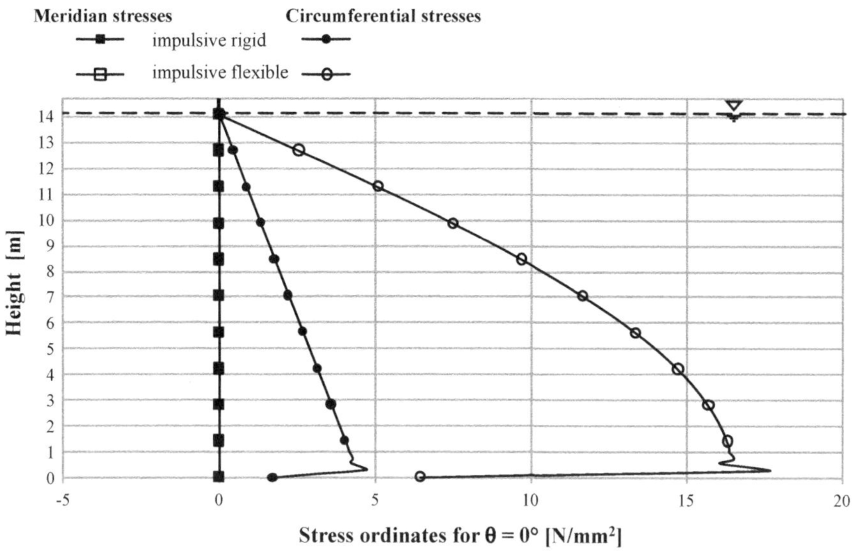

Fig. 5.54 Circumferential and meridian stresses of the pressure components due to vertical seismic excitation

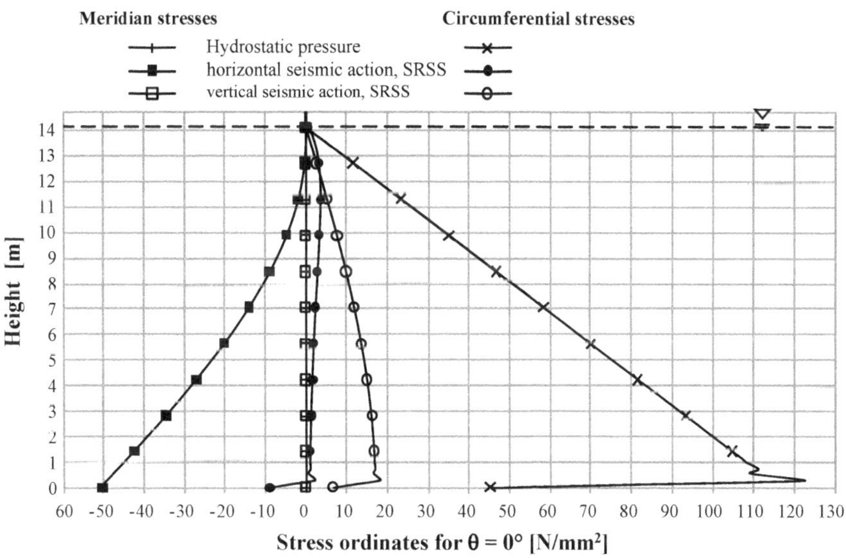

Fig. 5.55 Circumferential and meridian stresses of the pressure components due to horizontal and vertical seismic excitation superposed with the SRSS rule and combined with hydrostatic pressure

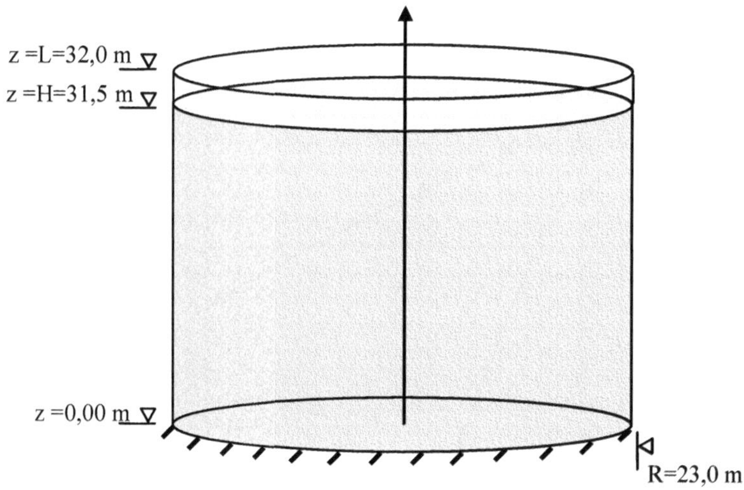

Fig. 5.56 Geometry of the tank investigated

5.6.10.1 Tank Description

The analysed tank has an inner radius of $R = 23.0\,m$, a height of $L = 32.0\,m$ and a filling height of $H = 31.5\,m$ (Fig. 5.56). This results in a tank slenderness of $\gamma = H/R = 1.37$. The tank wall is subdivided into segments with different wall thicknesses, but the calculations are carried out with an average thickness of $t = 11.3\,mm$. The material is steel with a Young's modulus of $E = 20.000\,\text{kN/cm}^2$ and the tank is used to store a liquid with a density of $\rho_L = 0.568\,\text{t/m}^3$. The importance factor is given as $\gamma_I = 1.6$. The design response spectra for the pressure components are determined according to the German National Annex DIN EN 1998-1/NA (2011) using the specific damping values proposed in Sect. 5.6.6. The following input values are used to compile the design response spectra:

Earthquake zone 3	$a_g = 0.8\text{m/s}^2$ (DIN EN 1998-1/NA 2011)
Ground type C-R	Soil factor $S = 1.5$
Control periods	$T_A = 0\text{s} \mid T_B = 0.05\text{s} \mid T_C = 0.3\text{s} \mid T_D = 2.0\text{s}$
Amplification factor	$\beta_0 = 2.5$
Damping for convective pressure	$0.5\,\% \Rightarrow \eta = 1.348$ after (5.70)
Damping for impulsive flexible pressure	$2.5\,\% \Rightarrow \eta = 1.155$ after (5.70).

5.6.10.2 Finite Element Modelling of the Tank

The modelling of the tank structure is carried out with shell elements using the FE program ANSYS. The tank support is assumed to be fully clamped at the bottom.

5.6.10.3 Base Shear and Overturning Moment with Tabulated Pressure Curves

Convective pressure component due to horizontal seismic excitation

The fundamental natural period for sloshing is calculated according to Eq. (5.42):

$$T_{k1} = \frac{2\pi}{\sqrt{\frac{g \cdot \lambda_1 \cdot \tanh(\lambda_1 \cdot \gamma)}{R}}} = \frac{2\pi}{\sqrt{\frac{9.81 \cdot 1.841 \tanh(1.841 \cdot 1.37)}{23.0}}} = 7.137\,s$$

The spectral acceleration is determined according to DIN EN 1998-1/NA (2011) for a spectrum with 0.5% damping:

$$\begin{aligned} a_{k1} &= a_g \cdot \gamma_I \cdot S \cdot \eta \cdot \beta_0 \cdot \frac{T_C \cdot T_D}{T^2} \\ &= 0.8 \cdot 1.6 \cdot 1.5 \cdot 1.348 \cdot 2.5 \cdot \frac{0.3 \cdot 2.0}{7.137^2} = 0.0762\,\text{m/s}^2 \end{aligned}$$

The participation factor Γ_{k1} of the fundamental natural vibration mode is calculated with $\lambda_1 = 1.841$ following Eq. (5.41):

$$\Gamma_{k1} = \frac{2 \cdot \sinh(\lambda_1 \cdot \gamma) \cdot [\cosh(\lambda_1 \cdot \gamma) - 1]}{\sinh(\lambda_1 \cdot \gamma) \cdot \cosh(\lambda_1 \cdot \gamma) - \lambda_1 \cdot \gamma} = 1.798$$

The tabulated coefficients are determined from Table 5.27 for $\gamma = 1.37$:

$C_{F,k}$	$C_{MW,k}$	$C_{MB,k}$	$C_{M,k}$
0.3477	0.2306	0.0429	0.4722

Using these coefficients, the resulting base shear and overturning moments are:

$$F_{b,k} = m_L \cdot a_k(t) \cdot \Gamma_k \cdot C_{F,k} = 29734.76 \cdot 0.0762 \cdot 1.798 \cdot 0.3477 = 1416.49\,\text{kN}$$

$$\begin{aligned} M_{W,k} &= \pi \cdot R^2 \cdot H^2 \cdot \rho_L \cdot a_k(t) \cdot \Gamma_k \cdot C_{MW,k} \\ &= 936644.93 \cdot 0.0762 \cdot 1.798 \cdot 0.2306 = 29592.32\,\text{kNm} \end{aligned}$$

$$\begin{aligned} M_{B,k} &= \pi \cdot R^4 \cdot \rho_L \cdot a_k(t) \cdot \Gamma_k \cdot C_{MB,k} \\ &= 499355.17 \cdot 0.0762 \cdot 1.798 \cdot 0.0429 = 2935.02\,\text{kNm} \end{aligned}$$

$$\begin{aligned} M_{G,k} &= \pi \cdot R^4 \cdot \rho_L \cdot a_k(t) \cdot \Gamma_k \cdot C_{M,k} \\ &= 499355.17 \cdot 0.0762 \cdot 1.798 \cdot 0.4722 = 32305.78\,\text{kNm} \end{aligned}$$

Impulsive rigid pressure component due to horizontal seismic excitation
The impulsive rigid pressure component due to horizontal seismic excitation is calculated using the spectral acceleration of the design response spectrum at period T = 0 s according to DIN EN 1998-1/NA (2011):

$$a_{is,h} = a_g \cdot \gamma_I \cdot S = 0.8 \cdot 1.6 \cdot 1,5 = 1.92\,\text{m/s}^2$$

The participation factor $\Gamma_{is,h}$ of the impulsive rigid pressure component is equal to 1.0 and the tabulated coefficients are determined using Table 5.27:

$C_{F,is,h}$	$C_{MW,is,h}$	$C_{MB,is,h}$	$C_{M,is,h}$
0.6501	0.2673	0.2071	0.7365

The resulting base shear and overturning moments are:

$$\begin{aligned} F_{b,is,h} &= m_L \cdot a_{is,h}(t) \cdot \Gamma_{is,h} \cdot C_{F,is,h} \\ &= 29734.76 \cdot 1.92 \cdot 1.0 \cdot 0.6501 = 37114.69\,\text{kN} \\ M_{W,is,h} &= \pi \cdot R^2 \cdot H^2 \cdot \rho_L \cdot a_{is,h}(t) \cdot \Gamma_{is,h} \cdot C_{MW,is,h} \\ &= 936644.93 \cdot 1.92 \cdot 1.0 \cdot 0.2673 = 480701.17\,\text{kNm} \\ M_{B,is,h} &= \pi \cdot R^4 \cdot \rho_L \cdot a_{is,h}(t) \cdot \Gamma_k \cdot C_{MB,is,h} \\ &= 499355.17 \cdot 1.92 \cdot 1.0 \cdot 0.2071 = 198559.60\,\text{kNm} \\ M_{G,is,h} &= \pi \cdot R^4 \cdot \rho_L \cdot a_{is,h}(t) \cdot \Gamma_{is,h} \cdot C_{M,is,h} \\ &= 499355.17 \cdot 1.92 \cdot 1.0 \cdot 0.7365 = 706128.16\,\text{kNm} \end{aligned}$$

Impulsive flexible pressure component due to horizontal seismic excitation
The fundamental natural period of the joint bending vibration of the tank and the liquid due to horizontal seismic excitation is calculated using the approximate Eq. (5.50):

$$T_{if,h,1} = 2 \cdot R \cdot F(\gamma) \sqrt{\frac{H \cdot \rho_L}{E \cdot s(\zeta = 1/3)}} = 2 \cdot 23.0 \cdot 3.15 \cdot \sqrt{\frac{31.5 \cdot 0.568}{200e6 \cdot 0.0113}} = 0.407\,s$$

with:

$$F(\gamma) = 0.157 \cdot \gamma^2 + \gamma + 1.49 = 3.15$$

For a better comparability of the numerical values and the results calculated with tabulated values, the numerically determined eigenperiod of 0.357 s is used in the following. The participation factor $\Gamma_{is,h}$ is calculated from Table 5.27 by interpolation:

$$\Gamma_{if,h} = 1.5797$$

The spectral acceleration according to DIN EN 1998-1/NA (2011) is calculated at the fundamental natural period $T_{if,h,1}$:

$$a_{if,h} = a_g \cdot \gamma_I \cdot S \cdot \eta \cdot \beta_0 \cdot \frac{T_C}{T} = 0.8 \cdot 1.6 \cdot 1.5 \cdot 1.155 \cdot 2.5 \cdot \frac{0.3}{0.357} = 4.66 \text{ m/s}^2$$

The tabulated coefficients are computed using Table 5.27 (sine function) and Table 5.28 (parametrized sine function):

Coefficient	$C_{F,if,h}$	$C_{MW,if,h}$	$C_{MB,if,h}$	$C_{M,if,h}$
Table 5.27	0.3588	0.1734	0.0798	0.4246
Table 5.28	0.3716	0.1778	0.0849	0.4204
Deviation (%)	3.44	2.47	6.01	0.99

The differences between the two methods are small. Thus, the tabulated coefficients according to Table 5.27 are used to calculate the base shear and overturning moments:

$$\begin{aligned}
F_{b,is,h} &= m_L \cdot a_{if,h}(t) \cdot \Gamma_{if,h} \cdot C_{F,if,h} \\
&= 29734.76 \cdot 4.66 \cdot 1.5797 \cdot 0.3588 = 78537.56 \text{ kN} \\
M_{W,if,h} &= \pi \cdot R^2 \cdot H^2 \cdot \rho_L \cdot a_{if,h}(t) \cdot \Gamma_{if,h} \cdot C_{MW,if,h} \\
&= 936644.93 \cdot 4.66 \cdot 1.5797 \cdot 0.1734 = 1195596.44 \text{ kNm} \\
M_{B,if,h} &= \pi \cdot R^4 \cdot \rho_L \cdot a_{kif,h}(t) \cdot \Gamma_k \cdot C_{MB,if,h} \\
&= 499355.17 \cdot 4.66 \cdot 1.5797 \cdot 0.0798 = 293341.14 \text{ kNm} \\
M_{G,if,h} &= \pi \cdot R^4 \cdot \rho_L \cdot a_{kif,h}(t) \cdot \Gamma_{if,h} \cdot C_{M,if,h} \\
&= 499355.17 \cdot 4.66 \cdot 1.5797 \cdot 0.4246 = 1560810.13 \text{ kNm}
\end{aligned}$$

5.6.10.4 Base Shear and Overturning Moment According to DIN EN 1998-4 (2007)

A simplified method for the calculation of the base shear and the overturning moments is proposed in Annex A.3.2.2 of Eurocode 8-4 (2006), as described in Sect. 5.6.7.2. At first, the fundamental natural period of the convective pressure component is calculated using Eq. (5.80):

$$T_k = C_k \sqrt{R} = 1.49 \cdot \sqrt{23} = 7.146 \ s$$

The spectral acceleration is determined according to DIN EN 1998-1/NA (2011) for a spectrum with 0.5% damping:

$$a_k = a_g \cdot \gamma_I \cdot S \cdot \eta \cdot \beta_0 \cdot \frac{T_C \cdot T_D}{T^2}$$
$$= 0.8 \cdot 1.6 \cdot 1.5 \cdot 1.348 \cdot 2.5 \cdot \frac{0.3 \cdot 2.0}{7.146^2} = 0.0760 \text{ m/s}^2$$

The fundamental natural period of the impulsive pressure component is calculated from Eq. (5.79):

$$T_i = C_i \cdot \frac{\sqrt{\rho_L} \cdot H}{\sqrt{s/R} \cdot \sqrt{E}} = 6.14 \cdot \frac{\sqrt{0.568} \cdot 31.5}{\sqrt{0.0113/23.0} \cdot \sqrt{200e6}} = 0.465\, s$$

The spectral acceleration according to DIN EN 1998-1/NA (2011) is calculated at the fundamental natural period T_i:

$$a_i = a_g \cdot \gamma_I \cdot S \cdot \eta \cdot \beta_0 \cdot \frac{T_C}{T} = 0.8 \cdot 1.6 \cdot 1.5 \cdot 1.155 \cdot 2.5 \cdot \frac{0.3}{0.465} = 3.577 \text{ m/s}^2$$

The total mass of the liquid content is equal to:

$$m = \pi \cdot R^2 \cdot H \cdot \rho_L = 29734.76\, t$$

The convective mass and the convective lever arms are calculated with Table 5.14:

$$m_k = 0.35 \cdot 29734.76 = 10417.17$$
$$h_k = 0.67 \cdot 31.5 = 21.11\, m$$
$$h_{ku} = 0.75 \cdot 31.5 = 23.63\, m$$

The impulsive mass and the impulsive lever arms are calculated using Table 5.14:

$$m_i = 0.65 \cdot 29734.76 = 19327.59\, t$$
$$h_i = 0.43 \cdot 31.5 = 13.55\, m$$
$$h_{iu} = 0.60 \cdot 31.5 = 18.90\, m$$

The convective and impulsive base shears are:

$$F_{b,k} = m_k \cdot a_k = 10407.17 \cdot 0.0760 = 790.95 \text{ kN}$$
$$F_{b,i} = m_i \cdot a_i = 19327.59 \cdot 3.577 = 69234.79 \text{ kN}$$

The convective overturning moments due to pressure on the tank wall and pressures on the wall and the bottom are calculated as follows:

$$M_{W,k} = m_k \cdot h_k \cdot a_k = 10407.17 \cdot 21.11 \cdot 0.0760 = 16696.84 \text{ kNm}$$

$$M_{G,k} = m_k \cdot h_{ku} \cdot a_k = 10407.17 \cdot 23.63 \cdot 0.0760 = 18690.03 \text{ kNm}$$

Finally, the impulsive overturning moments due to pressure on the tank wall and pressures on the wall and the bottom are given by:

$$M_{W,i} = m_i \cdot h_i \cdot a_i = 19327.59 \cdot 13.55 \cdot 3.577 = 936776.40 \text{ kNm}$$
$$M_{G,i} = m_i \cdot h_{iu} \cdot a_i = 19327.59 \cdot 18.90 \cdot 3.577 = 1306647.52 \text{ kNm}$$

5.6.10.5 Result Comparison and Discussion

The calculation results of the base shears and the overturning moments are summarised in Table 5.19. The results of the iterative solution with ANSYS and the calculation with tabulated coefficients show a good agreement, while the method according to DIN EN 1998-4 (2006) yields significantly lower base shears and overturning moments.

As the method according to DIN EN 1998-4 (2006) sums up the impulsive rigid pressure component and the impulsive flexible pressure component, the resulting pressure coefficients are not fully transparent. While the method itself is simple and easy to apply, the results are not always on the safe side. Therefore, the final tank design should be carried out using more accurate calculation methods.

It is also noticeable from the second sample calculation that the percentage of the overturning moment from the pressure components on the tank bottom has considerably increased compared to the slender tank analysed in Sect. 5.6.9. Disregarding the pressure parts on the bottom would result in a considerable underestimation of the seismic reactions and shell stresses.

5.6.11 Summary

With the simplified method of tabulated pressure coefficients, the user is provided with a tool for calculating the relevant seismic induced pressure components. These pressure components can be applied to an FE model as surface loads on the tank shell, which allows a linear elastic analysis to calculate reaction forces and shell stresses for design purposes. The use of tabulated pressure coefficients substitutes the implementation of mathematically sophisticated pressure formulas like those given in DIN EN 1998-4 (2006). This considerably simplifies the calculation of tank structures subjected to seismic load, both with regard to the program implementation and the required calculation time. The presented method is much closer to reality than the model according to Housner (1963) that does not take the joint bending vibration into account. His model leads to a considerable underestimation of stresses especially in the case of slender tanks. Also, the application of the simplified method

Table 5.19 Results for the tank with a slenderness of $\gamma = 1.37$

Pressure component	Result	Unit	ANSYS	Tables	DIN EN 1998-4 (2006)
Convective	Base shear	(kN)	1336.10	1416.49	790.95
	Overturning moment: wall pressure	(kNm)	27878.20	29592.32	16696.84
	Overturning moment: wall and bottom pressure	(kNm)	30573.80	32305.78	18690.03
	Fundamental natural period	(s)	7.137	7.137	7.146
	Acceleration	(m/s^2)	0.0762	0.0762	0.0760
	Participation factor	(–)	1.798	1.798	–
Impulsive rigid	Base shear	(kN)	37511.50	37114.69	Pressure part is considered in the impulsive flexible pressure component!
	Overturning moment: wall pressure	(kNm)	484732.80	480701.17	
	Overturning moment: wall and bottom pressure	(kNm)	686512.70	706128.16	
	Fundamental natural period	(s)	0.00	0.00	
	Acceleration	(m/s^2)	1.92	1.92	
	Participation factor	(–)	1.00	1.00	
Impulsive flexible	Base shear	(kN)	80199.00	78537.56	69134.79
	Overturning moment: wall pressure	(kNm)	1207926.90	1195596.44	936776.40
	Overturning moment: wall and bottom pressure	(kNm)	1515064.80	1560810.13	1306647.52
	Fundamental natural period	(s)	0.357	0.357	0.465

(continued)

Table 5.19 (continued)

Pressure component	Result	Unit	ANSYS	Tables	DIN EN 1998-4 (2006)
	Acceleration	(m/s^2)	4.658	4.658	3.577
	Participation factor	(–)	1.565	1.580	–
Superposed with SRSS rule	Base shear	(kN)	88548.16	86877.24	69139.31
	Overturning moment: Wall pressure	(kNm)	1301856.55	1288953.13	936925.19
	Overturning moment: Wall and bottom pressure	(kNm)	1663627.30	1713414.40	1306781.18

after DIN EN 1998-4 (2006), Annex A.3.2.2 may underestimate the reaction forces as demonstrated in the example of a tank with medium slenderness.

It should be pointed out that only anchored tanks were analysed here. In the case of unanchored tanks a partial uplift of the bottom can occur, which can cause significantly higher compression stresses on the opposite side of the tank. Furthermore, dynamic effects due to soil-structure interaction and elevations were not considered. Both aspects are object of current research activities and will certainly increase the level of complexity for tank calculation and design.

Annex: Tables of the Pressure Components

See Tables 5.20, 5.21, 5.22, 5.23, 5.24, 5.25, 5.26, 5.27 and 5.28.

Table 5.20 Coefficient $\mathbf{C_k}$ for the **convective pressure component** considering the fundamental natural mode for sloshing; to be multiplied with $\mathbf{R \cdot \rho_L \cdot \cos(\theta) \cdot a_k(t) \cdot \Delta_k}$, where $\mathbf{a_k}$ = horizontal spectral acceleration at the fundamental period T_{k1} according to Eq. (5.42)

$\zeta = \frac{z}{H}$	$\gamma = 0.2$	$\gamma = 0.4$	$\gamma = 0.6$	$\gamma = 0.8$	$\gamma = 1.0$	$\gamma = 1.5$	$\gamma = 2.0$	$\gamma = 2.5$	$\gamma = 3.0$	$\gamma = 3.5$	$\gamma = 4.0$	$\gamma = 5.0$	$\gamma = 6.0$	$\gamma = 7.0$	$\gamma = 8.0$	$\gamma = 9.0$	$\gamma = 10$
1.00	0.8371	0.8371	0.8371	0.8371	0.8371	0.8371	0.8371	0.8371	0.8371	0.8371	0.8371	0.8371	0.8371	0.8371	0.8371	0.8371	0.8371
0.95	0.8318	0.8183	0.8012	0.7838	0.7672	0.7300	0.6965	0.6650	0.6351	0.6065	0.5792	0.5283	0.4818	0.4395	0.4008	0.3656	0.3334
0.90	0.8268	0.8007	0.7679	0.7348	0.7039	0.6369	0.5796	0.5284	0.4819	0.4395	0.4008	0.3334	0.2774	0.2307	0.1919	0.1597	0.1328
0.85	0.8220	0.7841	0.7368	0.6898	0.6466	0.5560	0.4825	0.4198	0.3656	0.3184	0.2774	0.2104	0.1597	0.1211	0.0919	0.0697	0.0529
0.80	0.8176	0.7686	0.7080	0.6485	0.5947	0.4857	0.4017	0.3336	0.2774	0.2307	0.1919	0.1328	0.0919	0.0636	0.0440	0.0305	0.0211
0.75	0.8134	0.7541	0.6814	0.6107	0.5479	0.4247	0.3345	0.2651	0.2105	0.1672	0.1328	0.0838	0.0529	0.0334	0.0211	0.0133	0.0084
0.70	0.8095	0.7407	0.6569	0.5763	0.5057	0.3717	0.2788	0.2107	0.1597	0.1211	0.0919	0.0529	0.0305	0.0175	0.0101	0.0058	0.0033
0.65	0.8059	0.7283	0.6343	0.5449	0.4678	0.3259	0.2325	0.1676	0.1212	0.0878	0.0636	0.0334	0.0175	0.0092	0.0048	0.0025	0.0013
0.60	0.8026	0.7168	0.6137	0.5166	0.4339	0.2863	0.1941	0.1333	0.0920	0.0636	0.0440	0.0211	0.0101	0.0048	0.0023	0.0011	0.0005
0.55	0.7995	0.7063	0.5950	0.4910	0.4037	0.2522	0.1623	0.1062	0.0699	0.0461	0.0305	0.0133	0.0058	0.0025	0.0011	0.0005	0.0002
0.50	0.7967	0.6968	0.5780	0.4681	0.3768	0.2228	0.1361	0.0847	0.0531	0.0334	0.0211	0.0084	0.0033	0.0013	0.0005	0.0002	0.0001
0.45	0.7941	0.6883	0.5629	0.4477	0.3532	0.1978	0.1144	0.0676	0.0404	0.0243	0.0146	0.0053	0.0019	0.0007	0.0003	0.0001	0
0.40	0.7919	0.6806	0.5494	0.4298	0.3326	0.1765	0.0967	0.0542	0.0308	0.0176	0.0101	0.0033	0.0011	0.0004	0.0001	0	0
0.35	0.7899	0.6739	0.5377	0.4142	0.3148	0.1586	0.0822	0.0437	0.0236	0.0128	0.0070	0.0021	0.0006	0.0002	0.0001	0	0
0.30	0.7882	0.6681	0.5275	0.4009	0.2996	0.1437	0.0705	0.0355	0.0182	0.0094	0.0049	0.0013	0.0004	0.0001	0	0	0
0.25	0.7867	0.6632	0.5190	0.3897	0.2870	0.1315	0.0613	0.0292	0.0141	0.0069	0.0034	0.0008	0.0002	0.0001	0	0	0
0.20	0.7855	0.6592	0.5121	0.3806	0.2769	0.1219	0.0541	0.0244	0.0112	0.0052	0.0024	0.0005	0.0001	0	0	0	0
0.15	0.7846	0.6561	0.5067	0.3736	0.2690	0.1146	0.0487	0.0209	0.0091	0.0040	0.0018	0.0004	0.0001	0	0	0	0
0.10	0.7839	0.6539	0.5029	0.3686	0.2635	0.1094	0.0450	0.0186	0.0077	0.0032	0.0014	0.0002	0	0	0	0	0
0.05	0.7835	0.6526	0.5006	0.3657	0.2602	0.1064	0.0428	0.0172	0.0069	0.0028	0.0011	0.0002	0	0	0	0	0
0	0.7834	0.6521	0.4998	0.3647	0.2591	0.1054	0.0421	0.0168	0.0067	0.0027	0.0011	0.0002	0	0	0	0	0

Table 5.21 Coefficient $\mathbf{C_{is,h}}$ for the **impulsive rigid pressure component** due to horizontal seismic excitation; approximation of the Bessel function with a series of 200 terms; to be multiplied with $\mathbf{R} \cdot \rho_L \cdot \mathbf{cos(\theta)} \cdot \mathbf{a_{is,h}(t)} \cdot \mathbf{\Gamma_{is,h}}$, where $\mathbf{a_{is,h}}$ = horizontal spectral acceleration at period T = 0 (free field acceleration)

$\zeta=\frac{z}{H}$	$\gamma=0.2$	$\gamma=0.4$	$\gamma=0.6$	$\gamma=0.8$	$\gamma=1.0$	$\gamma=1.5$	$\gamma=2.0$	$\gamma=2.5$	$\gamma=3.0$	$\gamma=3.5$	$\gamma=4.0$	$\gamma=5.0$	$\gamma=6.0$	$\gamma=7.0$	$\gamma=8.0$	$\gamma=9.0$	$\gamma=10$
1.00	0	0	0	0	0	0	0	0	0	0	0	0	0	0	0	0	0
0.95	0.0277	0.0569	0.0864	0.1138	0.1390	0.1934	0.2401	0.2819	0.3202	0.3556	0.3885	0.4481	0.5007	0.5475	0.5893	0.6270	0.6609
0.90	0.0469	0.0965	0.1458	0.1916	0.2327	0.3184	0.3882	0.4481	0.5008	0.5476	0.5895	0.6611	0.7195	0.7674	0.8070	0.8397	0.8667
0.85	0.0625	0.1289	0.1948	0.2553	0.3088	0.4166	0.5002	0.5691	0.6272	0.6769	0.7196	0.7882	0.8397	0.8786	0.9079	0.9302	0.9470
0.80	0.0759	0.1569	0.2369	0.3098	0.3732	0.4970	0.5887	0.6610	0.7195	0.7675	0.8071	0.8669	0.9080	0.9364	0.9560	0.9695	0.9789
0.75	0.0876	0.1813	0.2738	0.3571	0.4286	0.5642	0.6601	0.7322	0.7882	0.8322	0.8669	0.9161	0.9471	0.9666	0.9789	0.9867	0.9916
0.70	0.0980	0.2030	0.3063	0.3987	0.4770	0.6211	0.7182	0.7879	0.8397	0.8786	0.9080	0.9471	0.9695	0.9825	0.9899	0.9942	0.9966
0.65	0.1072	0.2222	0.3352	0.4355	0.5193	0.6694	0.7658	0.8318	0.8786	0.9121	0.9364	0.9666	0.9825	0.9908	0.9952	0.9974	0.9986
0.60	0.1154	0.2394	0.3610	0.4680	0.5565	0.7107	0.8049	0.8664	0.9079	0.9364	0.9560	0.9789	0.9899	0.9952	0.9977	0.9989	0.9995
0.55	0.1227	0.2547	0.3838	0.4968	0.5392	0.7459	0.8371	0.8937	0.9301	0.9539	0.9696	0.9867	0.9942	0.9975	0.9989	0.9995	0.9998
0.50	0.1291	0.2683	0.4041	0.5222	0.6178	0.7759	0.8636	0.9153	0.9469	0.9666	0.9789	0.9916	0.9967	0.9987	0.9995	0.9998	0.9999
0.45	0.1348	0.2803	0.4220	0.5445	0.6427	0.8014	0.8854	0.9324	0.9596	0.9758	0.9854	0.9947	0.9981	0.9993	0.9998	0.9999	1
0.40	0.1399	0.2909	0.4377	0.5640	0.6644	0.8230	0.9033	0.9458	0.9692	0.9824	0.9899	0.9967	0.9989	0.9996	0.9999	1	1
0.35	0.1442	0.3000	0.4512	0.5808	0.6829	0.8411	0.9178	0.9563	0.9764	0.9872	0.9930	0.9979	0.9994	0.9998	1	1	1
0.30	0.1479	0.3078	0.4628	0.5951	0.6986	0.8561	0.9295	0.9645	0.9819	0.9906	0.9951	0.9987	0.9996	0.9999	1	1	1
0.25	0.1510	0.3143	0.4725	0.6070	0.7116	0.8684	0.9388	0.9708	0.9859	0.9931	0.9966	0.9992	0.9998	1	1	1	1
0.20	0.1535	0.3196	0.4803	0.6166	0.7221	0.8781	0.9460	0.9756	0.9888	0.9948	0.9976	0.9995	0.9999	1	1	1	1
0.15	0.1554	0.3237	0.4863	0.6240	0.7301	0.8854	0.9513	0.9791	0.9909	0.9960	0.9982	0.9997	0.9999	1	1	1	1
0.10	0.1568	0.3266	0.4906	0.6292	0.7358	0.8906	0.9550	0.9814	0.9923	0.9968	0.9986	0.9998	1	1	1	1	1
0.05	0.1576	0.3283	0.4932	0.6323	0.7392	0.8936	0.9572	0.9828	0.9931	0.9972	0.9989	0.9998	1	1	1	1	1
0	0.1579	0.3289	0.4940	0.6334	0.7403	0.8946	0.9579	0.9832	0.9933	0.9973	0.9990	0.9998	1	1	1	1	1

Table 5.22 Coefficient $\mathbf{C_{if,h}}$ for the **impulsive flexible pressure component** due to horizontal seismic excitation; approximation of the Bessel function with a series of 100 terms; to be multiplied with $\mathbf{R} \cdot \rho_L \cdot \cos(\theta) \cdot \mathbf{a_{if,h}(t)} \cdot \Gamma_{if,h}$, where $\mathbf{a_{if,h}}$ = horizontal spectral acceleration at the fundamental period $T_{if,h,1}$ calculated with an FE model or according to Eq. (5.50)

$\zeta = \frac{z}{H}$	$\gamma = 0.2$	$\gamma = 0.4$	$\gamma = 0.6$	$\gamma = 0.8$	$\gamma = 1.0$	$\gamma = 1.5$	$\gamma = 2.0$	$\gamma = 2.5$	$\gamma = 3.0$	$\gamma = 3.5$	$\gamma = 4.0$	$\gamma = 5.0$	$\gamma = 6.0$	$\gamma = 7.0$	$\gamma = 8.0$	$\gamma = 9.0$	$\gamma = 10$
	$f(\zeta) = \sin((\pi/2) \cdot \zeta)$								$f(\zeta) = \zeta$							$1 - \cos((\frac{\pi}{2}) \cdot \zeta)$	
1.00	0	0	0	0	0	0	0	0	0	0	0	0	0	0	0	0	0
0.95	0.0232	0.0472	0.0715	0.0951	0.1176	0.1701	0.2180	0.2621	0.2710	0.3063	0.3392	0.3989	0.4516	0.4986	0.5406	0.5505	0.5845
0.90	0.0373	0.0763	0.1156	0.1534	0.1892	0.2704	0.3422	0.4061	0.4014	0.4478	0.4895	0.5609	0.6192	0.6671	0.7066	0.6840	0.7108
0.85	0.0479	0.0983	0.1490	0.1974	0.2426	0.3429	0.4290	0.5032	0.4785	0.5275	0.5699	0.6385	0.6900	0.7289	0.7583	0.6989	0.7155
0.80	0.0561	0.1154	0.1748	0.2311	0.2831	0.3959	0.4896	0.5681	0.5211	0.5681	0.6073	0.6668	0.7079	0.7362	0.7558	0.6626	0.6715
0.75	0.0625	0.1287	0.1949	0.2572	0.3139	0.4341	0.5309	0.6094	0.5405	0.5832	0.6174	0.6663	0.6972	0.7168	0.7291	0.6071	0.6115
0.70	0.0673	0.1388	0.2102	0.2767	0.3364	0.4599	0.5560	0.6316	0.5428	0.5800	0.6086	0.6472	0.6695	0.6824	0.6898	0.5435	0.5453
0.65	0.0708	0.1463	0.2214	0.2907	0.3521	0.4756	0.5683	0.6388	0.5327	0.5641	0.5874	0.6169	0.6326	0.6409	0.6452	0.4790	0.4795
0.60	0.0732	0.1515	0.2290	0.2998	0.3617	0.4825	0.5695	0.6334	0.5133	0.5391	0.5573	0.5793	0.5900	0.5952	0.5977	0.4155	0.4152
0.55	0.0747	0.1546	0.2335	0.3049	0.3661	0.4819	0.5615	0.6179	0.4873	0.5076	0.5215	0.5373	0.5444	0.5475	0.5489	0.3550	0.3543
0.50	0.0752	0.1559	0.2353	0.3063	0.3661	0.4751	0.5460	0.5940	0.4564	0.4717	0.4818	0.4925	0.4970	0.4988	0.4995	0.2980	0.2971
0.45	0.0750	0.1557	0.2348	0.3046	0.3623	0.4630	0.5242	0.5634	0.4222	0.4329	0.4395	0.4461	0.4486	0.4495	0.4498	0.2452	0.2442
0.40	0.0742	0.1542	0.2323	0.3003	0.3553	0.4467	0.4976	0.5277	0.3858	0.3923	0.3959	0.3990	0.3998	0.4000	0.4001	0.1971	0.1960
0.35	0.0729	0.1517	0.2282	0.2938	0.3458	0.4271	0.4673	0.4881	0.3483	0.3508	0.3516	0.3515	0.3509	0.3505	0.3502	0.1537	0.1525
0.30	0.0712	0.1482	0.2228	0.2858	0.3344	0.4053	0.4348	0.4464	0.3108	0.3094	0.3076	0.3044	0.3024	0.3013	0.3007	0.1158	0.1145
0.25	0.0692	0.1442	0.2165	0.2767	0.3218	0.3824	0.4013	0.4038	0.2740	0.2690	0.2646	0.2582	0.2545	0.2525	0.2514	0.0830	0.0817
0.20	0.0670	0.1399	0.2098	0.2671	0.3088	0.3594	0.3684	0.3623	0.2392	0.2307	0.2237	0.2139	0.2082	0.2050	0.2030	0.0562	0.0549
0.15	0.0649	0.1356	0.2030	0.2576	0.2961	0.3376	0.3375	0.3234	0.2075	0.1956	0.1860	0.1725	0.1643	0.1593	0.1562	0.0349	0.0335
0.10	0.0629	0.1316	0.1969	0.2490	0.2848	0.3186	0.3107	0.2899	0.1805	0.1656	0.1535	0.1362	0.1252	0.1180	0.1131	0.0199	0.0185
0.05	0.0613	0.1284	0.1920	0.2422	0.2759	0.3041	0.2906	0.2646	0.1604	0.1431	0.1288	0.1079	0.0940	0.0843	0.0773	0.0105	0.0091
0	0.0605	0.1269	0.1897	0.2390	0.2718	0.2975	0.2815	0.2532	0.1515	0.1330	0.1176	0.0948	0.0791	0.0679	0.0594	0.0076	0.0062

Table 5.23 Coefficient $\mathbf{C_{if,h}}$ for the **impulsive flexible pressure component** due to horizontal seismic excitation; bending curve approximated with a parametrized sine function according to Eq. (**5.54**); to be multiplied with $\mathbf{R} \cdot \boldsymbol{\rho_L} \cdot \mathbf{cos(\theta)} \cdot \mathbf{a_{if,h}(t)} \cdot \boldsymbol{\Gamma}_{\mathbf{if,h}}$, where $\mathbf{a_{if,h}}$ = horizontal spectral acceleration at the fundamental period $T_{if,h,1}$ calculated with an FE model or according to Eq. (5.50)

$\zeta=\frac{z}{H}$	$\gamma=1.0$	$\gamma=1.2$	$\gamma=1.4$	$\gamma=1.6$	$\gamma=2.0$	$\gamma=2.4$	$\gamma=2.8$	$\gamma=3.0$	$\gamma=3.5$	$\gamma=4.0$	$\gamma=4.5$	$\gamma=5.0$	$\gamma=6.0$	$\gamma=7.0$	$\gamma=8.0$	$\gamma=10.0$	$\gamma=12.0$
1.00	0	0	0	0	0	0	0	0	0	0	0	0	0	0	0	0	0
0.95	0.1206	0.1400	0.1590	0.1776	0.2119	0.2416	0.2696	0.2834	0.3154	0.3446	0.3731	0.3993	0.4484	0.4923	0.5320	0.6013	0.6579
0.90	0.1956	0.2261	0.2549	0.2823	0.3293	0.3716	0.4087	0.4261	0.4655	0.5008	0.5330	0.5624	0.6129	0.6556	0.6910	0.7455	0.7835
0.85	0.2513	0.2898	0.3246	0.3562	0.4107	0.4564	0.4953	0.5126	0.5513	0.5844	0.6128	0.6377	0.6779	0.7085	0.7319	0.7631	0.7808
0.80	0.2945	0.3382	0.3764	0.4098	0.4661	0.5114	0.5479	0.5635	0.5970	0.6236	0.6453	0.6629	0.6891	0.7068	0.7185	0.7311	0.7359
0.75	0.3283	0.3747	0.4144	0.4481	0.5032	0.5444	0.5760	0.5889	0.6146	0.6334	0.6471	0.6574	0.6702	0.6766	0.6793	0.6794	0.6769
0.70	0.3542	0.4015	0.4411	0.4741	0.5248	0.5605	0.5856	0.5951	0.6123	0.6230	0.6292	0.6325	0.6336	0.6314	0.6277	0.6200	0.6136
0.65	0.3728	0.4199	0.4584	0.4894	0.5344	0.5632	0.5809	0.5869	0.5957	0.5985	0.5978	0.5953	0.5872	0.5785	0.5706	0.5580	0.5494
0.60	0.3849	0.4312	0.4676	0.4956	0.5337	0.5551	0.5653	0.5678	0.5684	0.5644	0.5578	0.5505	0.5354	0.5222	0.5114	0.4959	0.4861
0.55	0.3919	0.4362	0.4698	0.4943	0.5248	0.5383	0.5412	0.5404	0.5336	0.5236	0.5123	0.5012	0.4810	0.4648	0.4522	0.4349	0.4244
0.50	0.3944	0.4358	0.4660	0.4868	0.5089	0.5145	0.5106	0.5067	0.4936	0.4786	0.4635	0.4497	0.4259	0.4078	0.3941	0.3759	0.3650
0.45	0.3927	0.4308	0.4572	0.4739	0.4876	0.4855	0.4753	0.4686	0.4500	0.4311	0.4133	0.3975	0.3714	0.3522	0.3381	0.3195	0.3086
0.40	0.3872	0.4219	0.4442	0.4566	0.4620	0.4527	0.4367	0.4276	0.4044	0.3825	0.3627	0.3457	0.3185	0.2989	0.2847	0.2664	0.2556
0.35	0.3787	0.4097	0.4279	0.4359	0.4332	0.4173	0.3961	0.3851	0.3583	0.3341	0.3130	0.2954	0.2678	0.2485	0.2347	0.2169	0.2066
0.30	0.3680	0.3952	0.4091	0.4130	0.4027	0.3806	0.3549	0.3422	0.3125	0.2869	0.2652	0.2474	0.2203	0.2015	0.1883	0.1715	0.1618
0.25	0.3558	0.3790	0.3889	0.3889	0.3713	0.3439	0.3144	0.3004	0.2686	0.2420	0.2201	0.2026	0.1763	0.1586	0.1463	0.1308	0.1219
0.20	0.3424	0.3620	0.3681	0.3645	0.3407	0.3085	0.2760	0.2610	0.2276	0.2006	0.1789	0.1617	0.1368	0.1203	0.1090	0.0950	0.0871
0.15	0.3289	0.3453	0.3481	0.3412	0.3121	0.2761	0.2411	0.2253	0.1909	0.1638	0.1424	0.1259	0.1024	0.0873	0.0771	0.0647	0.0578
0.10	0.3164	0.3300	0.3301	0.3206	0.2874	0.2484	0.2117	0.1953	0.1604	0.1333	0.1124	0.0965	0.0744	0.0604	0.0513	0.0405	0.0346
0.05	0.3065	0.3180	0.3162	0.3049	0.2687	0.2280	0.1902	0.1735	0.1382	0.1113	0.0908	0.0754	0.0543	0.0414	0.0331	0.0233	0.0182
0	0.3018	0.3123	0.3096	0.2977	0.2605	0.2189	0.1807	0.1639	0.1285	0.1018	0.0815	0.0664	0.0458	0.0333	0.0252	0.0160	0.0112

Table 5.24 Coefficient $C_{if,v}$ for the **impulsive flexible pressure component** due to vertical seismic excitation; Bending curve: $\mathbf{f(\zeta)=\cos(\pi/2 \cdot \zeta)}$; including the correction factor to consider the degree of clamping at the tank bottom according to Eq. (5.66); values of the correction factor $\boldsymbol{\beta}$ for slenderness values $\gamma > 4$ have to be checked carefully; the coefficient $C_{if,v}$ is to be multiplied with $\mathbf{R \cdot \rho_L \cdot a_{if,v}(t) \cdot \Gamma_{if,v}}$, where $\mathbf{a_{if,v}}$ = spectral acceleration of the joint rotationally symmetric fundamental natural mode with the fundamental natural period $T_{if,v,1}$ calculated with an FE model or according to Eq. (5.63)

$\zeta=\frac{z}{H}$	γ=0.2	γ=0.4	γ=0.6	γ=0.8	γ=1.0	γ=1.5	γ=2.0	γ=2.5	γ=3.0	γ=3.5	γ=4.0	γ=5.0	γ=6.0	γ=7.0	γ=8.0	γ=9.0	γ=10
1.00	0	0	0	0	0	0	0	0	0	0	0	0	0	0	0	0	0
0.95	0.0107	0.0232	0.0386	0.0587	0.0878	0.1925	0.3468	0.5539	0.8161	1.1349	1.5118	2.4447	3.6227	5.0523	6.7390	8.6873	10.9016
0.90	0.0213	0.0463	0.0769	0.1171	0.1751	0.3838	0.6915	1.1044	1.6271	2.2628	3.0143	4.8744	7.2231	10.0735	13.4364	17.3211	21.7359
0.85	0.0318	0.0691	0.1147	0.1748	0.2613	0.5727	1.0318	1.6481	2.4281	3.3767	4.4983	7.2740	10.7790	15.0326	20.0510	25.8481	32.4362
0.80	0.0421	0.0914	0.1519	0.2314	0.3459	0.7581	1.3659	2.1817	3.2141	4.4698	5.9545	9.6287	14.2684	19.8990	26.5420	34.2157	42.9366
0.75	0.0522	0.1132	0.1881	0.2865	0.4283	0.9388	1.6915	2.7018	3.9803	5.5354	7.3740	11.9241	17.6699	24.6428	32.8693	42.3724	53.1722
0.70	0.0619	0.1343	0.2231	0.3399	0.5081	1.1137	2.0067	3.2052	4.7220	6.5668	8.7480	14.1460	20.9624	29.2345	38.9940	50.2678	63.0800
0.65	0.0712	0.1546	0.2568	0.3912	0.5848	1.2818	2.3095	3.6889	5.4345	7.5578	10.0681	16.2807	24.1257	33.6461	44.8782	57.8533	72.5989
0.60	0.0801	0.1739	0.2889	0.4401	0.6579	1.4420	2.5981	4.1498	6.1136	8.5021	11.3261	18.3150	27.1402	37.8502	50.4858	65.0821	81.6702
0.55	0.0885	0.1922	0.3192	0.4863	0.7269	1.5932	2.8706	4.5851	6.7549	9.3941	12.5143	20.2363	29.9874	41.8210	55.7821	71.9097	90.2380
0.50	0.0964	0.2092	0.3475	0.5294	0.7914	1.7347	3.1255	4.9922	7.3546	10.2281	13.6253	22.0329	32.6497	45.5339	60.7345	78.2940	98.2494
0.45	0.1037	0.2250	0.3737	0.5693	0.8511	1.8654	3.3611	5.3685	7.9090	10.9990	14.6523	23.6937	35.1107	48.9661	65.3125	84.1955	105.6551
0.40	0.1103	0.2394	0.3976	0.6057	0.9055	1.9847	3.5759	5.7117	8.4146	11.7022	15.5890	25.2084	37.3553	52.0963	69.4877	89.5779	112.4094
0.35	0.1162	0.2523	0.4190	0.6384	0.9543	2.0917	3.7687	6.0197	8.8683	12.3332	16.4296	26.5677	39.3695	54.9054	73.2346	94.4080	118.4707
0.30	0.1215	0.2636	0.4379	0.6671	0.9973	2.1858	3.9383	6.2905	9.2674	12.8881	17.1689	27.7631	41.1410	57.3760	76.5300	98.6561	123.8015
0.25	0.1259	0.2734	0.4540	0.6917	1.0341	2.2665	4.0836	6.5226	9.6093	13.3636	17.8023	28.7874	42.6589	59.4929	79.3535	102.2960	128.3691
0.20	0.1297	0.2814	0.4674	0.7121	1.0645	2.3331	4.2038	6.7145	9.8919	13.7568	18.3260	29.6342	43.9137	61.2429	81.6877	105.3051	132.1452
0.15	0.1326	0.2877	0.4779	0.7280	1.0883	2.3854	4.2980	6.8650	10.1136	14.0650	18.7367	30.2984	44.8979	62.6154	83.5184	107.6650	135.1066
0.10	0.1346	0.2922	0.4854	0.7395	1.1055	2.4230	4.3657	6.9731	10.2730	14.2866	19.0319	30.7757	45.6052	63.6018	84.8341	109.3612	137.2350
0.05	0.1359	0.2950	0.4899	0.7464	1.1158	2.4456	4.4065	7.0383	10.3689	14.4201	19.2097	31.0632	46.0313	64.1961	85.6268	110.3830	138.5174
0	0.1363	0.2959	0.4914	0.7487	1.1193	2.4532	4.4201	7.0600	10.4010	14.4647	19.2691	31.1593	46.1736	64.3946	85.8916	110.7244	138.9457

Table 5.25 Correction factor **β** to consider the clamping degree at the tank bottom. The factor is already considered in Table 5.24

γ=0.2	γ=0.4	γ=0.6	γ=0.8	γ=1.0	γ=1.5	γ=2.0	γ=2.5	γ=3.0	γ=3.5	γ=4.0	γ=5.0	γ=6.0	γ=7.0	γ=8.0	γ=9.0	γ=10
1	1	1	1	1.0780	1.1891	1.2679	1.3291	1.3790	1.4213	1.4578	1.5190	1.5689	1.6112	1.6478	1.6800	1.7089

Table 5.26 Participation factor $\Gamma_{if,v}$ for the **impulsive flexible pressure component due to vertical seismic excitation**

	γ=0.2	γ=0.4	γ=0.6	γ=0.8	γ=1.0	γ=1.5	γ=2.0	γ=2.5	γ=3.0	γ=3.5	γ=4.0	γ=5.0	γ=6.0	γ=7.0	γ=8.0	γ=9.0	γ=10
$\Gamma_{if,v}$	1.1892	1.0958	0.9896	0.8807	0.7807	0.5893	0.4650	0.3815	0.3224	0.2788	0.2453	0.1976	0.1653	0.1420	0.1244	0.1107	0.0997

Table 5.27 Coefficients $\mathbf{C_{F,j}}$, $\mathbf{C_{MW,j}}$, $\mathbf{C_{MB,j}}$, $\mathbf{C_{M,j}}$ and participation factors $\boldsymbol{\Gamma_j}$ for the **convective (j = k), impulsive rigid (j = is, h) and impulsive flexible pressure components (j = if, h)**

Coefficient	$\gamma = 0.2$	$\gamma = 0.4$	$\gamma = 0.6$	$\gamma = 0.8$	$\gamma = 1.0$	$\gamma = 1.5$	$\gamma = 2.0$	$\gamma = 2.5$	$\gamma = 3.0$	$\gamma = 3.5$	$\gamma = 4.0$	$\gamma = 5.0$	$\gamma = 6.0$	$\gamma = 7.0$	$\gamma = 8.0$	$\gamma = 9.0$	$\gamma = 10$
Convective pressure component																	
$C_{F,k}$	0.8704	0.7541	0.6360	0.5328	0.4493	0.3120	0.2355	0.1886	0.1572	0.1348	0.1179	0.0943	0.0786	0.0674	0.0590	0.0524	0.0472
$C_{MW,k}$	0.4434	0.3985	0.3520	0.3105	0.2758	0.2147	0.1762	0.1494	0.1297	0.1144	0.1023	0.0843	0.0717	0.0623	0.0551	0.0493	0.0447
$C_{MB,k}$	0.2311	0.1924	0.1474	0.1076	0.0764	0.0311	0.0124	0.0050	0.0020	0.0008	0.0003	0.0000	0.0000	0.0000	0.0000	0.0000	0.0000
$C_{M,k}$	0.2488	0.2561	0.2742	0.3063	0.3523	0.5143	0.7170	0.9388	1.1689	1.4023	1.6372	2.1084	2.5801	3.0520	3.5240	3.9963	4.4689
Γ_k	1.5101	1.5389	1.5830	1.6371	1.6954	1.8289	1.9173	1.9635	1.9847	1.9938	1.9975	1.9996	1.9999	2.0000	2.0000	2.0000	2.0000
Impulsive rigid pressure component																	
$C_{F,is,h}$	0.1148	0.2386	0.3591	0.4636	0.5478	0.6861	0.7630	0.8102	0.8418	0.8644	0.8813	0.9051	0.9209	0.9321	0.9406	0.9472	0.9524
$C_{MW,is,h}$	0.0459	0.0952	0.1435	0.1861	0.2214	0.2834	0.3224	0.3494	0.3694	0.3847	0.3969	0.4150	0.4278	0.4372	0.4445	0.4503	0.4549
$C_{MB,is,h}$	0.0172	0.0571	0.1024	0.1424	0.1736	0.2189	0.2376	0.2451	0.2480	0.2492	0.2497	0.2500	0.2500	0.2500	0.2500	0.2500	0.2500
$C_{M,is,h}$	0.0191	0.0723	0.1541	0.2615	0.3950	0.8565	1.5273	2.4290	3.5724	4.9623	6.6008	10.6261	15.6508	21.6749	28.6986	36.7218	45.7444
$\Gamma_{is,h}$	1.0000	1.0000	1.0000	1.0000	1.0000	1.0000	1.0000	1.0000	1.0000	1.0000	1.0000	1.0000	1.0000	1.0000	1.0000	1.0000	1.0000
Impulsive flexible pressure component																	
Approach	$f(\zeta) = \sin((\pi/2)\cdot\zeta)$								$f(\zeta)=\zeta$							$1-\cos((\frac{\pi}{2})\cdot\zeta)$	
$C_{F,if,h}$	0.0620	0.1286	0.1937	0.2510	0.2982	0.3801	0.4306	0.4647	0.3693	0.3846	0.3968	0.4149	0.4276	0.4371	0.4443	0.3151	0.3194
$C_{MW,if,h}$	0.0283	0.0586	0.0885	0.1157	0.1393	0.1854	0.2190	0.2448	0.2074	0.2211	0.2322	0.2493	0.2616	0.2708	0.2779	0.2204	0.2247
$C_{MB,if,h}$	0.0079	0.0258	0.0454	0.0614	0.0724	0.0823	0.0789	0.0713	0.0429	0.0377	0.0333	0.0269	0.0224	0.0192	0.0168	0.0022	0.0018
$C_{M,if,h}$	0.0090	0.0352	0.0772	0.1355	0.2118	0.4994	0.9547	1.6012	1.9094	2.7456	3.7493	6.2598	9.4396	13.2871	17.8012	17.8567	22.4761
$\Gamma_{if,h}$	1.6529	1.6581	1.6545	1.6417	1.6226	1.5646	1.5099	1.4656	1.7807	1.7401	1.7087	1.6642	1.6348	1.6141	1.5989	1.7553	1.7393

Table 5.28 Coefficients $\mathbf{C_{F,if,h}}$, $\mathbf{C_{MW,if,h}}$, $\mathbf{C_{MB,if,h}}$, $\mathbf{C_{M,if,h}}$ and participation factor $\boldsymbol{\Gamma_{if,h}}$ for the **impulsive flexible pressure component;** parametrized sine function of the bending curve according to Sect. 5.6.3.3

Coefficient	$\gamma = 1.0$	$\gamma = 1.2$	$\gamma = 1.4$	$\gamma = 1.6$	$\gamma = 1.8$	$\gamma = 2.2$	$\gamma = 2.6$	$\gamma = 3.0$	$\gamma = 3.5$	$\gamma = 4.0$	$\gamma = 4.5$	$\gamma = 5.0$	$\gamma = 6.0$	$\gamma = 7.0$	$\gamma = 8.0$	$\gamma = 10.0$	$\gamma = 12.0$
Impulsive flexible pressure component																	
Approach	Parametrized Sine function																
$C_{F.if.h}$	0.3215	0.3530	0.3749	0.3893	0.3983	0.4062	0.4065	0.4036	0.3984	0.3931	0.3883	0.3845	0.3788	0.3755	0.3737	0.3724	0.3727
$C_{MW.if.h}$	0.1482	0.1658	0.1799	0.1909	0.1996	0.2121	0.2203	0.2259	0.2308	0.2343	0.2369	0.2391	0.2427	0.2457	0.2483	0.2528	0.2564
$C_{MB.if.h}$	0.0800	0.0846	0.0850	0.0825	0.0782	0.0674	0.0563	0.0465	0.0366	0.0290	0.0233	0.0189	0.0131	0.0095	0.0072	0.0046	0.0032
$C_{M.if.h}$	0.2281	0.3234	0.4375	0.5711	0.7248	1.0940	1.5455	2.0800	2.8638	3.7770	4.8198	5.9962	8.7486	12.0473	15.9008	25.2863	36.9317
$\Gamma_{if.h}$	1.5374	1.5509	1.5623	1.5737	1.5849	1.6060	1.6246	1.6396	1.6536	1.6628	1.6684	1.6708	1.6707	1.6665	1.6608	1.6490	1.6386

References

American Society of Civil Engineers (ASCE): Minimum design loads for buildings and other structures. SEI/ASCE 7-05, ISBN 0-7844-0831-9, Reston, VA (2006)

API 650: Welded Steel Tanks for Oil Storage. American Petroleum Institute (Hrsg.) (2003)

Bachmann, H.: Neue Tendenzen im Erdbebeningenieurwesen. Beton- und Stahlbetonbau, vol. 99, Heft 5, S. 356–371 (2004)

BASF: Internetseite der BASF SE (online Pressefotos) (2010). http://www.basf.com

Bauer, E. Zum mechanischen Verhalten granularer Stoffe unter vorwiegend ödometrischen Beanspruchungen. Veröffentlichungen des Institutes für Bodenmechanik und Felsmechanik der University Fridericiana in Karlsruhe, No 130. Ph.D. Thesis, Universität Fridericiana zu Karlsruhe, Karlsruhe, Germany (1992)

Braun, A. Schüttgutbeanspruchungen von Silozellen unter Erdbebeneinwirkungen. Institut für Massivbau und Baustofftechnologie. Ph.D. Thesis, Karlsruhe Institute of Technology, Karlsruhe, Germany (1997)

Bronstein, I.N., Semendjajew, K.A.: Teubner-Taschenbuch der Mathematik. E. Zeidler (Hrsg.), Teubner Verlagsgesellschaft, ISBN 3-8154-2001-6 (1996)

Bruneau, M.: Building damage from the Marmara, Turkey earthquake of August 17, 1999, Multidisciplinary Center for Earthquake Engineering Research, and Department of Civil, Environmental and Structural Engineering, University at Buffalo, Buffalo, NY 14260, USA (2001)

Chopra, A.K., Goel, R.: Seismic code analysis of buildings without locating centers of rigidity. J. Struct. Eng. **119**(10), 3039–3055 (1993)

Clough, D.P.: Experimental Evaluation of Seismic Design Methods for Broad Cylindrical Tanks. Report No. UCB/EERC-77/10, University of California, Berkeley, California (1977)

Cornelissen, P.: Erarbeitung eines vereinfachten impulsiv-flexiblen Lastansatzes für die Berechnung von Tankbauwerken unter Erdbebenlast. Diplomarbeit, RWTH Aachen (2010)

DIN 4149: Bauten in deutschen Erdbebengebieten. Deutsches Institut für Normung (DIN), Berlin Beuth-Verlag, Berlin (2005)

DIN 1055-6: Einwirkungen auf Silos und Flüssigkeitsbehälter: Deutsches Institut für Normung (DIN), Beuth-Verlag, Berlin (2005)

DIN EN 1998-1: Eurocode 8: Auslegung von Bauwerken gegen Erdbeben—Teil 1: Grundlagen, Erdbebeneinwirkungen und Regeln für Hochbauten; Deutsche Fassung EN 1998-1:2004 + AC:2009, December 2010

DIN EN 1998-1/NA: Nationaler Anhang—National festgelegte Parameter—Eurocode 8: Auslegung von Bauwerken gegen Erdbeben—Teil 1: Grundlagen, Erdbebeneinwirkungen und Regeln für Hochbau, January 2011

Eurocode 0: Basis of structural design, European Standard, European Committee for Standardization, April 2002

Eurocode 1-1-4: Actions on structures—General actions—Wind actions, European Standard, European Committee for Standardization, January 2004

Eurocode 1-1-3: Actions on structures—General actions—Snow loads, European Standard, European Committee for Standardization, July 2003

Eurocode 1-4: Actions on structures—General actions—Silos and tanks, European Standard, European Committee for Standardization, May 2006

Eurocode 1-4: Actions on structures—General actions—Silos and tanks, European Standard, European Committee for Standardization, May 1995

Eurocode 3-1-1: Design of steel structures—General rules and rules for buildings, European Standard, European Committee for Standardization, May 2005

Eurocode 3-1-6: Design of steel structures—Strength and Stability of Shell Structures, European Standard, European Committee for Standardization, February 2007

Eurocode 3-4: Design of steel structures—Silos, European Standard, European Committee for Standardization, February 2007

Eurocode 8-1: Design of structures for earthquake resistance, General rules, seismic actions and rules for buildings, European Standard, European Committee for Standardization, May 2004
Eurocode 8-4: Design of structures for earthquake resistance, Silos, tanks and pipelines, European Standard, European Committee for Standardization, July 2006
Eurocode 8-5: Design of structures for earthquake resistance, Foundations, retaining structures and geotechnical aspects, European Standard, European Committee for Standardization, November 2004
El-Zeiny: Nonlinear Time-Dependent Seismic Response of Unanchored Liquid Storage Tanks. Dissertation, University of California, Irvine, CA (2000)
FEMA 450: Federal Emergancy Managament Agency: NEHRP recommended provisions for the development of seismic regulations for new buildings and other structures. 2003 Ed. (FEMA 450), American Society of Civil Engineers (2003)
Fischer, F.D., Rammerstorfer, F.G.: The Stability of Liquid-Filled Cylindrical Shells under Dynamic Loading. In: E. Ramm (Hrsg.): Buckling of Shells. S., pp. 569–597 (1982)
Fischer, F.D., Rammerstorfer, F.G., Scharf, K.: Earthquake Resistant Design of Anchored and Unanchored Liquid Storage Tanks under Three-Dimensional Earthquake Excitation. In: G.I. Schuëller (Hrsg.): Structural Dynamics. Springer, pp. 317–371 (1991)
Fischer, F.D., Seeber, R.: Dynamic response of vertically excited liquid storage tanks considering liquid-soil interaction. Earthq. Eng. Struct. Dynam. **16**, 329–342 (1988)
Freeman, S.A.: Review of the Development of the Capacity Spectrum Method. ISET J. Earthq. Technol. **41**(1), paper no. 438, 1–13 (2004)
Gehrig, H.: Vereinfachte Berechnung flüssigkeitsgefüllter verankerter Kreiszylinderschalen unter Erdbebenbelastung. Stahlbau, vol. 73, Heft 1 (2004)
Gudehus, G.: A comprehensive equation for granular materials. Soils Found. **36**(1), 1–12 (1996)
Niemunis, A., Herle, I.: Hypoplastic model for cohesionsless soils with elastic strain range. Mech. Cohesive-Frict. Mater. **2**, 279–299 (1997a)
Guggenberger, W.: Schadensfall, Schadensanalyse und Schadensbehebung eines Silos auf acht Einzelstützen. Stahlbau, Nr. 67, Heft 6 (1998)
Gupta, B., Eeri, M., Kunnath, K.: Adaptive Spectra-based pushover procedure for seismic evaluation of structures. Earthq. Spectra **16** (2000)
Haack, A., Tomas, J.: Untersuchungen zum Dämpfungsverhalten hochdisperser, kohäsiver Pulver. Chemie Ingenieur Technik, Band 75, Nr.11 (2003)
Habenberger, J.: Beitrag zur Berechnung von nachgiebig gelagerten Behältertragwerken unter seismischen Einwirkungen. Dissertation, Weimar (2001)
Holl, H.J.: Parameteruntersuchung zur Abgrenzung der Anwendbarkeit eines Berechnungskonzeptes für Erdbebenbeanspruchte Tankbauwerke. Heft ILFB—1/87 der Berichte aus dem Institut für Leichtbau und Flugzeugbau der Technischen Universität Wien, Diplomarbeit (1987)
Holler, S., Meskouris, K.: Granular Material Silos under dynamic excitation: Numerical simulation and experimental validation. J. Struct. Eng. **132**(10), 1573–1579 (2006)
Holtschoppen, B., Butenweg, C., Meskouris, K.: Seismic Design of Secondary Structures. Im Tagungsband Seismic Risk 2008—Earthquakes in North-Western Europe. Liege (2008)
Holtschoppen, B.: Beitrag zur Auslegung von Industrieanlagen auf seismische Belastungen. Dissertation, Lehrstuhl für Baustatik und Baudynamik, RWTH Aachen (2009a)
Holtschoppen, B., Butenweg, C., Meskouris, K.: Seismic Design of Non-Structural Components in Industrial Facilities. Int. J. Eng. Under Uncertain. Hazards Assessment and Mitigation (2009b)
Holtschoppen, B., Cornelissen, P., Butenweg, C., Meskouris, K.: Vereinfachtes Berechnungsverfahren der Interaktionsschwingung bei flüssigkeitsgefüllten Tankbauwerken unter seismischer Belastung. Tagungsband Baustatik-Baupraxis, Innsbruck (2011)
Housner, G.W.: The dynamic behaviour of water tanks. Bull. Seismol. Soc. Am. **53**, 381–387 (1963)
IBC: International Building Code. International code council, 2015
Janssen, H.A.: Getreidedruck in Silozellen. Z. Ver. Dt. Ing. **39**, 1045–1049 (1895)

Kettler, M.: Earthquake Design of Large Liquid-Filled Steel Storage Tanks. Diplomarbeit, TU Graz, 2004, Vdm Verlag Dr. Müller, ISBN 978-3639059588 (2008)

Kneubühl, F.K.: Repetitorium der Physik. 5. Auflage, Teubner Verlag (1994)

Kolymbas, D.: Ein nichtlineares viskoplastisches Stoffgesetz für Böden. Dissertation, Veröffentlichungen des Instituts für Bodenmechanik und Felsmechanik der Universität Fridericiana in Karlsruhe, Heft 77 (1978)

Luft, R.W.: Vertical accelerations in prestressed concrete tanks. J. Struct. Eng. **110**(4), 706–714 (1984)

Martens, P.: Silohandbuch. Wilhelm Ernst&Sohn Verlag, Berlin (1998)

Meskouris, K.: Baudynamik. Modelle, Methoden, Praxisbeispiele. Bauingenieur-Praxis. Berlin: Ernst & Sohn (1999)

Niemunis, A., Herle, I.: Hypoplastic model for cohesionsless soils with elastic strain range. Mech. Cohesive Friction. Mater. **2**, 279–299 (1997b)

Nottrott, Th: Schwingende Kamine und ihre Berechnung im Hinblick auf die Beanspruchung durch Kármán-Wirbel. Bautechnik Heft **12**, 411–415 (1963)

New Zealand Standard 1170.5: Structural Design Actions, Part 5: Earthquake Actions; Standards New Zealand: Wellington, New Zealand (2004)

Petersen, C.: Dynamik der Baukonstruktionen. Vieweg Verlag, Braunschweig/Wiesbaden (2000)

Pieraccini, L., Silvestri, S., Trombetti, T.: Refinements to the Silvestri's theory for the evaluation of the seismic actions in flat-bottom silos containing grain-like material. Bull. Earthq. Eng. **131**, 3493–3525 (2015)

Rammerstorfer, F.G., Fischer, F.D.: Ein Vorschlag zur Ermittlung von Belastungen und Beanspruchungen von zylindrischen, flüssigkeitsgefüllten Tankbauwerken bei Erdbebeneinwirkung. Neuauflage des Institutsberichtes ILFB-2/90, Institut für Leichtbau und Struktur-Biomechanik (ILSB) der TU Wien (2004)

Rammerstorfer, F.G., Scharf, K., Fischer, F.D.: Storage tanks under earthquake loading. Appl. Mech. Rev. **43**(11), 261–279 (1990)

Rammerstorfer, F.G., Scharf, K., Fischer, F.D., Seeber, R.: Collapse of earthquake excited tanks. Res Mechanica **25**, 129–143 (1988)

Rinkens, E.: Automatische Berechnung und Bemessung von Metallsilos mit der FE-Methode nach DIN 1055-6:2005. Diplomarbeit, RWTH-Aachen (2007)

Rotter, J.M., Hull, T.S.: Wall loads in squat steel silos during earthquake. Eng. Struct. **11**, 139–147 (1989)

Rotter, J.M.: Structures, stability, silos and granular solids: A personal adventure. In: Chen, J.F., Ooi, J.Y., Teng, J.G. (eds.) Structures and Granular Solids: From Scientific Principles to Engineering Application, pp. 1–20. London, UK, Taylor & Francis Group (2008)

Rotter, J.M.: Silos and tanks in research and practice: State of the art and current challenges. In: Proceedings of the International Association for Shell and Spatial Structures (IASS) Symposium 2009—Evolution and Trends in Design, Analysis and Construction of Shell and Spatial Structures, Valencia, Spain, 28 September–2 October 2009; Domingo, A., Lazaro, C., Eds.; Universitat Politècnica de València: Valencia, Spain (2009)

Sakai, F., Ogawa, H., Isoe, A.: Horizontal, vertical and rocking fluid-elastic responses and design of cylindrical liquid storage tanks. In: Proceedings of the 8th World Conference on Earthquake Engineering (1984)

Sakai, M., Matsumura, H., Sasaki, M., Nakamura, N., Kobayashi, M., Kitagawa, Y.: Study on the dynamic behavior of coal silos against earthquakes. Bulk Solids Handl. **1985**, 5 (1021)

Younan, A.H., Veletsos, A.S.: Dynamics of solid-containing tanks I: rigid tanks. J. Struct. Eng. **124**, 52–61 (1998)

Sasaki, K.K., Freeman, S.A., Paret, T.F.: Multimode Pushover Procedure (MMP)—a method to identify the effects of higher modes in a pushover analysis. In: Proceedings of the 6th U.S. National Conference on Earthquake Engineering, Seattle, Washington (1998)

Scharf, K., Rammerstorfer, F.G.: Probleme bei der Anwendung der Antwortspektrenmethode für Flüssigkeit-Festkörper-Interaktionsprobleme des Erdbebeningenieurwesens. Zeitschrift für angewandte Mathematik und Mechanik, ISSN 0044-2267, vol. 71, no. 4, Seiten T160-T165 (1991)

Scharf, K.: Beiträge zur Erfassung des Verhaltens von erdbebenerregten, oberirdischen Tankbauwerken. Dissertation, Wien, Fortschrittsbericht aus der VDI-Reihe 4 Nr. 97, ISBN 3-18-149704-5, VDI-Verlag (1990)

Shimamoto, A., Kodama, M., Yamamura, M.: Vibration tests for scale model of cylindrical coal storing silo. In: Proceedings of the 8th World Conference on Earthquake Engineering, San Francisco, CA, USA, 21–28 July 1984; vol. 5, pp. 287–294 (1984)

Schmidt, H.: Schalenbeulen im Stahlbau—Ein spannendes Bemessungsproblem. Essener Unikate (2004)

Seed, H.B., Lysmer, J.: Geotechnical Engineering in Seismic Areas. Bergamo (1980)

Shih, C.-F.: Failure of Liquid Storage Tanks due to Earthquake Excitation. Dissertation, California Institute of Technology (1981)

SIA 261: Einwirkungen auf Tragwerke. Schweizerischer Ingenieur- und Architektenverein, Zürich (2003)

SIA 261: Einwirkungen auf Tragwerke. Schweizerischer Ingenieur- und Architektenverein, Zürich (2014)

Sigloch, H.: Technische Fluiddynamik. 7. Auflage, Springer Verlag, ISBN 978-3-642-03089-5 (2009)

Silvestri, S., Ivorra, S., Chiacchio, L.D., Trombetti, T., Foti, D., Gasparini, G., Taylor, C.: Shaking table tests of flat bottom circular silos containing grain like material. Earthq. Eng. Struct. Dyn. **45**, 69–89 (2016)

Silvestri, S., Gasparini, G., Trombetti, T., Foti, D.: On the evaluation of the horizontal forces produced by grain-like material inside silos during earthquakes. Bull. Earthq. Eng. **65**, 69–89 (2012)

Singh, M.P., Moreschi, L.M.: Simplified methods for calculating seismic forces for non-structural components. In: Proceedings of Seminar on Seismic Design, Retrofit, and Performance of Non-structural Components (ATC-29-1), Applied Technology Council (1998)

Smoczynski, K., Schmitt, T.: Resultierende Erdbebeneinwirkung aus zwei Horizontalkomponenten. DGEB-Tagung in Hannover (2011)

Smoltczyk, U.: Grundbau Taschenbuch. Vierte Auflage, Teil 1, Ernst & Sohn, ISBN: 3-433-01085-4 (1991)

Stempniewski, L.: Flüssigkeitsgefüllte Stahlbetonbehälter unter Erdbebeneinwirkung. Dissertation, Karlsruhe (1990)

Taiwan earthquake code: Seismic Design Code for Buildings in Taiwan. Construction and Planning Agency, Ministry of the Interior (2005)

Tang, Y.: Studies of Dynamic Response of Liquid Storage Tanks. Dissertation, Rice University Houston, Texas (1986)

UBC 1997: Uniform Building Code, "UBC 97". Whittier, CA (1997)

Verband der Chemischen Industrie (VCI): Leitfaden: Der Lastfall Erdbeben im Anlagenbau. Edition 10/2012, www.vci.de

Verband der Chemischen Industrie (VCI): Erläuterungen zum Leitfaden: Der Lastfall Erdbeben im Anlagenbau. Edition 10/2012, www.vci.de

Veletsos, A.S., Younan, A.H.: Dynamics of solid-containing tanks II: flexible tanks. J. Struct. Eng. **124**, 62–70 (1998)

Villaverde, R.: Seismic design of secondary structures: state of the art. J. Struct. Eng. **123**(8), 1011–1019 (1997)

Wagner, R.: Seismisch belastete Schüttgutsilos. Ph.D. Thesis, Lehrstuhl für Baustatik und Baudynamik, RWTH Aachen University, Aachen, Germany (2009)

Wolf, J.P.: Foundation Vibration Analysis Using Simple Physical Models; PTR Prentice Hall Inc. Upper Saddle River, NJ, USA, p. 27 (1994)

Yanagida, T., Matchett, A., Asmar, B., Langston, P., Walters, K., Coulthard, M.: Damping characteristics of particulate materials using low intensity vibrations: effects of experimental variables and their interpretation. J. Chem. Eng. Jpn. **36**(11), 1339–1346 (2003)

Yokota, H., Sugita, M., Mita, I.: Vibration tests and analyses of coal-silo model. In: Proceedings of the 2nd International Conference on the Design of Silos for Strength and Flow, Stratford-upon-Avon, UK, 7–9 Nov. 1983, pp. 107–116 (1983)

Chapter 6
Structural Oscillations of High Chimneys Due to Wind Gusts and Vortex Shedding

Francesca Lupi, Hans-Jürgen Niemann and Rüdiger Höffer

Abstract During the last 2 decades, new code generations have been introduced which incorporate the achieved scientific and technological state of the art that has been proven its usefulness in practical application. The Eurocodes and the CICIND model codes are examples of this development. The Eurocode applies a modified gust response factor to model in-line wind loading and resonance due to turbulence through increasing the peak velocity pressure by a factor which depends on the individual size and dynamic features of the structure considered. The CICIND model codes for steel and concrete chimneys take account of the particular mechanical behaviour and the specific design requirements of these structures and utilizes the mean and the gust wind force, where the gust load defines an equivalent static load scaled to reproduce the real base bending moment of the chimney induced by wind gustiness. The cross-wind excitation of chimneys by vortex shedding is calculated in the CICIND model applying a negative aerodynamic damping to incorporate the motion induced forces, and a bandwidth factor to account for the reduction of the lift force spectrum caused by wind turbulence. Contrarily, the Eurocode relies on an empirical concept. In addition, it contains a further method, where the aerodynamic damping parameter is given for zero turbulence only. The principal issues of the chapter are to identify the merits and the drawbacks of the different concepts and to identify their dominant fields of application.

Keywords Structural resonance due to wind turbulence · Vortex resonance Aerodynamic damping · Code models

During the last 2 decades, a change of paradigm has taken place in modelling the wind effects and the wind risk in the design of buildings and structures. A new code generation is being introduced which incorporates the scientific and technological state of the art that has been achieved in science and proven its usefulness in practical application. The Eurocodes EN 1990 "Basis of Design" and EN 1991-1-4 "Actions on Structures—Wind Actions" (both 2010) are examples of this development.

In-line wind loading and resonance due to turbulence. The Eurocode model applies a modified gust response factor, increasing the peak velocity pressure by a factor which depends on the individual size and dynamic features of the structure con-

K. Meskouris et al., *Structural Dynamics with Applications in Earthquake and Wind Engineering*, https://doi.org/10.1007/978-3-662-57550-5_6

sidered. The profile of the peak velocity pressure envelopes non-simultaneous local peaks. The resulting equivalent static wind force is generally accepted as an appropriate approximation. The CICIND model codes (CICIND 2010, 2011) for steel and concrete chimneys contain technical specifications for wind actions taking account of the particular mechanical behaviour and the specific design requirements of these structures. The CICIND approach utilizes instead a sum, namely of the mean and of the gust wind force. Both summands are in the same order of magnitude. The mean part is free from approximations and realistic. The gust load defines—similarly as the Eurocode—an equivalent static load scaled to reproduce the real base bending moment of the chimney induced by wind gustiness. In this manner, the CICIND model seems to be closer to reality since only half of the load is approximate. The differences between the CICIND approach and the Eurocode stipulations shall be highlighted in view of the response induced in steel and concrete chimneys.

Vortex excitation and vortex resonance. In the CICIND model, the cross-wind excitation of a chimney by vortex shedding is calculated applying an approach which was initially developed by Vickery and Basu (1983a, b, Part I and Part II). It applies a negative aerodynamic damping to incorporate the motion induced forces, and a bandwidth factor to account for the reduction of the lift force spectrum caused by wind turbulence. Contrarily, the Eurocode relies on an empirical concept developed by Ruscheweyh (method 1, Ruscheweyh 1982). In addition, the latest version of the code contains as method 2 an extension of the Vickery and Basu model by Hansen (1998). In this extension, the aerodynamic damping parameter is given for zero turbulence only. This procedure is recommended as a conservative approach in a cold climate where atmospheric stratification may largely cancel wind turbulence. The principal issues are to identify the merits and the drawbacks of the two concepts, to identify their dominant fields of application, and to consider how harmonisation of the CICIND model codes with other related codes can be achieved in this field of vortex excitation.

6.1 Gust Wind Response Concepts

6.1.1 Models for Gust Wind Loading

6.1.1.1 Davenport's Gust Response Factor G

All wind force models presently utilized in modern wind loading codes go back to the classical Davenport approach. It amplifies the mean wind force applying the gust response factor G. The wind force may be given as one or several point loads F_w, or as a line load w_m. The latter is appropriate for slender, line-like structures such as chimneys:

$$w(z) = G \cdot w_m(z) \tag{6.1}$$

In Eq. (6.1), z is the height above the ground; G is the gust response factor. The mean wind force is based on the mean wind velocity pressure q_m:

$$w_m(z) = C_D(z) \cdot d(z) \cdot q_m(z) \tag{6.2}$$

in which C_D—aerodynamic drag coefficient; d—diameter of the chimney; $q_m(z)$—velocity pressure of the 10-min mean wind speed at level z. The mean wind force is a real physical quantity, whereas the wind force Eq. (6.1) is an equivalent static load, intended to reproduce the effects of the stochastic wind loading process on the most important structural stressing.

The aerodynamic drag coefficient C_D results from wind tunnel experiments. Equation (6.2) provides the definition to be applied in evaluating test results: the mean load measured at level z is referred to the mean velocity pressure at the same level and the local diameter or width of the chimney ("mean-to-mean").

The line load varies along the chimney axis predominantly according to the profile of the natural wind. In a 2-dimensional flow, the variation of w_m is therefore largely compensated and C_D in this definition is approximately constant. In fact, the real flow is 3-dimensional for two main reasons: (i) The wind profile causes pressure differences along the cylinder axis which induce a corresponding secondary flow; (ii) For finite cylinders the free end flow reduces the wind force in general but may also increase it in the tip region, depending on the slenderness.

Davenport (1961) developed the concept of the gust response factor, G. He considered cantilevered, vertical structures and their response to the wind action, namely the mean (static) and time dependent (quasi-static and dynamic) components. He defined the gust response factor as the ratio of the peak to the mean response, E_p to E_m:

$$G = \frac{E_p}{E_m} = \frac{E_m + k_p \sigma_E}{E_m} = 1 + k_p \frac{\sigma_E}{E_m} \tag{6.3}$$

where σ_E is the standard deviation (or rms-value) of the fluctuating response; k_p is the peak factor, which is the ratio of the peak response fluctuation to its standard deviation, σ_E. The load fluctuations due to wind turbulence provide a broad band excitation of the structure. Part of it will cause resonance which may become important when the chimney is flexible and the 1st natural frequency n_1 is too low. A rough estimate for limiting resonance in a beam mode to <10% is

$$n_1 \geq \left(\frac{70}{h}\right)^{0.75} \tag{6.4}$$

where h is the chimney height in m. The rms response is split into resonant and quasi-static components, σ_{ER} and σ_{EB} (since the latter is also called background response, the subscript B is introduced):

$$\frac{\sigma_E^2}{E_m^2} = \frac{\sigma_{EB}^2}{E_m^2} + \frac{\sigma_{ER}^2}{E_m^2} \tag{6.5}$$

As Davenport showed, both components are proportional to the intensity of turbulence $I_v = \sigma_v/V$ if the structural behaviour is linear and the aerodynamic transmittance is quasi-stationary. Introducing factors R and B, the following expression for the gust response factor is obtained:

$$G = 1 + 2k_p I_v(z_s)\sqrt{B^2 + R^2}$$
$$\text{where B} = \frac{\sigma_{EB}}{E_m}\frac{1}{2I_v} \text{ and } \text{R} = \frac{\sigma_{ER}}{E_m}\frac{1}{2I_v} \tag{6.6}$$

In Eq. (6.6), z_s is the height at which the reference flow parameters are utilized for determining G. The gust response factor is a structural, not a load parameter. In it, B takes into account the reduction of the load effect due to the non-simultaneous occurrence of the load peaks over the structural surface whereas R accounts for the amplification of the load effect due to resonance with wind turbulence. It depends on one hand on the wind flow parameters e.g. the mean wind profile, the turbulence intensity, and the integral length scales of the oncoming flow; it accounts on the other hand for structural parameters such as the size of the structure and its dynamic behaviour, in particular the lowest natural frequency, the related mode shape, and the damping. It is obvious that the gust response factor is a statistical quantity and has a considerable scatter. Davenport's equivalent static wind load is distributed along the cylinder axis according to the profile of the mean velocity pressure; it reproduces the base bending moment correctly. This means that the bending moments at other heights will be underestimated.

6.1.1.2 The Eurocode Model

The Eurocode EN 1993-3-2 (2006) dealing with the design of steel chimneys does not give guidance on the wind loading but refers to the general wind loading code EN 1991-1-4 (2010). In the latter code document, Davenport's concept has been modified. The code introduces a wind force acting on a total structure or a structural component as:

$$F_w(z_e) = c_s(z_s)c_d(z_s) \cdot c_f \cdot A_{ref}q_p(z_e) \tag{6.7}$$

in which the following symbols apply: $c_s c_d$—so-called structural factor; z_s—reference height for determining the structural factor; c_f—force coefficient; q_p—peak velocity pressure; z_e—reference level; $A_{ref} = 1 \cdot d$—reference area. This wind force is an equivalent static load in the same sense as in Davenport's concept, the main difference being that the force is based on the peak velocity pressure. The structural factor is determined from the wind flow parameters at a reference height z_s.

Before deriving the interrelation between the concepts, two aspects shall be considered.

Strictly speaking, F_w is not a line load—as is typically applied in the design of slender structures, but a point load. However, the Eurocode introduces sections of length l of structural parts which allows to subdivide the total length arbitrarily and in this way to obtain a line load.

From Eq. (6.7), a different definition of the aerodynamic shape factor results: c_f is the ratio of the peak wind force measured in a turbulent flow (boundary layer wind tunnel) to the peak velocity pressure (Eq. (6.8)) times the structural coefficient. If the latter is 1, c_f is the ratio of "peak-to-peak". The measured force depends on the structure of the turbulence in the wind tunnel flow. The application to other boundary layers than the wind tunnel simulation will remain approximate.

As previously mentioned, an important difference to the Davenport approach is that the Eurocode bases the wind force on the peak (or gust) velocity pressure q_p. It is defined as the short-term maximum within the extreme mean wind. The peak velocity pressure is the sum of the mean q_m and the standard deviation σ_q amplified by the peak factor k_q as in Eq. (6.8):

$$q_p(z) = q_m + k_q \sigma_q \tag{6.8}$$

$$q_p = (1 + 7I_v(z)) \cdot \frac{\rho_a \cdot V^2(z)}{2} \tag{6.9}$$

Assuming $(\sigma/q_m) = 2I_v$ and $k_q = 3.5$, the approximate expression (6.9) results, where $I_v(z)$ is the turbulence intensity at height z. In the structural factor Eq. (6.10a, b), c_s is the size factor covering the reduction of the load effect due to non-simultaneous gust peaks; c_d accounts for resonant amplification of the gust effect. These factors are given by:

$$c_d = \frac{1 + 2k_p I_v(z_s)\sqrt{B^2 + R^2}}{1 + 2k_p B I_v(z_s)} \qquad c_s = \frac{1 + 2k_p B I_v(z_s)}{1 + 7I_v(z_s)} \tag{6.10a, b}$$

The dynamic factor c_d is always >1 in principle. However, it is close to 1 for stiff structures when the resonant contribution R is much smaller than the background contribution B. The background contribution B is ≤ 1; it decreases as the loaded areas become larger.

The background factor B^2 and the resonant response factor R^2 are calculated according to Eqs. (6.11) and (6.12), respectively.

$$B^2 = \frac{1}{1 + 0.9\left(\frac{b+h}{L(z_s)}\right)^{0.63}} \tag{6.11}$$

where b and h are the width and height of the structure, respectively; $L(z_s)$ is the turbulent length scale at the reference height z_s.

$$R^2 = \frac{\pi^2}{2\delta} \mathrm{S_L}(\mathrm{z_s}, \mathrm{n_{1,x}}) R_h(\eta_h) R_b(\eta_b) \tag{6.12}$$

where δ is the total logarithmic decrement of damping, S_L is the non-dimensional power spectral density function and R_b and R_h are the aerodynamic admittance functions. The aerodynamic admittance functions for a fundamental mode shape may be approximated using Eqs. (6.13) and (6.14):

$$R_h = \frac{1}{\eta_h} - \frac{1}{2\eta_h^2}\left(1 - e^{-2\eta_h}\right); \quad R_h = 1 \ for \ \eta_h = 0 \tag{6.13}$$

$$R_b = \frac{1}{\eta_b} - \frac{1}{2\eta_b^2}\left(1 - e^{-2\eta_b}\right); \quad R_b = 1 \ for \ \eta_b = 0 \tag{6.14}$$

where:

$$\eta_h = \frac{4.6h}{L(z_s)} f_L(z_s, n_{1,x}) \ \text{ and } \ \eta_b = \frac{4.6b}{L(z_s)} f_L(z_s, n_{1,x}) \tag{6.15}$$

The total logarithmic decrement of damping δ is the sum of structural damping δ_s and aerodynamic damping δ_a, plus, in case, the damping due to special devices. The aerodynamic damping is generally positive and thus increases the total damping. It can be estimated, for the fundamental bending mode of vibration, by Eq. (6.16):

$$\delta_a = \frac{c_f \rho V_m(z_s)}{2n_1 \mu_e} \tag{6.16}$$

where c_f is the force coefficient for the wind action and μ_e is the equivalent mass per unit length.

The non-dimensional power spectral density function $S_L(z, n)$, expressed by Eq. (6.17), represents the wind distribution over frequencies. f_L is the non-dimensional frequency ($f_L = nL(z)/V_m(z)$) corresponding to the first natural vibration mode of the structure $n_{1,x}$.

$$S_L(z, n) = \frac{nS_v(z, n)}{\sigma_v^2} = \frac{6.8 f_L(z, n)}{(1 + 10.2 f_L(z, n))^{5/3}} \tag{6.17}$$

The peak factor k_p is calculated according to Eq. (6.18):

$$k_p = \sqrt{2\ln(\nu T)} + \frac{0.6}{\sqrt{2\ln(\nu T)}} \quad \left(k_p \geq 3\right) \tag{6.18}$$

where ν is the up-crossing frequency (Eq. (6.19), being $n_{1,x}$ the first natural frequency of the structure) and T = 600 s is the averaging time for the mean wind velocity. The limit of ν = 0.08 Hz corresponds to a peak factor of 3.0.

$$\nu = n_{1,x}\sqrt{\frac{R^2}{B^2 + R^2}} \quad (\nu \geq 0.08\,\text{Hz}) \tag{6.19}$$

Then, by substitutions into Eq. (6.7), Eq. (6.20) is obtained:

$$F_w(z_e) = \frac{G}{1 + 7I_v(z_s)}(1 + 7I_v(z_e)) \cdot \frac{\rho V^2(z_e)}{2} c_f A_{ref} \tag{6.20}$$

The Eurocode includes the size and resonance effects in the same manner as in Davenport's classical approach. The main difference is that instead of the mean velocity pressure, the profile of the peak velocity pressure, see Eq. (6.9) is introduced. The base bending moments calculated by the two models will be the same, whereas the values at other levels along the chimney axis will differ. Finally, as mentioned earlier, the aerodynamic coefficients differ, too. They are determined nowadays in boundary layer wind tunnels. For stiff wind tunnel models the dynamic factor is $c_d = 1$. Then, Eqs. (6.2) and (6.6) yield the relation:

$$c_f = \frac{F_w / q_p}{F_m / q_m} C_D \tag{6.21}$$

6.1.1.3 The CICIND Approach

The CICIND (International Committee for Industrial Chimneys, www.cicind.org) wind load model is depicted in Eq. (6.22). It is unique in that it gives individual expressions for the mean wind load w_m and the gust wind load w_g, and their variation with respect to the height z above the ground:

$$w(z) = w_m(z) + w_g(z) \tag{6.22}$$

Regarding the mean load, the model is realistic. The mean component amounts to about 50% of the total load. The gust load component has to account for the worst case structural response induced by the multiple stochastic load process of wind gustiness. It is typically an equivalent static load which is less realistic and has a larger uncertainty than the mean load. By separating both components, the gust load may be given any profile appropriate to reproduce the distribution of the real bending response due to gustiness.

In the CICIND model the mean wind load is given by:

$$w_m(z) = C_D \cdot d(z)\frac{\rho V^2}{2} \tag{6.23}$$

In a turbulent flow, the mean velocity pressure is in fact greater than the velocity pressure of the mean wind speed V. A factor of $(1 + Iv^2)$ is obtained. The true mean wind force is accordingly higher than given by Eq. (6.23). The difference is however small and neglected in wind loading codes. Furthermore, the aerodynamic shape factor varies over the chimney height, $C_D = C_D(z)$. It is approximately constant over the tower height from the bottom to 8 times the diameter. It increases close to the chimney top. This observation is known as the tip effect (see Sect. 6.1.2).

The mean value of the bending moment at the base z = 0 is obtained as:

$$M_m(0) = \int_0^h w_m(z) \cdot z \cdot dz \tag{6.24}$$

Amplification of the mean by the gust response factor G gives the peak base moment and from it the contribution induced by the gust load, $M_g(0)$, is derived as

$$M_g(0) = M_p(0) - M_m(0) = (G - 1) \cdot M_m(0) \tag{6.25}$$

The gust load component should reproduce (i) the base moment of Eq. (6.25), (ii) the distribution of the gust moment over the tower height, by an appropriate gust load profile. Vickery has proposed a linear profile with $w_g = 0$ at z = 0. On this basis, the CICIND stipulation of the gust load component becomes:

$$w_g(z) = \frac{3}{h^2}\frac{z}{h}(G - 1) \cdot M_m(0) \tag{6.26}$$

where h = chimney height; G = gust response factor of the base bending moment M(0); $M_m(0)$ = mean base bending moment. The profile does not account for variations of the diameter d and the drag coefficient C_D.

The Gust Response Factor is calculated in the CICIND according to Eq. (6.6). The background and the resonant contributions (B^2 and R^2, respectively) are calculated as in the EN (Eqs. (6.11) and (6.12)). In this regard, the only one difference in the formulation between the EN and the CICIND is in the evaluation of the aerodynamic damping. In the EN it shall be calculated according to Eq. (6.16), while the CICIND for concrete chimneys (CICIND 2011) recommends the value $\xi_s = 0.016$ (corresponding to $\delta_s = 2\pi \cdot 0.016 = 0.1$) and the CICIND for steel chimneys (CICIND 2010) uses Eq. (6.27), see Sect. 7.2.7. The CICIND Code uses in the calculation the critical damping ratio ξ, while the EN uses the logarithmic decrement δ. In the following, δ is used, unless differently specified.

$$\xi_a = 2.7 \times 10^{-6} \frac{V}{f_1 t} \tag{6.27}$$

where V is the design wind speed at the top of the chimney, f_1 is the fundamental natural frequency and t is the thickness of the wall in the top third.

6.1.2 Aerodynamic Coefficients

On a circular cylinder, the aerodynamic coefficients (also referred to as shape factors or drag/pressure coefficients) are function of several parameters, such as the Reynolds number Re (Re), the turbulence intensity (I_v), the ratio of surface roughness of the stack and the stack diameter (k/d), the aspect ratio of the stack (h/d) and three-dimensional effects due to the cylinder free-end and the presence of the ground. The atmospheric boundary layer also has an influence, as the pressure gradient enhances vertical flow movements. The approach of the Codes, like the Eurocode and the CICIND, is that of expressing the shape factor as the product of a basic value (which might depend on Re and k/d and applies to a cylinder without a free-end) and an end-effect correction factor which accounts for the reducing effect due to h/d on small aspect ratios.

6.1.2.1 The Approach of the Eurocode

The Eurocode EN 13084-1:2007 for free-standing chimneys recalls in Sect. 5.2.3.2 the EN 1991-1-4 to calculate the wind action. The Eurocode EN 1991-1-4:2010-12 for wind loading defines the aerodynamic force coefficient c_f according to the Eq. (6.7). On the basis of Eq. (6.20), the factor $(1 + 7I_v(z_e))/(1 + 7I_v(z_s))$ is in the order of 1 and can be interpreted as a correction to Davenport's approach. It then follows from Davenport's approach that the force coefficient c_f, defined in Eq. (6.7), is a mean force coefficient ("mean to mean").

The definition of the pressure coefficient C_{pe} (where the suffix "e" means on the external surface) is instead somewhat different, as it is expressed—according to Eq. (6.28)—as the ratio of the peak wind pressure and the peak velocity pressure. Consequently, C_{pe} is a peak coefficient ("peak to peak").

$$w_e = q_p(z_e)c_{pe} \tag{6.28}$$

The force coefficient c_f for a finite circular cylinder is defined in the Eurocode as the product of the basic drag coefficient $c_{f,0}$ and the end-effect factor ψ_λ:

$$c_f = c_{f,0}\psi_\lambda \tag{6.29}$$

The basic drag coefficient is the force coefficient for cylinders without free-end and it accounts for both surface roughness k/d and Reynolds number Re.

The Reynolds number is referred to the gust velocity, i.e. $\sqrt{2q_p/\rho}$. The Eurocode does not specify at which height the Reynolds number should be calculated. The exact calculation would then imply the evaluation at each level along the tower height. Instead, the CICIND refers to the representative height z = 0.75H.

Values for the equivalent roughness k are given in the Table 7.13 of the Code. For concrete k ranges from 0.2 mm for smooth surface until 1.0 mm for rough surface. For metals, k ranges from 0.002 to 0.2 mm depending on the surface treatment.

The basic drag coefficient can be determined according to Fig. 7.28 of EN 1991-1-4 (2010). In particular, above the critical range of Re, the following formula applies:

$$c_{f,0} = 1.2 + \frac{0.18\log(10k/d)}{1 + 0.4\log\left(\mathrm{Re}/10^6\right)} \tag{6.30}$$

The end-effect factor ψ_λ is a function of the effective slenderness λ and it can be determined by a linear interpolation in the log-linear plane, depending on the solidity ratio φ, as shown by Fig. 7.36 of EN 1991-1-4 (2010). The effective slenderness is equal to the slenderness h/d if $h < 15$ m and decreases up to 0.7 h/d when $h > 50$ m. In any case, for circular cylinders $\lambda < 70$, which means $\psi_\lambda < 0.91$. The Eurocode does not specify at which height the slenderness ratio should be calculated. The exact calculation would then imply the evaluation at each level along the tower height. Instead, the CICIND refers to the representative height $z = 0.75$ h.

The pressure coefficients c_p for circular cross-sections are defined in the Eurocode depending on the Reynolds number. The latter is referred the peak wind velocity at the height $z_s = 0.6$ h.

The external pressure coefficients c_{pe} for a finite circular cylinder is defined as the product of the external pressure coefficient for a cylinder without a free-end ($c_{p,0}$) and the end-effect factor $\psi_{\lambda\alpha}$:

$$c_{pe} = c_{p0}\psi_{\lambda\alpha} \tag{6.31}$$

The external pressure coefficient for a cylinder without a free-end ($c_{p,0}$) is a function of the angle and depends on Re, as shown by Fig. 3.3. Depending on the circumferential angle α, three ranges are identified along the circumference by α_{min} (position of the minimum pressure) and α_A (position of the flow separation). c_{p0min} and c_{p0h} are the minimum pressure coefficient (at $\alpha = \alpha_{min}$) and the base pressure coefficient, respectively. Figure 7.27 of EN 1991-1-4 (2010) is based on the equivalent roughness k/d less than 5×10^{-4} and any further information for different values of surface roughness is not provided. For chimneys of relatively small diameter, this only applies for small values of equivalent surface roughness (Table 6.1).

Moreover, it should also be observed that once the in-wind pressure components ($C_p\cos\varphi$) are integrated along the circumference, the resulting in-wind force coefficient is generally higher (in the order of 7–8%) than the force coefficient c_f (Table 6.1)

The end-effect factor $\psi_{\lambda\alpha}$ is a function of α in the three ranges and is based on the end-effect factor ψ_λ for the force coefficient.

Table 6.1 Aerodynamic coefficients and turbulence intensity of the flow in wind tunnel tests

z/H	Iv	C_D	σ, C_D	If = σ, C_D/C_D	If/Iv
0.99	0.077	0.724	0.120	0.166	2.143
0.95	0.079	0.800	0.111	0.139	1.751
0.91	0.081	0.770	0.115	0.149	1.850
0.89	0.081	0.716	0.113	0.158	1.951
0.85	0.082	0.607	0.100	0.165	2.006
0.75	0.086	0.501	0.080	0.160	1.851
0.65	0.093	0.486	0.078	0.161	1.735
0.55	0.105	0.486	0.084	0.172	1.639
0.52	0.110	0.493	0.089	0.180	1.640
0.505	0.112	0.483	0.090	0.186	1.661
0.495	0.114	0.482	0.092	0.190	1.671
0.48	0.116	0.504	0.092	0.183	1.581
0.45	0.119	0.509	0.094	0.184	1.540
0.35	0.137	0.555	0.116	0.209	1.525
0.25	0.153	0.604	0.139	0.229	1.501
0.15	0.157	0.699	0.163	0.233	1.486
0.05	0.157	0.801	0.187	0.233	1.484

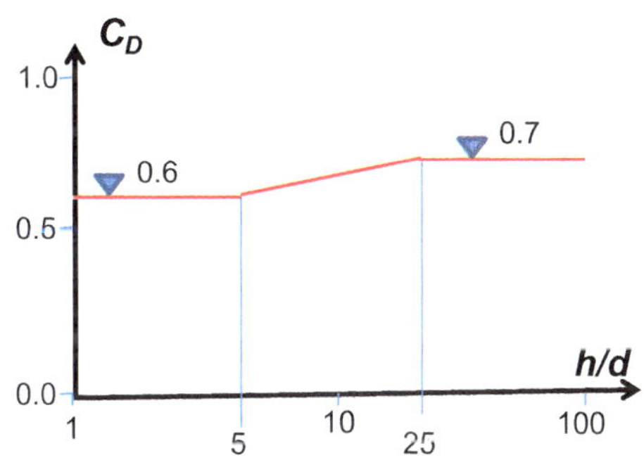

Fig. 6.1 Basic shape factor as a function of the aspect ratio for concrete chimneys

6.1.2.2 The Approach of the CICIND Concrete Chimney Model Code

For isolated chimneys (no interference effects), the shape factor C_D equals the basic shape factor $C_{D,0}$. The latter is a function of the aspect ratio and ranges between 0.6 (h/d < 5) and 0.7 (h/d ≥ 25). Intermediate values are linear interpolations in the linear-log plane, where the log axis applies to aspect ratio. So, the effect of slenderness is included in the definition of the basic shape factor (Fig. 6.1).

For concrete chimneys, the influence of Reynolds number, turbulence intensity and surface roughness is ignored. As explained in the Commentaries, this is because the Reynolds number is generally larger than 10^6, i.e. in the post-critical regime,

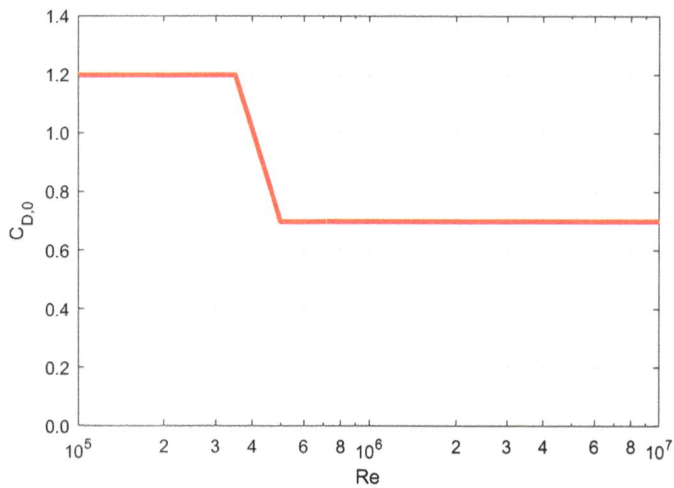

Fig. 6.2 Basic shape factor as a function of the Reynolds number, compare to Fig. 7.3 of the CICIND Steel Code

where the influence of the turbulence intensity and the surface roughness is much smaller than in other Re regimes.

Table 6.1 shows that $C_{D,0}$ proposed by the CICIND for h/d ≥ 25 (infinitely long cylinder) is between 10 and 20% (depending on surface roughness k/d) lower than the corresponding force coefficient calculated according to the Eurocode. However, once the end-effect factor in the EN is applied to c_f, the CICIND force coefficient for the chimney samples considered is on the safe side (Table 6.1).

6.1.2.3 The Approach of the CICIND Steel Chimney Model Code

The shape factor C_D for a chimney is defined in the CICIND Steel Code as the product of the basic shape factor $C_{D,0}$ and the end-effect factor k_a. So, differently from the Concrete Code, end-effects are considered separately.

$$C_D = k_a C_{D,0} \tag{6.32}$$

The basic shape factor $C_{D,0}$ is a function of Re and ranges between 1.2 (Re < 2.5×10^5) and 0.7 (Re > 3.5×10^5), as shown in Fig. 6.2, red line. This is a simplified representation, which also accounts for the influence of the turbulence intensity and surface roughness, although they do not explicitly enter as parameters in the calculation.

The main influence of turbulence is to shift the curve to lower Re, so instead of a single curve there exists a curve for each value of the turbulence intensity. These curves can be approximated by a single curve by using the so-called effective

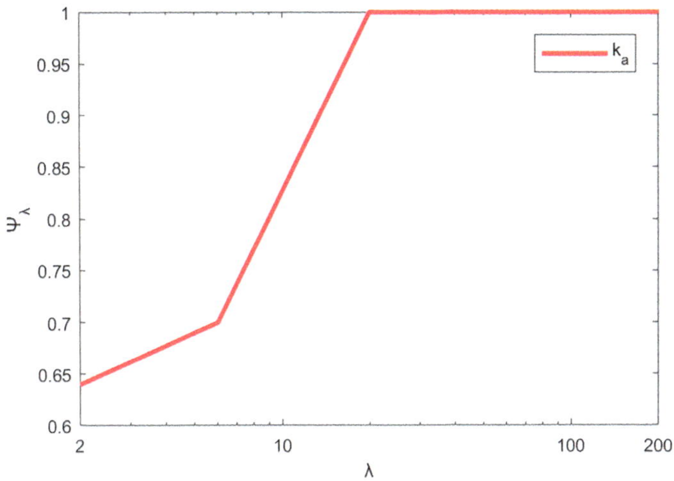

Fig. 6.3 End-effect factor as a function of the aspect ratio (and of the solidity ratio φ in the EN), compare to Fig. 7.4 of the CICIND Steel Code

Reynolds number, that is the actual Re multiplied by a turbulence parameter λ_T. It is especially in the supercritical range, that the shape factor of a chimney can have a much different value for low and high turbulence intensity. So, in order not to underestimate the shape factor, the Model Codes applies the post-critical value also in the supercritical range.

For increasing surface roughness, the drag-versus-Re curve moves to the left i.e. to lower Re and the drag coefficient in the supercritical and post-critical range increases. This is accounted by introducing a roughness parameter λ_R in the expression of the effective Reynolds number. However, since for industrial chimneys the ratio k/d is typically (much) less than 10^{-4}, the correction factor results close to 1 and can be ignored without underestimation of the shape factor. The goodness of this approximation shall be estimated in Sect. 6.1.3

The correction factor k_a takes into consideration the end-effect for chimneys with aspect ratio less than 20. Its value is less than or equal to one, as the resulting mean shape factor (averaged over the height) of a cantilevered cylinder is smaller than its 2D value. This reduction decreases for large aspect ratios, so that $k_a = 1$ for H/D > 20 (Fig. 6.3). Figure 6.3 also shows the more conservative end-effect factors for the CICIND with respect to the EN. As the aspect ratio for industrial steel chimneys is usually larger than 20, for most steel chimneys the end-effect factor is equal to one.

6.1.2.4 Non-uniform Distribution by Wind Tunnel Tests

The flow around a finite length circular cylinder (h/d = 6.7) has been recently investigated in atmospheric boundary layer flow at WIST wind tunnel at Ruhr-University

Bochum (see Lupi 2013). External pressures are measured at 17 levels along the height and at an angular spacing of 20°. Pressure coefficients are calculated by Eq. (6.33), i.e. by using the local velocity pressure $q_m(z)$ at each level.

$$c_{pe} = \frac{p - p_0}{q_m(z)} \tag{6.33}$$

Circular cylinders belong to the class of bluff bodies with rounded shape, characterized by a separation point which can move and adjust itself in response to the flow structure in the separated region. The state of the flow is largely dominated by the Reynolds number. Most of chimneys lie in the range of transcritical Re and it is difficult to perform tests in the wind tunnel at such high Re on scaled models. It is therefore often used the concept of effective Reynolds number (ESDU 81017). This is a modified Re ($Re_e = Re\lambda_R\lambda_T$), which incorporates a factor λ_R depending on surface roughness and a factor λ_T depending on incoming turbulence. Re_e reproduces, at lower Reynolds numbers, the same effects in the flow that would occur at higher Reynolds numbers.

Within this work, the wind tunnel tests are performed on a rough cylinder ($k/d = 0.25/150 = 1.67 \times 10^{-3}$, ribs at spacing of 20°) at $Re = 2.5 \times 10^5$. Detailed information about the effects of Re, surface roughness and turbulence intensity can be found in Lupi (2013). The target full-scale condition, typical of an industrial chimney, is a relatively smooth circular cylinder (i.e. without vertical ribs) in the transcritical range of Re.

Beside Re, surface roughness and turbulence of the incoming flow, the flow around circular is also affected by the aspect ratio and end-effects (ground-wall effects and free-end effects). The latter are even enhanced in presence of atmospheric boundary layer, due to the vertical flow movements created by velocity and pressure gradients.

The non-uniform distribution of drag coefficients measured in the wind tunnel along the height of the chimney (mean and rms values) is reported in Fig. 6.4. In particular, three issues should be observed:

(1) The mean drag coefficient in the tip region is 60% higher than the drag coefficient at middle height. This is due to the tip effect. In the tip region, the pressure induced by the flow on the forward-facing surface is significantly higher than that on the rearward facing surface, therefore a flow is induced over the tip of the cylinder from front to rear. The separated flow over the tip creates a region of very low pressure, which induces a spanwise flow towards the tip of the cylinder. This flow sweeps up the separated shear layers from intermediate heights. At short distance below the free-end (about d/2), vortex sheets roll up into a pair of trailing (tip) vortices, which form because of the interaction between upward-directed separated flow at the sides of the cylinder and downward-directed flow over the tip. The tip vortices are counter-rotating open vortex loops with their axis perpendicular to both the free-stream direction and the longitudinal axis of the cylinder (Gould et al. 1968). The free-end effect is limited to the upper two or three diameters and it is not affected by the boundary layer conditions. A typical feature of the mean surface pressure distribution in the tip region

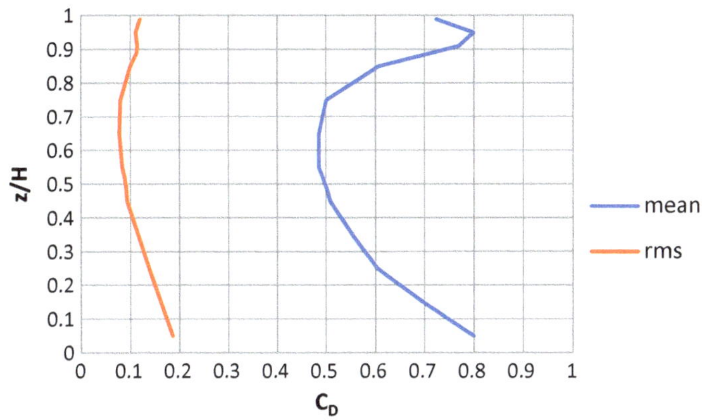

Fig. 6.4 Shape factors from wind tunnel tests (h/d = 6.67)

is the displacement of the separation point to higher angles. A typical islet of minimum pressure appears at $\varphi \approx 135°$ and $z'/d \approx 1/3$ from the free-end. This arises considerably the local drag near the free-end.

(2) The drag coefficient at middle height of a finite length circular cylinder is comparatively smaller than the 2D value (infinitely long cylinder). In fact, below the tip region, the downwash flow leads to an increase of the wake pressure. The mean drag coefficient increases as h/d increases (Okamoto et al. 1992), because the strength of the downwash is reduced as the distance from the tip increases and it tends to the nominal 2D value. For this reason, the Codes apply a smaller end-effect correction factor as the slenderness ratio of the chimney increases.

(3) The drag coefficient in the lower half is affected by ground effects, such as horseshoe vortices and base vortices. In fact, in the ground-wall region, a three-dimensional separation of the boundary layer occurs upstream of the cylinder and the cylinder end is submerged in a retarded wall boundary layer. The adverse pressure gradient causes the three dimensional boundary layer separation at some distance upstream of the body, followed by a roll-up of the separated boundary layers into a system of swirls. This swirl system is swept around the base of the cylinder and assumes a characteristic shape, which is responsible for the name horseshoe or necklace vortex (Baker 1980). Moreover, a weak upwash flow can be seen in the rear of the cylinder near the ground, which moves towards the central region of the wake. This creates a vortex, namely base vortex, near the cylinder wall-junction. As a result of the horseshoe and base vortices, the separation point moves downstream and the wake suction increases. This increases the drag coefficient in the bottom region of the chimney. Such an increase is even more evident in atmospheric boundary layer flow, due to the vertical velocity and pressure gradients, which enhance vertical flows. However, the drag distribution in the lower half of the tower has only a small effect on the wind-induced bending moments.

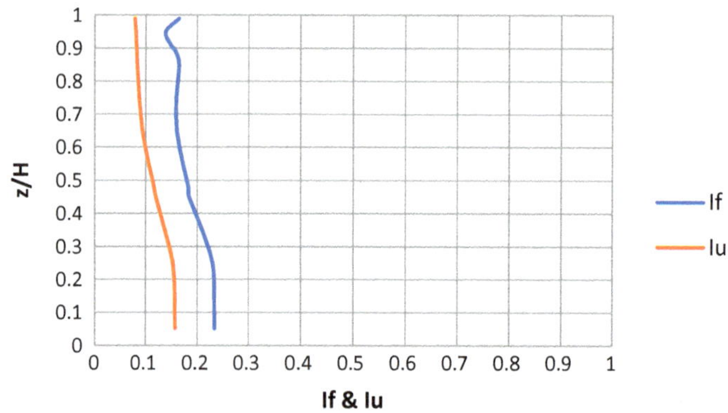

Fig. 6.5 Turbulence intensity of the incoming flow and intensity of the force ($Iv(z) = \sigma v(z)/V(z)$ and $If = \sigma,CD/CD$)

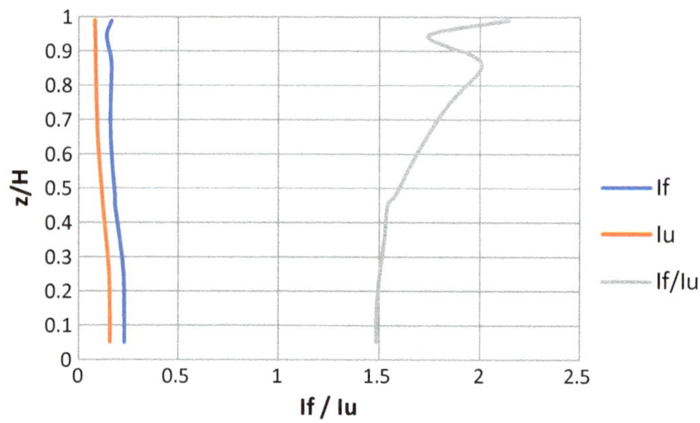

Fig. 6.6 Ratio intensity of force-to-intensity of turbulence: wind tunnel tests and quasi-stationary approach

The Codes propose constant or nearly constant aerodynamic coefficients and do not account for their non-uniform distribution due to three-dimensional effects. However, the shape factors proposed by the Codes are generally higher than the actual C_D at middle height. This globally leads to a conservative design (see Sect. 6.1.2).

The aforementioned wind tunnel results refer to the condition of zero-efflux out of the chimney. This situation represents in common practice the most critical design condition for a stack. In fact, the presence of efflux disturbs the tip vortices and reduces the tip effect.

Figures 6.5 and 6.6 report the turbulence intensity of the incoming flow measured in the wind tunnel. It should be observed that the ratio intensity of force-to-intensity of turbulence is generally lower than 2, that is the value which is usually assumed in the quasi-stationary approach. This confirms that the quasi-stationary approach is conservative.

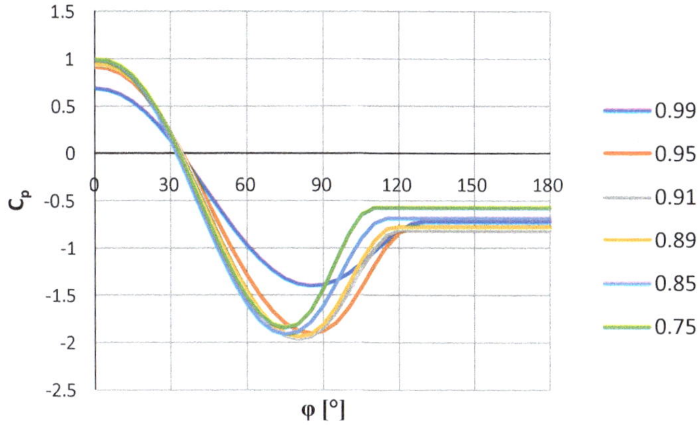

Fig. 6.7 Pressure coeff. distribution, tip region (wind tunnel tests, h/d = 6.67)

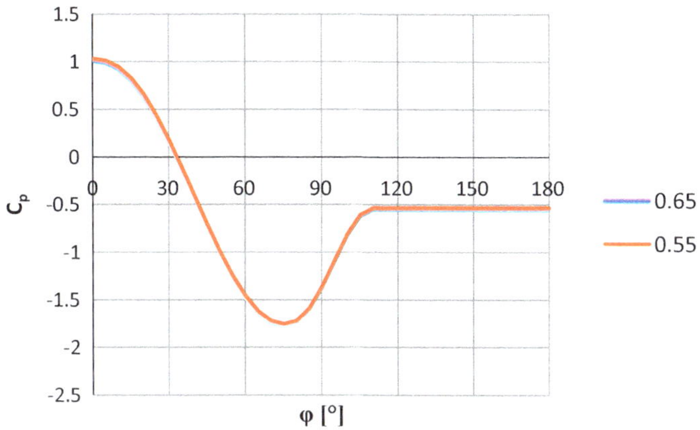

Fig. 6.8 Pressure coeff. distribution, middle region (wind tunnel tests, h/d = 6.67)

Numerical values are then summarized in Table 6.1.

Figures 6.7, 6.8 and 6.9 show the non-uniform pressure coefficient distribution corresponding to the drag force in Fig. 6.4, as it results from wind tunnel tests.

6.1.3 Comparison of the Models Based on Cantilever Responses

The aim of this section is a comparison of beam models with regard to the following issues: (1) Models of the gust wind load; (2) Non-uniform distribution of aerodynamic coefficients. Beam models are applied to the chimney samples.

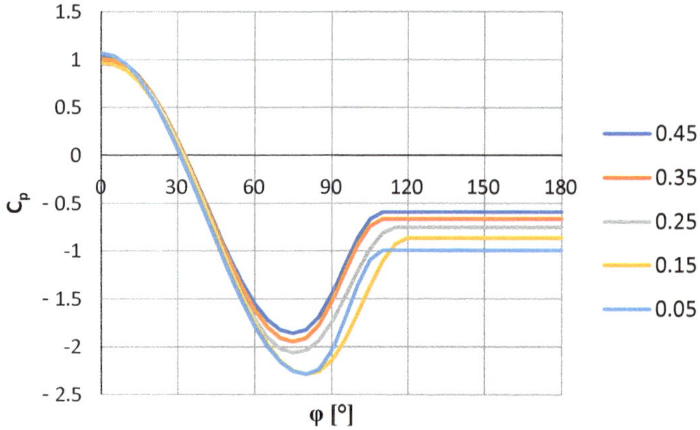

Fig. 6.9 Pressure coeff. distribution, low region (wind tunnel tests, h/d = 6.67)

The wind action is calculated for a terrain of the second EN category ($z_0 = 0.05$ m, $\alpha = 0.16$). The power law (CICIND Code) is used for the mean wind profile instead of the log law (EN Code). The basic wind velocity at z = 10 m is 25 m/s.

6.1.3.1 Differences Due to the Characteristics of the Models

The aim of this sub-section is to investigate the effect on the response of different models for the gust wind load. The latter were presented in Sect. 6.2.1. The calculation applies the same distribution of force coefficients, that is the one measured in the wind tunnel.

The GRF is calculated according to the CICIND or EN, as the two procedures are identical (see Sect. 6.1.1). The GRF by applying the two Codes would be only slightly different due to the different values of the damping that are suggested by the EN and the CICIND. The calculation of the GRF for the chimney samples is shown in Table 6.2 (according to the EN) and Table 6.3 (according to the CICIND). The values in Table 6.2 are negligibly different in case of smooth or rough surface, due to the different c_f which enters in Eq. (6.16). In the following figures (from Figs. 6.10, 6.11, 6.12, 6.13 and 6.14) it is referred to the GRF corresponding to the surface roughness k = 0.2 mm. The value of the damping which is recommended by the CICIND is generally higher than the EN, this implies lower G by applying the CICIND code (see Table 6.3). This issue will be addressed in Sect. 6.2.4. Here, it is always referred to the value of the aerodynamic damping as it is calculated in the Eurocode (Eq. (6.16)).

The following figures show a comparison of the beam models for the gust wind load on the chimneys presented in Sect. 6.1.1. These are Davenport's model, the EN approach (i.e. modified Davenport's model) and the CICIND approach (i.e. triangular gust load after Vickery). The aerodynamic coefficients are the same and derived from wind tunnel tests. The equivalent static wind load is defined by the aforementioned gust load models to reproduce the same base bending moment

Table 6.2 Calculation of gust response factor according to EN for chimney samples

EN	Concrete								Steel	
	Chimney n.1		Chimney n.2		Chimney n.3		Chimney n.4		Chimney n.5	
	Smooth	Rough	Smooth	Rough	Smooth	Rough	Smooth	Rough	Smooth	Rough
k (mm)	0.2	1.0	0.2	1.0	0.2	1.0	0.2	1.0	0.02	0.2
H (m)	130	130	213	213	177	177	148	148	40	40
z_s (m)	78.0	78.0	127.8	127.8	106.2	106.2	88.8	88.8	24.0	24.0
$V_m(z_s)$ (m/s)	34.73	34.73	37.59	37.59	36.47	36.47	35.47	35.47	28.76	28.76
$I_v(z_s)$	0.14	0.14	0.13	0.13	0.13	0.13	0.13	0.13	0.17	0.17
$L_{ux}(z_s)$ (m)	211.36	211.36	240.41	240.41	228.90	228.90	218.73	218.73	155.57	155.57
$b(z_s)$ (m)	5.77	5.77	23.35	23.35	11.60	11.60	7.15	7.15	0.81	0.81
$\lambda(z_s)$	22.54	22.54	9.12	9.12	15.26	15.26	20.70	20.70	49.58	49.58
$c_f(z_s)$	0.5870	0.6491	0.5339	0.5825	0.5554	0.6097	0.5801	0.6400	0.5284	0.6601
B^2	0.5949	0.5949	0.5290	0.5290	0.5566	0,5566	0.5797	0.5797	0.7208	0.7208
f_L	2.5379	2.5379	1.7267	1.7267	1.3179	1.3179	1.7021	1.7021	2.7047	2.7047
S_L	0.0715	0.0715	0.0898	0.0898	0.1046	0.1046	0.0906	0.0906	0.0688	0.0688
η_D	0.3186	0.3185	0.7715	0.7715	0.3072	0.3072	0.2559	0.2559	0.0645	0.0645
η_H	7.1806	7.1806	7.0373	7.0373	4.6877	4.6877	5.2977	5.2977	3.1990	3.1990
R_D	0.8175	0.8175	0.6357	0.6357	0.8231	0.8231	0.8492	0.8492	0.9583	0.9583
R_H	0.1296	0.1296	0.1320	0.1320	0.1906	0.1906	0.1709	0.1709	0.2638	0.2638

(continued)

Table 6.2 (continued)

EN	Concrete								Steel	
	Chimney n.1		Chimney n.2		Chimney n.3		Chimney n.4		Chimney n.5	
	Smooth	Rough	Smooth	Rough	Smooth	Rough	Smooth	Rough	Smooth	Rough
m_e (kg)	11309	11.309	65.430	65.430	24.920	24.920	14.311	14.311	275	275
δ_s	0.0300	0.0300	0.0300	0.0300	0.0300	0.0300	0.0300	0.0300	0.0140	0.0140
δ_a	0.0156	0.0172	0.0166	0.0181	0.0281	0.0308	0.0233	0.0257	0.0556	0.0695
δ	0.0456	0.0472	0.0466	0.0481	0.0581	0.0608	0.0533	0.0557	0.0696	0.0835
R^2	0.8202	0.7915	0.0.7986	0.7735	1.3949	1.3319	1.2177	1.1652	1.2331	1.0283
ν (Hz)	0.3175	0.3151	0.2094	0.2081	0.1775	0.1764	0.2272	0.2255	0.3972	0.3834
k_p	3.4183	3.4161	3.2947	3.2928	3.2444	3.2424	3.3193	3.3171	3.4831	3.4729
C_s	0.8881	0.8881	0.8720	0.8720	0.8789	0.8789	0.8846	0.8846	0.9190	0.9190
C_d	1.2151	1.2081	1.1923	1.1864	1.2978	1.2860	1.2790	1.2683	1.3163	1.2703
C_sC_d	1.0791	1.0730	1.0397	1.0346	1.1407	1.1302	1.1314	1.1219	1.2097	1.1675
G	2.1124	2.1003	1.9594	1.9497	2.1805	2.1605	2.1919	2.1736	2.6083	2.5172
$G/(1 + 7 * Iu(z_s))$	1.0791	1.0730	1.0397	1.0346	1.1407	1.1302	1.1314	1.1219	1.2097	1.1675

Table 6.3 Calculation of Gust Response Factor acc. to CICIND for chimney samples

CICIND	Concrete				Steel
	Chimney n.1	Chimney n.2	Chimney n.3	Chimney n.4	Chimney n.5
k (mm)	Any	Any	Any	Any	Any
H (m)	130	213	177	148	40
z_s (m)	78.0	127.8	106.2	88.8	24.0
$V_m(z_s)$ (m/s)	34.73	37.59	36.47	35.47	28.76
$I_v(z_s)$	0.14	0.13	0.13	0.13	0.17
$L_{ux}(z_s)$ (m)	211.36	240.41	228.90	218.73	155.57
$b(z_s)$ (m)	5.77	23.35	11.60	7.15	0.81
$\lambda(z_s)$	22.54	9.12	15.26	20.70	49.58
B^2	0.5949	0.5290	0.5566	0.5797	0.7208
f_L	2.5379	1.7267	1.3179	1.7021	2.7047
S_L	0.0715	0.0898	0.1046	0.0906	0.0688
η_D	0.3186	0.7715	0.3072	0.2559	0.0645
η_H	7.1806	7.0373	4.6877	5.2977	3.1990
R_D	0.8175	0.6357	0.8231	0.8492	0.9583
R_H	0.1296	0.1320	0.1906	0.1709	0.2638
δ_s	Not entering	Not entering	Not entering	Not entering	0.0125
δ_a	Not entering	Not entering	Not entering	Not entering	0.1681
δ	0.**1005**	0.**1005**	0.**1005**	0.**1005**	0.**1806**
R^2	0.3719	0.3700	0.8057	0.6454	0.4754
ν_T	155.1769	103.9294	96.8987	120.1918	189.1296
k_p	3.3580	3.2369	3.2152	3.2813	3.4162
G	**1.9032**	**1.7756**	**1.9774**	**1.9728**	**2.2343**

induced by gustiness. The comparison aims at investigating the goodness of the approximation in the response at different levels. It can be observed that the CICIND triangular gust loading model reproduces correctly the high load in the tip region, as well as the bending moments along the height. In fact, even if the gust load is calculated with reference only to the base bending moment, the differences along the height in all the examples are lower than (or in the order of) 10%. Furthermore, the CICIND approach is always on the safe side. Therefore Vickery's CICIND approach represents a simple conservative approximation for calculating the gust wind load on chimneys.

6.1.3.2 Non-uniform Distribution of Aerodynamic Coefficients

At first, the values of the constant or nearly constant aerodynamic coefficients calculated according to the Codes (EN and CICIND) are compared. As mentioned in

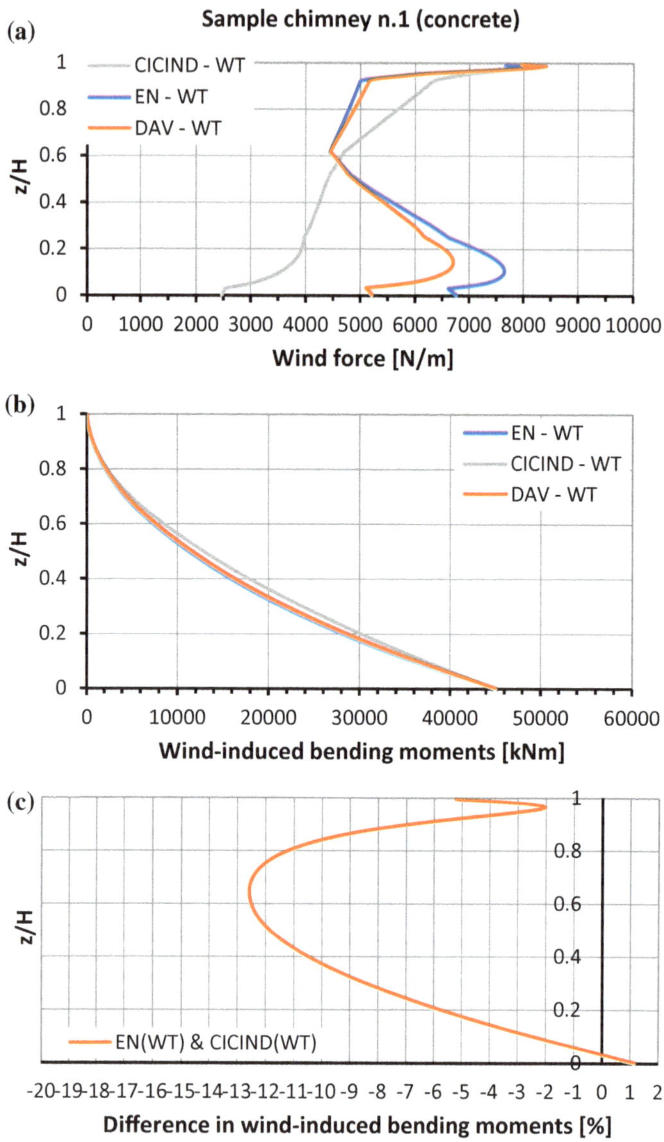

Fig. 6.10 Load and response to different Gust Loading Models—sample n.1 (Concrete, h/d = 23.0)

Sect. 6.1.2, the CICIND proposes a simplified approach, which neglects the influence of surface roughness and turbulence intensity. The Eurocode, instead, includes the dependency on k/d and Re. The dependency on Re is introduced in the CICIND only for steel chimneys (Fig. 6.2); for concrete chimneys the Reynolds number is always assumed larger than 10^6, i.e. in the post-critical regime.

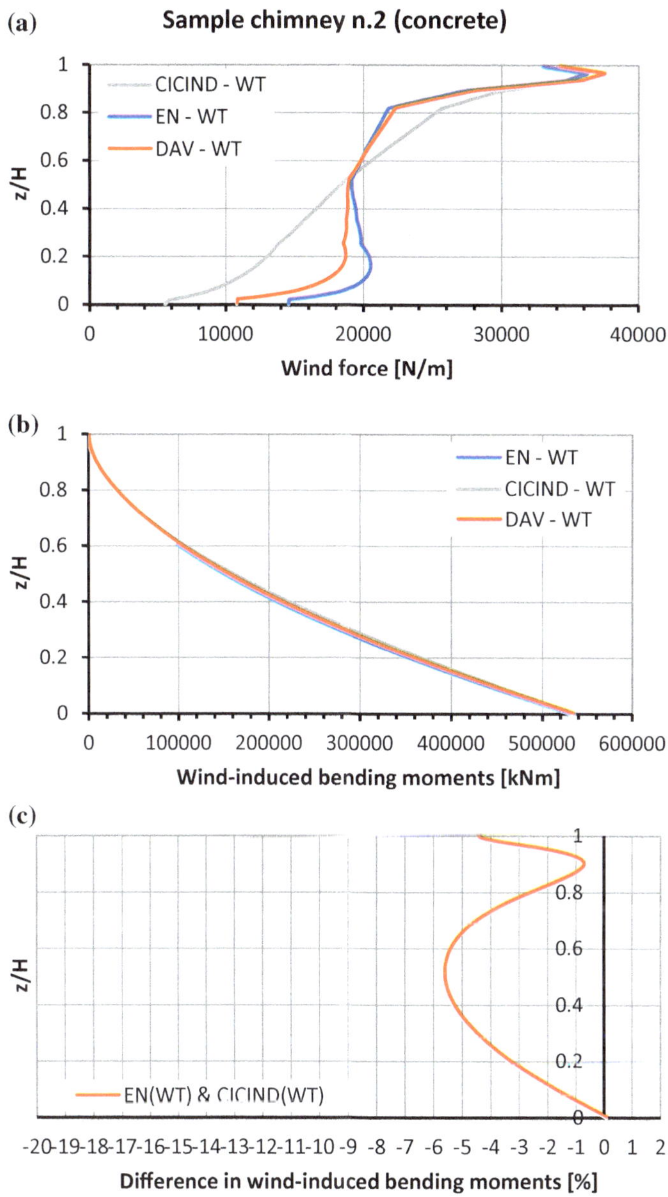

Fig. 6.11 Load and response to different Gust Loading Models—sample n.2 (Concrete, h/d = 9.1)

For the purpose of comparison, the force coefficients are calculated at the height 0.75H, which is the reference height indicated by the CICIND. Results are reported in Table 6.4.

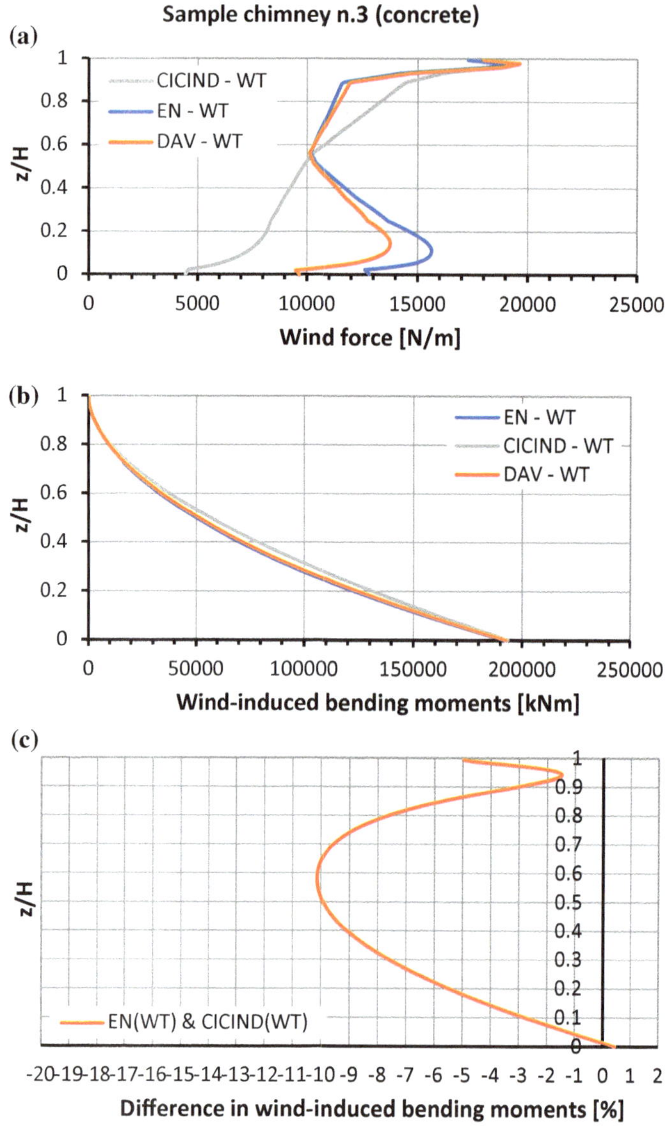

Fig. 6.12 Load and response to different Gust Loading Models—sample n.3 (Concrete, h/d = 15.3)

Table 6.4 shows that $C_{D,0}$ in the Eurocode is higher than the value suggested by the CICIND for h/d > 20 (infinitely long cylinders). However, the end-effect factor applied by the EN is more pronounced than the reduction applied by the CICIND, so that in the end the C_D calculated according to the CICIND is on the safe side. It should be also observed that the integration of C_p (EN) in the in-wind direction is

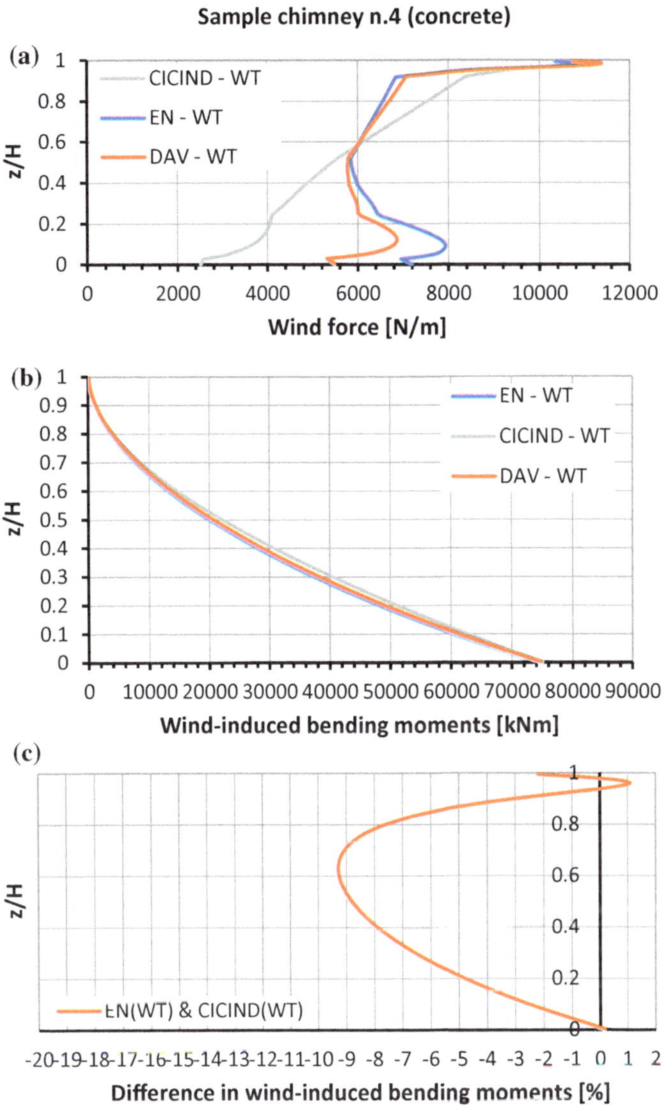

Fig. 6.13 Load and response to different Gust Loading Models—sample n.4 (Concrete, h/d = 20.7)

not related to the force coefficient c_f. The values are reported only in the k/d range in which the pressure distribution of the EN applies, i.e. $k/d < 5 \times 10^{-5}$.

Furthermore, the effect of different and, in case, not-uniform distributions of force coefficients is evaluated on the response. Differences are both in the magnitude of the force coefficient (see C_D-CICIND and c_f-EN, Table 6.4) and in the spanwise distribution. The c_f-EN present a weak height-dependency, due to the variation of

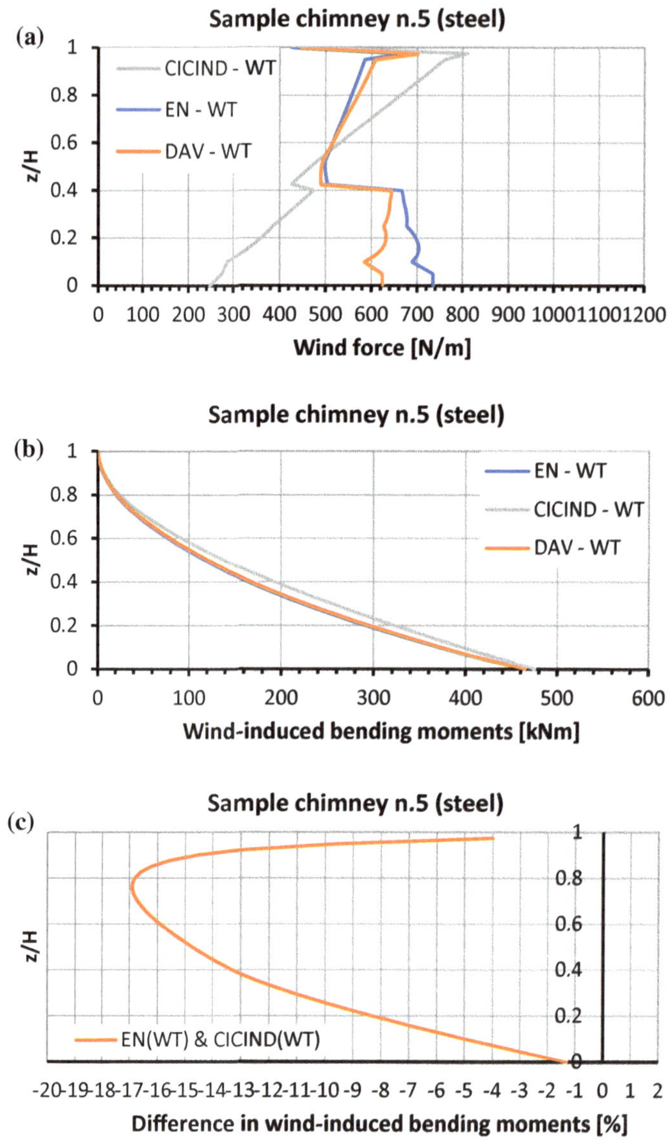

Fig. 6.14 Load and response to different Gust Loading Models—sample n.5 (Steel, h/d = 49.6)

diameter and wind profile along the height (thus variation of Re(z), of k/d(z) and of h/d(z)).

It should be observed that the results presented here also include the differences in the gust wind load (i.e., whenever the CICIND model is applied in the following graphs, both C_D and w_g are calculated according to the CICIND). In any case, the

Table 6.4 Drag coefficients at zref = 0.75H, EUROCODE and CICIND

	h (m)	d (m)	h/d	Re	k (mm)	k/d	EUROCODE			CICIND	
							$C_{D,0}$	C_D	ΣCpcosφ	$C_{D,0}$	C_D
n.1 CONCRETE	130	5.65	23.01	1.88E+07	0.2	3.54E−05	0.7885	0.5896	0.8465	0.7000	0.6948
					0.4	7.08E−05	0.8244	0.6164			
					0.6	1.06E−04	0.8454	0.6321			
					0.8	1.42E−04	0.8603	0.6433			
					1.0	1.77E−04	0.8719	0.6519			
n.2 CONCRETE	213	23.35	9.12	8.26E+07	0.2	8.57E−06	0.7856	0.5346	0.8465	0.7000	0.6374
					0.4	1.71E−05	0.8163	0.5555	0.8465		
					0.6	2.57E−05	0.8342	0.5677	0.8465		
					0.8	3.43E−05	0.8470	0.5764	0.8465		
					1.0	4.28E−05	0.0.8568	0.5831	0.8465		
n.3 CONCRETE	177	11.60	15.26	4.01E+07	0.2	1.72E−05	0.7873	0.5563	0.8465	0.7000	0.6693
					0.4	3.45E−05	0.8203	0.5796	0.8465		
					0.6	5.17E−05	0.8396	0.5932			
					0.8	6.90E−05	0.8533	0.6029			

(continued)

Table 6.4 (continued)

	h (m)	d (m)	h/d	Re	k (mm)	k/d	EUROCODE			CICIND	
							$C_{D,0}$	C_D	$\Sigma C_p \cos\varphi$	$C_{D,0}$	C_D
					1.0	8.62E−05	0.8639	0.6104			
n.4 CONCRETE	148	7.28	20.34	2.46E+07	0.2	2.75E−05	0.7882	0.5796	0.8465	0.7000	0.6872
					0.4	5.50E−05	0.8230	0.6052			
					0.6	8.25E−05	0.8434	0.6202			
					0.8	1.10E−04	0.8578	0.6308			
					1.0	1.37E−04	0.8690	0.6391			
n.5 STEEL	40	30.00	0.81	4.96E+01	0.02	2.48E−05	0.6342	0.5303	0.8465	0.7000	0.7000
					0.2	2.48E−04	0.7911	0.6615			

Table 6.5 Base bending moment Mm(0) due to different distributions of aerodynamic coefficients

	Concrete				Steel
	n.1 (kNm)	n.2 (kNm)	n.3 (kNm)	n.4 (kNm)	n.5 (kNm)
EN-WT	45.119	530.948	192.676	74.474	465
EN-smooth	48.560	494.730	193.249	79.873	485
EN-rough	53.358	537.046	210.100	87.362	583
CICIND	58.327	594.754	235.214	95.641	634

differences due to the choice of the gust load model are not significant, as previously shown.

It can be concluded that the constant force coefficient proposed by the CICIND is higher than the actual wind tunnel distribution of C_D in the middle region of the cylinder. This compensates the increase of load in the tip region, which is not accounted for by the CICIND. It results that the CICIND approach is always on the safe side. However, in case of slender cylinders, the overestimation in the base bending moment due to the CICIND C_D distribution with respect to the wind tunnel C_D distribution might even be in the order of 30% (Table 6.5) (Figs. 6.15, 6.16, 6.17, 6.18 and 6.19).

6.1.4 Conclusions and Future Outlook

In Sect. 6.1, two main issues were addressed: the modelling of the gust wind load, and the aerodynamic coefficients.

As regards the first issue, i.e. the modelling of the gust wind load, three main approaches are reviewed. They apply to beam structures. The basic approach is developed by Davenport and a slightly modified version is applied by the Eurocode (Eurocode 1, 2010). Davenport's approach accounts for gust effects by a Gust Response Factor G, which amplifies the mean wind load. Therefore, the load pattern which is used to calculate the effects of gustiness is still the mean load pattern. Davenport derived the gust response factor for a global response, i.e. the base bending moment. But G can be determined for any response which in case may be the leading stress in a specific design condition. The Eurocode introduces the structural factor and applies a modified version of Davenport's approach, which differs by the factor $(1 + 7Iv(z_e))/(1 + 7Iv(z_s))$, in the order of 1. Instead, the CICIND approach (after Vickery) is somewhat different, as the gust load has a triangular (not realistic) shape, which reproduces the gust wind-induced bending at the base $(G - 1)M_{mb}$. The effect on the response of the different gust loading models is investigated in Sect. 6.1.3. It results that the CICIND proposes a simple conservative approach in the design of chimneys. Even though it is derived to reproduce the gust wind-induced bending moments at the base, the differences along the height are less than (or in the order of) 10% for all the chimney samples.

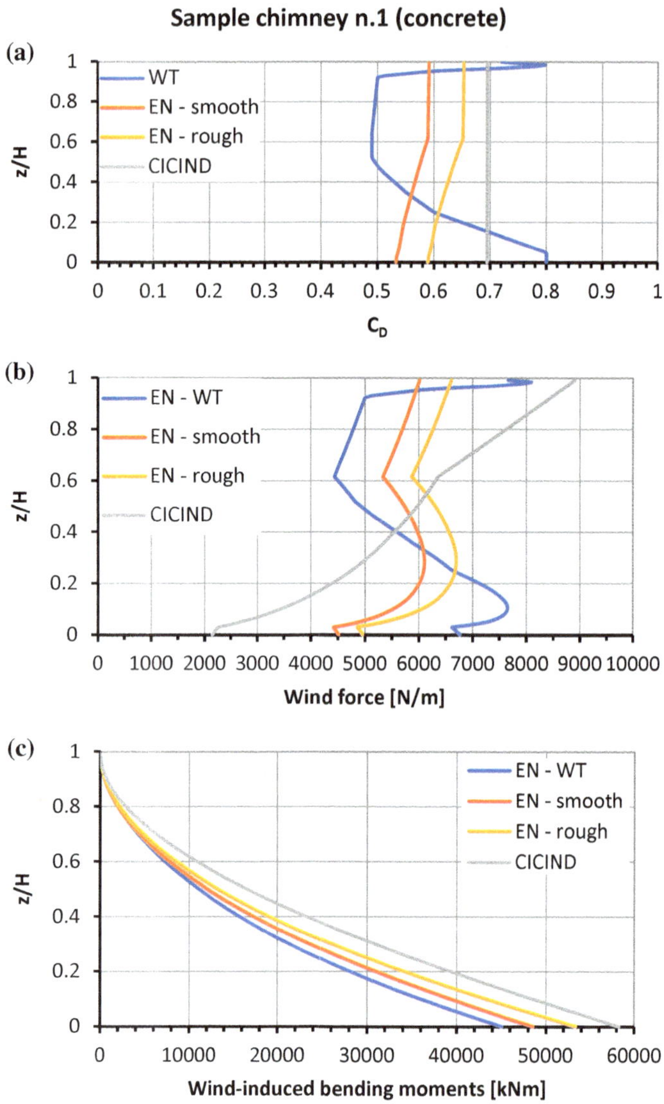

Fig. 6.15 Load and response to different aerodynamic coefficient distributions—sample n.1 (Concrete, h/d = 23.0)

The second issue which is addressed in this chapter concerns the aerodynamic coefficients and their non-uniform distribution along the height of the chimney. The aerodynamic coefficients around circular cylinders depends on the Reynolds number, on surface roughness, on turbulence intensity of the incoming flow and on the aspect ratio. These parameters are considered in the Eurocode formulation, while the CICIND proposes a more simple approach. It assumes concrete chimneys to lie in the

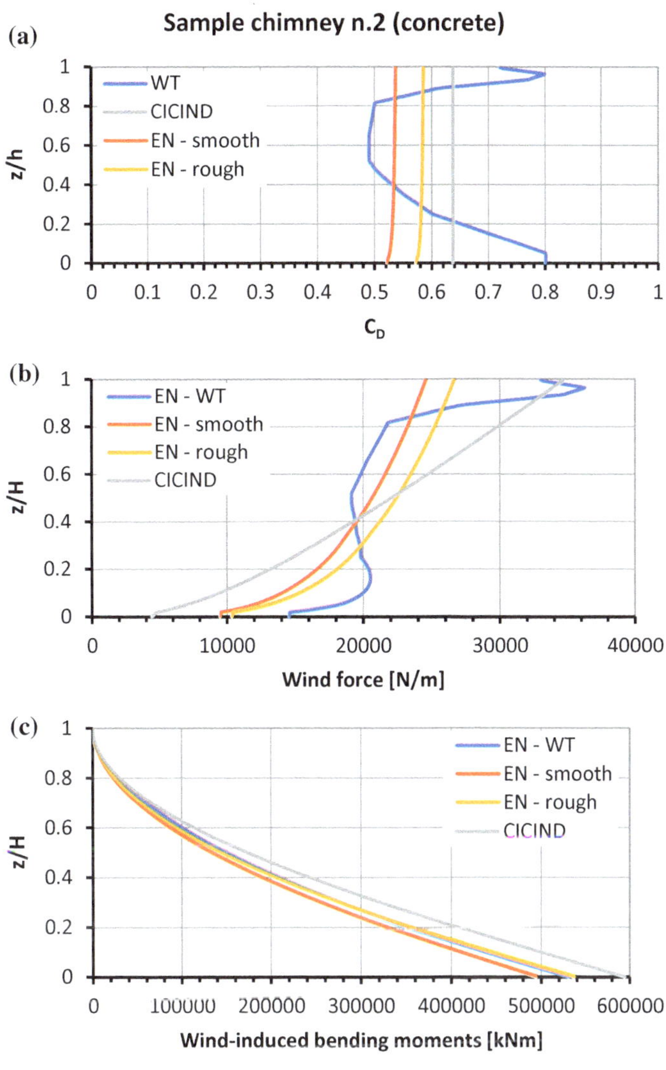

Fig. 6.16 Load and response to different aerodynamic coefficient distributions—sample n.2 (Concrete, h/d = 9.1)

transcritical range and ignores effect of surface roughness and turbulence intensity. For steel chimneys a simplified dependency on Re is considered. In all the codes, the aspect ratio enters in an end-effect factor for finite lengths cylinders. In any case, the calculation applied to the five chimney samples showed that the prediction of the CICIND is conservative. This is still true also with regard to the non-uniform distribution of force coefficients as measured by wind tunnel tests. The reason is that the constant force coefficient proposed by the CICIND is higher than the C_D in the

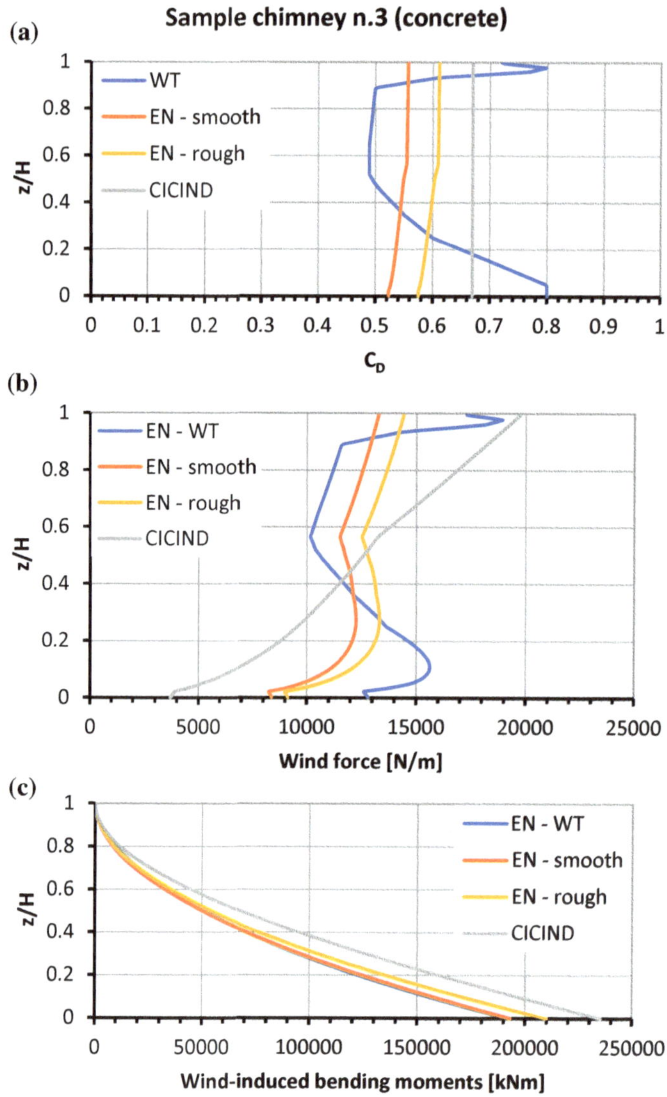

Fig. 6.17 Load and response to different aerodynamic coefficient distributions—sample n.3 (Concrete, h/d = 15.3)

middle region of the cylinder measured in the wind tunnel. This compensates the increase of load in the tip region, which is in fact not accounted for by the CICIND. However, in case of slender cylinders, the overestimation in the base bending moment due to the CICIND C_D distribution with respect to the wind tunnel C_D distribution might be in the order of 30%.

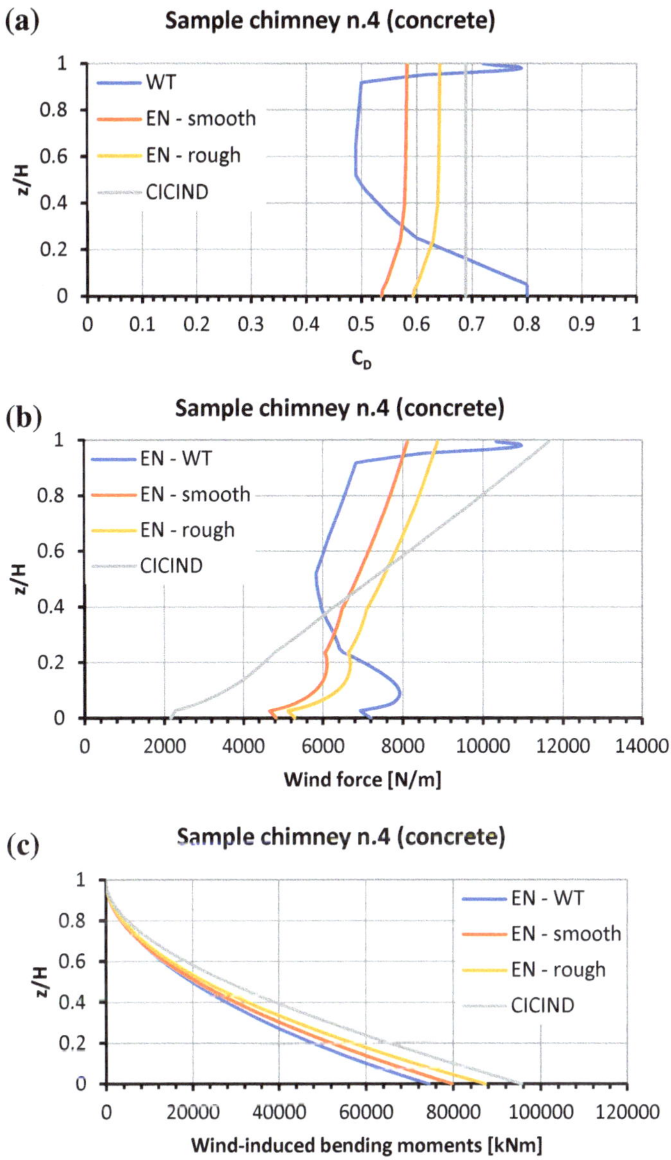

Fig. 6.18 Load and response to different aerodynamic coefficient distributions—sample n.4 (Concrete, h/d = 20.7)

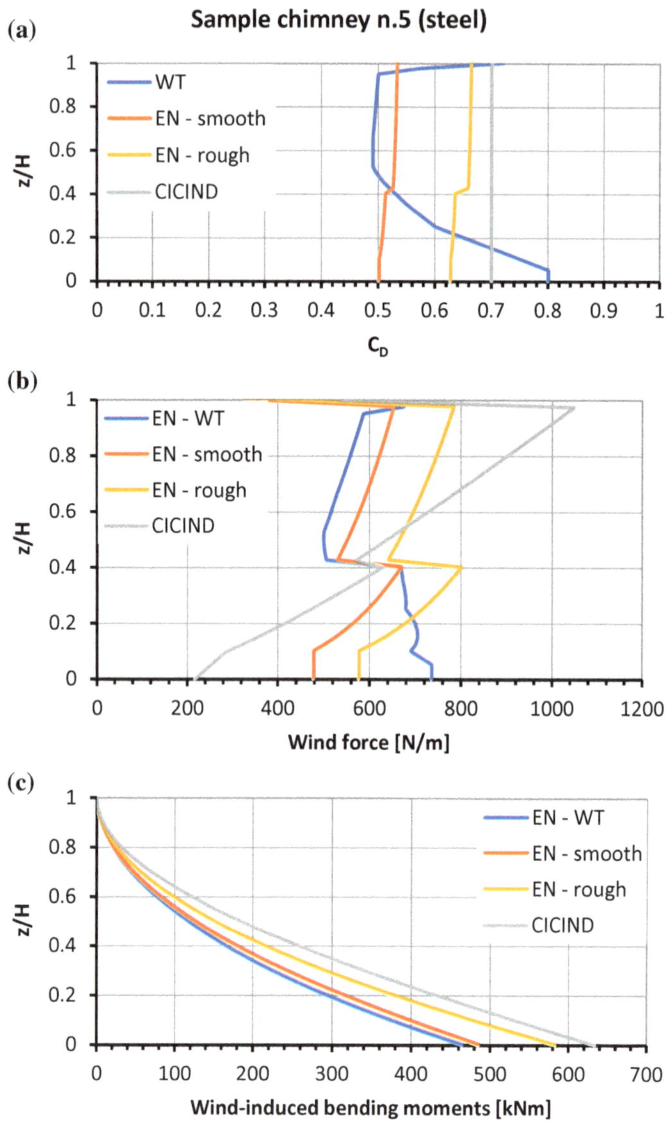

Fig. 6.19 Load and response to different aerodynamic coefficient distributions—sample n.5 (Steel, h/d = 49.6)

6.2 Vortex Excitation and Vortex Resonance Using the Example of High Chimneys

Francesca Lupi, Hans-Jürgen Niemann, Rüdiger Höffer

6.2.1 Introduction

Vortices are shed regularly from the sides of a slender body and form a periodic vortex street in its wake. Theodor von Kármán was the first to develop a theory predicting the frequency of vortex separation and the travelling speed of the vortices. When a vortex is formed on one side of the structure, the wind speed is increased on the other side, and according to Bernoulli's theory, this results in reduced pressure. Thus, the structure is subjected to a lateral force away from the side where a vortex is formed. The result is a dynamic load oscillating with the separation frequency. The frequency of the cross-wind load caused by vortex shedding is proportional to V/d. The factor of proportionality is the Strouhal number St:

$$f_s = St\frac{V}{d} \tag{6.34}$$

The in-wind component is small and oscillates at double frequency. It is superimposed on the mean wind load and the turbulent load fluctuations. For circular cylinders, the role of Reynolds number on the organization of the wake has been studied intensively. Several parameters dominate the load frequency and intensity. (1) Considering a 3-dimensional body, the vortices form cells in which the load oscillations are highly correlated. (2) In natural wind, the approaching flow is highly turbulent. Under these circumstances, the vortex separation becomes a random process showing in the spectral density a pronounced peak at the Strouhal frequency. At the same time, the force correlation along the beam axis decays in turbulent flow. (3) The variability of the mean wind velocity along the axis of a vertical beam, which occurs in natural wind, induces a shift of the peak frequency. (4) The features aforementioned relate to fixed bodies. For a flexible body vibrating at sufficiently high amplitudes, a pronounced aeroclastic effect arises in the loading process, which has an important influence on the load. Large vibrations may occur if the dominating frequency of vortex shedding f_s is in the same range as the natural frequency of a cross-wind mode of the structure. The critical velocity for the nth mode is defined by imposing $f_s = f_n$. If the body is flexible, vortex separation is tuned to the structural vibration frequency over a relatively wide range of wind speeds, that is the lock-in effect. The load correlation along the beam axis improves, the random character of the loading process decreases, and the structural response becomes more or less sinusoidal.

Cross-wind loading can generally be attributed to three contributions: (1) net gust load caused by lateral fluctuations (gusts); (2) loads caused by vortex shedding (net vortex shedding load), which occurs whether or not the structure is moving, but may be strongly dependent on the size of the motion; (3) motion-induced forces.

The net load caused by vortex shedding is characterized by aerodynamic parameters such as the lift coefficient, the Strouhal number, the load spectral bandwidth and the spanwise load correlation. All of them depends on the cross-sectional shape, the Reynolds number, the turbulence scale and intensity, the aspect ratio.

As regards aeroelastic effects, experience has shown that in structures with Scruton number lower than 10 the risk of lock-in is very pronounced. Instead, there is no risk of lock-in if Sc > approximately 20 (Dyrbye and Hansen 1997). Besides the Scruton number, that is related to the mechanical damping, also low Reynolds numbers (and thus low critical velocity, small chimney diameter and natural frequency) are indicators of the proneness of the structure to vortex resonance. In fact, low Reynolds numbers are associated to high (negative) aerodynamic damping. In addition to that, low turbulence conditions increase the risk of violent vortex shedding structural vibrations. Smooth air flow occurs for instance in the stable atmosphere. For example, large vibrations of chimney were observed during periods of cold weather with temperatures of approx. −10 to −5 °C (Hansen 1998). These vibrations occur primarily in the morning and/or in the evening, indicating that the air flow may be characterized by extremely low turbulence levels due to stable stratification of the atmosphere. Rare events with large vibrations may also occur during weather conditions with temperatures well above 0 °C. However, the probability of low turbulence situations in cold weather is larger than in normal weather situations (Hansen 1998).

Vortex resonance has also caused structural failures of towers and chimneys in particular by fatigue. Fatigue of the material is determined by the combination of the number of load cycles during the operational life time of the chimney and the stress range occurring. The fatigue check shall ascertain that the movement due to vortex shedding does not result in the initiation and gradual propagation of cracks in the shell material. The Palmgren-Miner sum is used for the fatigue check, i.e. the chimney is expected to develop a crack if the factored Palmgren-Miner sum (i.e. the Palmgren-Miner sum multiplied by a partial safety factor) is equal or larger than 1.

6.2.2 Models, Methods and Parameters—The Eurocode Models 1, 2 and the CICIND Model Codes

Cross-wind deflections can be evaluated by using two methods, which are named Method 1 (effective correlation length method) and Method 2 (spectral method) in the Eurocode. Method 1 relies on an empirical concept developed by Ruscheweyh (1982) at the Technical University of Aachen (Germany), whereas Method 2 is based on the Vickery&Basu model (1983a, b) (University of Western Ontario, Canada) and revised in the Eurocode by S.O. Hansen. The substantial difference lies in the formulation of the lift force: Method 1 uses a pure deterministic formulation, whereas Method 2 applies a spectral formulation.

The CICIND approach is a modified version of the Vickery&Basu model (after Hansen 1998 and Dyrbye and Hansen 1997), which accounts for the influence of turbulence intensity on both the bandwidth of the lift spectrum and the aerodynamic damping parameter. This represents a more realistic design approach, which should take rare as well as frequent events into account. Moreover, the full formulation given in the CICIND Commentaries allows to consider the height-dependency of the diameter (for chimneys with complicated shape) and of the wind profile, as well as the dependency of the vibration amplitude on V/Vcr and the mode shape. The introduction of the mode shape in the calculation of the variance of the displacement (CICIND Commentaries, Eq. (6.39)) might be interpreted as an extension of Method 2 of the Eurocode to other mode shapes than the first one and even to non-cantilevered structures. Moreover, the possibility to calculate the vibration amplitude for a certain range of wind speeds and not only in resonant conditions is important for fatigue analyses.

Method 1 and Method 2 in the Eurocode, as well as the CICIND approach, are reviewed in the following. The starting point for the two last approaches goes back to Vickery and Clark (1972) and to Basu's Ph.D. dissertation (1982). In particular, for a chimney approximated as a system with one degree of freedom, the variance of the cross-wind deflection is given by:

$$\sigma_y^2 = \int_0^\infty |H(f)|^2 \cdot S_L(f) df \tag{6.35}$$

where f is the natural frequency, H(f) is the response function of a system with one degree of freedom and S_L is the spectrum of the generalized lift force (Verboom and van Koten 2010). If the damping is small, Eq. (6.35) can be approximated by:

$$\begin{aligned} \sigma_y^2 &\approx S_{L,n}(f_n) \int_0^\infty |H_n(f)|^2 df = S_{L,n}(f_n) \frac{1}{4 \cdot K_n \cdot C_n} \\ &= S_{L,n}(f_n) \frac{1}{8 \cdot \omega_n^3 \cdot \xi_n \cdot M_n^2} \end{aligned} \tag{6.36}$$

The formula is applied to the nth mode shape, so that: ω_n is the natural frequency of the chimney in the nth mode ($\omega_n^2 = K_n/M_n$), K_n and C_n are the generalized stiffness and the generalized damping of the chimney ($C_n = 2\omega_n M_n \xi$), respectively; and M_n is the generalized mass (Eq. (6.37)).

$$M_n = \int_0^h m(z) u_n^2(z) dz \tag{6.37}$$

The form taken by the spectrum of the lift force in turbulent flow is suggested by the nature of atmospheric turbulence. In particular, large scale turbulence can be regarded as a slowly varying mean wind speed, which has the effect of altering the shedding frequency f_s. Thus, if the fluctuating lift force were sinusoidal in smooth flow, Gaussian turbulence would cause the spectrum of the lift to take a Gaussian form (Vickery et al. 1983a, b). Because of that, by assuming that the longitudinal turbulent component of the flow follows a normal distribution, the auto-spectrum of the normalized lift force at level z is expressed in Gaussian form:

$$\frac{S_{C_L}(z)f}{C_L^2} = \frac{f}{\sqrt{\pi}B(z)f_s(z)}\exp\left[-\left(\frac{1-f/f_s(z)}{B(z)}\right)^2\right] \tag{6.38}$$

where B is the bandwidth of the lift spectrum, $f_s(z)$ is the vortex shedding frequency and C_L^2 is the variance of the lift coefficient.

For a line-like structure in the nth mode of vibration, the function $g_n(f_n, z)$ is defined as the square root of the spectral density of the lift force at level z (evaluated at $f = f_n$) multiplied by the nth mode shape $u_n(z)$ at the same level:

$$\begin{aligned} g_n(f_n, z) &= \sqrt{S_L(f_n, z)} \cdot u_n(z) \\ &= \frac{1}{2}\rho V_m^2(z)d(z) \cdot \sqrt{S_{C_L}(f_n, z)} \cdot u_n(z) \\ &= \frac{1}{2}\rho V_m^2(z)d(z)\frac{C_{L,n}u_n(z)}{\sqrt{\sqrt{\pi}B(z)f_s(z)}}e^{-\frac{1}{2}\left(\frac{1-f_n/f_s(z)}{B(z)}\right)^2} \end{aligned} \tag{6.39}$$

being ($f_n = \omega_n/2\pi$).

Dyrbye and Hansen (1997) proposed the approximation (6.40) for the modal spectrum of the lift force, by introducing λ as the load correlation length in diameters. This describes the vortex-induced wind load on the non-vibrating structure, as explained in Hansen (1998).

$$S_{L,n}(f_n) \approx 2\lambda d_1 \int_0^h g_n^2(f_n, z)dz \tag{6.40}$$

Equation (6.40) is an approximation of the exact solution, which would require the knowledge of the cross-spectra of the lift force for an integration along the height:

$$\begin{aligned} S_{L,n}(f_n) = \int_0^h\int_0^h & \sqrt{S_L(f_n, z_1)}u_n(z_1) \cdot \sqrt{S_L(f_n, z_2)}u_n(z_2) \\ & \cdot Coh_{12}^2(f_n, z_1, z_2) \cdot dz_1 dz_2 \end{aligned} \tag{6.41}$$

By neglecting the frequency-dependence of the cross-spectra and considering just the correlation between two levels, one can write:

$$S_{L,n}(f_n) = \int_0^h \int_0^h g_n(z_1) \cdot g_n(z_2) \cdot \rho(z_1, z_2) \cdot dz_1 dz_2 \tag{6.42}$$

A further simplification (see Dyrbye and Hansen 1997, p. 122) results from assuming that the correlation depends only on the distance between two levels ($r = |z_1 - z_2|$) and not on their coordinates. Consequently, the double integral can be expressed as two single integrals:

$$S_{L,n}(f_n) = \int_0^h \left(\int_0^{h-r} g_n(z) g_n(z+r) \right) dr \cdot \rho(r) \cdot dr \tag{6.43}$$

Then, Dyrbye and Hansen (1997) assume a coherence function with a negative exponential decay and by integration calculate a non-dimensional correlation length λ close to unit (this is also the value for λ which is recommended in the CICIND). So, the correlation length for net vortex shedding load is relatively small (about one diameter) and this allows to write Eq. (6.40).

Even though the vibration of the structure increases the correlation length of the load—due to the aeroelastic effect—this is not included in λ, which only refers to the non-vibrating structure. It is a typical feature of the spectral method to account for aeroelastic effects only by aerodynamic damping forces. This is a basic difference with Method 1 of the Eurocode, which is then an iterative method, as the effective correlation length (much higher than λ) depends on the amplitude.

The **rms of the lift coefficient** $C_{L,n}$ is a function of the Reynolds number Re_n (i.e. corresponding to the critical velocity of the nth mode) and in the CICIND it is defined as:

$$\begin{aligned} C_{L,n} &= 0.7 \text{ for } Re_n \leq 2 \times 10^5 \\ C_{L,n} &= 0.2 \text{ for } Re_n > 5 \times 10^5 \end{aligned} \tag{6.44}$$

with log-linear interpolation for intermediate Re.

The **width of the lift spectrum** B is a dimensionless measure of the band of frequencies, within which vortices are shed. It depends on turbulence intensity and in the CICIND it is given by:

$$B(z) = 0.1 + Iv(z) \leq 0.35 \tag{6.45}$$

Vickery and Clark (1972) were the first who approximated the lift spectrum by a Gaussian function with width B. Then, Basu (1982) introduced the influence of Iv on B and Vickery (1995) proposed B = 0.1 + 2Iv. In fact, it was clear from wind tunnel experiments that B is a function of Iv and B = 0.1 can be assumed as limit in smooth flow. However, the formulation of B is amenable to discussion and from the limitation of B in all full-scale applications one may suspect that the factor of 2 is too large (Verboom and van Koten 2010). Based on these observations, the CICIND adopted Eq. (6.45).

The damping ξ in Eq. (6.36) is the sum of the structural damping ξs and the negative aerodynamic damping ξa. **Aerodynamic damping** accounts for motion-induced wind loads and depends on the ratio between the wind velocity and the critical velocity, the Reynolds number, the scale and the intensity of the incoming turbulence. It may be characterized by the dimensionless parameter Sa:

$$Sa_n = \frac{C_{a,n}}{\rho_a d_1^2 f_n} = \frac{2\delta_{a,n} m_{o,n}}{\rho_a d_1^2} = \frac{4\pi \xi_{a,n} m_{o,n}}{\rho_a d_1^2} = 4\pi K_{a,n} \tag{6.46}$$

where $K_{a,n}$ is the aerodynamic damping parameter, $m_{0,n}$ is the equivalent mass, d_1 is the chimney diameter, f_n is the natural frequency and ρ_a the density of air, as it will be explained later.

The **Scruton number** Sc is defined in a similar way, but by using structural damping. That is, therefore, a mechanical damping:

$$Sc_n = \frac{4\pi \xi_s m_{o,n}}{\rho_a d_1^2} \tag{6.47}$$

The effective structural damping is thus proportional to the sum of the Scruton number Sc and the parameter Sa. Aerodynamic damping depends on structural vibrations and also structural damping increases with vibration amplitude. However, the last issue is not taken into account by using a constant Scruton number value.

In the spectral model by Vickery and Basu (1983a, b), aerodynamic damping is expressed in the form $a\dot{y} - b\dot{y}^3$. The first linear term introduces negative aerodynamic damping, while the last non-linear term gives positive damping and ensures that the response is self-limiting. The CICIND and the Eurocode (Method 2) formulations use this basic principle and define the total damping (after Vickery&Basu) as:

$$\xi = \xi_s - \frac{K_{a,n}\rho_a d_1^2}{m_{o,n}}\left(1 - \left(\frac{\sigma_{y,n}}{a_{L,n} d_1}\right)^2\right) \tag{6.48}$$

Similarly, by introducing the Scruton number (Eq. (6.47)), it results:

$$\xi \cdot \frac{m_{o,n}}{\rho_a d_1^2} = \frac{Sc_n}{4\pi} - K_{a,n}\left(1 - \left(\frac{\sigma_{y,n}}{a_{L,n} d_1}\right)^2\right) \tag{6.49}$$

i.e.:

$$\begin{aligned} \xi &= \xi_s - \frac{K_{a,n}\rho_a d_1^2}{m_{o,n}}\left(1 - \left(\frac{\sigma_{y,n}}{a_{L,n} d_1}\right)^2\right) \\ &= \xi_s - \xi_a = \frac{\rho_a d_1^2}{m_{o,n}}\left[\frac{Sc_n}{4\pi} - K_{a,n}\left(1 - \left(\frac{\sigma_{y,n}}{a_{L,n} d_1}\right)^2\right)\right] \end{aligned} \tag{6.50}$$

where $m_{o,n}$ is the equivalent mass per unit length of the nth mode, $a_{L,n}$ is the limiting deflection amplitude as a fraction of d and $K_{a,n}$ is the aerodynamic damping parameter. $m_{o,n}$ and $a_{L,n}$ are defined as:

$$m_{o,n} = \frac{\int_0^h m(z)u_n^2(z)dz}{\int_0^h u_n^2(z)dz} \tag{6.51}$$

$$a_{L,n} = \frac{0.4 f_1}{f_n} \tag{6.52}$$

Equation (6.52) is an approximation introduced by the CICIND Model Code, while Eq. (6.51) is the definition of the modal mass.

Only few information is available on the **aerodynamic damping parameter** $K_{a,n}$. Based on theoretical arguments (see Verboom and van Koten 2010), K_a is a function of the Reynolds number, the turbulence intensity, the mean wind velocity and also the aspect ratio.

The values of $K_{a,n}$ given by the CICIND are the result of the following piece of information:

- Vickery (1978) presented a compilation of wind tunnel data (Nakamura et al. 1971; Schmidt 1965.; Szechenyi et al. 1975; Yano and Takahara 1971) for chimneys at given Re in smooth flow. The values of K_a are normalized with respect to the maximum value (which is attained at $V \approx V_{cr}$) and plotted in the graph $K_a/K_{a,max}$ versus V/V_{cr}. This graph is also reported in Vickery et al. (1983a, b).
- As these wind tunnel measurements were obtained in smooth flow, Vickery (1978) and Basu (1982) integrated the product of the smooth flow value and the Gaussian probability density function of the fluctuating wind to yield K_a as a function of Iv. In fact, as explained in Hansen (1998) and already mentioned in Vickery et al. (1983a, b), the effect of turbulence is mainly imputed to large-scale turbulence, which may be interpreted as a slowly varying mean wind velocity. Therefore, the influence of large-scale turbulence is estimated by integrating the aerodynamic damping parameter measured for different mean wind velocities and weighted with a Gaussian distribution describing the variation of the longitudinal turbulent component (Hansen 1998). Once the influence of turbulence intensity is introduced, it results $K_a < K_{a,max}$ for $I_v > 0$.
- In order to obtain K_a for an arbitrary Reynolds number, measurements by Szechenyi and Loiseau (1975) (reported in Basu 1982) are used by Verboom and van Koten (2010). Their result (Fig. 6.21) is included in the CICIND Commentaries (Fig. C.3.3.4).

According to the CICIND Approach of the Commentaries, $K_{a,n}$ is expressed as:

$$K_{a,n} = K_{a,\max,n}(\mathrm{Re}_n)\langle K_{a,0}(V/V_{cr}, Iv)\rangle \tag{6.53}$$

where the symbol $\langle\rangle$ stands for average of the aerodynamic damping parameter over the large scale turbulence.

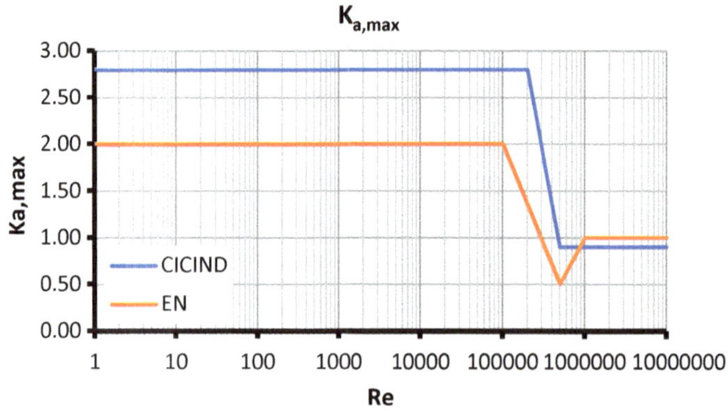

Fig. 6.20 Aerodynamic damping parameter at Iv = 0 ($K_{a,max}$) versus Re

In the CICIND Model Code, the expression (6.53) is simplified and only the maximum value of $\langle K_{a,o}(V, Iv)\rangle$, at V $\approx$ V_{cr}, is considered. It results in:

$$\langle K_{a,o}(Iv)\rangle = 1.0 - 3Iv(z) \geq 0.25 \tag{6.54}$$

The formulation in the Eurocode (Method 2) is even simpler. In fact, being $K_a \leq K_{a,max}$ for $I_v \geq 0$ (the pedix n is omitted because the Eurocode refers always to n = 1), the Eurocode recommends to use $K_a = K_{a,max}$ for every I_v. This approach is, in principle, always on the safe side. However, by comparing results from the Eurocode and the CICIND Commentaries, differences and even underestimations of the Eurocode with respect to the CICIND may result in certain ranges of Scruton number-to-aerodynamic damping parameter. They are due to the actual value of $K_{a,max}(Re_n)$ which is used in the calculation. In this regards, differences between the two Codes are as follows (see also Fig. 6.20):

$$\text{CICIND} \quad \begin{array}{l} K_{a,\max} = 2.8 \text{ for } \mathrm{Re}_n \leq 2 \times 10^5 \\ K_{a,\max} = 0.9 \text{ for } \mathrm{Re}_n > 5 \times 10^5 \end{array} \tag{6.55}$$

$$\text{EN} \quad \begin{array}{l} K_{a,\max} = 2.0 \text{ for } \mathrm{Re}_n \leq 10^5 \\ K_{a,\max} = 0.5 \text{ for } \mathrm{Re}_n = 5 \times 10^5 \\ K_{a,\max} = 1.0 \text{ for } \mathrm{Re}_n \geq 5 \times 10^5 \end{array} \tag{6.56}$$

The value of $K_{a,max}$, as well as its variation with height, are crucial to estimate reliable results especially in the rare events of meteorological conditions with smooth air flow. This issue will be addressed later on, also by using a practical example in Sect. 6.2.3.

Equations (6.36), (6.39) and (6.40) lead to the **full formulation for the variance of deflection** (Eq. C3.3.3 in the **CICIND Commentaries**), i.e. (6.57):

$$\sigma_{y,n}^2 = \frac{2\lambda d_1 \int\limits_0^h g^2(f_n, z)dz}{8 \cdot \omega_n^3 \cdot \mathrm{M}_n{}^2 \cdot \frac{\rho_a d_1^2}{m_{0,n}} \left[\frac{Sc_n}{4\pi} - K_{a,n}\left(1 - \left(\frac{\sigma_{y,n}}{a_{L,n} d}\right)^2\right)\right]} \tag{6.57}$$

A more common form than Eq. (6.57) for $\sigma_{y,n}^2$ is (CICIND C3.3.8):

$$\sigma_{y,n}^2 = \frac{C_{a,n}}{\frac{Sc_n}{4\pi} - K_{a,n}\left(1 - \left(\frac{\sigma_{y,n}}{a_{L,n} d}\right)^2\right)} \tag{6.58}$$

where $C_{a,n}$ assumes in the full formulation (CICIND Commentaries) the following expression:

$$\begin{aligned} C_{a,n} &= \frac{2\lambda d_1 \int_0^h g_n^2(f_n, z)dz}{8 \cdot (2\pi f_n)^3 \cdot \frac{\rho_a d_1^2}{m_{o,n}} \cdot \left(m_{o,n} \int_0^h u^2(z)dz\right)^2} \\ &= \frac{2\lambda \int_0^h \frac{\rho^2 \cdot V_m^4(z) \cdot d^2(z) \cdot C_{L,n}^2 \cdot u_n^2(z)}{4\sqrt{\pi} \cdot B(z) \cdot \frac{St \cdot V_m(z)}{d(z)}} e^{-\left(\frac{1 - f_n/f_s(z)}{B(z)}\right)^2} dz}{8 \cdot (2\pi f_n)^3 \cdot \rho \cdot d_1 \cdot m_{o,n} \cdot \left(\int_0^h u^2(z)dz\right)^2} \end{aligned} \tag{6.59}$$

The Strouhal number at the power of 4 can be extracted from Eq. (6.59), in case $f_n = f_s = St * V/d$.

By referring to the interpretation of Hansen (1998), the coefficient $C_{a,n}$ in Eq. (6.58) results from the product of two contributions. The first contribution describes the vortex-induced wind load on non-vibrating structures and contains the rms of lift coefficient, the load correlation length and the width of the spectrum. This contribution depends on turbulence intensity and Reynolds number. The second contribution depends primarily on the mode shape, on the velocity profile and on the variation of diameter with height. In this regard, in the full formulation of the CICIND Commentaries, the function g_n^2 (Eq. (6.39)) is introduced, because the velocity profile of the wind and a decreasing width of the structure affect the net vortex shedding load. In fact, the vortex shedding $f_s(z)$ increases if the mean wind velocity increases, and/or if the structural width d is reduced, so that the net vortex shedding load occurs on a height-dependent band of frequencies. Moreover, as previously said, the explicit use of the mode shape in Eq. (6.39) allows the extension of the method to an arbitrary, even non-cantilevered, structure.

In case the variability with height is negligible, Eq. (6.59) can be rather simplified. In particular, if only the mode shape is kept as function of height, while the wind velocity, the diameter and the turbulence intensity are fixed at their value at the top of the chimney, the formulation of the CICIND Model Code is obtained. This is shown in the following.

The starting point is the combination of Eqs. (6.58) and (6.59), i.e. the full formulation of the CICIND Commentaries (Eq. (6.60)), which includes: (1) profile of mean wind, (2) profile of Iv, (3) variable diameter with height, (4) mode shape. The first step (that results in Eqs. (6.65) and (6.66)) is to understand the formulation of the CICIND Model Code by progressive simplification of the full procedure (Equations from (6.60) to (6.66)). The latter still includes the mode shape u(z) in the expression and it is therefore applicable to the higher mode shapes of a cantilever beam and it might even be extended to non-cantilevered structures.

$$\sigma_{y,n}^2 = \frac{2\lambda \int_0^h \frac{\rho^2 \cdot V_m^4(z) \cdot d^2(z) \cdot C_{L,n}^2 \cdot u_n^2(z)}{4\sqrt{\pi} \cdot B(z) \cdot \frac{St \cdot V_m(z)}{d(z)}} e^{-\left(\frac{1-f_n/f_s(z)}{B(z)}\right)^2} dz}{8 \cdot (2\pi f_n)^3 \cdot \rho \cdot d_1 \cdot m_{o,n} \cdot \left(\int_0^h u^2(z) dz\right)^2} \cdot \frac{1}{\frac{Sc_n}{4\pi} - K_{a,n}\left(1 - \left(\frac{\sigma_{y,n}}{a_{L,n} d}\right)^2\right)} \tag{6.60}$$

Then the variability with z of d(z), B(z), $V_m(z)$ is neglected and the parameters are taken out of the integral.

$$\sigma_{y,n}^2 = \frac{\frac{2}{4\sqrt{\pi}} \frac{\rho^2 \cdot V_m^4 \cdot d^2 \cdot \lambda \cdot C_{L,n}^2}{B \cdot \frac{St \cdot V_m}{d}} e^{-\left(\frac{1-f_n/f_s}{B}\right)^2} \int_0^h u_n^2(z) dz}{8 \cdot (2\pi f_n)^3 \cdot \rho \cdot d \cdot m_{o,n} \cdot \left(\int_0^h u_n^2(z) dz\right)^2} \cdot \frac{1}{\frac{Sc_n}{4\pi} - K_{a,n}\left(1 - \left(\frac{\sigma_{y,n}}{a_{L,n} d}\right)^2\right)} \tag{6.61}$$

i.e.:

$$\sigma_{y,n}^2 = \frac{\frac{2}{4\sqrt{\pi}}}{8 \cdot (2\pi)^3} \frac{\rho \frac{V_m^3 \cdot d^3}{St} \frac{\lambda \cdot C_{L,n}^2}{B} e^{-\left(\frac{1-f_n/f_s}{B}\right)^2}}{\left(\frac{V_{cr,n} \cdot St}{d}\right)^3 \cdot d \cdot m_{o,n}} \frac{\int_0^h u_n^2(z) dz}{\left(\int_0^h u_n^2(z) dz\right)^2} \cdot \frac{1}{\frac{Sc_n}{4\pi} - K_{a,n}\left(1 - \left(\frac{\sigma_{y,n}}{a_{L,n} d}\right)^2\right)} \tag{6.62}$$

The equation is then further simplified in a more readable way:

$$\left(\frac{\sigma_{y,n}}{d}\right)^2 = \frac{1}{2\sqrt{\pi} 8 \cdot (2\pi)^3} \frac{\rho \cdot d^3}{m_{o,n}} \frac{\lambda \cdot C_{L,n}^2}{B \cdot St^4} \cdot \frac{V_m^3}{V_{cr,n}^3} e^{-\left(\frac{1-f_n/f_s}{B}\right)^2} \frac{1}{\int_0^h u_n^2(z) dz} \cdot \frac{1}{\frac{Sc_n}{4\pi} - K_{a,n}\left(1 - \left(\frac{\sigma_{y,n}}{a_{L,n} d}\right)^2\right)} \tag{6.63}$$

at $f_n = f_s$, $V_m = V_{cr}$ and the exponential function equals 1:

$$\left(\frac{\sigma_{y,n}}{d}\right)^2 = 1.42 \times 10^{-4} C_{L,n}^2 \frac{\lambda}{BSt^4} \frac{\rho d^3}{m_{o,n}} \frac{1}{\int_0^h u_n^2(z)dz} \cdot \frac{1}{\frac{Sc_n}{4\pi} - K_{a,n}\left(1 - \left(\frac{\sigma_{y,n}}{a_{L,n}d}\right)^2\right)} \quad (6.64)$$

Finally, the expression of the CICIND Model Code is obtained (6.65). As it can be seen by Eq. (6.66), the dependency on the mode shape is maintained.

$$\left(\frac{\sigma_{y,n}}{d}\right)^2 = \frac{C_{a,n}}{\frac{Sc_n}{4\pi} - K_{a,n}\left(1 - \left(\frac{\sigma_{y,n}}{a_{L,n}d}\right)^2\right)} \quad (6.65)$$

where $C_{a,n}$ is dimensionless and equal to:

$$C_{a,n} = \frac{1.42 \times 10^{-4} \cdot \lambda \cdot C_{L,n}^2}{St^4 \cdot B \int_0^h u_n^2(z)dz} \frac{\rho_a \cdot d_1^3}{m_{o,n}} \quad (6.66)$$

It should be observed that $C_{a,n}$ in (6.59) is not a non-dimensional parameter (it has the dimension of a length squared), while $C_{a,n}$ in (6.66) is a non-dimensional parameter.

A further step is now to highlight the direct comparison between the formulations of the CICIND Model Code (6.65), (6.66) and the EN Method 2 (E.14), reported in (6.67).

$$\left(\frac{\sigma_y}{d}\right)^2 = \frac{1}{St^4} \frac{C_c^2}{\frac{Sc}{4\pi} - K_a\left(1 - \left(\frac{\upsilon_y}{a_{L,n}d}\right)^2\right)} \frac{\rho d^2}{m_o} \frac{d}{h} \quad (6.67)$$

In the Eurocode, C_c is given as a number, which depends on Re:

$$\begin{aligned} C_c &= 0.020 \ \mathrm{Re} \le 10^5 \\ C_c &= 0.005 \ \mathrm{Re} = 5 \times 10^5 \\ C_c &= 0.010 \ \mathrm{Re} \ge 10^6 \end{aligned} \quad (6.68)$$

In fact, the EN Method 2 is a further simplification of the CICIND Model Code by using the coefficient C_c. This assumes $u(z) = \text{const} = 1$ (so that the integration of $u^2(z)$ along the height is equal to h), $B(Iv) = \text{const}$ (whereas in the CICIND Eq. (6.45) holds and B ranges from 0.1 for $Iv = 0$ and $B_{max} = 0.35$) and it maintains the dependency on Re that was in C_L (rms of the lift coefficient). By comparing Eqs. (6.65) and (6.67), it can be seen that the coefficient C_c given by the Eurocode (6.68) results from:

$$\frac{C_c^2}{h} = 1.42 \times 10^{-4} \cdot C_L^2 \frac{\lambda}{B} \frac{1}{\int_0^h u_n^2(z)dz} \quad (6.69)$$

By approximating u(z) = const = 1, it results:

$$C_c = C_L \sqrt{1.42 \times 10^{-4} \frac{\lambda}{B}} \quad (6.70)$$

By assuming:

$\lambda = 1$;
B = 0.1 (at Iv = 0, see (6.45))
C_L is ranging from 0.7 at low Re and 0.2 at high Re (see, for example Eq. (6.44), coming from the CICIND, and Figure E.2 in the Eurocode, where C_{lat} for Re > 10^7 is somewhat higher, i.e. 0.3), it results:

$$C_c = C_L \sqrt{1.42 \times 10^{-4} \frac{\lambda}{B}} = 0.7 \sqrt{1.42 \times 10^{-4} \frac{1}{0.1}} = 0.026 \quad (6.71)$$

$$C_c = C_L \sqrt{1.42 \times 10^{-4} \frac{\lambda}{B}} = 0.2 \sqrt{1.42 \times 10^{-4} \frac{1}{0.1}} = 0.0075 \quad (6.72)$$

$$C_c = C_L \sqrt{1.42 \times 10^{-4} \frac{\lambda}{B}} = 0.3 \sqrt{1.42 \times 10^{-4} \frac{1}{0.1}} = 0.011 \quad (6.73)$$

These values are reasonably comparable to those in Eq. (6.68).

In order to introduce the dependency of the mode shape in the EN—Method 2, i.e. in order to extend the EN—Method 2 to higher modes, one should use a modified C_c coefficient, according to Eq. (6.82):

$$C_c = C_{c,EN} \sqrt{\frac{h}{\int_0^h u_n^2(z)dz}} \quad (6.74)$$

The same (practically speaking) results would be obtained by applying either the CICIND Model Code or the Eurocode Method 2 extended to the second mode shape, once the $K_{a,max}$ has the same distribution, e.g. like in Eq. (6.55). Otherwise, results will differ especially in the coefficient c_1. This issue is proved with a practical example in Sect. 6.2.3 (Fig. 6.33).

Equation (6.57) (i.e. Eq. (6.60)) from the CICIND Commentaries is consistent with Dyrbye & Hansen's formulation (1997). This is also true for Eqs. (6.40)–(6.43). Even though the final result does not change, it should be observed that in Dyrbye & Hansen's approach the function g(z, f) is referred to a reference height level and the so-called joint acceptance function $|J(f)|^2$ (frequency-dependent) is introduced in the calculation. The function g(z, f) and $|J(f)|^2$ are defined as follows:

Table 6.6 Code stipulations for coefficients c1 and c2

	EN	CICIND Model Code	CICIND Commentaries
c_1	$\frac{a_L^2}{2}\left(1-\frac{Sc_1}{4\pi K_{a,1}}\right)$	$\frac{a_L^2}{2}\left(1-\frac{Sc_n}{4\pi K_{a,n}}\right)$	$\frac{a_L^2}{2}\left(1-\frac{Sc_n}{4\pi K_{a,r,n}}\right)$
c_2	$\frac{\rho_a d^2}{m_{o,n}}\frac{a_L^2}{K_{a,1}}\frac{C_C^2}{St^4}\frac{d}{h}$	$\frac{a_L^2 C_{a,n}}{K_{a,n}}$	$\frac{a_L^2 C_{a,n}}{K_{a,r,n}d^2}$

$$g(f,z)=\frac{q(z)\cdot d(z)\cdot C_L(z)\cdot u(z)}{(q\cdot d\cdot C_L\cdot u)_{ref}}\sqrt{\frac{B_{ref}}{B(z)}}\sqrt{\frac{f}{f_s(z)}}e^{-\frac{1}{2}\left(\frac{1-f/f_S(z)}{B(z)}\right)^2} \tag{6.75}$$

$$|J(f)|^2=\frac{1}{h^2}\int_0^h\int_0^h g(z_1,f)\cdot g(z_2,f)\cdot\rho(z_1,z_2)\cdot dz_1dz_2$$
$$\approx\frac{1}{h^2}2\lambda d_{ref}\int_0^h|g(z,f)|^2dz \tag{6.76}$$

Equation (6.58) (and, similarly, Eq. (6.67)) can be solved as a second degree equation in the unknown $(\sigma_y/d)^2$. The positive solution is the physically correct one.

$$\left(\frac{\sigma}{d}\right)^4-2c_1\left(\frac{\sigma}{d}\right)^2-c_2=0\rightarrow\left(\frac{\sigma}{d}\right)^2=c_1\pm\sqrt{c_1^2+c_2} \tag{6.77}$$

The exact expressions of the coefficients c_1 and c_2 depend on the formulation which is used (either (6.58) like in the CICIND or (6.67) like in the Eurocode). It should also be observed that the coefficient c_2 (dimensionless) is defined differently in the CICINC Model Code and in the Commentaries. This is due to the fact that the full formulation of $C_{a,n}$ (6.59) has the dimension of a square length, whereas the simplified formulation of the Model Code (6.66) is dimensionless. In summary, it is:

Table 6.6 shows that there is general agreement in the definition of the coefficient c_1. In most cases, its sign governs the response. In fact, the either small or large cross-wind vibration amplitude $\sigma_{y,n}$ of the structure depends on the ratio between two dimensionless parameters: the Scruton number Sc_n and the aerodynamic damping parameter (4π) $K_{a,n}$. The value of such a ratio (either bigger or smaller than 1) changes the sign of the coefficient c_1, so that:

$$\frac{Sc_n}{4\pi K_{a,n}}\ll 1\quad(c_1>0)\quad\text{Large vibration amplitude(lock - in regime)} \tag{6.78}$$

$$\frac{Sc_n}{4\pi K_{a,n}}\gg 1\quad(c_1<0)\quad\text{Small vibration amplitude(forced vibration regime)} \tag{6.79}$$

Therefore, in order to avoid the range of large deflections one must assure that $c_1 < 0$, i.e. $Sc_n > 4\pi K_{a,n}$, i.e. the mechanical damping should be larger than the (negative) aerodynamic damping.

In most cases $|c_{1,n}| \gg c_{2,n}$ and therefore the result of expression (6.77) strongly depends the sign of $c_{1,n}$. In particular, c_1 alone governs the response in the range of large deflections, while the value of c_2 becomes more important in the range of small deflections, i.e.:

$$\frac{\sigma_y}{d} \approx \sqrt{2c_1} \quad if \quad c_1 > 0 \ and \ c_1 \gg c_2 \tag{6.80}$$

$$\frac{\sigma_y}{d} \approx \sqrt{\frac{c_2}{2c_1}} \quad if \quad c_1 < 0 \ and \ |c_1| \gg c_2 \tag{6.81}$$

It should be noted that the Scruton number enters in c_1 and therefore affects the response in both the small and the large amplitude range; whereas the St enters at the power of 4 only in c_2, not in c_1, and therefore it mainly influences the response in the range of large deflections. However, the Strouhal number enters in the critical velocity and thus in the Reynolds number and therefore it affects indirectly K_a and thus c_1, too.

A parenthesis is now open on the sensitivity of the model to the values of the aerodynamic damping parameter K_a. At $I_v = 0$, $K_{a,n} = K_{a,max}$. The values of $K_{a,max}$ given in the CICIND and in the Eurocode are directly comparable. The comparison, as a function of the Reynolds number, in shown in Figs. 6.20, 6.21, 6.22 and 6.23. The figure shows that $K_{a,max}(EN) < K_{a,max}(CICIND)$ for $Re < 9 \times 10^5$. The values of $K_{a,max}$ provided by the Eurocode result from Basu (1982). For instance, at low Re ($Re < 10^5$) $K_{a,max} = 2.0$. Instead, the CICIND decides to adopt at low Re the higher value $K_{a,max} = 2.8$. Verboom and van Koten (2010) explains the reason for this choice. Basu's value of 2 is multiplied by the factor 1.4 in order to translate the wind tunnel experiments for a cylinder with small aspect ratio (ca. 11) to full-scale dimension. This is also documented in Vickery (1973), Fig. 15.18. Therefore, the prediction of the Eurocode at $Iv = 0$ might not always be on the safe side for $Re < 9 \times 10^5$, due to the uncertainty on $K_{a,max}$. In fact, at the same value of Re, a lower value of $K_{a,max}$ implies higher ratio $Sc/(4\pi K_{a,max})$ and it may significantly underestimate large vibration amplitudes in smooth flow. Further documentation should be necessary to assess the correct value of $K_{a,max}$ to be used in the calculation.

Figure 6.21 shows the variation of the turbulence mean aerodynamic parameter $\langle K_{a,0} \rangle$ with Iv and V/V_{cr}, according to the approach of the CICIND Commentaries. As the velocity profile varies with height, it is basically a dependency on the height. In case $Iv > 0$, the height dependency is also due to the profile of Iv(z). The parameter $K_{a,r,n}$, which enters in the full calculation of the commentaries in place of $K_{a,n}$, is derived by integration of $K_{a,n}(V(z)/V_{cr})$ (see Eq. (6.82)).

$$K_{a,r,n} = \frac{\int_0^h K_{a,n}(z) d^2(z) u_n^2(z) dz}{d_1^2 \int_0^h u_n^2(z) dz} \tag{6.82}$$

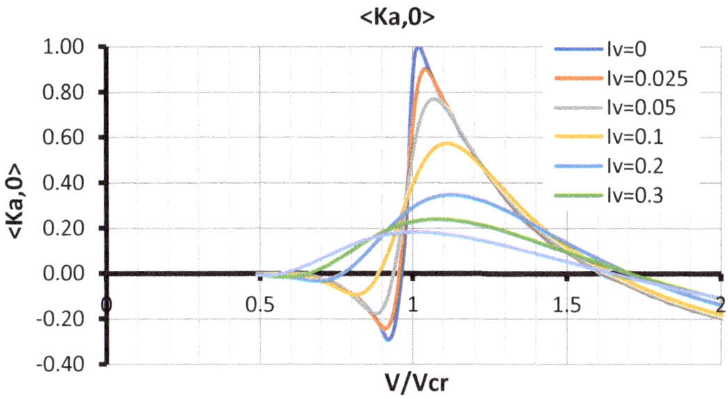

Fig. 6.21 Turbulence mean aerodynamic damping parameter

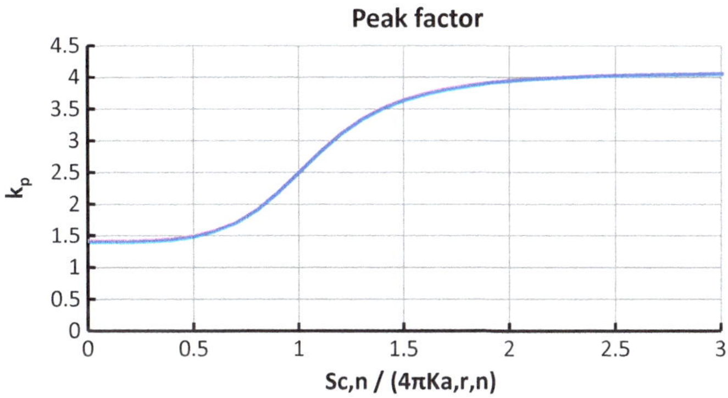

Fig. 6.22 Peak factor

Particular attention should be drawn to the range $Sc_n \approx 4\pi K_{a,r,n}$, i.e. $c_1 \approx 0$, where very different solutions, sensitive to small changes of turbulence intensity, Scruton number and Reynolds number may occur. In fact, in this range even small variations of input parameters may be responsible for switching between $Sc_n > 4\pi K_{a,r,n}$ or $Sc_n < 4\pi K_{a,r,n}$. The response will then turn to be significantly different in the two cases because of the change of sign of c_1 (Table 6.6) and it may lie either in the small- or large-amplitude range. This issue will be addressed in more detail on practical chimney example in Sect. 6.2.3. The effect is especially crucial to estimate the response in low turbulence conditions, where $K_a = K_{a,max}$, which may occur in rare events of meteorological conditions with cold weather and stable atmosphere. In fact, as documented in Hansen (1998), large vibrations of chimneys have been observed in these conditions.

The maximum amplitude of vibration at each height in the nth mode of vibration is given by:

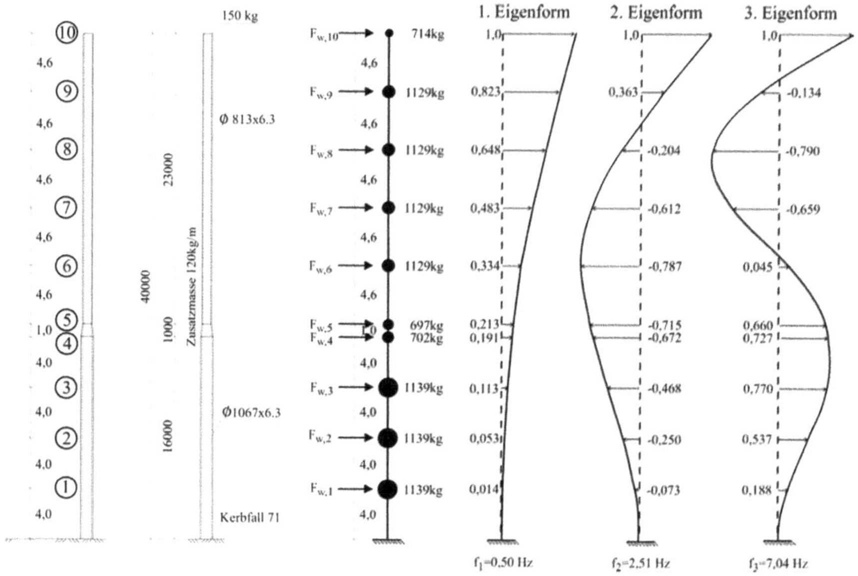

Fig. 6.23 Steel chimney sample, distribution of masses and mode shapes (Clobes M., A)

$$y_n(z) = k_{p,n}\sigma_{y,n}u_n(z) \tag{6.83}$$

where $k_{p,n}$ is the peak factor, $\sigma_{y,n}$ is the standard deviation of deflection in the nth mode and $u_n(z)$ is the nth mode itself. Both the Eurocode and the CICIND Model Code calculate the response only at the critical wind speed, whereas the CICIND Commentaries also introduce the dependency on V/Vcr.

Both in the Eurocode as well as in the CICIND, the **peak factor** is a continuous function of $Sc_n/(4\pi Ka_n)$:

$$k_{p,n} = \sqrt{2}\left[1 + 1.2\arctan\left(0.75\left(\frac{Sc_n}{4\pi K_{a,n}}\right)^4\right)\right] \tag{6.84}$$

This formulation was given by Hansen (1998) and it is an approximation of the discontinuous result reported by Basu (1982). For small amplitudes (approx. 1–2% of the diameter), the peak factor is approx. 3.5–4 depending on the natural frequency of the structure. It corresponds to a stochastic type of vibration. For large amplitudes, the peak factor is equal to $\sqrt{2}$ and it corresponds to a sinusoidal vibration with nearly constant amplitude. For intermediate amplitudes, the peak factor increases gradually with decreasing amplitude (Fig. 6.22).

Eurocode Method 1 (after Ruscheweyh), on the other hand, is based on the concept of effective correlation length and calculates the maximum cross-wind vibration amplitude by the following equation:

$$\frac{y_{F,\max}}{d} = \frac{1}{St^2} \cdot \frac{1}{Sc} \cdot K \cdot K_W \cdot c_{lat} \tag{6.85}$$

K_w is the effective correlation length factor, which accounts for aeroelastic forces, K is the mode shape factor and c_{lat} is the lateral force coefficient.

For each mode of vibration which is considered, the effective correlation length L_j is positioned in the range of the antinodes. As L_j depends on the amplitude, the procedure is iterative. In particular, L_j ranges between 6 times the diameter for small amplitudes of vibration (i.e. below 10% of the diameter) and 12 times the diameter in case of large amplitudes, i.e. higher than 60% of the diameter.

The lateral force coefficient depends on the Reynolds number and assumes the maximum (basic) value, equal to $c_{lat,0}$, when the critical velocity $V_{cr,i}$ is less than 83% of the mean wind velocity at the centre of the effective correlation length ($V_{m,Lj}$). Then, it progressively decreases to zero as $V_{cr,i}/V_{m,Lj}$ increases. For $V_{cr,i} > 1.25V_{m,Lj}$ the force becomes zero.

The effective correlation length factor is the percentage of the area subtended by the eigenmode in the regions where vortex excitation occurs at the same time, with respect to the whole area subtended by the eigenmode. In other words, it is:

$$K_W = \frac{\sum_{j=1}^{n} \int_{Lj} \left|u_{i,y}(z)\right| dz}{\sum_{j=1}^{m} \int_{lj} \left|u_{i,y}(z)\right| dz} \leq 0.6 \tag{6.86}$$

In particular, the numerator contains the summation and integration only on the n regions where the vortex shedding occurs simultaneously, while the denominator integrates over the whole length of the structure between two nodes.

The coefficient K is named mode shape factor and it is defined as:

$$K = \frac{\sum_{j=1}^{m} \int_{lj} \left|u_{i,y}(z)\right| dz}{4 \cdot \pi \cdot \sum_{j=1}^{m} \int_{lj} u_{i,y}^2(z) dz} \tag{6.87}$$

Finally, it results:

$$\frac{y_{F,\max}}{d} = \frac{1}{St^2} \cdot \frac{1}{Sc} \cdot \frac{\sum_{j=1}^{m} \int_{lj} \left|u_{i,y}(z)\right| dz}{4 \cdot \pi \cdot \sum_{j=1}^{m} \int_{lj} u_{i,y}^2(z) dz} \cdot \frac{\sum_{j=1}^{n} \int_{Lj} \left|u_{i,y}(z)\right| dz}{\sum_{j=1}^{m} \int_{lj} \left|u_{i,y}(z)\right| dz} \cdot c_{lat} \tag{6.88}$$

Equation (6.88) shows some similarities with Eqs. (6.65) and (6.66), even thought the latter provide the variance of the amplitude and the peak factor must be applied to obtain the maximum value. In particular, the Strouhal number enters at the same power, the rms of the force coefficient acts in the same manner and also the mode shape enters as a integration of $u^2(z)$ along the height. It should be observed, instead, that the Scruton number plays a different role. In the amplitude, it enters under square root in Method 2 and linearly in Method 1.

Table 6.7 Variation of structural damping and Scruton number for the chimney sample n.5 (steel)

δs	Sc,1	
0.010	6.350	E.g. un-insulated chimney without lining (δs min, see Table 6.13)
0.015	9.525	E.g. un-insulated chimney without lining (ca. typical value, see Table 6.13)
0.020	12.700	E.g. unlined chimney with external thermal insulation (see Table 6.13)
0.025	15.875	E.g. two or more liners + external thermal insulation (see Table 6.13)
0.030	19.050	E.g. with internal concrete shell (see Table 6.13)
0.040	25.400	E.g. guyed stack without liner (see Table 6.13)
0.070	44.451	E.g. with internal brick liner (see Table 6.13)

6.2.3 *Worked Examples for Vortex Resonance*

6.2.3.1 Steel Chimney Sample

The maximum cross-wind deflection is calculated in this section by applying to a steel chimney the CICIND Model Code and Commentaries, as well as the methods 1 and 2 of the Eurocode. Distributions of masses and mode shapes are sketched in Fig. (Clobes xxx) and result from a real world example. In this section the calculation is repeated for different values of structural damping (and thus Scruton number), according to Table 6.7. The actual calculation on the real world sample of the chimney (i.e. by using $\delta_s = 0.014$) will then be presented in Sect. 6.2.3.

6.2.3.2 The CICIND Approach (Model Code and Commentaries)

The CICIND Model Code allows to evaluate the response of the chimney at different values of turbulence intensities. The result is generally accurate for chimneys of simple shape and far from the range $Sc \approx 4\pi K_a$, where the coefficient c_1 changes sign. In those cases, the full procedure of the Commentaries should be applied. The maximum deflection in the first cross-wind mode of vibration is plotted in Fig. 6.24, for the range of Iv from 0 to 25%. Different Scruton numbers are considered, by changing the structural damping according to Table 6.7.

Figure 6.24 proves that the higher is the value of the Scruton number, the lower is the turbulence intensity at which the transition regime between the lock-in region and the forced vibration region occurs. Moreover, at a given Scruton number, the effect of turbulence intensity may act in the reduction of amplitude of one order of magnitude. The reason is that turbulence of the flow reduces the (negative) aerodynamic damping, because Iv enters in the calculation of the aerodynamic damping parameter K_a (Fig. 6.21). On the other hand, being for this chimney K_a in any case

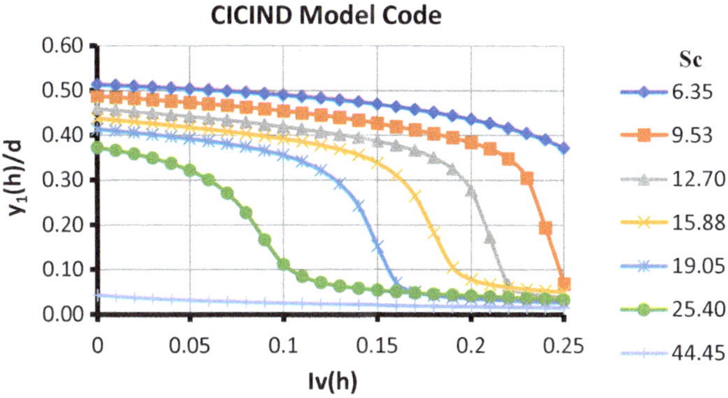

Fig. 6.24 Maximum cross-wind deflection (1st mode) versus Iv at different Sc, CICIND Model Code

high due to the low Re, a relatively high structural damping is necessary to attain a safe chimney design out of the lock-in regime, preferably at any turbulence intensity of the incoming wind.

The CICIND Commentaries, with respect to the Model Code, allow to evaluate the cross-wind response of the chimney as a function of the wind velocity. It results that the maximum response amplitude occurs at the ratio V/Vcr $\approx$ 1.05–1.10. In particular, the lower is the turbulence intensity, the lower is the ratio V/Vcr (≥ 1) at which the maximum response occurs. This is mainly attributed to the behaviour of $<K_{a,0}>$, as shown in Fig. 6.21, where the curves show their maximum at progressively higher values as Iv increases.

The results of the full procedure of the CICIND Commentaries, plotted for different V/V_{cr} ratios, is shown in Fig. 6.25 for a Scruton number fully in the lock-in region and in Fig. 6.26 for a Scruton number four times larger. In the latter case, the lock-in region is limited at a narrow range around resonance in laminar flow and it is suppressed at higher turbulence intensities. However, even though in Fig. 6.26 the Scruton number is relatively high, the lock-in regime still occurs at low turbulence. In fact, the response lies either in the lock-in region or in the forced-vibration region depending not only on Sc, but on the ratio $Sc/(4\pi K_a)$, and $K_{a,max}$ is high, in any case, due to the low Reynolds number.

By applying the complete calculation procedure of the CICIND Commentaries, at each Scruton number the maximum cross-wind response is plotted versus Iv in Fig. 6.20. By comparing with Fig. 6.27, it can be seen that results differ especially in the range Sc $\approx 4\pi K_{a,r,n}$. In fact, in this range (where c_1 changes sign) the CICIND recommends to apply the full procedure in the Commentaries. At Iv = 0, this range occurs at Sc about 25.40 (Fig. 6.28).

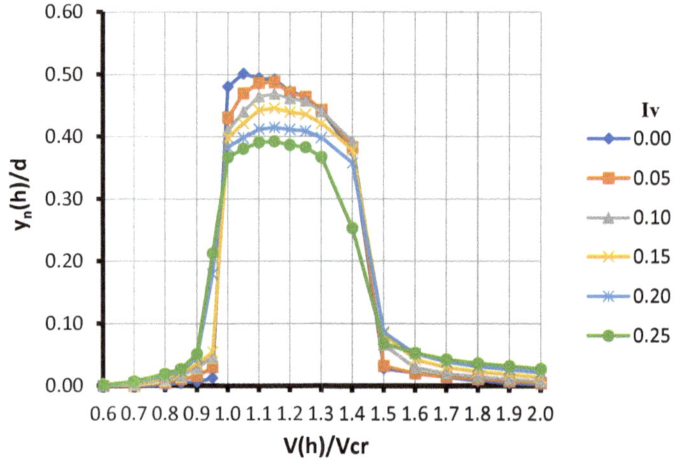

Fig. 6.25 Cross-wind deflection at different Iv (Sc = 6.35) after CICIND Commentaries

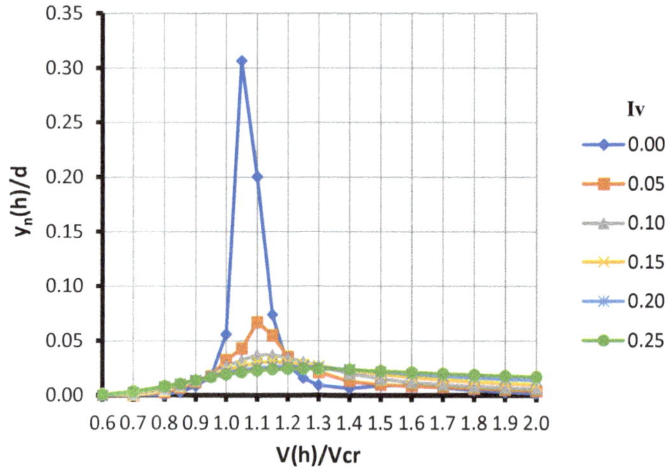

Fig. 6.26 Cross-wind deflection at different Iv (Sc = 25.40) after CICIND Commentaries

6.2.3.3 Comparative Calculations After the CICIND Models and the Eurocode Methods

The aim of this sub-section is to compare oscillation amplitudes predicted by the CICIND (Model Code and Commentaries) and the Eurocode (Method 1 and Method 2) on a real world chimney example, in three different possible response regimes:

1. Lock-in regime (Sc = 6.35 and Sc = 15.87; $Sc/(4\pi K_a) \ll 1$, Fig. 6.29 and Fig. 6.30, respectively);
2. Forced-vibration regime (Sc = 44.45 and $Sc/(4\pi K_a) \gg 1$, Fig. 6.31);
3. Transition regime (Sc = 25.40 and $Sc/(4\pi K_a) \approx 1$, Fig. 6.32).

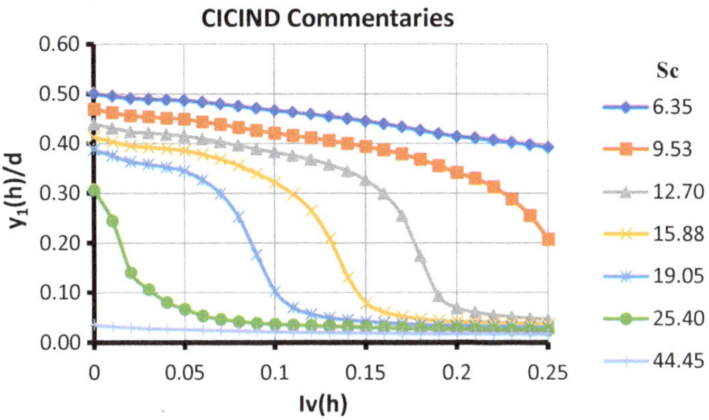

Fig. 6.27 Maximum cross-wind deflection (1st mode) versus Iv at different Sc, CICIND Commentaries

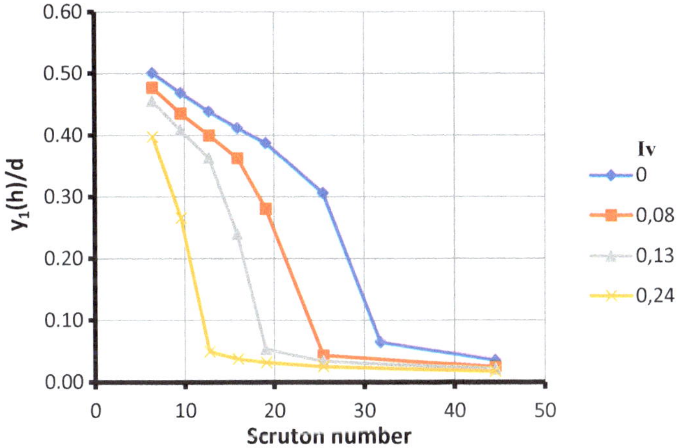

Fig. 6.28 Cross-wind deflection as a function of the Scruton number, CICIND Commentaries

Figure 6.29 shows that in the lock-in regime ($Sc/(4\pi Ka) \ll 1$) the response is similarly evaluated by both the CICIND (either Model or Commentaries) and the Eurocode, Method 2. Method 1 underestimates the response considerably. However, further studies should be necessary in order to assess how realistic are the large amplitudes predicted by Vickery & Basu-based models on existing chimneys.

Moreover, in Fig. 6.29 it can be seen that due to the relatively low Scruton number and above all due to the small ratio $Sc/(4\pi K_a)$, aggravated by high K_a, the effect of turbulence intensity is not so pronounced. Instead, in Fig. 6.30 (higher Sc) the effect of Iv in the range 10–15% reduces the amplitude of about one order of magnitude. This reduction is not considered in the Eurocode, which prediction is by the way well matching the one of the CICIND at Iv = 0. The differences at Iv = 0 in the Eurocode

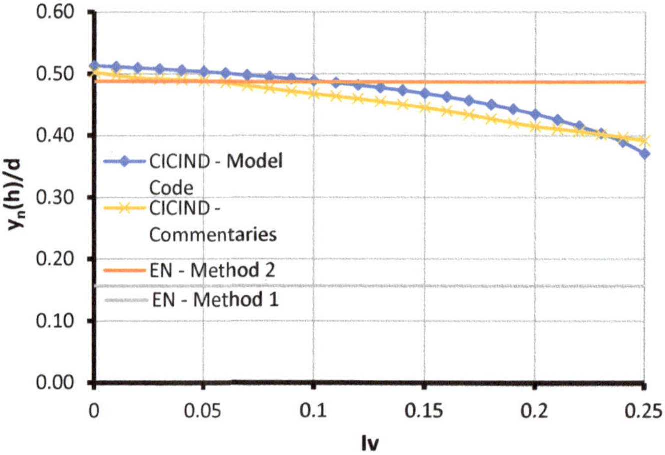

Fig. 6.29 Cross-wind vibration amplitudes in the lock-in regime: comparison CICIND and Eurocode (Sc = 6.35)

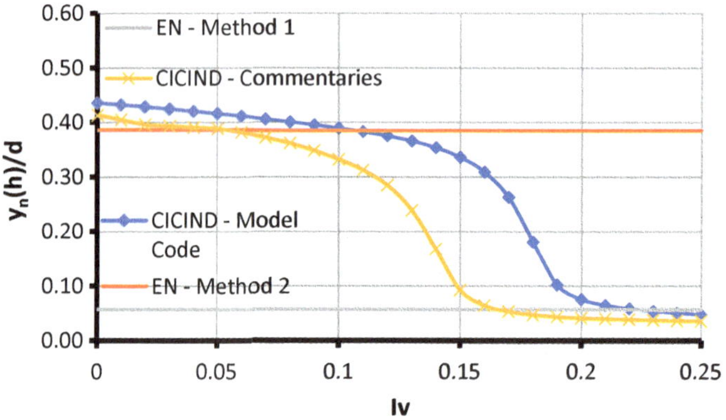

Fig. 6.30 Cross-wind vibration amplitudes in the lock-in regime: comparison CICIND and Eurocode (Sc = 15.87)

calculation with respect to the CICIND Model Code are due to the different values of $K_{a,max}$ adopted (see Eqs. (6.55) and (6.56)). The difference with respect to the Commentaries are also due to the integration all over the height (see (6.74)).

Figure 6.31 shows that in the forced-vibration range ($Sc/(4\pi Ka) \gg 1$ and $y/d < 0.1$), the response is well estimated by all the Codes and Methods, even though with some differences. As it is the small amplitude range, the influence of turbulence intensity is not so important. The reason is that the mechanical damping is in any case much higher than the aerodynamic damping and the further reduction of aerodynamic damping by turbulence intensity becomes unimportant.

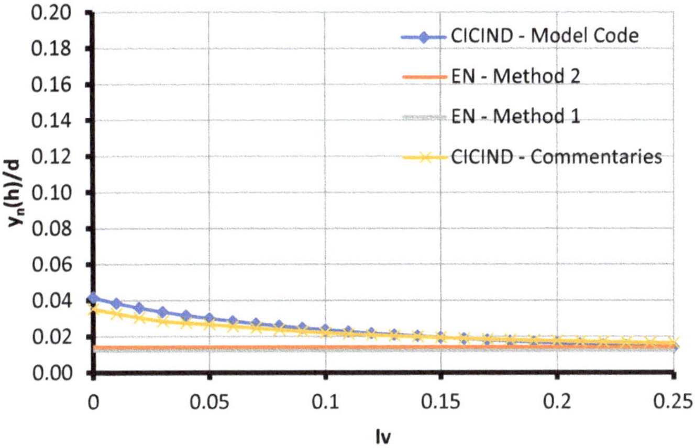

Fig. 6.31 Cross-wind vibration amplitudes in the forced-vibration regime: comparison CICIND and Eurocode (Sc = 44.45)

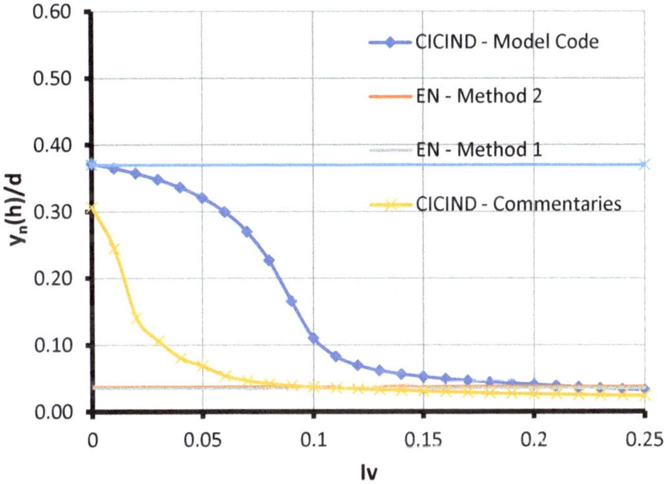

Fig. 6.32 Cross-wind vibration amplitudes in the transition regime: comparison CICIND and Eurocode (Sc = 25.40)

Instead, Fig. 6.32 shows that in the range $Sc/(4\pi K_a) \approx 1$ the amplitude of vibration is very sensitive to input model parameters and very different results may occur. In this range, the CICIND Model Code suggests to use the full procedure of the Commentaries, because difference up to 25% are to be expected. This is consistent with the results in the figure.

At zero turbulence intensity, $K_a = K_{a,max}$. The most relevant result of Fig. 6.32 is the significant underestimation of the Eurocode, Method 2, even and especially at Iv = 0. It is due to the value of $K_{a,max}$ which is used in the calculation

Table 6.8 Results on the steel chimney at different Sc and Iv = 0, according to the CICIND Model Code

CICIND—Model Code (n = 1)						
f_n (Hz)	0.5					
St	0.2					
d (m)	0.813					
$V_{cr,n}$ (m/s)	2.033					
$R_{e,n}$	1.10E+05					
δ_s	0.0100	0.0150	0.0200	0.0300	0.0400	0.0700
$m_{0,n}$ (kg/m)	262.326					
$S_{c,n}$	**6.350**	**9.525**	**12.700**	**19.050**	**25.400**	**44.451**
Iv(h)	0.00					
$k_{p,n}$	1.416	1.421	1.436	1.523	1.755	3.262
$a_{L,n}$	0.4					
$K_{a,n}$	2.800					
δ_a	−0.05541					
δ_{tot}	−0.04541	−0.04041	−0.03541	−0.02541	−0.01541	0.01459
$C_{a,n}$	0.00012					
$S_c/(4\pi K_{arn})$	0.18047	0.27071	0.36095	0.54142	0.72189	1.26331
$c_{1,n}$	0.06556	0.05834	0.05112	0.03669	0.02225	−0.02106
$c_{2,n}$	0.00001	0.00001	0.00001	0.00001	0.00002	0.00001
$\sigma_{y,n}$	0.29446	0.27779	0.26005	0.22036	0.17224	0.01039
$y_{n,max}/d$	0.51269	0.48554	0.45927	0.41292	0.37185	0.04168

(the value given by the EN is smaller than the value given by the CICIND). The result contradicts the remark in the comma (4) in the section E.1.5.3 of the Eurocode, according to which a conservative design is always guaranteed by using $K_{a,max}$ (Table 6.8).

The substantial difference in the results is numerically explained by Table 6.9 and Table 6.10. At Sc = 25.40, and at the Re of the chimney, $K_{a,r,n} = 2.327$ according to the CICIND (Eqs. (6.53) and (6.82)) and $K_{a,max} = 2.0$ according to the Eurocode (Eq. (6.56)). Consequently, $Sc/(4\pi K_{a,r,n}) = 0.87 < 1$ according to the CICIND and $Sc/(4\pi K_{a,r,n}) = 1.06 > 1$ according to the Eurocode. This means that $c_1 > 0$ in the first case and $c_1 < 0$ in the second case. The change of sing of the coefficient c_1 has a significant effect on the response, because the response range switches between large amplitude and small amplitude (Fig. 6.31).

In the critical range of $Sc/(4\pi K_{a,r,n}) \approx 1$ and $Iv \approx 0$, the substantial disagreement between the CICIND and the Eurocode is only attributable to the different values of $K_{a,max}$ which are adopted (Eqs. (6.55) and (6.56)). Once the same value, e.g. the one which is given by the CICIND (that is higher and thus on the safe side) is assumed and the Eurocode procedure is applied, the two models would agree on the same response

Table 6.9 Results on the steel chimney at different Sc and Iv = 0, according to the CICIND Commentaries

CICIND—Commentaries (n = 1)						
fn (Hz)	0.5					
St	0.2					
d (m)	0.813					
$V_{cr,n}$ (m/s)	2.033					
$R_{e,n}$	1.10E+05					
δ_s	0.0100	0.0150	0.0200	0.0300	0.0400	0.0700
$m_{0,n}$ (kg/m)	262.326					
$S_{c,n}$	**6.350**	**9.525**	**12.700**	**19.050**	**25.400**	**44.451**
Iv(h)	0.00					
$k_{p,n}$	1.417	1.429	1.460	1.642	2.099	3.665
$a_{L,n}$	0.4					
$K_{a,r,n}$	2.327					
δ_a	−0.04604					
δ_{tot}	−0.03604	−0.03104	−0.02604	−0.01604	−0.00604	0.02396
$C_{a,n}$	0.00007					
$Sc/(4\pi K_{arn})$	0.21719	0.32578	0.43438	0.65157	0.86876	1.52033
$c_{1,n}$	0.06262	0.05394	0.04525	0.02787	0.01050	−0.04163
$c_{2,n}$	0.00001	0.00001	0.00001	0.00001	0.00001	0.00001
$\sigma_{y,n}$	0.28780	0.26711	0.24469	0.19220	0.11882	0.00784
$y_{n,max}/d$	0.50163	0.46935	0.43928	0.38823	0.30684	0.03532

amplitude at Iv = 0. This is shown by the light blue curve (labelled "EN—Method 2, modified $K_{a,max}$") in Fig. 6.32.

It follows that further studies should be necessary in order to assess how realistic are the large amplitudes predicted by the CICIND model and how reliable are the values of the aerodynamic damping given by the Codes.

Moreover, as the results differ significantly at very low turbulence intensities, a further issue would be to investigate whether they should be realistically expected in the atmosphere.

A further issue to be better investigated would be the extension of the calculation procedure to the second mode shape. As the expression of the mode shape is maintained in the formulation of the CICIND (both in the Model Code and in the Commentaries) the CICIND procedure can be applied to the higher mode shapes. The Eurocode Method 2 is instead only applicable to the first mode shape. The reason lies in the coefficient C_c in (6.67). This maintains only the dependency on Re. It might be attempted to introduce in the EN Method 2 the dependency on the mode

Table 6.10 Results on the steel chimney at different Sc and Iv = 0, according to the Eurocode, Method 2

EN—Method 2 (1st mode of vibration)						
f_n (Hz)	0.5					
St	0.18					
D (m)	0.813					
V_{cr} (m/s)	2.258					
Re	1.10E+05					
δ_s	0.0100	0.0150	0.0200	0.0300	0.0400	0.0700
m_0 (kg/m)	262.326					
S_c	**6.350**	**9.525**	**12.700**	**19.050**	**25.400**	**44.451**
Iv(h)	–	–	–	–	–	–
k_p	1.420	1.446	1.514	1.905	2.696	3.888
a_L	0.4					
K_a	1.910					
δ_a	−0.03779					
δ_{tot}	−0.02779	−0.02279	−0.01779	−0.00779	0.00221	0.03221
C_a	–	–	–	–	–	–
Sc/(4πKa)	0.26459	0.39689	0.52919	0.79378	1.05838	1.85216
c_1	0.05883	0.04825	0.03766	0.01650	−0.00467	−0.06817
c_2	0.00000	0.00000	0.00000	0.00000	0.00000	0.00000
σ_y	0.27890	0.25258	0.22317	0.14780	0.01136	0.00301
$y_{1,max}/d$	0.48728	0.44917	0.41558	0.34639	0.03768	0.01437

Table 6.11 Structural damping and Scruton number in the first two modes

δs	Sc,1	Sc,2
0.010	6.350	6.451
0.015	9.525	9.676
0.020	12.700	12.902
0.025	15.875	16.127
0.030	19.050	19.353
0.040	25.400	25.803
0.070	44.451	45.156

shape by using the modified version of the C_c in Eq. (6.74). This is done in Fig. 6.33. In order to check the validity of the assumption the same values of $K_{a,max}$ as in the CICIND are used. As it can be seen, the results are well matching the Model Code. The differences in the Commentaries are due to the vertical variation of wind profile, turbulence intensity and tower diameter (Fig. 6.34 and Table 6.11).

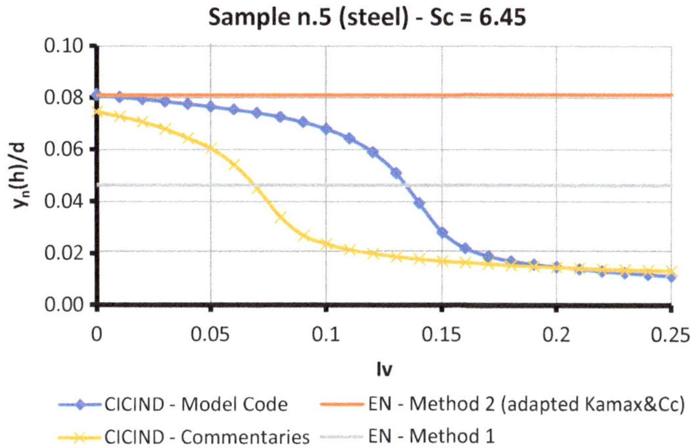

Fig. 6.33 Vibration amplitude in the second mode shape (Sc,2 = 6.45)

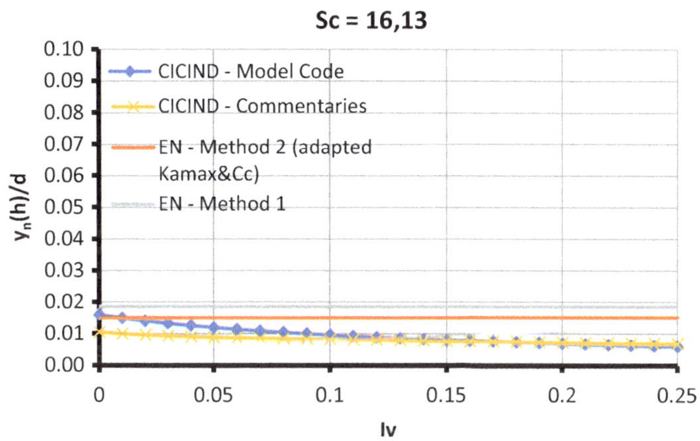

Fig. 6.34 Vibration amplitude in the second mode shape (Sc,2 = 16.13)

6.2.3.4 Application to a Real World Chimney Sample

The real world chimney sample, which is taken as reference in this sub-section, is the sample n.5 (made of steel). The following value of structural damping is used (after Eurocode):

$\delta s = 0.014$ (one liner + external thermal insulation, h/d > 26)
Sc = 8.89 (1st natural mode)

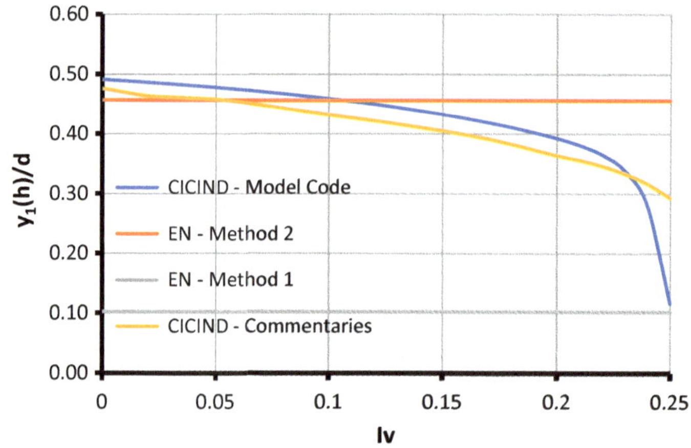

Fig. 6.35 Cross-wind vibration amplitudes (Sc,1 = 8.89)

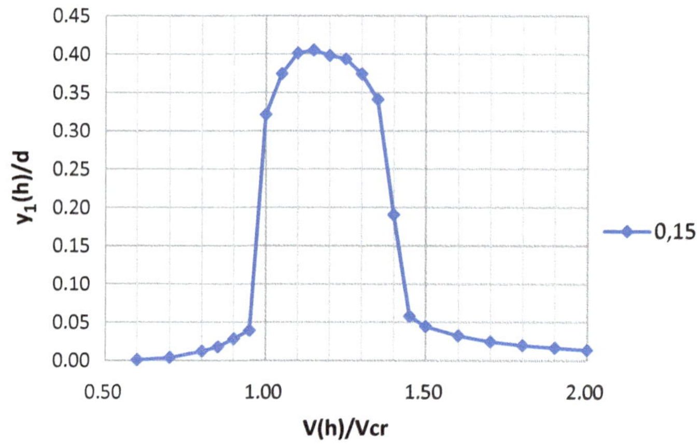

Fig. 6.36 Cross-wind response Iv(h) = 0.15 (Sc = 8.89) after CICIND commentaries

The design condition for turbulence intensity, in normal climate, is:
Iv(10 m) = 0.19 → Iv(h) = 0.15.

Results by applying different models are plotted in Fig. 6.35 and listed in Table 6.12 (first natural mode). Figure 6.36 shows the lock-in range, between 0.95 and 1.45 V(h)/Vcr.

The core of the Vickery&Basu method lies in the concept of a negative aerodynamic damping, which accounts for the interaction of the moving chimney and the air flow, whereas the lift force spectrum acts on the non-vibrating structure. These are the fundamental model assumptions. The model also introduces a self-limiting oscillation amplitude, that limits the vibration to a maximal value.

Table 6.12 Results in the 1st mode shape on the steel chimney n.5 (Sc = 8.89) after CICIND and EN

	CICIND		EN		
	Commentaries	Model	Method 2	Method 1	
f_1	0.5	0.5	0.5	0.5	Hz
St	0.20	0.20	0.18	0.18	[-]
d	0.813	0.813	0.813	0.813	m
V_{cr}	2.03	2.03	2.26	2.26	m/s
Re	1.10E+05	1.10E+05	1.22E+05	1.22E+05	[-]
δ_s	0.014	0.014	0.014	0.014	[-]
m_0	262	262	262	262	kg/m
Sc	8.89	8.89	8.89	8.89	[-]
Iv(h)	0.15	0.15	–	–	[-]
kp	1.55	1.47	1.44	–	[-]
a_L	0.4	0.4	0.4	–	[-]
Ka,r	1.23	1.54	1.91	–	[-]
δa	−0.024	−0.055	−0.038	–	[-]
δtot	−0.010	−0.041	−0.024	–	[-]
Ca	3.61E−05	4.83E−05	–	–	[-]
$Sc/(4\pi K_a)$	0.575	0.459	0.370	–	[-]
c_1	3.4027E−02	4.3249E−02	5.0365E−02	–	[-]
c_2	7.0944E−06	5.0220E−06	1.8634E−06	–	[-]
σy	0.21	0.24	0.26	–	m
$c_{lat,0}$	–	–	–	0.7	[-]
c_{lat}	–	–	–	0.7	[-]
L/d	–	–	–	6.03	[-]
Kw	–	–	–	0.32	[-]
K	–	–	–	0.13	[-]
$y_{1,max}/d$	0.41	0.43	0.46	0.10	[-]

The large-scale turbulence of the incoming flow is responsible for the modulation of the shedding frequency. Its effect is included in the model by the bandwidth B of the lift spectrum. Furthermore, the CICIND extension of the Vickery&Basu model takes into account the reducing effect of turbulence on the aerodynamic damping parameter. In particular, according to the CICIND approach, the model of the aerodynamic damping is split up into the value of $K_{a,max}$ and the reducing factor $<K_a>$. $K_{a,max}$ depends on the Reynolds number and is calculated in smooth flow, whereas $<K_a>$ accounts for the influence of turbulence intensity and in minor importance also for the variation with height of the mean wind profile.

As regards the influence of the aspect ratio, it is taken into account in the CICIND approach by applying a safety factor to Basu's experimental results. At low Reynolds numbers, a safety factor of 1.4 is applied. This is not applied in the Eurocode.

The CICIND model is theoretically founded and presents several strong points with respect to the Eurocode-Method 2. First, as previously said, the CICIND model allows to consider the effect of incoming turbulence and therefore it allows to distinguish rare and frequent events. Secondly, it includes the mode shape in the formulation and because of that it is applicable to higher modes of vibration and, in principle, even to non-cantilevered structures.

The application of the CICIND model to a steel chimney sample showed significant vibrations at low values of the turbulence intensity and a wide lock-in region even for relatively high values of the Scruton number. Apart from the situation $Sc \approx 4\pi Ka$ (in which the response is extremely sensitive to the values of the input parameters and the model lack of robustness), the results of the CICIND model at $Iv = 0$ are in substantial agreement with the results of the Eurocode Method 2. Compared to the results of the Eurocode Method 1, the CICIND gives much more conservative predictions. It is known that the Eurocode Method 1 refers to normal conditions of the atmosphere, so that a direct comparison at very low Iv should not be done, as also cited by the Eurocode itself. In any case, from the CICIND results in the previous graphs, it would follow that the number of existing chimneys, which would not be safe according to the CICIND model, would be rather high. Therefore, further studies should be necessary to assess how realistic are the models for predicting vortex excitation.

The structural response is governed by the relative importance of aerodynamic damping and mechanical damping. The concept of the negative aerodynamic damping is theoretically founded, but it is based on the modelling and on the estimation of the aerodynamic damping parameter K_a, which is—in fact—an uncertain and delicate key input parameter. Few information is available on it and it does depend on several factors like the Reynolds number, the turbulence intensity, the mean wind velocity related to the critical value and also the aspect ratio.

The mechanical damping is another general feature of uncertainty. However, it is not an uncertainty of the model, but an uncertainty of the parameter δ_s (structural damping). The scatter and the uncertainty in the values of structural damping will be addressed in Sect. 3.4.4.3.

As regards the steel chimney sample which has been investigated in this chapter, it is a typical example of chimney which is very prone to vortex excitation. In particular, in the specific case study, it emerged that the main reason for which this chimney experiences large vibration amplitudes even at relatively high Scruton numbers (e.g. $Sc = 25.40$ at $Iv = 0$, Fig. 6.26) is attributable to a relatively high value of K_a. According to the definition (6.53), such a high value is basically governed by $K_{a,max}$, which acts in smooth flow and depends on the Reynolds number. In particular, due to the small tower diameter and the relatively low natural frequency, the chimney goes in resonance in the first mode at a small critical velocity (about 2 m/s). This low value of critical velocity, combined again with the small tower diameter (which then acts in the power of 2), produces a small Reynolds number. Because of that,

$K_{a,max,1}(Re_1)$ lies on the highest plateau of 2.8 (Fig. 6.20). In the CICIND, this plateau is—by definition—even higher than the one given be the Eurocode, where $K_{a,max,1}$ = 2. As previously explained, the motivation for increasing $K_{a,max}$ from the original Basu's value ($K_{a,max}$ = 2 at low Re) to the actual CICIND value ($K_{a,max}$ = 2.8 at low Re) is due to the translation of the wind tunnel experiments for a cylinder with small aspect ratio to full-scale dimension (Verboom and van Koten 2010). Due to the importance of $K_{a,max}$ especially for chimneys with small Re (which are more prone to experience lock-in effects), it might be wondered whether the choice $K_{a,max}$ = 2.8 is too conservative. In fact, by comparison with the Eurocode, the CICIND method always predicts larger values of amplitudes, which are mainly attributable to the choice of $K_{a,max}$ as aerodynamic damping parameter.

As a general rule of thumb, in order to reduce the sensitivity of the chimney to vortex resonance, the low range of Re should be avoided. In other words, it is always recommended that the Re of the chimney lies on the lowest plateau of $K_{a,max}$ in Fig. 6.20 (i.e. Re > 5 × 10^5). This requirement can be translated in:

$$\mathrm{Re}_1 = \frac{V_{cr,1} \cdot d}{\nu} = \frac{\frac{f_1 d}{St} \cdot d}{\nu} = \frac{f_1 \cdot d^2}{St \cdot \nu} > 5 \times 10^5 \tag{6.89}$$

The kinematic viscosity of air ν is equal to 15 × 10^{-6} m^2/s, therefore:

$$\mathrm{Re}_1 = \frac{f_1 \cdot d^2}{St \cdot 15 \times 10^{-6}} > 5 \times 10^5 \rightarrow \frac{f_1 \cdot d^2}{St} > \frac{15}{2} \tag{6.90}$$

if it assumed St = 0.2, it results:

$$f \cdot d^2 > 1.5 \tag{6.91}$$

Equation (6.91) assures a low value of the aerodynamic parameter. In fact, as it is a negative aerodynamic damping (see (6.48)), a small value is preferable. Besides that, the mechanical damping of course plays a role. In fact, it is always advisable that Sc ≫ 4πKa.

6.2.4 *Structural Damping*

6.2.4.1 Basics

The damping is basically the dissipation of mechanical energy, which decreases the amplitudes of real vibrating systems. There are different energy transformation mechanisms; mostly, mechanical energy is converted into caloric energy, e.g. viscous damping and friction. Then there is also, for example, radiation damping, in which there is not any loss of mechanical energy, but distribution in an infinite system (Krätzig et al. 1996).

In the linear vibration theory for one degree of freedom system, damping is usually introduced as viscous damping, i.e. the damping force is proportional to the velocity through the damping factor c:

$$m\ddot{x} + c\dot{x} + kx = F \tag{6.92}$$

In the free vibrations of a damped linear system, the fading vibration is itself a measure of damping. Being ξ the ratio of the damping parameter to the critical damping ($\xi = c/c_{cr}$), i.e. the critical damping ratio, the relation between two following amplitudes can be evaluated as:

$$\frac{x_i}{x_{i+1}} = \exp\left\{\frac{2\pi\xi}{\sqrt{1-\xi^2}}\right\} \tag{6.93}$$

The logarithmic of this ratio is the so-called logarithmic damping decrement:

$$\delta = \ln\left(\frac{x_i}{x_{i+1}}\right) = \frac{2\pi\xi}{\sqrt{1-\xi^2}} \tag{6.94}$$

For small values of damping, the following approximation can be assumed:

$$\delta \approx 2\pi\xi \tag{6.95}$$

Multi-degree linear vibration systems are often analyzed by modal techniques, which allow to uncouple the system of equations into uncoupled one-dimensional ordinary differential equations, that can be solved independently. By introduction of modal coordinates, it is possible to obtain diagonal matrices for both mass and stiffness, thanks to the orthogonality of the eigenmodes. In order to express modal damping as a diagonal matrix, the Rayleigh form is often used:

$$\{u\}^T[C]\{u\} = \alpha[K] + \beta[M] \tag{6.96}$$

Damping is then expressed as the sum of two contributions, one which is proportional to the mass matrix, the other one which is proportional to the stiffness matrix. The values of α and β are calculated from modal damping ratios of two modes of vibrations, (i = j, k). It is assumed that the sum of α and β terms is nearly constant over the range of frequencies between ω_j and ω_k.

$$\xi_i = \frac{\alpha}{2\omega_i} + \frac{\beta\omega_i}{2} \tag{6.97}$$

The actual value of structural damping which enters the calculation decides on the dynamic sensitivity of the structure. In Sect. 6.2.3, the response of a steel chimney to vortex excitation has been calculated for different values of structural damping (Table 6.7). In this regard, the relative importance of mechanical damping (i.e. Scruton number) on aerodynamic damping characterizes the structural response, which may lie either in the lock-in regime or in the forced vibration one.

Flexible mechanical structures which are moving in fluids, experience a contribution to the damping properties which is called aerodynamic damping. It results from the relative flow forces between the moving structure and the fluid. These aerodynamic forces are thus expressible as products of fluid-dependent constants and time derivatives of the structural displacements. Therefore, aerodynamic damping is usually treated as a viscous contribution to the total damping (Krätzig et al. 1996).

As explained in Sect. 3.4, negative aerodynamic damping is the core of Vickery&Basu's model and it is the sum of a linear term plus a non-linear one, which allows to the vibration to be self-limited.

Aerodynamic damping, usually positive in sign, also acts in the in-wind direction. It increases the total damping of the structure and its contribution enters in the resonant factor R^2.

The interaction between the structure and the soil is another source of damping. The knowledge of the force settlement behaviour of the subsoil is necessary to describe the damping behaviour of the system. Further information can be found in (Krätzig et al. 1996); this issue is not further investigated in this report.

6.2.4.2 Code Stipulations

Table 6.13 summarizes the code stipulations on structural damping for both the CICIND and the Eurocode, concrete and steel chimneys. Whereas there is substantial agreement for steel chimneys, relatively different values are given for concrete chimneys. In particular, the Eurocode stipulation for structural damping of concrete chimneys is much lower than the CICIND one:

$$\delta_s(\mathrm{EN}) = 0.03 \ll \delta_s(\mathrm{CICIND}) = 0.101.$$

6.2.4.3 Comparison to Literature

In order to have a perception of the code stipulations regarding the structural damping, data from literature about measured values of structural damping of chimneys can be taken into account. It can be seen that the values of the Eurocode for the structural damping of steel chimneys derive directly from the measured values reported in Verwiebe et al. (1999).

Petersen (2000) distinguishes three mechanisms of damping: (1) damping of the material (δ_1); (2) damping of structural components and connections (δ_2); (3) damping of soil and foundation (δ_3). The logarithmic decrement of structural damping can be interpreted as the sum of the three contributions, Eq. (3.194). Values of δ_1, δ_2, and δ_3 for different types of steel and concrete chimneys are reported in Petersen (2000). In case, the structural damping can be thought as a weighted sum of the three

Table 6.13 Code stipulations for structural damping (CICIND A and B)

Code	Material	Characteristics		Aspect ratio	ξs	δs
CICIND	Concrete				0.016	0.101
	Steel	**Unlined**	Un-insulated		0.002	0.013
			Externally insulated		0.003	0.019
		Lined	With refractory concrete		0.005	0.031
			With brickwork		0.015	0.094
			Steel liners	$\lambda < 26$	0.006	0.038
				$\lambda > 28$	0.002	0.013
		Coupled group			0.004	0.025
		With tuned mass damper			0.02 min	
		Soft foundation (decrease of the f1 more than 10%)			+0.0005	
EN	Concrete				0.005	0.030
	Steel	**Unlined**	Without ext. Thermal insulation		0.002	0.012
			With ext. Thermal insulation		0.003	0.020
		Lined	One liner + ext. Thermal insulation	$\lambda < 18$	0.003	0.020
				$20 < \lambda < 24$	0.006	0.040
				$\lambda > 26$	0.002	0.014
			Two or more liners + ext. Thermal insulation	$\lambda < 18$	0.003	0.020
				$20 < \lambda < 24$	0.006	0.040
				$\lambda > 26$	0.004	0.025
			Internal brick liner		0.011	0.070
			Internal concrete shell		0.005	0.030
		Coupled stacks without liner			0.002	0.015
		Guyed steel stack without liner			0.006	0.040

contributions. However, no enough information and knowledge is available to define the weight factors.

$$\delta_s = \delta_1 + \delta_2 + \delta_3 \tag{6.98}$$

Material damping is not, in general, a constant, but it depends on the stress level. In metals, the damping becomes stronger due to fatigue and especially due to yielding. In concrete, the damping depends on the water-concrete factor and decreases with increasing dehydration. In uncracked concrete, the friction effects in the matrix and in the aggregate predominate. With increasing of cracks, damping increases, too, due to friction effects.

As regards damping in the soil, two causes should be distinguished. Damping can occur in the soil due to deformation and due to dissipation of energy in an infinite space. The latter is named geometric damping.

The value $\delta_s = 0.03$ for concrete chimneys, which is given by the Eurocode, is basically considering only the damping of the material, in the un-cracked situation. The concrete material damping is typically the predominant contribution in the summation, but its value can be more than doubled due to the onset of cracks, presence of lining and soft soil. According to Petersen's collection of data, even for the less damped case in concrete chimneys, i.e. un-cracked concrete chimney without lining and on rock, the minimum measured damping ($\delta_s = 0.039$) is already higher than the given value of the Eurocode. On the other hand, the CICIND value ($\delta_s = 0.101$) is somewhat optimistic, being in the order of the highest maximum damping measured by Petersen, for concrete chimneys with cracks, liners and on piles ($\delta_s = 0.103$).

As regards steel chimneys, there is also substantial agreement between Petersen's results and the Codes. In particular, the values of the Eurocode 0.012 and 0.020 (or, similarly, of the CICIND: 0.013 and 0.019) for unlined chimneys (un-insulated or insulated, respectively) are in the same order of $\delta_{s,min}$ in Petersen (2000) for a structure with the same characteristics. The value, about 0.040, for lined chimneys, is also consistent with Petersen's values. In case of internal brick liner, the damping is much higher: the CICIND gives 0.095, the Eurocode 0.070 and Petersen ranges about the same values.

References

Baker, C.J.: The turbulent horseshoe vortex. J. Wind Eng. Ind. Aerodyn. **6**(1–2), 9–23 (1980)

Basu, R.I.: Across-wind response of slender structures of circular cross-section to atmospheric turbulence. Ph.D. thesis, University of Western Ontario, London, Ontario, 1982

CICIND: Model Code for Steel Chimneys. The CICIND Chimney Standard. www.cicind.org (2010)

CICIND Steel Chimney Model Code—Commentaries and Appendices. www.cicind.org (2010)

CICIND: Model Code for Concrete Chimneys, Part A: The Shell. www.cicind.org (2011)

Clobes, M.: Böen und Wirbelerregung eines Schornsteins Berechnungsbeispiel nach DIN 1055-4. Internal Report, Institut für Stahlbau, TU Braunschweig (2014)

Davenport, A.G.: The application of statistical concepts to the wind loading of structures. ICE Proc. **19**(4), 449–472 (1961)
Dyrbye, C., Hansen, S.O.: Wind Loads on Structures. Wiley (1997)
ESDU 81017: Mean forces, pressures, and moments for circular cylindrical structures: finite-length cylinders in uniform and shear flow, incl. Amendment (A), 01 May 1987. Published in Release 2000-03, Engineering Sciences Data Unit, London
European Committee for Standardization: EN 1990 Basis of Design (2010)
European Committee for Standardization: EN 1991 Actions on Structures, Part 1–4: General Action-s—Wind Actions (2010)
European Committee for Standardization: EN 1993 Design of Steel Structures, Part 3–2: Towers, Masts and Chimneys—Chimneys (2006)
Gould, R.W.E., Raymer, W.G., Ponsford, P.J.: Wind tunnel tests on chimneys of circular section at High Reynolds Number. In: Proceedings of the Symposium on Wind Effects on Buildings and Structures, Loughborough (1968)
Hansen, S.O.: Vortex-induced vibrations of line-like structures. CICIND Rep. **15**(1), 15–23 (1998)
Krätzig, W.B., Niemann, H.-J.: Dynamics of Civil Engineering Structures. Balkema, A.A (1996)
Lupi, F.: A new aerodynamic phenomenon and its effects on the design of ultra-high cylindrical towers. Dissertation, University of Florence/TU Braunschweig (2013)
Nakamura, Y., Kaku, S., Mizota, T.: Effect of mass ratio on the vortex excitation of a circular cylinder. In: Proc. 3rd Int. Conf. on Wind Effects on Building and Structures, Tokyo, 1971
Okamoto, S., Sunabashiri, Y.: Vortex shedding from a circular cylinder of finite length placed on a ground plane. Trans. ASME J. Fluids Eng. **114**(4), 512–521 (1992)
Ruscheweyh, H.: Dynamic Wind Effects on Buildings, vols. I and II. Bauverlag, Wiesbaden, (1982)
Petersen, C.: Dynamik der Baukonstruktionen. Vieweg, 2000
Schmidt, L.V.: Measurements of fluctuating air loads on a circular cylinder. J. Aircr. **2**, 49–55 (1965)
Szechenyi, E., Loiseau, H.: Portance Instationnaires sur un Cylindre vibrant dans un Ecoulement supercritique. La Recherche Aerospatiale **1**, 45–57 (1975)
Verboom, G.K., van Koten, H.: Vortex excitation: three design rules tested on 13 industrial chimneys. J. Wind Eng. Ind. Aerodyn. **98**, 145–154 (2010)
Verwiebe, C., Berger, G.W.: Gemessene Dämpungsdecremente von Stahlschornsteinen und deren Bewertung in Hinblick auf die Bauart. Stahlbau **69**(1) (1999)
Vickery, B.J.: A model for the prediction of the response of chimneys to vortex shedding. In: Proceedings of the 3rd International Symposium on Design of Industrial Chimneys, Munich, pp. 157–162 (1978)
Vickery, B.J.: The response of chimneys and tower-like structures to wind loading. In: Proceedings of the 9th International Conference on Wind Engineering, India (1995)
Vickery, B.J., Basu, R.I.: Across-wind vibrations of structures of circular cross-section. Part 1: Development of a mathematical model for two-dimensional conditions. J. Wind Eng. Ind. Aerodyn. **12**, 49–73 (1983a)
Vickery, B.J., Basu, R.I.: Across-wind vibrations of structures of circular cross-section. Part 2: Development of a mathematical model for full-scale application. J. Wind Eng. Ind. Aerodyn. **12**, 75–97 (1983b)
Vickery, B.J., Clark, A.W.: Lift of across-wind response of tapered stacks. Proc. Am. Soc. Civil Eng. J. Struct. Div. **1**, 1–19 (1972)
Yano, T., Takahara, S.: Study on unsteady aerodynamic forces acting on an oscillating cylinder. In: Proceedings 3rd International Conference on Wind Effects on Building and Structures, Tokyo (1971)

Printed by Printforce, the Netherlands